MACROMOLECULAR MECHANOCHEMISTRY

MACROMOLECULAR MECHANOCHEMISTRY

Polymer Mechanochemistry
Volume One, Part 2

Cleopatra Vasiliu-Oprea
Florin Dan

Viva Books

New Delhi | Mumbai | Chennai | Kolkata | Bangalore | Hyderabad | Kochi

First Indian Edition 2008

VIVA BOOKS PRIVATE LIMITED

- 4737/23 Ansari Road, Daryaganj, New Delhi 110 002
  E-mail: vivadelhi@vivagroupindia.net, Tel. 42242200
- Plot No. 76, Service Industries, Shirvane, Sector 1, Nerul, Navi Mumbai 400 706
  E-mail: vivamumbai@vivagroupindia.net, Tel. 27721273, 27721274
- Jamals Fazal Chambers, 26 Greams Road, Chennai 600 006
  E-mail: vivachennai@vivagroupindia.net, Tel. 28290304, 28294241
- B-103, Jindal Towers, 21/1A/3 Darga Road, Kolkata 700 017
  E-mail: vivakolkata@vivagroupindia.net, Tel. 22836381, 22816713
- 7, Sovereign Park Apartments, 56-58, K. R. Road, Basavanagudi, Bangalore 560 004
  E-mail: vivabangalore@vivagroupindia.net, Tel. 26607409, 26607410
- 101-102 Mughal Marc Apartments, 3-4-637 to 641, Narayanguda, Hyderabad 500 029
  E-mail: vivahyderabad@vivagroupindia.net, Tel. 27564481, 27564482
- First Floor, Beevi Towers, SRM Road, Kaloor, Kochi 682 018, Kerala
  E-mail: vivakochi@vivagroupindia.net, Tel: 0484-2403055, 2403056

www.vivagroupindia.com

Published by arrangement with

Cambridge International Science Publishing
7, Meadow Walk, Great Abington
Cambridge CB1 6AZ, UK

ISBN-10: 81-309-0939-1 (Set of Two Volumes)
ISBN-13: 978-81-309-0939-4

Published by Vinod Vasishtha for Viva Books Private Limited, 4737/23, Ansari Road, Daryaganj, New Delhi-110 002. Printed and bound by Anand Sons, Delhi - 110 092.

The authors

Cleopatra VASILIU-OPREA held the post of Professor of Monomer and Polymer Technology and Polymer Mechanochemistry at the Gh. Asachi Technical University, Iasi, Romania. She received a chemical engineer degree at the Polytechnical Institute of Iasi, Romania, and PhD degree from the Techniche Hochschule für Chemie, Merseburg, Germany. Prof. Vasiliu-Oprea is PhD supervisor in the fields of Polymer Chemistry and Technology and Material Science.

During 45 years of scientific activity she published over 450 professional papers and holds more than 40 patents, with special emphasis on polymer degradation and mechanochemical synthesis, mechanochemistry of polymer and composites fracture and deformation. She is the author or coauthor of over 20 monographs, books, and book chapters, including *Mechanochemistry of Macromolecular Compounds; Polymers. Theory and Applications; Polymers. Structure and Properties; Polymers Fracture. Theory and Applications; Mechanochemical Polycondensation and Polycomplexation* in Polymer Kinetics and Technology, *Mechanochemistry of Polymer Deformation and Fracture Process* in Elastomer Technology Handbook, *Monomer and Polymer Technology*. She is a member of the Romanian Society of Polymer Science, the Romanian Society of Chemistry and Chemical Engineering, Romanian Society of Inventors, and of the International Federations of Inventors Associations, IFIIA-WIN, Geneva. For her scientific activity Prof. Vasiliu-Oprea received several awards, as follows: Gh. Spacu Prize of Romanian Academy, 1981 (for research in the field of Polymer Mechanochemistry), title of Evidenced Professor of the Romanian Education Minister, 1984, Elite Inventor award of Romanian Society of Inventors, 1991, 1992, 1994, Henry Coanda gold and silver medals of Romanian Society of Inventors, gold medals and diplomas from the EUREKA-Brusseles, 1994–1996. In 2003, she has been honored with Opera Omnia prize and medal for excellence in scientific research by Romanian Minister of Education and Research.

Florin DAN is a lecturer of Monomer and Polymer Technology and Polymer Processing at the Gh. Asachi Technical University, Iasi, Romania. Prior to this post he served as a chemical engineer in the rubber and plastic processing industry, UAMT-Oradea, Romania. He received his B.Sc. and Ph.D (supervisor C. Vasiliu-Oprea) degrees from the Gh. Asachi Technical University. He then spent a year as a Postgraduate Fellow in Polymer Science at the Institute of Macromolecular Chemistry of Prague, Czech Republic, with J. Stehlicek. This fellowship was followed by a year and a half as

Postdoctorate Fellow at the TotalFinaElf and a year and a half as an associate researcher at the National Research Council of France at the Blaise Pascal University of Clermont-Ferrand, France, with J.P. Grolier. His research interests focus on anionic polymerization of lactams in organic media, electroactivated stimuli responsive polymer gels, combined techniques for on-line monitoring of chemical processes, with an emphasis on the use of reaction calorimetry. Dan has authored more than 30 research papers, two invited book chapters, and holds one patent.

Preface

Mechanochemical phenomena are implied both in fundamental processes of life and in many branches of science and technology. Notable progress in areas such as solid body resistance, physico-chemical mechanics of materials, colloidal chemistry, polymers physics and chemistry, biophysics, chemical kinetics, theory of molecular capacity of reactions, molecular acoustics, shock waves detonation and propagation, friction and transfer, heterogeneous catalysis and adsorption, inorganic technology, metal corrosion, generally allow us to define and organise the mechanochemistry object, particularly in polymer mechanochemistry.

Presently, within the specialised scientific medium, the capacity of mechanical energy to modify into a specific manner, in accordance with the nature of co-existent environment, the relation between polymer structure and properties is unanimously recognised. The mechanisms of irreversible mechanochemical processes are clearly proved. In polymers, the mechanisms of deformation and fracture are studied and discussed in accordance with their physical state, as these ones are concretised in polymer synthesis, processing, and exploitation. Unfortunately, in industrial practice these results are not yet fully utilised.

Very promising results have been obtained in the last two decades in the field of reversible mechanochemical processes. The exploitation of the principle of stimulation of conformational changes from the natural or artificial macromolecular structures and the release of chemical and physical processes producing the mechanical work, as well as the assurance of the required feed-back of these processes, and not in the last instance the obtaining of new polymeric materials, characterised by very short time response at the applied stimuli, constitute major contributions to the further implementation of this new source of energy in practice. The high efficiency of conversion of chemical energy into mechanical work, and the ability of self-adapting of waste-free and quiet chemo-mechanical systems are some of significant advantages, allowing the fulfillment of one of the major objectives: the discover and use of some alternative energies.

These contributions determine the contour of macromolecular mechanochemistry.

The names of well-known schools, such as those led by

H. Staudinger, H. Grohn, W.F. Watson and R.J. Ceresa, N.K. Baramboin, W. Kuhn and A. Katchalsky, R.S. Porter, and not in the last instance by U. Murakami and Y. Osada made possible the progress in this field. The contributions of the Romanian school of polymer mechanochemisty are ranging among the trajectory of these researches. For of sixteen years, Professor C.I. Simionescu initiated and stimulated the investigations in this field and an optional discipline with this object of study was introduced the education programme of the undergraduate students of Department of Technology of Macromolecular Compounds, Faculty of Industrial Chemistry-Iasi.

Our book Macromolecular Chemistry presents from the theoretical and experimental point of view the main problems of this field, including the results obtained in more than a century of research. It is organised in two volumes: Polymer Mechanochemistry and Polymers with Chemomechanical Functions, respectively. The present volume (which is in two parts) deals with: Chained Multistage Character of Mechanochemical Process (1), Mechanochemistry of Polymers Deformation (2); Mechanochemistry of Polymer Fracture (including also the Fracture of Composite Materials) (3), and Mechanochemical processes for Energy Conversion (4). In this framework, the theoretic and experimental material is organised in correlation to the reaction mechanism, the type of mechanical solicitation, and the nature of environmental medium.

This book is to be considered as a guideline and not an encyclopedia. The treatment is not exhaustive and it is opened to adsorption and integration of new data, as well as to the critique analysis and suggestions, for which the authors will be indebted.

The book is addressed to professors, students, and researchers involved in the field of polymer science, to the engineers from the industry of synthesis and processing of plastic materials, elastomers, and fibres, as well as to the specialists from all technical domains that exploit the polymer-based materials. They will find inside treatment of the theoretical, experimental, and applicative problems and a wide access to the basic literature of this field.

The authors are grateful to Mrs. eng. Brandusa Vahnoveanu for the constant help during the make-up of this book.

Contents

3.8.2. Mechanochemical Reactions during Polymer Processing

3.8.2.1. Mechanochemical Reactions of Solid State Polymer Processing

Semicrystalline polymers below T_m, as well as the amorphous ones below T_g, are characterized by increased rigidity.

In the case of semicrystalline polymers, the high degree of macromolecule packing, known as reducing their freedom of movement and increasing the intensity of their interactions, is responsible for increased rigidity.

In the case of amorphous polymers below T_g, the amplitude of the oscillatory movements of the carbon atoms around the simple covalent bonds, in which they are involved, is quite low.

The macromolecules with a low energetic level cannot overcome the energetic activation barrier of free rotation, so that they cannot change their conformation easily. Consequently, they will lack flexibility.

The stress relaxation time in such polymers is long, consequently an intense, short time stress will not permit the macromolecule accommodation with the stress, so that they will finally split.

Therefore, degradation of rigid, crystalline or amorphous polymers requires application of some shock stresses on them, such as vibratory milling.

Sometimes, polymer processing involves a grinding phase, aimed to make uniform or even to decrease, for example, the dimensions of some polymer particles. Also, the grinding process constitutes a widely applied method in the processing of plastics or even fibre waste, in view of their recycling and reutilization.

A characteristic feature of the crystalline or amorphous polymers under T_g, processed by vibratory milling, is that the process involves not only the formation of new surfaces by the mechanochemical mechanism of fracture, but also a continuous increase of the material specific surface, as due to mechano dispersion.

In such a case, the process of mechanodestruction occurs simultaneously with mechanodispersion.

3.8.2.1.1. Mechanochemical Reactions of Solid State Semicrystalline Polymers Processing

In highly pure macromolecular compounds, with a regular structure of the chain which assures long-range ordering. the macromolecules tend to pack themselves, thus forming high-density regions, with the maximum ordering degree, defined as crystals.

As a result of their higher dimensions, comparatively with low molecular weight compounds, no macromolecules as such will be found in the nodes of the crystalline network, but only their constituting atoms. The larger dimensions of the chains reduce drastically the mobility of their individual atoms and, consequently, the speed at which they move for getting their position in the crystal. This explains the polymers' poorer and only partial crystallization, imperfect crystal formation, comparatively with the low molecular weight compounds.

The main element of a crystalline polymer's structure is the crystalline cell, with various shapes and dimensions, as determining its morphology.

Due to their higher dimensions, the macromolecules may successively penetrate several crystalline domains, thus interconnecting them. The spaces between these crystalline regions do not have a ordered structure, any longer, appearing as amorphous regions. In practice, there are no polymers with 100% crystallinity degree. The correct and common definition of these polymer types is 'semicrystalline polymers'.

In a semicrystalline polymer, the amorphous regions form mainly defective 'joining' regions in which mechanical energy is preferentially concentrated.

However, structural defects may also occur in the crystalline regions, an example being the chain ends of some macromolecules.

Several structural models have been proposed for semicrystalline polymers, such as the Hosemann–Bonnart model [691], in a polystructure subjected to mechanical stresses.

In semicrystalline polymers, scission occurs by a mechanism typical of the amorphous phase which results in crystal destruction. The crystallinity index is therefore reduced, an effect actually recorded with all semicrystalline polymers subjected to grinding.

As a conclusion, grinding of semicrystalline polymers results in polymer mechanodegradation (manifested as a reduction of its molecular weight), concomitantly with its mechanodispersion (resulting in the increase of the specific surface, by the formation of some extremely reactive new surfaces) and reduction of crystallinity index.

The mechanical stress in semicrystalline polymers, especially the shock and friction forces developed through vibratory milling, results, as an effect of the presence of macroradicals, in a continuous formation of some new surfaces, extremely rich in energy and, therefore, highly reactive.

At the level of such surfaces, several chemical reactions, such as grafting and block-copolymerization, polycondensation, complexation, etc., are developed.

Therefore, in spite of the fact that the main consequence – evidenced macroscopically – of semicrystalline polymer vibratory milling is mechanodispersion, this one is induced by mechanodegradation. The latter phenomenon is evidenced by the decrease of the molecular weight as well as by the increase in concentration of the final functional groups formed by macroradicals' stabilization.

Vibratory milling or CEL, PAN, polyesters and polyamides induces a continuous decrease of the molecular weight, which tends-in time-towards a limit value, characteristic to the mechanical treatment applied, simultaneously with the increase in the number of amino, carboxyl, hydroxyl groups or of the terminal unsaturated ones. Such changes at molecular level are also reflected in the modification of properties such as: solubility, swelling capacity and crystallinity.

Apparently, the crystallinity index should increase, as the mechanical energy is concentrated mainly on the amorphous domains, known as forming the defective regions of the crystalline polymer structure. Nevertheless, under conditions of intense shock and friction, the forces developed during vibratory milling and the stress supported even by the crystalline domains are very high. The stress rate being also high, the macromolecular crystals respond in a rigid manner to the stress applied, having no time to redistribute it over the amorphous domains. As a result, the shock energy concentrated on the crystal defect zones, will cause its destruction.

The formation of macroradicals, accompanying permanently the process of mechanodispersion and mechanodegradation of semicrystalline polymers, may be evidenced through some specific reactions of theirs; also, it may be utilized for obtaining modified polymers with new physical and chemical properties.

One of the methods of the macroradical's evidencing involves their generation in the presence of a vinyl (or diene) monomer, which causes its polymerization. Consequently, grinding of the PAN in the presence of vinyl chloride or butadiene, leads to a block-or grafted-copolymer, as well as of some minor amounts of homopolymer [356].

This type of mechanochemical synthesis has also been performed by Baramboim and coworkers in the modification of PE

[692], and polyamides [611, 612].

PeCL was either grafted or block-copolymerized with various vinylic monomers, such as styrene and vinyl chloride [124–126]. The first observation made was that mechanodegradation of polyamide through vibratory milling is characterized by a negative temperature coefficient. The grafting degree increases slightly with decreasing temperature (Figure 3.219), while homopolymerization is in no way affected by such a parameter.

A thorough study devoted to mechanochemical synthesis and, equally, to the behavior of PET under conditions of vibratory milling in the presence of some vinyl monomers, has been initiated by Simionescu and Vasiliu-Oprea.

Polyester mechanochemical destruction begins after a long induction period (minimum 12 hours), grafting occurring over the 12–48 h interval (at T = 18 °C). The highest values of the grafting degree have been recorded for gaseous-state monomers. In all situations, the grafting processes were seen as prevailing, with homopolymerization occurring only to a minor extent.

Such synthesis processes may be also exemplified for CEL triacetate-CEL [351], and, respectively, PAN-PVC mixtures [693].

Among the factors influencing the process and worth mentioning are: duration, the gaseous atmosphere, installation's building material, and mainly temperature; all are seen as affecting both the composition and properties of the products thus synthesized.

Macroradical formation, accompanying permanently mechanodispersion and mechanodestruction of semicrystalline polymers during vibratory milling, may be also evidenced by the utilization of

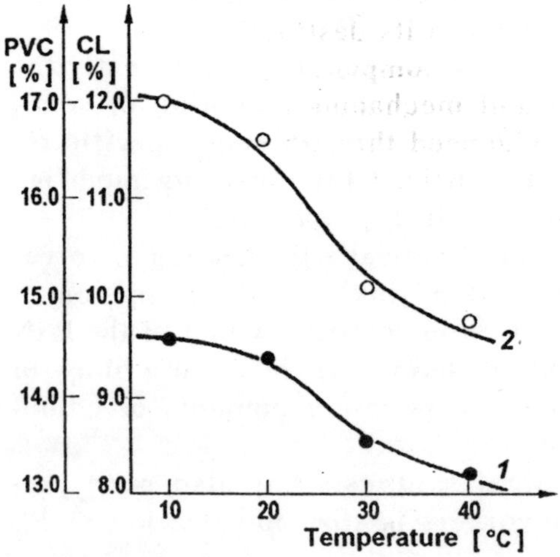

Figure 3.219. The effect of temperature on the grafting degree of P-e-Cl with vinyl chloride. 1) %Cl in graft copolymer; 2) %PVC in graft copolymer.

some radicalic acceptors of the aromatic diamines type; thus new modified polymers, possessing special chemical and physico-mechanical properties, may be obtained.

An the same time, such reactions as polycondensation and mechanochemical complexation, have been performed [83,85,300, 593,668–67].

Simionescu and Oprea have observed that in the case of P-ε-CL complexation may also be obtained, too, in the absence of ethylene diamine as a condensation agent, through mechanical activation of the polypeptidic groups from polymer's structure [668–671].

The presence of the metallic ions confers to the polychelates thus synthesized semiconducting and paramagnetic properties, along with increased thermostability.

CEL's vibratory milling in the presence of some transitional metals salts (Ni, Fe, Mn, Co, Cu, Ti, V) leads to macromolecular complexes capable of coupling enzymes [92, 313].

The Cu-containing complex was found as highly efficient in coupling of invertase, a situation in which an immobilized enzyme with an activity close to that of the free one is obtained, which may be nevertheless reused in several hydrolytic cycles.

3.8.2.1.2. Mechanochemical Reactions of Solid State Amorphous Polymers under T_g

Macromolecular compounds are characterized by a continuous transition, depending on the ordering degree, from an ideal order (single crystal) to an absolutely disordered state. Using the SAXS method, it was found that, in the so called 'crystalline' polymers, reflections evidencing the existence of an advanced order (Figure 3.220b) are obtained, while in the case of amorphous polymers, such reflections are absent (Figure 3.220a), [714].

Measurement of some parameters characteristic of non-crystalline polymers has evidenced the presence, within their structure, of not only some structural domains with dimensions of tens of angstroms, but also of some structural formations with considerable sizes.

It was therefore suggested that there is a possible aggregation of the domains into 'superdomains', similarly with the growth of spherulites in crystalline polymers, inducing formation of lamellar and fibrillar crystals (a model proposed by Kabanov [694]). Such a quasi-crystalline structure may have a characteristic period of ordering of 1 μm.

The microheterogeneity of the structure of non-crystalline poly-

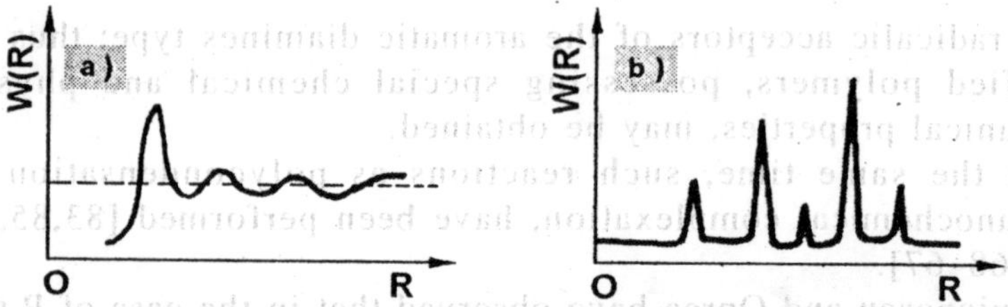

Figure 3.220. Variation of the $W_{(R)}$ function (radial distribution of the interatomic distances) for an amorphous (a) and crystalline, (b) polymer.

mers, characterized by the existence of either domains of 'superdomains' of relative ordering and of the totally amorphous zones, influences, of course, their properties, thus determining a certain anisotropy of such polymers.

This situation may constitute one of the reasons explaining the high discrepancies observed between the theoretical and experimental value of the, e.g., mechanical properties. The wholly amorphous regions act as factors of stress concentration, thus reducing the polymer's strength. In the same manner, an explanation may be provided for the fact that, frequently, with several samples belonging to the same batch, quite different values of the breaking strength up to deviations of 20% versus the main value were experimentally determined.

It is obvious that, in such situations, the anisotropy of the non-crystalline polymers, usually – and improperly defined as amorphous ones, has a significant influence.

The amorphous polymers are processed only in the visco-elastic or flowing state. However, quite frequent are the situations in which, prior to their processing the polymers are subjected, at inferior temperatures, either to vibratory milling or ultrasonic irradiaton, in view of their homogenization from the viewpoint of the particle size. Also, vibratory milling under T_g is applied for the recovery of wastes; that is why, knowledge of the chemical transformations suffered – under such conditions – by the amorphous polymers and, respectively, activation of certain chemical reactions under the action of mechanical stresses, may result in significant practical applications.

The intense mechanical stress of rigid polymers, especially by vibratory milling, determines morphological changes materialized in the considerable dispersion of the material. The process consists in the reduction of the grain's geometrical dimensions and the formation of new surfaces, a phenomenon accompanied by the appearance of free macroradicals due to the splitting of the chemical bonds from the basic chain.

In the case of amorphous polymers under T_g, the dispersion takes place up to very small geometrical limits of the grains (microns size), being accompanied by a molecular weight decrease. The process is influenced by the chemical nature of the polymer, the mechanical processing regime, the reaction medium nature, etc. Generally speaking, milling under these conditions produces particles with the size of 1 to 3 μm, after which the dispersion degree does not practically change, but instead the reduction of the molecular weight still continues as a result of the mechanochemical destruction process.

During vibratory milling, the polymer stores up the elastic energy received as shock and friction forces. The physico-mechanical changes produced under the action of the shock waves determined by the impact with the milling bodies as well as with the walls of the equipment take place extremely quickly (in less than 10^{-10} s). It is considered that, at the pressures, which are developed in the front of the shock wave, the polymer is transformed into an activated complex similar to the metallic state, characteristic of high degrees of compression. A relaxation effect takes place after the passing of the shock wave and a specific state of matter, corresponding to the dilatation degree attained at the moment the activated complex decomposition is established.

The anisotropy of the physico–mechanical properties constitutes an important factor, which controls the mechanochemical destruction. Thus, the polymers characterized by isotropic properties are milled, forming particles with a reduced degree of asymmetry, so that a determined direction of the destruction process could not be noticed. This fact evidently comes out from the increase of the polymer anisotropy, implicitly characterized, as well by the increase of the asymmetry degree of the produced particles.

The nature of the liquid medium and their interaction with the polymer will greatly determine the evolution of the mechanodispersion process. In the first stages of the process, the liquid penetrates in the faulty zones of the polymer structure (preexisting cracks, interfibrillar spaces, nascent microcracks, etc.) giving rise to pressures, which lead to the intensification of the material destruction. The nature of the gaseous atmosphere is very important, too. The higher the geometrical dimensions of the gas molecules are, the greater the 'wedge' effect exerted by them on the top of the main crack will be, thus determining the increase of the mechano-dispersion and mechanodegradation efficiency. An 'active' gaseous atmosphere will determine as well reactions with the radical centers appeared on the material fracture surfaces, forming some new functional groups.

The most important phenomenon, a result of the mechanodispersion process, is the increase of the specific surface by forming new surfaces. The mechanism according to which the effect is accomplished covers all the stages of the fracture mechanism being based on the mechanochemical act. There is a relation between the attained degree of milling (reflected by the specific surface) and the molecular weight, $\overline{M}_w$, a relation proved to be true for many polymers, the amorphous ones included

$$\overline{M}_w = f(S)$$

Thus, in vibratory milling of the PS in a gaseous medium, the molecular weight decrease is produced only at the same time with specific surface increase and with the grain dimensions decrease, respectively, while the established function is determined by the working temperature, the frequency of the mechanical stress, and the nature of the medium [694].

A similar effect appears during PVC vibratory milling either in a dry medium or in the presence of methanol [696,697].

Analysis of the results presented in Figures 3.221 and 3.222 shows that the molecular weight is reduced (according to an exponential low characteristic to the mechanical processes) at the same time with the increase of the vibratory milling duration (Figure 3.221 curve 1), while the specific surface registers a relative continuous increase (Figure 3.221 curve 2).

One can note a higher efficiency of the mechanodispersion process in the presence of methanol (Figure 3.222), a consequence of this polar liquid corrosive effect, which penetrates in microcracks and in the main crack accelerating the nucleation process and the propagation of the later one. Accordingly, the molecular weight of the polymer, milled under these conditions, registers a more pronounced decrease as compared to that in the dry medium. The increase of the specific surface area as well as the appearance and multiplication of the functional groups from the newly formed surfaces increase the polymer reactivity and improve some physicochemical properties. Thus, plasticizer sorption increases during dry milling from 95 % (the initial polymer) to

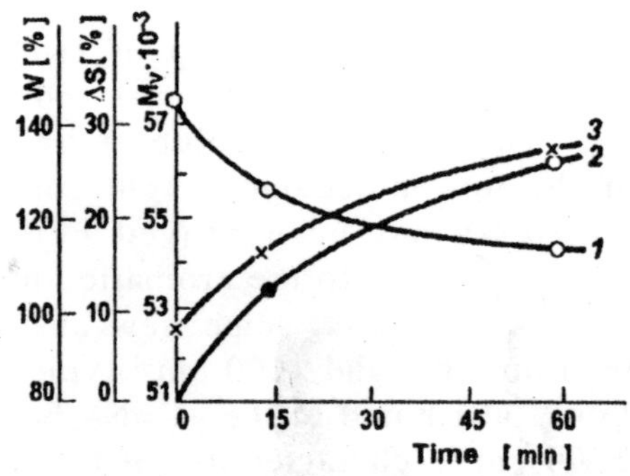

Figure 3.221. Influence of vibratory milling time (in dry medium) on some characteristics of PVC-S. 1) M_η; 2) specific surface (ΔS, %); 3) plasticizer sorption (W, %).

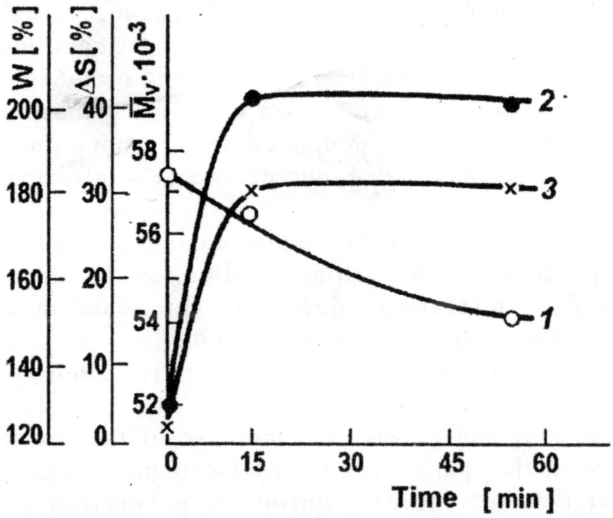

Figure 3.222. Influence of vibratory milling duration (in methanol) on some characteristics of PVC-S. 1) M_η; 2) specific surface (ΔS, %); 3) plasticizer sorption (W, %).

135 % after 60 min processing (curve 3, Figure 3.221). This coefficient has a much higher value in methanol, reaching the 220 % value only after 15 min of milling (curve 3, Figure 222). All these changes are based on the mechanochemical act starting when obtaining a critical amount of elastic energy stored up in the polymer during milling, thus, macroradicals are generated; those present during the entire mechanical stressing period can be pointed out by reactions with specific radical acceptors or aromatic diamines.

Therefore, by the PVC vibratory milling in the presence of

benzidine and *p*-PhDA, products made up of two fractions of reaction products (one soluble in cyclohexanone and quantitatively dominant – over 80 % – the other one insoluble) are obtained. The chemical bonding of the low molecular weight compound to the PVC destruction fragments has been proved both by spectral methods and by reactions specific to the aromatic amino group. In the IR spectra of soluble fractions, some new absorption bands appear, such as those from 1490 and 1600 cm^{-1} typical of the bond –CN and from 1700 cm^{-1} attributed to the –C=N– bond. A marked maximum value at 3400 cm^{-1}, characteristic of the primary imido and amino group is to be noted, as well.

Considering that the aromatic diamine works as a radical acceptor through one of the functional groups, the other one remaining free on the destruction fragments end, it has proceeded to accomplish the diazotization reaction and subsequent coupling with phenolic components. In this way it was obtained products with a color that is specific to the newly formed azoic groups, the color intensity being a function of the coupled diamine amount, an indirect reflection of the mechanical regime parameter's influence on the reaction.

The analysis of the reacted diamine amount has therefore pointed out its dependence on the milling duration, on the diamine quantity from the PVC mixture subjected to the reaction and on the micromolecular compound chemical nature.

The influence of the milling duration is reflected by the results given in Table 3.79.

One can note the continuous increase of the reacted diamine quantity with this parameter of the mechanical regime, a consequence of the free radical continuous generation at the newly formed surfaces.

The diamine amount from the milling–subjected mixture (10 g of PVC) also influences the reaction efficiency (Table 3.80) which increases with an increase in the amount of diamine.

From the above-given data we can conclude that the PVC reaction with aromatic diamine is conditioned by the chemical nature of the micromolecular compound. The most active seems to be benzidine, whose content in synthesis products is constantly higher, providing, as well, deeper colors of the diazotizing and coupling products.

It must be mentioned that the new products synthesized by PVC vibratory milling in the benzidine and *p*-PhDA presence are characterized by their particular thermostability pointed out by the

Table 3.79. The influence of milling time on the amount of chemically bonded diamine (10 g of PVC + 2 g of aromatic diamine)

	Duration of mechanical activation (h)		
	24	72	96
	Chemically bonded diamine amount (%)		
Benzidine	1.60	4.80	5.46
p-PhDA	1.58	3.39	5.40

Table 3.80. The influence of the diamine quantity on the reaction efficiency with the mechanoradicals formed by PVC vibratory milling (10 g, 24 h)

	Quantity of diamine from the initial mixture , (g)		
	1	2	3
	Chemically bonded diamine amount (%)		
Benzidine	1.60	4.07	4.60
p-PhDA	1.58	2.78	3.30

losses in weight with the temperature, losses which are inferior to the PVC witness, and by the calcination residue, which is superior (at 1200 °C) demonstrating the formation of some structures with a remarkable specific thermal stability at temperature.

The molecular weight decrease and the possibility of obtaining some reactions of the PVC with the aromatic diamine-type radical acceptors are conclusive arguments for a radical destructive mechanism of the polymers under vibratory milling conditions. Consequently, mechanodispersion is a mechanochemical-type process, developing itself according to the general laws of polymer fracture under stress.

During their processing or exploitation, polymers may be subjected either to some stresses the intensity of which increases in time (e.g., traction, compression etc.) or to the periodical, cyclical action of some stresses of low or average intensity; in this latter case, the process is also known as 'fatigue'.

The mechanism controlling breaking and formation of new surfaces in amorphous polymers under T_g subjected to cyclic stress, involves the same stages discussed in section 3.3, although with some specific peculiarities.

The appearance of the microcracks suggests an induction period of the fatigue fracture which, however, can be absent in the pre-existing crack polymers. Whether this period exists or not, the process develops continuously by incremental increases as shown by Andrews [1] and not by the catastrophic propagation of the main crack as it happens for other types of stresses. Thus, the fatigue implies the non-catastrophic fracture propagation through the material, the main crack increasing with each cycle but only to a small extent, insufficient to produce the macroscopic failure of the polymer. The mechanism of the gradual incremental propagation of the main crack in cyclically stressed polymers has been checked out for a wide range of macromolecular compounds [698]. On studying, for instance, the PVC behavior under fatigue, regular groves are evident on the surface of the samples, the distance between them being correlated with the application rate of stress [699,700]. The number of the groves represents about 2 % of the number of cycles, which lead to the conclusion that the fatigue fracture of this material is a discontinuous process, the main crack propagating once for every two cycles. Similar results have been obtained by Constable and coworkers [699], in the case of PMMA.

A particular case of mechanical fatigue stress is the ultrasonic treatment when reduced intensity but very high frequency stresses act on the polymer. Some data concerning the molecular mechanism and the structural morphological changes caused by the ultrasonic treatment are revealed by the study of PVC [309,701]. Ultrasonic treatment was applied to a PVC/water slurry, the frequency being 40 kHz and the duration ranging between 10 and 120 h. The PVC used was obtained by the suspension process, having a molecular weight of about 55000 and the size distribution ranging between 40 and 250 μm.

Macroscopically, the ultrasonic action on polymers is materialized in intense dispersing and the increase of weight of the 80 μm diameter fraction. This effect is, obviously, the result of the formation of new surfaces by the cleavage of the monolith structure of the grains due to the accumulation of mechanical stresses on the faults by the mechanism of crack appearance and propagation. The longer the duration of the ultrasonic treatment, the more intense the dispersion will be (see Figure 3.144, Chapter 3.6.2.2.), as manifested by the prevailing accumulation of the 80 μm diameter fraction in the material.

The mechanochemical act of the homolytical splitting of chemical bonds with the formation of macroradicals is the basis of the mechanodispersion process; this act is demonstrated both by continuous diminishing of the molecular weight and by the increase of the number of split chemical bonds during the ultrasonic treatment (Figure 3.223).

The shape of the variation curve of the molecular weight is typical for the mechanodestructive processes, tending to a limiting value which, however, has not been attained on the experimental interval. It is also evident that a maximum intensity of the destruction is registered at the 200 μm diameter fraction, upon which the mechanical energy is especially concentrated.

The mechanodispersion process in the given conditions has, at the same time, led to the appearance of new functional groups by the reciprocal stabilization of the active destruction fragments or by their reaction with radical acceptors. The IR spectra have proved that, as a result of the ultrasonic treatment, the concentration of the end non-saturations increases and numerous oxidized functions of the hydroxyl, carboxyl or peroxy types appear. All these bring about the increase of the newly formed surface reactivity, an increase rendered evident by the growth of the plasticizer sorption processes or by the acceleration of the reactions in the heterogeneous system. Therefore, the non-ultrasonically treated 200 μm diameter fraction has a soaking period ranging between 92 and 105°C; after the ultrasonic treatment, even after the first 10 h, this frac-

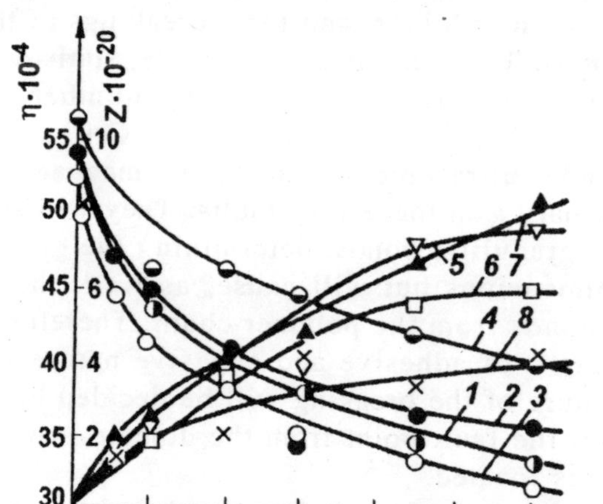

Figure 3.223. Variation of the molecular weight M_η and of the split bonds number (Z) as a function of ultrasonic treatment duration of PVC in aqueous suspension. 1) M_η in unfractionated polymer; 2) M_η in the 80 μm fraction; 3) M_η in the 200 μm fraction; 4) M_η in the 40 μm fraction; 5) Z in the unfractionated polymer; 6) Z in the 80 μm fraction; 7) Z in the 200 μm fraction; 8) - Z in the 40 μm fraction.

tion practically absorbs instantaneously the plasticizer, but swelling and dissolution are produced in the 25 to 70°C interval.

In the light of these results, mechanodispersion has to be considered not only as a process of forming new surfaces, but also as a process which ensures the increase of the system free energy entails the displacement of the chemical and phase equilibrium. The existence in the solid body of an excess of free energy leads to the appearance of an unstable thermodynamic state.

One can consider that the ultrasonic treatment produces in every polymer grain a structural reaction consiting of two inter-conditioning components, namely, mechanodegradation (or mechanocracking) in microdomains which constitutes the active seeds of mechanodispersion. In every moment of the process equilibrium is established between the number of split molecules defining the decrease of the molecular weight and the corresponding specific surface, which is an evaluation criterion of mechanodispersion. The interdependence of these effects brought about by the mechanical energy is clearly made evident for the 80 μm diameter fraction (Figure 3.224).

Each polymer grain is an imperfect conglomerate of small solid bodies kept together by adhesive forces (as result from suspension polymerization technology), containing numerous faults such as pores, air bubbles, embeddings, and local breakings of the cohesive and adhesive bonds. These faults are, as a rule, statistically distributed inside the granulitic formations; the number of faults is proportional to the dimensions of the grains. Critical stresses appear because, under ultrasonic treatment, the mechanical energy is concentrated primarily on these very faults. They will strain the adhesive-type intergranulitic bonds, determining the grain dispersion up to certain dimensions but will cause, as well, the splitting of some chemical bonds from the polymer chain. Therefore, two types of bonds splitting – of adhesive and cohesive nature – occur. The place and the nature of the breaking will be decided by the position and the nature of the fault point from the domain in which the mechanical energy is located.

Taking into account the broad size-distribution of PVC particles nature and the higher number of faults existing in the higher dimensioned grains, one could expect that both the mechanodestruction and the mechanodispersion should take place predominantly in these grains. The conclusion is supported by the distribution curves (see Figure 3.144, section 3.6.2.2) which indicate polymer uniformisation by its concentration in fractions characterized by an

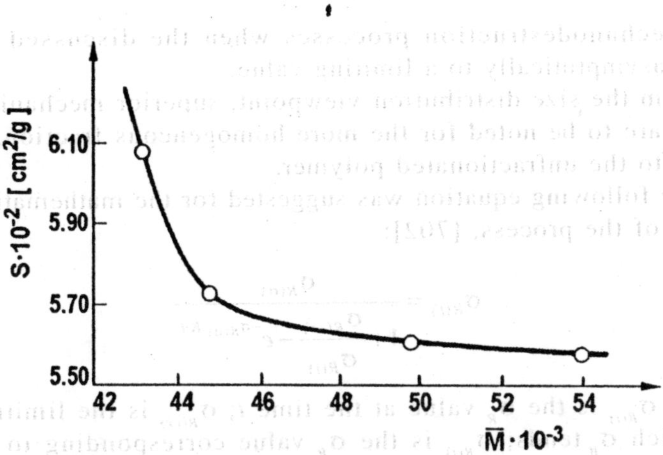

Figure 3.224. Specific surface variation (S) with molecular weight for different durations of ultrasonic treatment (0, 60, 90, 120 h).

optimum diameter for a given mechanical regime (80 μm diameter for the discussed case).

An important role in accomplishing the mechanodispersion is played by the liquid; as a function of its nature, it can loosen some physical interactions, especially under the conditions of penetrating in microcracks or gaps where it acts just like a 'wedge', thus favoring both the mechanodestruction and mechanodispersion.

It should be emphasized that both processes are characterized by a limit determined by the polymer chemical nature and by the mechanical regime (nature of medium, working parameters, temperature). The decrease in the size of the polymer grains to an equilibrium value takes place at the beginning, after which the dispersion stops but the molecular phenomenon still goes on in the stressed material, the result being a decrease in molecular weight.

The molecular weight reduction, which is a first consequence of the mechanochemical act, is macroscopically reflected in the change of some mechanical properties, among them being the tensile strength [701]. Accordingly, the variation of σ_R with the parameters of ultrasonic treatment, especially with the time, can constitute an estimation criterion for the intensity of the mechanochemical processes which develop in the mechanically stressed material.

The evolution of σ_R along with the ultrasonic treatment duration both for the unfractioned PVC and for the 45 and 100 μm diameter fractions, respectively, is shown in Figure (Figure 3.153, Chapter 3.6.2.2.). The shape of the curves is identical and typical for the mechanodestruction processes when the discussed property tends asymptotically to a limiting value.

From the size distribution viewpoint, superior mechanical properties are to be noted for the more homogeneous fraction, as compared to the unfractionated polymer.

The following equation was suggested for the mathematical modelling of the process, [702]:

$$\sigma_{R(t)} = \frac{\sigma_{R(\sigma)}}{1 + \frac{\sigma_{R(cr)}}{\sigma_{R(i)}} - e^{-\sigma_{R(\sigma)} \cdot K \cdot t}} \tag{3.118}$$

where $\sigma_{R(t)}$ is the σ_R value at the time t; $\sigma_{R(cr)}$ is the limiting value to which σ_R tends; $\sigma_{R(i)}$ is the σ_R value corresponding to the non-ultrasonically treated polymer; t is ultrasonic treatment time (h); and K a coefficient whose value is determined from experimental data by means of the least square method:

$$K = \frac{\nu}{\nu_{R(cr)}} \cdot \frac{\Sigma y_i N_i}{\Sigma N_i^2} \tag{3.119}$$

where ν is the frequency of the ultrasonic field (Hz) and N_i number of stress cycles at the time t,

$$y_i = \ln \frac{\frac{\sigma_{R(cr)}}{\sigma_{R(i)}} - 1}{\frac{\sigma_{R(cr)}}{\sigma_{R(t)}} - 1} \tag{3.120}$$

The theoretically obtained curves are practically identical to the experimental ones, thus demonstrating that the proposed mathematical model has been properly chosen.

The above presented results has led to the conclusion that the fatigue process of the amorphous polymers under T_g has a mechanochemical basis, evolving practically according to the same laws as the fracture process of the polymers in a constant or variable monotonously increasing field of forces. From a macroscopic viewpoint, this fatigue process is manifested by forming new surfaces, which are rich in energy and with increased reactivity.

3.8.2.2. Mechanochemistry of highly elastic state polymers processing

The highly elastic state is exclusively specific to the polymers being a consequence of the linear macromolecule flexibility.

Taking into account the macromolecule's great dimensions as well as their capacity of aggregation in various supramolecular forms, the polymer viscosity becomes so high that the flowing process does not occur even after long duration of stress application. Consequently, the tension developed in the material cannot relax and causes the deformation of the flexible high macromolecular chain, in the sense of their uncoiling and orientation according to the mechanical stress direction, and which appears as a reversible high-elastic deformation. Therefore, this property appears only when the macromolecule deformation rate is much higher than the displacement one, so that the irreversible deformation can be neglected. That is why even if we often speak about an equilibrium high-elastic deformation of a linear polymer; in reality this equilibrium is only partial. Referring to the highly elastic polymers we must admit that low-intensity relaxation phenomena – more pronounced in the transition domains – may be always noticed, especially at deformations close to the critical one.

By virtue of the polymer polydisperse character, the more reduced length macromolecules will reach the strained conformation faster, thus supporting a higher fraction of the local stress, a fact that entails their preferential splitting. However, at the same time, due to the chain's statistically coiled ball appearance, there exist numerous 'knots' between them which act as real crosslinkings. When stress is applied, the knotted macromolecules will undergo strong stresses in their contact points, a fact that entails some chemical bond homolytical splitting. Also, at very high deformations, when the mechanical vitrifying phenomenon or even the crystallization process take place within the elastomer, the mechanochemical processes become very frequent due to the strong orientation of the macromolecules and to the intensification of the interactions between them, having as a result the macroradical formation.

Therefore, the mechanochemical phenomena are also present in the case of highly elastic state polymers processing even if their intensity is lower in comparison with vitreous state polymers.

Usually, elastomers are processed through extrusion, forming in matrices, mixing. The first observations on the cold degradation of rubber by two roll-mixing, in the last century, refers to the fact that, under prolonged processing, natural rubber becomes sticky [703–705].

Subsequent investigations established, beyond any doubt, that such effects, known as 'plasticization' are induced by the modifications suffered by polymers in their molecular structure, i.e. reduction of their molecular weight.

Degradation of elastomers by two roll-mixing has several practical applications in processing industries. From the synthesis, usually polymers with high molecular weight are obtained, which makes difficult their processing and especially their homogenization with micromolecular ingredients (such as carbon black, vulcanization agents, accelerators and activators etc.). A previous processing of the elastomer, consisting in the reduction of its molecular weight, which brings about decrease of its viscosity, becomes therefore quite necessary.

During two-roll mixing or mastication, applied not only to elastomers but also to plastomers, under the action of some strong shearing stresses occurring in the material, intense breaking processes are developed. As a result, new surfaces are formed, yet, without any increase in the polymer's specific surface, as in the case of vibratory milling of rigid polymers. The new surfaces are extremely reactive, as containing macromolecules in a mechanically-excited state, capably of easily reacting, and even active centers of radicalic nature, formed through homolytic splitting of the chemical bonds, capable of reacting either with atmospheric oxygen or with the polymer's ingredients.

The medium in which elastomer degradation on rollers occurs represents a key factor in the evolution of the process. In an active medium (oxygen), polymer structuring may be observed in either branched or crosslinked reaction products thus resulting, as due to the reactions of macroradicals' stabilization preferentially through transfer to the polymer.

The observation made a long time ago in the case of natural rubber, i.e. the appearance of branchings and crosslinkings during mastication, was found out for PεCL [705] and also for the copolymer of chloroprene with 2,3-dichloro-1,3 butadiene at various temperatures [706].

In the two roll-mixing process in the presence of some inert gases (He, N_2) a continuous decrease in time of the molecular weight may be observed. In such a reaction medium, stabilization of the macroradicals formed through chains' homolytical splitting occurs preponderantly through disproportionation, with no branchings or crosslinkings.

Examination of the evolution of Mooney's viscosity – an indirect measure of molecular weight, allowed accurate estimation of the evolution, during mastication, of this property, at various temperatures, as a function of the nature of the environment (Figure 3.225).

Temperature is therefore seen as playing a significant part in such processes, more reduced values of temperature favour the destruction.

Kempermann [707] studied the plasticization of a wide range of elastomers, thus establishing the dependence of the destruction degree on a series of parameters of the mechanical regime, such as: duration, temperature, distance between cylinders.

In his turn, Baramboim followed the degradation of some copolymers of butadiene with styrene and acrylonitrile, in the conditions of mastication on a Brabender plastographer [708].

Rubber processing was carried out in the presence of a thermoreactive resin (of the resole type), a chemical reaction between the two component being thus produced. The results obtained have a special applicative value, evidencing the possibility to synthesise new products, under intense shearing regimes, with improved operation properties comparatively to the inital rubbers.

The formation of elastomer macroradicals during the processing led to the idea of a possible development of homopolymerization reactions between certain vinylic monomers and, respectively, grafting and block-copolymerization reactions.

Natural rubber's processing by two roll-mixing, in the presence of styrene, at temperatures ranging within the 75–130°C interval permits a rapid polymerization of the monomer. The reaction is influenced by the elastomer/styrene weight ratio, being observed that its decrease causes increase of the polymerization reaction rate, which comes to exceed the one corresponding to styrene's thermal polymerization.

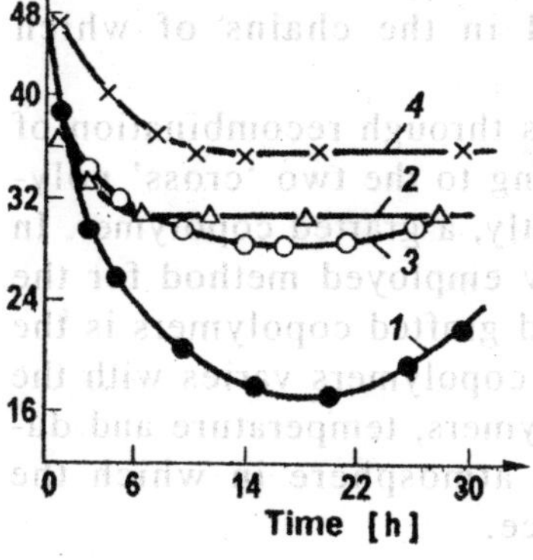

Figure 3.225. Dependence of Mooney viscosity of poly(chloroprene-co-2,3-dichloro-1,3-butadiene) on the duration of plasticization.

Under such circumstances, responsible for the initiation of polymerization are not the primary macroradicals from rubber, but the secondary ones, resulted from the reaction with atmospheric oxygen.

If, however, the objective is obtainment of block- or grafted copolymers, two variants may be applied, as follows:

- synthesis in the polymer–monomer system;
- synthesis in the polymer–polymer system, [659,709].

Regardless of the method of mechanical stress applied and of the initial reaction system (polymer–monomer or polymer–polymer), grafting and block-copolymerization involve a radicalic mechanism. The macroradical are formed through the basic homolytic splitting of the chains acting as initiators.

In this manner, natural rubber modified with maleic anhydride was obtained, two roll-mixing being applied as the processing method [710].

More frequent are the syntheses of block and grafted copolymers, in a polymer–polymer system. Usually, they are obtained as a mixture, which make impossible controlling of the synthesis for the obtainment of an unique type of copolymer.

However, the chemical nature of the polymers from the initial mixture may contribute to obtaining, predominantly, one of the two products.

Consequently, if the rates of the homopolymers' mechanical degradation are sensible equal, the possibility that two types of macroradicals should be formed in the system is the same.

The main way for their stabilization should be, in this situation, the 'crossed' recombination, with block-copolymers formed in most cases. If one of the polymers shows a high degradation rate in comparison with another polymer, macroradicals of this type will predominantly appear in the reaction medium; their stabilization will partially involve chain transfer, in which the macromolecules of the non-degraded polymer take part, and in the chains of which radicalic active centers will be formed.

In the next stage, stabilization occurs through recombination of the terminal macroradicals, corresponding to the two 'cross' polymers, the reaction product being, evidently, a grafted copolymer. In a viscoelastic state, the most frequently employed method for the mechanochemical synthesis of block and grafted copolymers is the two roll-mixing; the composition of the copolymers varies with the chemical structure of the processed polymers, temperature and duration of processing. In this case the atmosphere in which the process occurs has a negligible influence.

Some polymer–polymer systems, which led through mastication to block and graft copolymers, include: NR–PB or NBR [626], NR– SBR [659], PB–SBR [711], PB–ER [628], PB–PS [632], PIB–PE, PVC–PU [712].

The criterial factor for the degree of conversion is here viscosity, which induces increase of the shearing stresses and, consequently, the synthesis' efficiency.

These types of synthesis have applications in the industry of elastomer and thermoplastic polymer processing, permitting the obtainment of modified polymers, with higher physicochemical and mechanical properties, under conditions of their processing in classical installations.

3.8.2.3. Mechanochemistry of the polymers processing in the fluid-viscous state

3.8.2.3.1. The supramolecular structure and peculiarities of polymers processing in fluid viscous state

Tackling the problem of the morphology of the polymers in a fluid-viscous state is useful for two reasons: first of all, the physical and chemical properties can be explained in a correct way only on the basis of a structural model as close to reality as possible, secondly, because the impact of this structure on the properties of the product may be analyzed. This reaction product is obtained by changing the state of the polymer in a elastic state or solid one.

The structure of the polymeric fluid is given by the space arrangement of the entities that form it. In the case of the polymers in fluid state the disposition of the macromolecules depends on the chemical structure of the chain and on the thermomechanical conditions in which the macromolecules evolve.

When the thermomechanical conditions are stable in time and space, the structure of the polymeric fluid will be dictated by the intra and intermolecular interactions generated by the chemical structure of the macromolecules. The structure of the polymers in the fluid state has not been completely elucidated yet, but it is intensely studied.

J.P. Flory considered that in a melt the macromolecules takes the shape of a statistical coil, being dispersed between the macromolecules of the same type, [713].

If the chemical structure of the chain contains elements that do not hinder rotation around one bond and the flexibility is high, the macromolecule in the fluid state takes an isometric shape known under the name of globule [714]. The thermodynamic conditions reduce the interpenetration of the macromolecules so that, in a fluid state the molecules of the polymer will be placed in irregular rolls (Figure 3.226b) [715].

If the elements of the structural units hinder the rotation and rigidity of the chain, the macromolecules will take linear conformations, (Figure 3.227a). An experimental proof was presented for the globular structures as well as for the linear ones.

Using ED [716] and EM [717] certain ordered domains were found at distances of 50–100 Å in polymer melts. Thus, the linear conformations taken by the macromolecules in the melt are explained.

In another context, Fisher [718] have already presented dates based on the LS, SAXS and the SANS to demonstrate the globular conformation of the macromolecules in polymer melts.

Between the two limiting conformations, there are structure of

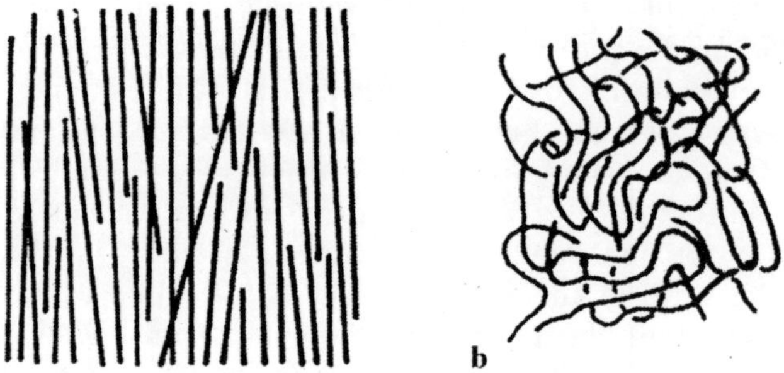

b

Figure 3.226. Supramolecular arrangement in: (a) conventional molten polymer and (b) liquid crystal polymer melt.

certain polymers in which the orientation can be induced as a result of intermolecular effects, whose size depends on the thermomechanical conditions.

The primary structure of the melt, resulted from synthesis, may suffer changes which hint at associations and orientations, generically referred to with the term structure.

The globular formations may be structured either through the aggregations of the globules forming structures that are more or less compact, or at a certain flexibility of the macromolecules the spreading of the globules takes place followed by the generation of some supramolecular filamentary structures.

The appearance of the linear aggregates takes place in polymer melts as the result of the intermolecular forces, when folding one

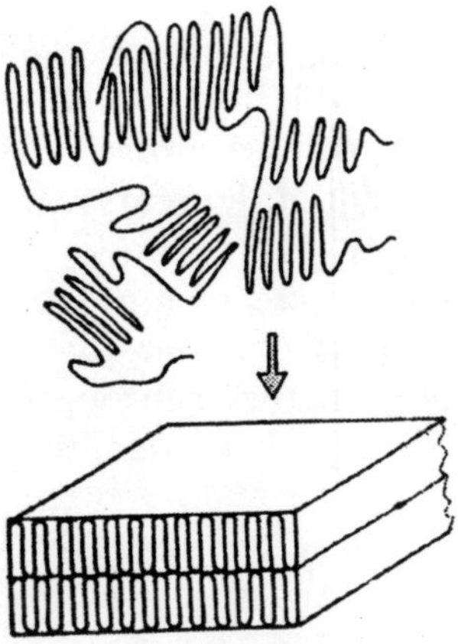

Figure 3.227. The scheme of the packing of chains segments and the ulterior structuring of these formations.

macromolecule or one part of it, or by coupling some isolated chains, Figure 3.227, [84].

The structure of the polymeric fluid can be realized through the connections rendered by the cohesion of secondary forces. The value of the cohesion energy of polymers in a viscous state exceeds the value of the energy of principal valence bonds, generating specific properties in the polymeric liquid.

Considering as a criterion the value of the cohesion energy for the chain of 5 Å, at the melt spanned polymers, the value of E_{5A} > 5 kcal/mole has been found [719].

Another way to structure the polymeric fluid is the formation of the temporary tridimensional network using entanglements.

Several models have been proposed for the structure of the polymer melts. Pechhold describes the polymer melt as a phase formed by the parallel macromolecules with a big concentration in defects, [720]. The free enthalpy of such a system is lowered through the summing of the defects in planes, which leads to the change in direction of the macromolecular fascicles and the appearance of some meanders. Such a model concords to the nodular structure in polymer melts, which show the presence of high anisotrope domains.

According to Lodge's theory, the polymeric fluid is considered to be a liquid elastomer, in which the cross-links are of a physical nature [721].

A 'physical cross-link' is defined as: "two points of two macromolecular chains which are moving together for a duration which is longer than the period of fluctuation τ^*.

The fluctuation period τ^*, is the minimal time required by a macromolecular chain to pass through the majority of its thermo-dynamical states" [721].

A series of models for the macromolecules which are in fluid state were elaborated by Curtis, Bird [722], Doi and Edwards [723] taking into consideration the principle of entanglements and replation.

Ferry considers that the temporary crosslinks are of topological nature and they are not obtained due to intermolecular forces [724]. The coupling of the entanglements can be realized through : 1 – adhesion or temporary network; 2 – local knots: 3 – superpositions of large loops, Figure 3.228.

The formation of entanglements between the macromolecules of a polymer melt is influenced by the dimensions and the structure of the chain.

The models that have been presented refer to the macromolecules in fluid state in absence of external mechanical stresses.

Above the flowing temperature the polymers are characterized by a high mobility of the macromolecules which results in the appearance of the macrobrownian motion. The cancellation in a rather great extent of the physical interactions between macromolecules facilitates their shifting to one another so that the flowing is the principal effect of the mechanical energy action.

But the flowing temperature depends on the polymers molecular weight. For this reason, by virtue of the polymer polydispersity, the fractions of molecular weight higher than the average one, will be characterized by a lower mobility so that they will suffer a homolytical splitting under the action of the shearing forces developed during processing.

The unavoidable presence of atmospheric oxygen accelerates the degradation process thus initiated so that the small-sized neighboring macromolecules are also affected which finally results in the decreasing in the average molecular weight of the polymer

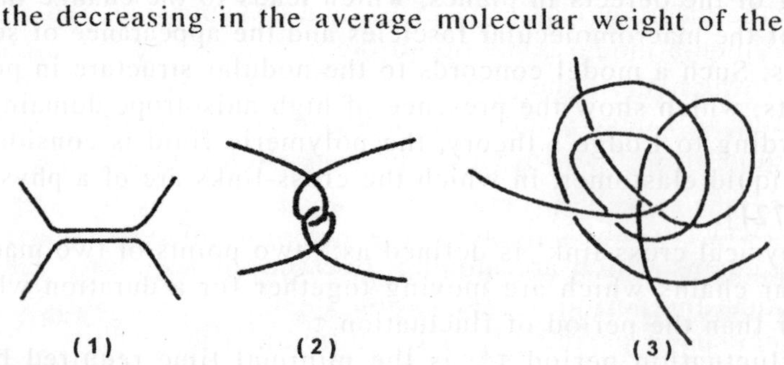

Figure 3.228. Types of entanglement coupling: (1) temporary cross-link; (2) local knot; (3) long-range contour loop.

as well as in the lowering of the polydispersity degree.

The study of the mechanochemical aspects involved in the processing of the polymer melts is rather difficult due to the superposition of the thermal and oxidative degradations. Thus, the effect of a single component of the overall process of thermal–mechanical oxidative degradation is almost impossible to separate.

The continuous stress of the polymer melt by shearing forces brings about the appearance of some discontinuities within the material, which causes the formation of new surfaces. The formation process of the new surfaces obey the general laws of the fracturing mechanism. Even if the relaxation times would be short suggesting thus the easy and rapid adaptation of the macromolecules to the stress, the high rate of the stress would result in the appearance of some strains exceeding the material resistance and determining the fracturing. A characteristic of the polymer fracturing in the flowing state is the fact that the total surface of the material does not modify although new surfaces continuously appear due to their strong adhesion.

If a melted polymer is forced to pass through a capillary the phenomenon of 'melt fracture' appears which is noticed with the non-Newtonian liquids only. This phenomenon appears along the whole capillary but it can be made evident especially at the outlet, which can be explained as follows.

During the flow through the capillary the macromolecules are orientated to the direction of the force action. According to the Boltzmann's theory, any process of molecular orientation corresponds to an entropycal state lower than that characteristic of the entirely random state. Hence, the polymer entropy at the capillary outlet is lower than the initial one. On the other hand, the brownian motion tends to disorder the system and in case of a slow flow this process can prevail preventing from the orientation. But in case of the rapid flow of very big macromolecules the orientation effect is quite marked.

When the Brownian motion does not manage to maintain the macromolecules disordered and the stress field is continuously increased the appearing effects could be explained by two alternatives:

– the main stress field changes itself so that the molecular orientation is diminished or even canceled (which is practically impossible with a steady state flow);

– the longer macromolecules are splinted into smaller fragments which are much more mobile and, hence, more liable to perform disordering Brownian motion.

The last explanation seems to be more plausible suggesting thus the mechanochemical aspects involved in 'melt fracture' [725–727].

By applying a field of external forces, the macromolecules are deformed and can attend extended conformations, in which only the entropy is reduced, this time the enthalpy suffering little changes, thus the melt temperature calculated from relation $T_m = \Delta H/\Delta S$, is higher [728].

In the polymeric fluid, which is mechanically stressed, linear formations called lines of nucleations, which have the same orientation as the force, appear.

In conclusion, the molecules of a polymeric liquid in flow, depending on the gradient rate, can be oriented and deformed significantly, compared to the isotropic structure, which prevails in the stationary liquid. If the macromolecules have already been partially oriented into liquid state, the change of the entropy is reduced and the crystals are easily formed. The kinetics of the crystallization process is considerably accelerated and the structure of the crystals that are produced is affected.

The mechanical deformation of the polymers in fluid state may produce a high axial orientation of the chains. The interchain junctions tend to separate the crystals, forming regions with a weak orientation and forbidden crystallization, Figure 3.229 curves c and d) [729].

This kind of defect reduces the total anisotropy and limits the level of the axial properties.

If the orientation is induced in the liquid crystal polymer, when there are fewer junctions, the orientated state is even more uniform and is characterized by high axial resistances and modules, Figure 3.230, curves a and b).

The fluid resistance to mechanical shearing or its internal friction is characterized by viscosity, being a reflection of the specific structure of the polymer in the fluid state.

The viscosity of the polymers in the fluid state illustrates a series of dependencies on the molecular weight, molecular weight distribution, chain structure and superstructure.

The viscosities of the polymers in the fluid state can be compared by varying one factor and maintaining the others constant.

Since the mechanical factor may influence significantly the viscosity of a fluid polymer the viscosities were compared under the

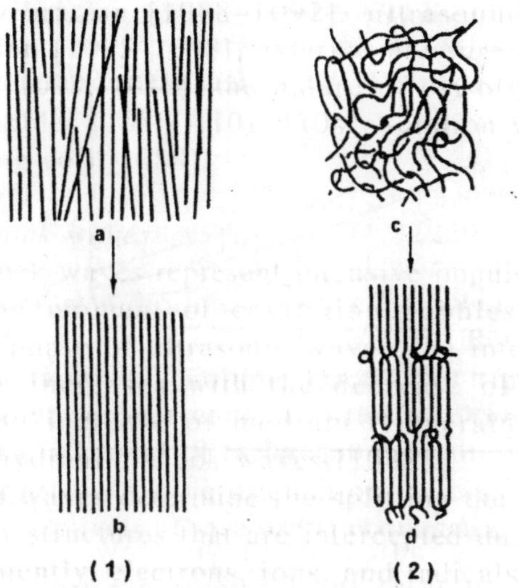

Figure 3.229. Crystallization of a liquid crystal polymer melt (1) and of a flexible chain polymer melt (2).

conditions of absence of shearing, the so-called 'initial viscosity' η_0.

This parameter was correlated with the characteristics of the macromolecular chain by the relationship derived by Bueche [730] and completed by Allen and Fox [731].

The dependence of polymer viscosity on the temperature is particularly important to be known for the good understanding of the flowing process mechanism and also for making evident the relation between the macromolecular fluid structure and its behavior to the mechanical stress [732].

The temperature dependence of the Newtonian viscosity of the polymer melts is given by the WLF equation:

$$\log \eta_0 = \log \eta_g + \frac{C_1 (T - T_g)}{C_2 + (T - T_g)} \qquad (3.121)$$

which is valid within the $T_g < T < T_g + 100$ °C domain. Andrade's law holds above this temperature:

$$\eta = A \cdot e^{\frac{E}{RT}} \qquad (3.122)$$

where E denotes the activation energy of the viscous flow.

In the case of polymer melts the log η versus $1/T$ relation is linear over a 50–60 °C range only [733,734].

This fact is indicative of a modification of the polymer fluid structure within a larger temperature range.

The second important aspect for the polymer processing in the fluid state is connected with the system behavior under the action of a mechanical factor.

The response of the flowing fluid to the stress is controlled by the correlation between the resultant tension and the normal and tangential tension bringing at the system motion.

The fluid polymers may be processed in a flow field with simple shearing, in a flow field with elongation, at compression, or in a combination of such fields.

In the case of a simple shearing flow of a non-compressible homogeneous fluid polymer, the components of the tension tensor, τ, can be expressed by three independent functions [735]:

$$\tau_{1,2} = \nu(\dot{\gamma})\dot{\gamma};\ \tau_{11} - \tau_{22} = \Psi_1(\dot{\gamma})\dot{\gamma}^2;\ \tau_{22} - \tau_{33} = \Psi_2(\dot{\gamma})\dot{\gamma}^2 \qquad (3.123)$$

where the index 1 is indicative of the flow direction, 2 for a direction perpendicular to the flowing direction and 3 for the neutral direction.

From the relation of the first difference of the normal tension

$$\tau_{11} - \tau_{22} = f(M, T, \eta_0)\dot{\gamma}^2 \qquad (3.124)$$

it is evident that the elements of elasticity have an important weight for a fluid polymer submitted to a simple shearing flow [735].

The fluids with measurable differences in the normal tension show non-Newtonian behavior.

In the case of high shearing tensions the first difference of the normal tension was shown to be proportional to the shearing tension. For the PE, PA melts, the appearing normal tension are by a magnitude order higher than the shear stresses developed during flowing [735,736].

The normal tension stresses the polymer fluid to elongation. The final results of the simple flow will be given by the shearings and elongations in the processing route. In the case of elongation, the polymer fluid shows elongational viscosity.

According to Trouton's rule, the elongational and the simple shearing viscosities are correlated by the relation:

$$\eta_E = 2\eta_S \qquad (3.125)$$

when the deformation rate tends to zero. For the PET melt the value of the factor of proportionality is ~4 [737].

The processes of polymer degradation in the flowing state develop inevitably during processing by injection, rolling, calandering, extrusion, and spinning.

For instance, Saveliev [738] processed the PVC by injection and found that the polymer viscosity was rather low at 190–195°C, the mechanical degradation being absent.

By decreasing the working temperature by 15–20 °C the viscosity increases and so do the high pressures necessary for the flowing of the melted material thus causing the splitting of chemical bonds. At temperatures above 195 °C, the degradation process again intensifies due to the contribution of thermal oxidation destruction.

It should also be mentioned that beside temperature the overall degradation process is also influenced by the pressure created for the material entering into the matrix and by the rolling rate of the helical conveyer.

The flowing of a fluid polymer through media whose geometrical characteristics (perimeter and area of the cross section) modify along the route will determine the appearance of a complex flowing field where shearing, elongation and twisting stresses are present [739–741].

There are several geometry combinations of the media through which the polymer flowing is performed. The following combination of geometries is often encountered in the working processes; a duct with a circular cross section of a high radius followed by a route with either circular or rectangular cross section of a small area.

In the case of extrusion, spinning, injection the fluid polymer is forced to flow from a space of a high cross section through a nozzle of a much smaller cross section. This technological situation may be illustrated as in Figure 3.230.

Between the reservoir and the capillary outlet a pressure drop ΔP_T appears which is plotted versus route length (Figure 3.230).

Four flow żones are distinct:

Ist zone: simple flow with shearing;

IInd zone: complex convergent flow with shearing and elongation

IIIrd zone: simple flow with shearing;

IVth zone: complex divergent flow with elongation and shearing.

For the fluid polymers the pressure drop at a certain flow rate is due to the viscous dissipation as well as to the elastic and inertial dissipation [742].

The elastic dissipation is responsible for some effects such as the great pressure losses, the swelling of the extruded polymer, melt fracture, elastic turbulence.

The inertial dissipation is low in comparison to the other two components and it is usually neglected when the flow behavior of

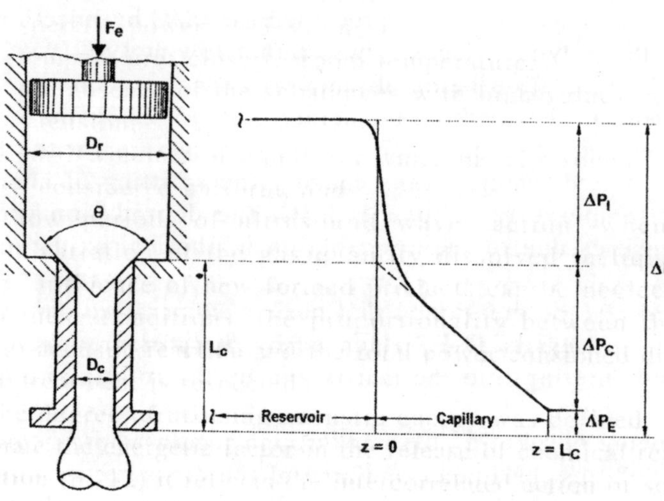

Figure 3.230. Geometry (a) and the diagram of the pressure (b) for the flow of a polymer in fluid state from a reservoir with the diameter D_r, trough a capillary with the diameter D_c.

a fluid polymer is studied. But the relative contributions of the viscous and elastic dissipations to the pressure drop cannot be separated and they are considered to change as a function of the route geometry.

The drop pressure in the reservoir is negligible with respect to the values of the other three zones, so that the total pressure drop, ΔP_T, is given by the relation:

$$\Delta P_T = \Delta P_I + \Delta P_C + \Delta P_E \tag{3.126}$$

where I, C, E denote the inlet, capillary route and exit, respectively.

The inlet pressure drop may be more or less important than that inside the capillary [743].

For the fluid polymer the inlet total pressure drop may be theoretically separated into viscous pressure drop, ΔP_{Iv}, and elastic pressure drop, ΔP_{Ie} [735].

$$\Delta P_I = \Delta P_{Iv} + \Delta P_{Ie} \tag{3.127}$$

For the polymer melts ΔP_{Iv} was found to represent less than 5% of ΔP_I which indicates the inlet drop pressure to be determined by the melt elasticity.

Practically, one may consider:

$$\Delta P_I \cong \Delta P_{Ie} \tag{3.128}$$

The mechanochemical phenomena connected with the polymer fluid at inlet are determined to the greatest extent by elongation and to a lesser extent by shearing.

Since the inlet pressure drop, ΔP_l, is strongly dependent on the behavior of the melt a change in the inlet geometry would be expected to influence the pressure drop as well as the elastic and shearing strains.

The experimental data obtained by Han [735] with PE at 154°C revealed the inlet pressure drop to increase with increasing shearing rate, at a given entrance angle. In addition to that, the inlet pressure loss was found to decrease with increasing entrance angle from 15 to 60 °C remaining than constant.

The entrance pressure drop is higher for small entrance angles. Practically the capillaries with small entrance angles are chosen since the critical shearing rate at which the melt fracture begins, is higher and thus high flow rates can result in processing [735,744].

By assuming a behavior according to the power law for both shearing and elongation flows and applying Cogswell's law [745], the inlet pressure drop can be estimated by means of the relationship:

$$\Delta P_l = \dot{\gamma}_w \left[\frac{2}{3} \eta \frac{1}{\tan \Theta} + \frac{\lambda}{3} \tan \Theta \right] \tag{3.129}$$

where η and λ denote the viscosities for simple shear and for elongation, respectively; Θ is the entrance angle of the capillary. This relation holds for the Newtonian or apparent-Newtonian behavior, $n \leq 1$.

Relation (128) indicates that the viscous component prevails for small entrance angles and the elongation component for large entrance angles.

The pressure drop in the capillary is given by:

$$\Delta P_c = 2 \eta \dot{\gamma}_w \frac{L}{D_c} \tag{3.130}$$

Using the $\Delta P_l / \Delta P_C$ ratio, it may be estimated which of the capillary zones influences the flow as well as the magnitude of the tensions. This ratio may be of sub-unitary, unitary or supra-unitary values.

With low L/D_C ratios, the effect of the pressure drop in the capillary is rather small and the inlet effect is prevailing.

At high L/D_c ratios, the inlet pressure drop is much lower than that in the capillary. The end effects can be neglected.

The total pressure drop changes with the convergence angle and shows minimum values for a certain angle, Θ.

But for the flow of the fluid polymers through a capillary with a high angle at the capillary inlet, a natural convergence angle, Θ′, forms, which is smaller than the capillary angle, Θ.

Under these conditions, the inlet pressure drop will be lower than the theoretical one corresponding to angle Θ and the elastic component will decrease. At the same time, the viscous component increases [746]. The occurrence of some high elongations strains at the capillary inlet results in the formation of some crystalline formations within the polymer melt and, hence, of some temporary networks [747].

The molecular and supermolecular structures of the polymer submitted to extension can be influenced even from the capillary inlet zone [748–750].

The analysis of the most important external factors involved in the fluid polymer processing, namely the temperature and the mechanical forces, reveals that the macromolecules can suffer thermal and mechanical stresses whose values can sometime exceed the limit of the fluid state existence. Irreversible transformations of the fluid molecular and supermolecular structures occur. When the mechanical factor prevails, the transformation is the mechanochemical nature.

Since the thermal factor is high during polymer processing in the fluid state the transformations are of the thermo-mechanochemical type.

The micromolecular fluids flow by the relative shifting of the molecules to one another.

The macromolecular components can flow by the shifting of some chain fragments. Sometimes the intermolecular interactions are strong and the rapid movement of these fragments cannot assure the shifting of the whole macromolecule. In these cases, by a prolonged action of the external strains the energy concentrates on some zones with structural defects and the chain is broken. Thus, radicalic fragments of size smaller than that of the initial macromolecule are formed. These fragments can act on the polymer as a plasticizer facilitating the much more rapid shifting of the moving macromolecules as well as the shifting of the motionless ones, which do not change their position due to the strong intermolecular interactions.

This procedure of initiating the flow and of amplifying the polymer flowing was called 'the chemical flow' by Kargin.

It was noticed that along with the common flow based on the shifting of the macromolecular segments a chemical flow also takes place involving the shifting of the newly formed fragments and their recombination as new chains.

Under the condition of the industrial processing of the polymers the viscous flow and the 'chemical flow' proceed simultaneously and their ratio varies over a wide range.

When the working temperature is higher and the molecular weight is rather low the flowing process is not accompanied by the scission of the chemical bonds. In case of the high molecular weights and lower temperatures when an intense mechanical stress is applied the role of the 'chemical flow' becomes important. Under the conditions of strong mechanical stress, high deformation rates, the chemical flow may prevail. This fact is particularly important for the mechanical processing of the crosslinked polymers. In this case the macromolecules fixed in an unique network cannot flow unless the chemical bonds are simultaneously broken.

Under the action of strong enough mechanical stresses the macromolecular network decomposes into diradicalic fragments which within a short time behave as molecules of a micromolecular fluid, suffering a relative shifting under the action of the mechanical stresses and fixing then again in a spatial network of a structure similar to the first but which differs from these by a new macroscopic form of the polymeric product.

The mechanical stress on the polymers in fluid state induces typical mechanochemical reactions proceeding unavoidable together with those caused by the thermal factor, being also promoted by active media (water, oxygen).

During the fluid-state processing the polymer material may be submitted to shearing and elongation stresses which may modify the supramolecular and even the molecular structures.

Important studies on the mechanochemistry of the polymer melt processing have been reported by Porter [164, 272, 273].

The importance of the coiling between macromolecules for their scission was made evident.

Under the conditions of shearing stresses within the melt mass the maximum mechanical stress concentrates on the macromolecule middle or in its neighboring but above all on the chemical bonds in the 'knot' regions. One of these possibilities prevails in function of the stress duration, working temperature, probability of the relative slipping of the macromolecules, resistance of the chemical bonds in the polymer main chain. Although the Buech's theory admits the preferential splitting of the macromolecules near their center, the Porter's experimental data do not entirely support this mechanism. The distribution curves of molecular weights (based on the HPLC and GPC data) indicate the accumulation of rather great amounts of fractions of low molecular weight which would suggest the chain breaking also takes place to a great extent at the chain ends, in the knot zone, probably.

Intensive processes of mechanical destruction also develop during melt spinning, rolling or calandering. The formation of new strongly activated surfaces makes possible the reaction of the polymer with other macro- or micromolecular compounds which supports the radicalic mechanism of the process.

3.8.2.3.2. Mechanochemical reactions during polymer processing by two roll-mixing

Two roll-mixing is a widely used working process in the elastomer and plastic industry. It consists in submitting a polymer, polymer mixtures or polymer-micromolecular compound mixtures to intensive shearing stresses by passing them repeatedly through metallic cylinders having either smooth or rough surfaces and rolling in the opposite directions at different speeds.

The peripheral speed of the cylinders affords to define the friction ratio, which varies between 1 and 2, taking usually the value 1.2. Due to the rolling of the cylinders, the adhesion and the friction, the material to be processed is passed through the slot between cylinders and taken over by one of them in function of their differences of tours and peripheral speed. Due to the difference in the rolling speed of the cylinders strong shearing stresses are generated within the interval between them which causes the unavoidable mechanical degradation of the polymer. The intensity of these stresses also depends on the distance between the two cylinders [751]. The occurrence of some strong shearing stresses due to the mechanical and thermal solicitation of the polymer during the two roll-mixing process causes firstly the appearance of the mechanoactivated state followed by the splitting of the chemical bonds in the main chain (according to the mechan-ochemical mechanism of the polymer breaking discussed above). Consequently, high concentrations of radicals appear which are located on the new surfaces occurring continuously under the action of the shearing stresses. The polymer becomes extremely reactive being able to interact chemically with other micro- or macromolecular compounds in the medium.

The possibility of activating some reactions of the polymer with certain compounds under these conditions constitutes a good argument supporting the radicalic mechanism of the process. At the same time, modified polymers can be obtained showing properties superior to those of the initial ones by the proper selection of the compounds able to react with the macroradicals formed by cracking. This way of chemical modification shows several advantages, the most important one being the possibility of its practical performing with the equipment commonly used for polymer processing.

PVC is one of the most important polymers, largely studied as regards the mechanochemical transformations occurring during its processing. Kargin and Sokolova [752,753] have advanced the theory of the 'chemical flow' consisting of the continuous splitting and remaking of th. chemical bonds. By means of this concept the behavior of PVC during its processing by injection was described.

At high rates of the extruder helical conveyer the properties of the final productare affected [738,754]. With increasing temperature the obtained working polymer shows, surprisingly, higher molecular weights as well as branched structures. At enough high temperatures the cross-linking reaction prevail; consequently some mechanical properties improve. The process of 'chemical flow' is negligible for the PVC of low molecular weight and it is completely absent in the melts with low viscosity.

The fact deserves mention that, surprisingly, by recycling and reprocessing of PVC under the same conditions the mechanical properties improve as a result of a better distribution of the ingredients used and of the higher homogeneity of the material.

Hence the 'chemical flow' involves the splitting of the covalent bonds in the main chain and their remaking in another part of the melt, where the macroradicals arrive. Their presence within the processed polymer melt was made evident by the PVC processing on rolls in the presence of aromatic diamines known for their property of accepting radicals.

Thus, by chemical and spectral methods, the products resulted by the simultaneous two roll-mixing of PVC with benzidine, *o*- or *p*-PhDA were proved to contain aromatic rings on the main chain [301,308]. By the photocolorimetric dosage of the unreacted diamine, the percentage of the diamine reacted with PVC (with respect to the initial amount) could be estimated, Figure 3.231.

The aromatic diamine links by a radicalic mechanism:

The benzidine seems to be the most reactive since it reacts al-

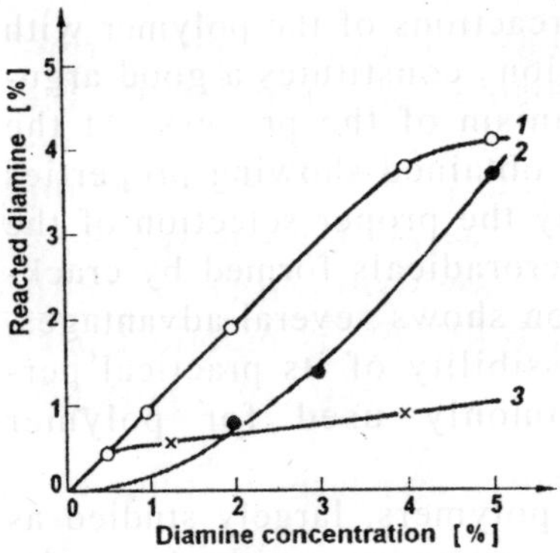

Figure 3.231. Variation of the reacted diamine amount with its content in the initial reaction mixture. 1) benzidine; 2) *p*-PhDA; 3) *m*-PhDA.

most entirely (Figure 2.232, curve 1). The *m*-PhDA is less reactive

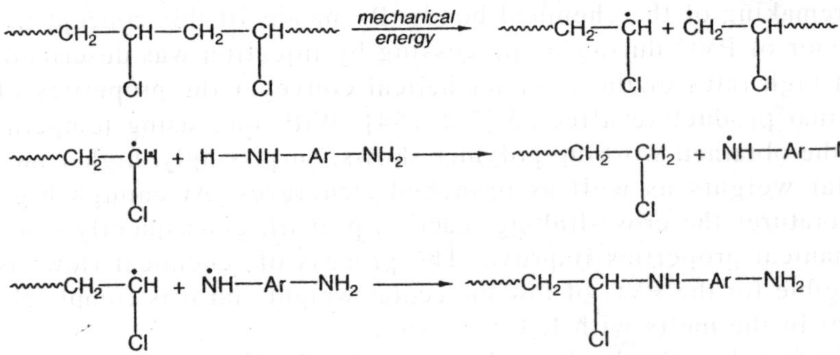

due to the position of its amino groups. The free amino groups can be diazotized and coupled with phenols or a mixture tertiary amines when various colors result.

The coloring capacity decreases with increasing process duration, which indicates a consumption of the free amino groups in time (an effect less evident with *m*-PhDA). This could be explained, by the fact that with both benzidine and *m*-PhDA, the free amino group can play the same role of chain-transfer agent linking by chemical bonds other radicalic fragments.

This hypothesis was experimentally confirmed by following the variation of molecular weight of the reaction products in time, Figure 3.232.

Hence in the presence of benzidine and, especially, of *p*-PhDA the molecular weight practically increases in time as a consequence of the formation of structures of the type:

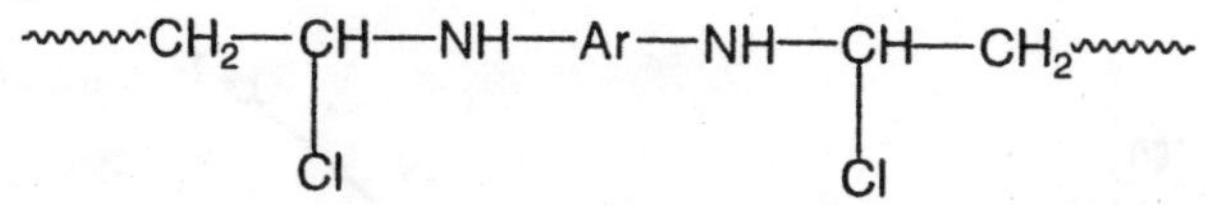

As regards *m*-PhDA, its structure induces steric hindrances, which do not allow for reaction of the second functional group, so that the molecular weight decreases continuously and the colors resulting by diazotization and coupling are intensified.

The chemical bonding of the micromolecular compound to the PVC chains causes important structural changes of the polymer, which significantly affects some properties. The most spectacular one is the tensile strength, which increases considerably with increasing diamine amount in the initial mixture, Figure 3.233, [311].

The best results are obtained with benzidine when a new material is obtained, showing a tensile strength higher by 140% than that of the initial polymer.

The tensile strength increases with duration and especially with diamine amount, correlated with the continuously decreasing in the dissolution capacity of the resulting products (for 5% *p*-PhDA in the initial mixture the polymer becomes practically insoluble) requires to take a crosslinking reaction into account which proceeds by hydrogen chloride elimination between the diamine and the chlorine in PVC.

An important effect for the practical processing of PVC is noticed in case of its two roll-mixing in the presence of benzidine which could partially substitutes the classical thermal stabilizer

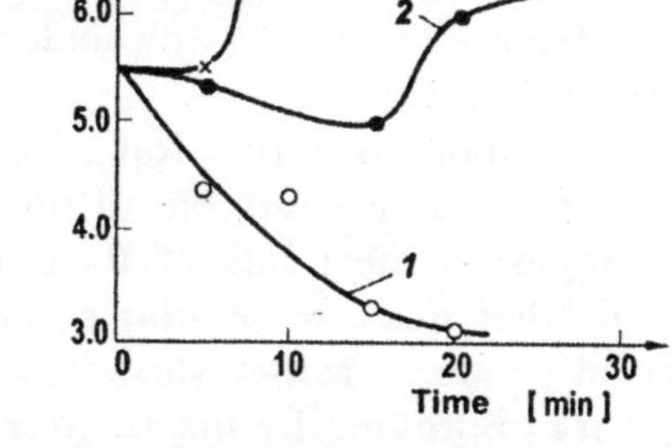

Figure 3.232. Variation in time of molecular weight of PVC by two-roll mixing in the presence of aromatic diamines: 1) *m*-PhDA; 2) benzidine; 3) *p*-PhDA.

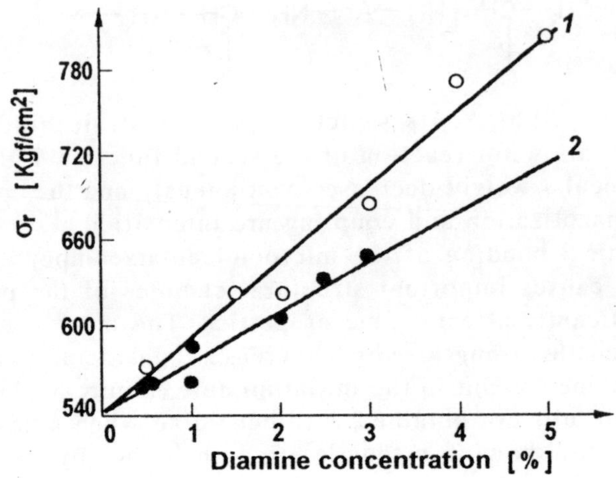

Figure 3.233. Variation of tensile strength of PVC with diamine concentration in the initial mixture. 1) benzidine; 2) *p*-PhDA.

(lead salts) when its concentrations exceed 3%. Its content with respect to the polymer may be diminished until 1% [312].

The efficiency of the PVC-benzidine reaction as reflected in the tensile strength increase depends not only on the micromolecular compound concentration (C_d) but also on other factors among which the concentration of the thermal stabilizer (C_s) and the two roll-mixing duration and temperature are the most important.

The mathematical modelling of the two roll-mixing process of the PVC-benzidine mixture, based on the above mentioned variables afforded the following function of tensile strength variation:

$$\sigma_r = 703 - 16C_s + 12.2C_d - 5C_s^2 - 5.3C_d^2 + 7.6tT + 9.4tC_s + 11.4tC_d - 9.3TC_s - 11.8TC_d \tag{3.131}$$

The material resulting from PVC processing under these conditions has a tensile strength of 722 Kgf/cm^2.

The possibility of mechanical activation of some reactions of PVC with micromolecular compounds based on the formation of free macroradicals suggested the trials of its grafting and block-copolymerization with other macromolecular compounds.

Since the PVC rigid products rather show low impact strength, the trial was made of its improving by mastication with elastomers.

The processing of PVC in the presence of PU (based on dibenzyl diisocyanate) with different alcohols (ethylene glycol, triethylene glycol, glycerol) resulted in products of higher impact strength in comparison with the starting PVC [605].

The IR spectral measurements as well as the analysis of the curves of turbidimetric titration confirmed the formation of block- or graft-copolymers.

The IR spectra show absorption bands at 1520 and 1580 cm^{-1}, ascribed to the –CO–NH– group, as well as those at 1690 cm^{-1} corresponding to the –CO–O– group, in the products resulted by co-mastication, after previous extraction of the starting homopolymer in adequate solvents. These products are not physical mixtures of PVC with PUs as shown by the curves of turbidimetric titration, having a single level and proving thus the existence of an unique compound resulting, probably, from the reaction of the two polymers, Figure 3.234.

The structure of the reaction product is of the type:

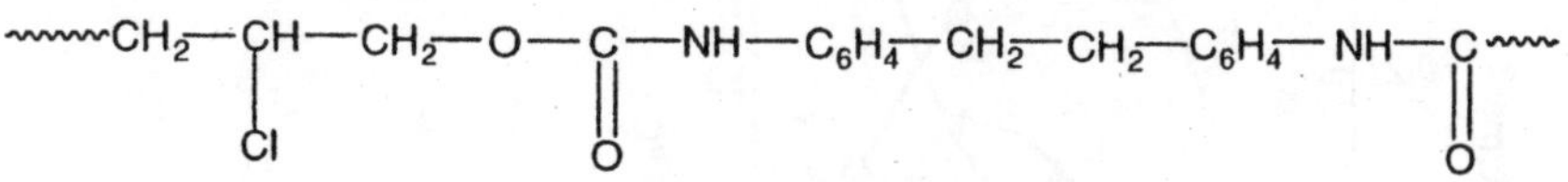

Block-copolymer

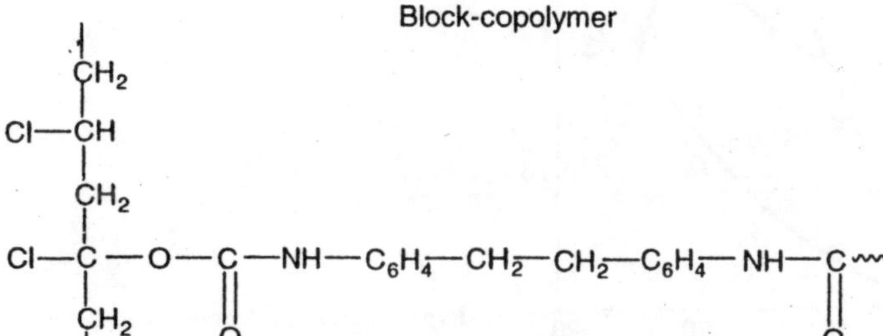

Graft-copolymer

The presence of the elastomer fragment either in the main chain of the reaction products or as a branch results in the increase in the macromolecule flexibility which determines at the macroscopic level the increase in the impact strength. A close correlation exists between the mechanical properties and the elastomer concentration in the mixture under processing (Figure 3.235), the impact strength improves with increasing concentration.

The reaction efficiency and, hence, the impact strength is also influenced by the processing temperature and duration.

The chemical nature of the diol in the PU structure also influences the mechanical properties under identical working conditions. Thus, with the glycerol based PU (curve 3) the impact strength shows a maximum at 170°C. Its decreasing above this two roll-mixing temperature is due to some crosslinking reactions of the de-

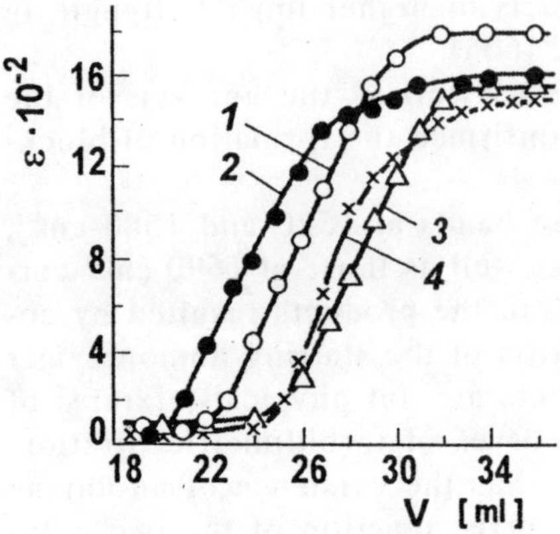

Figure 3.234. Curves of turbidimetric titration of some products obtained by co-mastication of PVC with polyurethanes. 1) standard sample; 2) PU with ethylene glycol; 3) PU with triethylene glycol; 4) PU with glycerol.

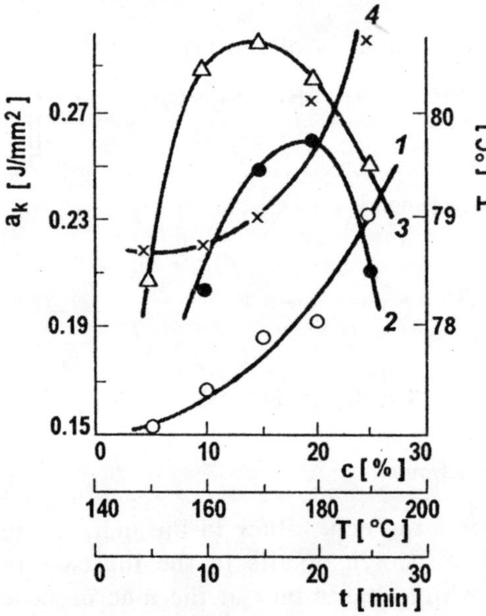

Figure 3.235. Variation of impact strength (a_x) with reaction parameters. 1) with concentration (PU with glycerol); 2) with duration (PU with glycerol); 3, 4) with temperature (PU with glycerol and ethylene glycol, respectively).

struction fragments either by recombinations or by means of the side hydroxyl groups; the macroscopic result is the polymer strengthening. The same effect is noticed with increasing two roll-mixing time (curve 2). The increasing in the impact strength for two roll-mixing duration of 20 min is due to the presence of continuously increasing amounts of elastic component in the reaction product structure, in spite of some few crosslinking reaction. However, the crosslinking reactions intensify at longer working duration, which results in the formation of a rigid structure and the decreasing in the impact strength.

The crosslinking does not proceed with increasing temperature in case of the ethylene glycol based PU (curve 4) so that the impact strength increases continuously.

The influence of the two roll-mixing process parameters on the mechanical properties of the product resulting by the simultaneous processing of PVC with an elastomer, was thoroughly studied with a PU containing dibenzyl diisocyanate-ethylene glycol and adipic acid as an elastic component [712].

The functions correlating the process parameters with the tensile strength (σ_r) and impact strength (a_k) have been derived by mathematical modelling:

$$\sigma_r = 635.9 + 14.8\times C + 8.6\times C\times T \quad (3.132)$$

$$a_k = 3.5\times10^{-2} + 2.8\times10^{-3}\times T - 5.2\times10^{-3}\times C\times T - 4.6\times10^{-3}\times t\times T \quad (3.133)$$

where C is the elastomer concentration; t the processing time; T the processing temperature.

The tensile strength increases with temperature due to the structuring reaction, Figure 3.236.

The impact strength shows a different variation for different processing temperature ranges, Figure 3.237.

At lower temperatures (150°C) the impact strength increases continuously with increasing two roll-mixing time certainly, by the increase of the elastomer content in the reaction product in the absence of the crosslinking reactions. At higher temperatures the effect is quite opposite (Figure 3.237b). The decrease in the impact strength with increasing two roll-mixing time and PU concentration is connected with the intensification of the crosslinking reaction, proceeding predominantly for the ethylene glycol-based elastomer.

The optimum conditions of processing the mixture of the two polymers for obtaining a product with mechanical properties much improved in comparison with those of PVC have thus been settled.

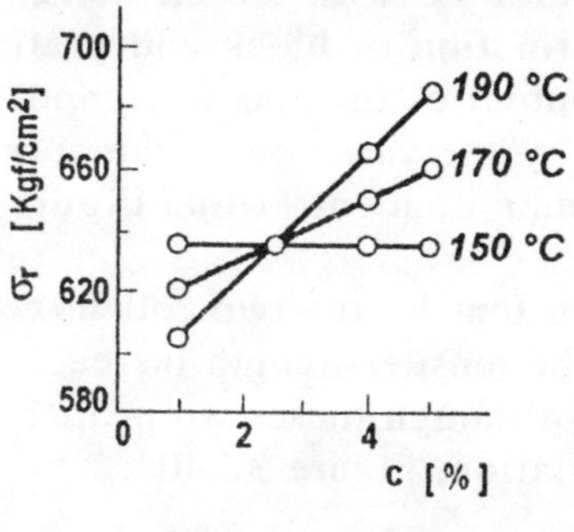

Figure 3.236. Tensile strength versus polyurethane concentration for different two roll-mixing temperatures.

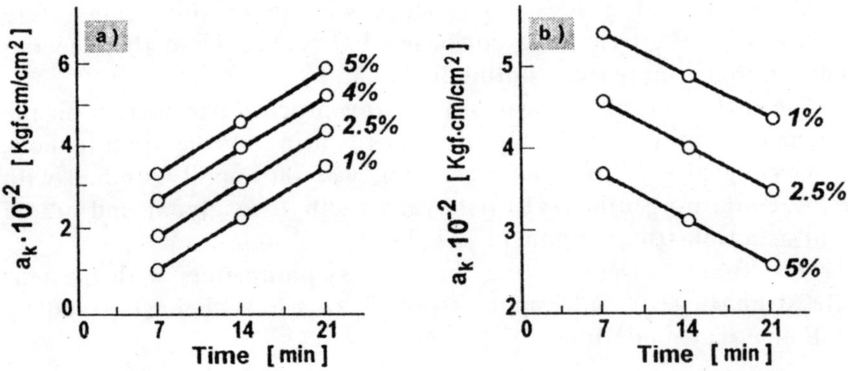

Figure 3.237. Impact strength versus working duration for different polyurethane concentrations: (a) T = 150 °C; (b) T = 190 °C.

Thus, a high tensile strength and low impact strength can be achieved with 5% PU concentration, short two roll-mixing time (7 min) and high temperature (190°C). A high impact strength and a tensile strength close to that of PVC can be obtained with short two roll-mixing times (7 min.), high temperatures (190°C) and low PU concentrations (1–2%) for preventing from the cross-linking reactions.

Finally, products of high impact strength and tensile strength can be obtained with high PU concentrations, long durations and lower temperature (150°C).

By means of co-mastication processes some reactions of PVC with nitrile rubber could be activated which proved to be of a practical importance [303]. In this case, the temperature, duration and elastomer concentration are also the main parameters influencing the process. The dependence of the impact strength on concentration and duration is illustrated by the constant level curves in the plane of the two variables, Figure 3.238.

The increase in the impact strength of PVC with the elastomer initial concentration is the result of formation of block-and graft copolymers with increasingly higher content of the elastic component. This conclusion is also supported by the variation of the softening point VICAT, with the elastomer concentration, Figure 3.239.

The interesting fact deserves mention that by the two roll-mixing processing of this polymer mixture the tensile strength increases concomitantly with the impact strength although these two properties are known to show an opposite variation, Figure 3.240.

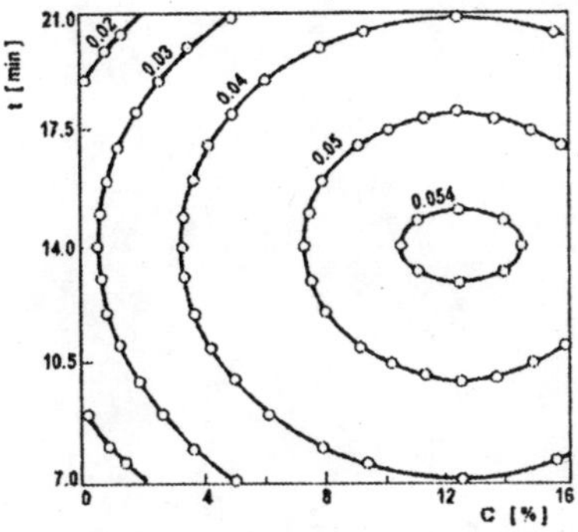

Figure 3.238. Constant level curves of impact strength in the plane of two variables: concentration – time (T = 170 °C).

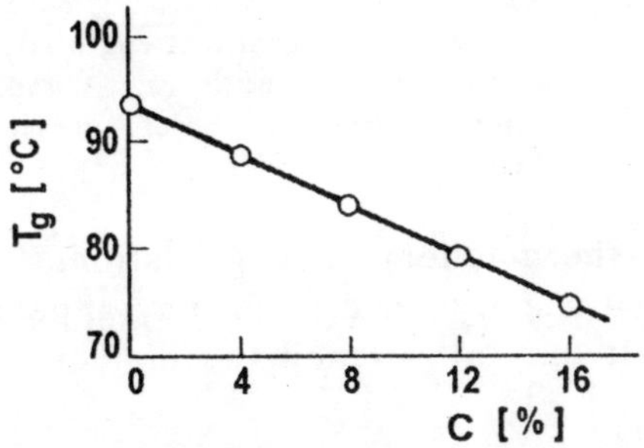

Figure 3.239. Softening point VICAT versus elastomers concentration (T = 170 °C; t = 14 min).

The maximal values corresponding to these two properties corresponds for an elastomer concentration of about 12%, in the starting mixture.

The conclusion can be drawn that due to the activation of some reactions as a consequence of formation of free radicals, the two roll-mixing process of polymers could be a simple and efficient method for performing chemical modifications of the macromolecular compounds as well as for synthesizing new compounds showing properties superior to the starting ones.

3.8.2.3.3. Mechanochemical transformations during polymer processing by extrusion and spinning

The main principle of the mechanochemical degradation is the scission of the inter and intramolecular links due to the action of mechanical energy. The scission of chemical bonds takes place as a result of its non-uniform distribution when the stress reaches critical values higher than the strength of the links in different places of the polymeric fluid.

The difficulties that arise in elucidating the mechanochemical reaction mechanisms in the polymers in a fluid state are related to the limited possibilities of separating the effects of the mechanical

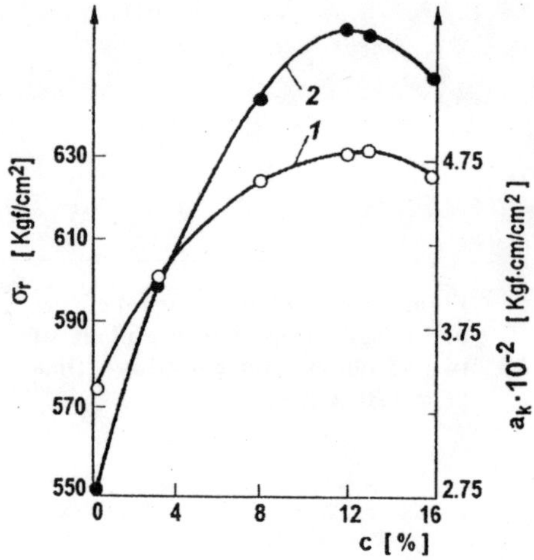

Figure 3.240. Tensile strength (σ_R) and impact strength (a_k) versus concentration of the nitrile elastomer.

stresses from the ones caused by thermochemical degradation.

The mechanical degradation of the polymeric fluid may appear at the supramolecular level, but it may also reach the molecular level.

3.8.2.3.3.1. The influence of some factors on mechanochemical transformation

It is considered that the mechanical degradation predominates in the processing through injection, extrusion, for shear gradients of more than $10^3\ s^{-1}$ [755].

The values of critical shear stress τ_{cr} or of critical shear rate $\dot{\gamma}_{cr}$, from which the mechanical degradation of the polymers in a fluid state begins, are determined by a series of factors, structurals or internal, thermomechanical or external, specific for every polymer and for every process.

In order to identify these values, the factor temperature, should be taken into consideration since it appears in all operations of polymer processing in the fluid state.

If the temperature is constant, the mechanochemical transformations of the polymers in the fluid state are influenced by a series of variables, some of them considered theoretical, others technological, and by correlations between them.

(a) Molecular weight

It is known that the viscosity of the polymer melt depends on the molecular weight and on the distribution of the molecular weight.

From the equation of the viscosity at the zero shear rate ($\eta_0 = kM^a$) it was noticed that if the molecular weight is higher than critical value M_c the value of a suddenly becomes 3.4 or 3.5.

If the viscosity measurements in melt are taken when the shear rates are increasing, there is a value of molecular weight M_D for which the constant a decreases, Figure 3.241.

M_c is the critical molecular weight at which the slope modifies from 1 to 3.4.

M_D is the molecular weight above which the polymer will be degraded at the respective shear rate.

The value of molecular weight at which degradation takes places decreases through the increase of the shear rate; $M_{Dn} < M_{Di}$, for $\dot{\gamma}_n \geq \dot{\gamma}_i$.

In general, mechanical degradation takes place in the case of molecular weights higher than the ones corresponding to the intersection point of the line, log η against log M, for a constant shear rate, with the line with slope 1, representing the viscosity of a polymer with a molecular weight smaller than critical value M_c.

Using the values obtained for M_D, the diagram of the degrading shear rate against molecular weight is plotted, Figure 3.242.

The increase of the molecular weight determines a grater number junctions between the macromolecule in a fluid state. In the case of a large number of connection points, the probability to break the macromolecule is high, even when the shear rate is low.

Thus, the polymers in the fluid state, having high molecular weights, are mechanically broken more easily than the ones with low molecular weights.

The above-mentioned behaviour, related to the initial molecular weight influencing the scission process, has been confirmed for melts of PS [273], PP [756] and PET [757].

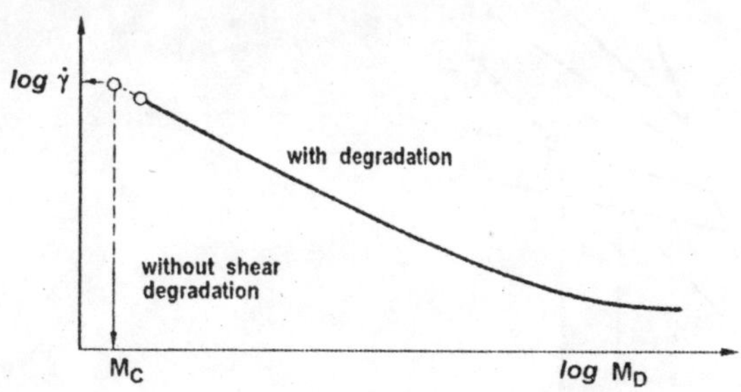

Figure 3.241. Relationship between apparent melt viscosity and molecular weight of the polymer.

At PP at the initial minimal molecular weight, the scission of the links becomes independent of the initial molecular weight, even for high shear rates [756].

(b) Geometry of the Processing Device

Another aspect of mechanochemical degradation refers to the discovery and evaluation of the influence of the mechanical factor during the processing of the polymers in the fluid state.

Overall degradation is very important for the quality of the final product and it represents the sum of mechanical, thermal, oxidative and hydrolytic degradation.

In order to show the influence of the mechanical factor, the other types of degradation have to be reduced or even eliminated.

In most cases, the effect of the thermal factor cannot be ignored and the polymer in the fluid state suffers thermomechanical degradation.

On a duct of cylindrical geometry the overall degradation of the molten PET was obtained as the sum of mechanical and thermal degradation in relation to the diameter of the duct [757].

The results obtained by Terrier show that in function of the diameter of the duct (its characteristic dimension) the thermal or mechanical degradation is predominant.

At small values of the diameter, the mechanical degradation prevails but thermal degradation cannot be eliminated.

Mechanical degradation occurs in processing ducts of the polymers in the fluid state which have small cross sections (capillary, dies, spinnarets, slits).

In order to simulate the processing conditions in the laboratory

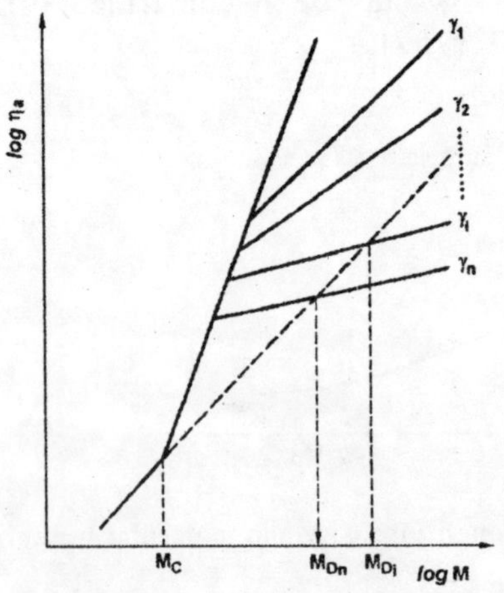

Figure 3.242. The degrading shear rate for different molecular weights.

and to separate the different types of degradation (mechanical, thermal, hydrolytic) the use of capillary rheometers is recommended.

A piston pushes the polymer through the fixed capillary at the end of a reservoir. The movement of the piston is realized by connection through a moving crosshead. The shear rate is adjusted through the crosshead speed. The extrusion force is recorded using a load cell cónnected to the upper part of the piston [284].

Capillary extrusion has the possibility to produce a large spectrum of molecular weights, on the transversal section of the extrudate because of the complex profile of the shear stresses.

Polymer degradation may take place in the reservoir, at the entrance in the capillary and in the cylindrical part of the capillary. It is difficult to separate the influence of each zone, in order to indicate the mechanochemical transformation of the polymer melt. When a series of characteristics remain invariable, the influence of one factor can be explained.

(*b1*) *Entrance angle*

For a PET with a given molecular weight and at constant, extrusion temperature, shear rate, capillary diameter, there were made determinations on capillaries with different entrance angles. Some of the characteristics of the obtained extrudates with the length between 5 and 8 cm were determined.

Changes in the Young modulus of the extrudate appear as a function of the entrance angle in the capillary in Table 3.81.

By increasing the entrance angle in the capillary from 90° to 180° the characteristics of the extrudate do not change in conformity with a continuous linear function. In the case of the Young modulus the maximum appears for an angle of 120° Reaching the maximum in the value of the modulus indicates the occurrence of an orientation process, especially for a certain value of the entrance angle.

Passing from the high diameter of the reservoir to the small diameter of the capillary, the polymer has a convergent flow in the entrance zone in the capillary. The direct effect of the convergent flow is the uniaxial orientation of the macromolecular chains [758]. In the polymer melt, as a result of the mechanical stresses from the entrance zone, linear structures called nucleation lines appear. The degree of ordering of fluid's macromolecules increases as compared to the initial one which is not deformed in the convergent entrance zone.

Table 3.81. The influence of the entrance angle in the capillary on the Young modulus of the extrudate

Entrance angle (°)	Young modulus (MPa)	
	L_c = 20 mm	L_c = 40 mm
90	1767	1498
120	1900	1856
180	1783	1388

T_e = 280 °C, D_c = 1 mm; $\dot{\gamma}$ 1 = 2400 s^{-1}; CCR-Reoscope 1000, for extrusion, ITTM-4301 for tensile test

The ordering increase facilitates the developing of a fibrillar crystallization process, after which the paralleled macromolecule can fuse into a fibrillar crystal.

In this way, temporary networks are formed at the entrance in the capillary under the action of normal or tangential stress [759,760].

The high normal or shear stress from the entrance zone or in the capillary destroys the temporary networks and even the macromolecular chains [761]. The orientation of the macromolecule is changed and the molecular weight and the young modulus of the extrudate decrease.

If the transformations from the entrance zone affected only the orientation of the macromolecule, some extrudates with different values of the Young modulus but with equal molecular weight should result. Determining the viscosity values of the solutions of the extruded PET, the results presented in Table 3.82 were obtained.

The values of the viscosities of extruded PET solutions indicate the developing of some degradations at molecular level which affect the chemical structure and which cause the transformation at supramolecular level, at certain entrance angle in the capillary.

Reaching the maximum at about 120°C indicates the existence of a minimal value of the pressure loss at the entrance in the capillary as the model given by relation (128) shows.

The influence of the value of the entrance angle in the capillary on the extrusion process was also showed for other polymer melts: PE, PP, PB, PI [747], polyester resins [746,762], PS [763]. As the thermal and oxidative factor were kept constant and some of the geometrical characteristics had constant values, the process of mechanochemical degradation can develop in the entrance zone in

Table 3.82. Influence of the entrance angle in the capillary on the viscosity of extruded PET solutions

Entrance angle (°)	Relative viscosity (in Ph/TCE 3:2)
90	1.766
120	1.773
180	1.749

T_e = 280 °C, D_c = 1 mm; L_c = 40 mm; $\dot{\gamma}_1$ = 2400 s^{-1}; CCR-Reoscope 1000 for extrusion

capillary and its intensity depends on the entrance angle.

(*b2*) *Capillary length*

Using a CEAST capillary rheometer-Reoscope 1000, a series of determinations on capillaries of different length was made. The entrance angle in the capillary, 90 °C, the diameter of the capillaries, 1 mm, the shear rate 2430 s^{-1} and the extrusion temperature 280 °C had constant values. For the extruded polymer was followed the initial viscosity in the melt.

The results of the measurements taken on the rotational rheometer, Rheometrics RDS 2 with a cone and plate geometry are presented in Figure 3.243 and Table 3.83.

The values of the initial viscosities of the extruded polymer are smaller that ones of the unextruded polymer on the whole range of the shear rate included in the testing program.

When the capillary length grows the diminishing of the initial viscosity of the extrudate amplifies. As the initial viscosity of the melt is proportional with the molecular weight, result that the molecular weight of the extruded polymer becomes smaller. The degradation at the molecular level appears at the length of 20 mm and at 40 mm.

With the temperature being constant, the reasons of the differences in the values of initial viscosity must be related in this case to the contributions of the mechanical factors.

For this purpose, the shear stress at the capillary wall and the end effect were calculated fo both lengths, Table 3.84. For both lengths the shear stresses which appear in the melt at the capillary wall reach values that are sufficient to cause the scission of the macromolecular chains.

For a 20 mm capillary, in the entrance zone, normal stresses and shear stresses appear and their value is not sufficient to cause the total orientation of the macromolecules. So, the normal stresses

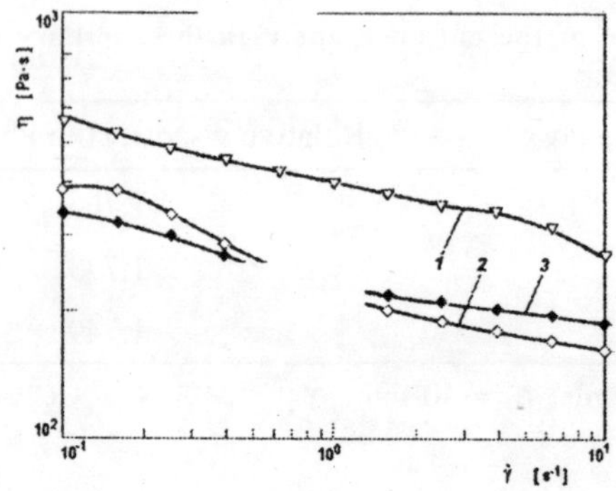

Figure 3.243. Rheograms at 280 °C for melts of PET: 1) unextruded polymer; 2) extruded polymer through a capillary with L_c = 20 mm; 3) L_c = 40 mm.

Table 3.83. The initial viscosity of extruded PET as a function of the length of the capillaries [167]

Capillary length *L* (mm)	Initial viscosity η_0 (Pa.s)
initial polymer	566
20	381
40	339

will produce a considerable end effect. When the capillary length is 40 mm, the stresses from the entrance zone determine the ordering and induce the flow crystallization of the macromolecules. The elasticity of the melt diminishes whereas the viscosity grows. The end effect will have negative values.

When passing through the capillary, the temporary network formed in the entrance zone is partially or totally destroyed, as a function of the stress level, both for the short and long capillaries. The macromolecules taken out of the network suffer thermal and mechanical degradation, providing in the extruded macromolecules with a smaller weight than the ones in the initial polyester.

According to the formula $\tau = k\dot{\gamma}_w^n$, the shear stress at a constant shear rate, should have constant values, no matter how long the capillary is. On the other hand, Han showed that ΔP_I is independent of the *L*/*D* ratio, [735]. This rule applies only for ratios *L*/*D* smaller than 40. For high values of the *L*/*D* ratio, the pressure loss at the entrance into the capillary diminishes, going towards zero.

Subtracting from the ΔP_T the pressure loss at the entrance, ΔP_C is obtained, for the length of 40 mm, higher than for the length of

Table 3.84. The shear stress at wall and the end effect for capillaries with different lengths

Capillary length, L (mm)	Shear stress at wall (dyne/cm^2) × 10^{-6}	End effect n
20	2.677	4.26
40	2.973	- 0.143

20 mm.

Calculating the shear stress at the wall, were obtained values which confirm the rapport of the two shear stresses, (Table 3.84). Superposing the effect of the shear stress at the capillary wall on the changes produced in the entrance zone, there results a series of phenomena, which determine the characteristics of the extruded polymer.

The flow crystallization takes place near the top of the entrance cone in the capillary because in this region the maximum of the elongational stress is attained.

The material from the top of the cone and from the first millimeters of the capillary will be under higher pressure, whereas the region from the lower part of the capillary will be not under stress. If crystallization did not take place in the entrance region, the macromolecules of the melt will relax in the lower region of the capillary, producing a loss of the orientation of the extruded polyester.

The cylindrical extrudate has a central zone which contains a certain degree of order induced at the entrance in the capillary and in passing through the capillary. It also has a peripheric zone with a small degree of order where the molecular weight diminishes as a result of thermal and mechanical degradation.

Information concerning the molecular and supramolecular structure of the extruded and on the nature of the changes that took place when the fluid polymer pass through the capillary may be obtained using thermal analysis.

For the unextruded polymer and for the extruded one through capillaries of different lengths, the results of DSC are presented in Figure 3.244 and Table 3.85.

If the shape of the thermograms is not modified a lot in the range of the glass transition, the melting and "cold" crystallization peaks show evident changes, in function of the length of the capillaries.

The range of the glass transition of the extruded polyester

moves towards smaller temperatures, so, interchains links were destroyed, in the case of the short capillaries and a series of intermolecular links were formed through crystallization during the flow in the case of long capillaries. For any length of the capillary the glass transition takes place in the extruded at smaller values than in the initial polymer. In short capillaries the degradation process of the interchain links is higher than in the long capillaries. Due to the crystallization induced in the convergent region of the long capillaries, at least 40 mm, the interchain links are remade and the glass transition temperature of the extruded is higher than the one recorded for the short capillaries.

In the case of polyesters, the phenomenon of cold crystallization appears, explained by the approaching of segments from the chain in the amorphous regions of the polymer, without any molecular rearrangements [765]. The exothermal peak characteristic of this phenomenon disappears or gets smaller, if the polymer sample suffers a treatment which increases its crystallinity.

The changes of the characteristics of the 'cold' crystallization range of the extruded polymer in function of the capillary length are presented in Table 3.85. The decrease of H_{CC} is a proof of the appearance of orientation during the extrusion through the capillaries of the melted PET, [766,767]. The orientation induced in the polymer in the entrance zone of the capillary is more accentuated in the case of extrusion through long capillaries.

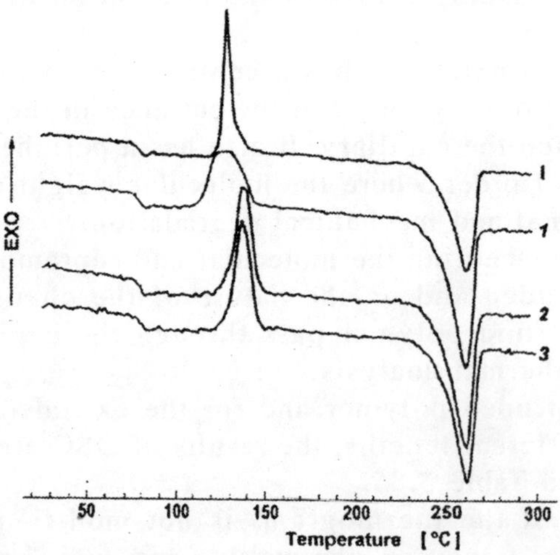

Figure 3.244. The influence of the capillary length on the thermal behaviour of the extruded PET; I – initial polymer; 1) L_c = 10 mm; 2) 20 mm; 3) 40 mm; (T_e = 280 °C); DSC-Mettler TC 11 sample of 10 mg; s.r. = 5 °C/min.

Table 3.85. The thermal characteristics of the extruded PET through capillaries with different lengths [167]

Capillary length L (mm)	T_g (°C)	T_{cc} (°C)	ΔH_{cc} (J/g)	T_m (°C)	ΔH_m (J/g)
initial polymer	81.9	132.5	29.6	261.1	43.2
10	78.5	126.7	29.3	261.3	46.6
20	76.1	124.2	29.3	260.2	49.0
40	78.5	128.9	28.4	260.1	48.2

Decrease of T_{cc} shows a growth of the degree of orientation in the extruded polymer.

When short capillaries are used for the extrusion of the PET melt, the orientation of the macromolecules in the direction of the flow is smaller than when long capillaries are used.

The appearance of the 'shoulder' in the exothermic peak of the cold crystallization for the PET extrudates obtained on the capillaries of 20 and 40 mm compared to the peak of the initial polymer, indicates the increase of the degree of crystallinity.

The orientation of the macromolecules in one direction will affect the global crystallinity. After the extrusion of the PET melt through short capillaries the crystallinity of the extruded is less affected.

10 and 20 mm capillaries were used to produce extrudates of PET with crystallinities that were higher than the ones from 40 mm capillaries. The end effects, in the form of which normal stress and shear stress are involved, generate a tridimensional structure in the extruded polymer, which gives a higher global crystallinity than the one of the initial sample.

The effect of the length of the capillary on the crystallinity of the extruded polymer can also be evaluated from the melting endothermal peak (Table 3.85).

The decrease of the melt temperature in extrudates obtained from capillaries of 20 and 40 mm shows the development of a degradation process of molecular weight, the fact which confirms the information given by melt viscosity measurements. After extrusion, the melt enthalpy of the polymer increased as a proof of the increase of crystallinity and of the decrease of molecular weight. The melt enthalpy and consequently the global crystallinity show a similar evolution, like the one indicated by T_{cc}.

By using long capillaries for the extrusion of the PET melt, the crystalline structure of the extruded matter is affected both in the entrance zone and inside the capillary.

Finally, this results in the formation of a macromolecular structure with the global crystallinity smaller than the one produced in extrusion through short capillaries.

(b3) Capillary diameter

The diameter of the capillary through which the extrusion of the PET melt is realized, influences the values of the mechanical stresses and deformation rates.

Table 3.86 shows the shear rate and extrusion force values for the usage of some capillaries with different diameters.

A far higher value of the shear rate in the case of the capillary with the diameter of 0.5 mm is noticed. Of course, the shear influences the flow, the pressure loss at the entrance and in the cylindrical part of the capillary.

In order to verify this theory, the calculation of the total pressure loss and of the end correction is used, on the basis of the extrusion force. The end effects are calculated using the method indicated by Brydson [747].

The results of the estimations are presented in Table 3.87. It is noticeable the fact that through the diminishing of the capillary diameter the crystallization in the entrance zone of the capillary is accentuated.

The shear stress at the wall of the capillary should have increased with an increase of the diameter according to the relation:

$$\tau_c = \Delta P_c \frac{D_c}{4} L \qquad (3.134)$$

In fact, the shear stress along the capillary drops when the diam-

Table 3.86. The influence of the diameter of the capillary on the extrusion force and on the shear rate [167]

Capillary diameter (mm)	Extrusion force (N)	Shear rate (s^{-1})
0.5	7500	4864
1.0	1110	609
2.0	500	76

L_c= 40 mm; entrance angle = 90°; T_{ext} = 280°C; CCR-Rheoscope 1000 for extrusion.

Table 3.87. The influence of the capillary diameter on the flow of the molten PET [167].

Capillary diameter (mm)	Total pressure loss in capillary $(N/m^2) \times 10^{-7}$	End effect	Shear stress in capillary $(N/m^2) \times 10^{-7}$
0.5	10.58	-33.0	1.1275
1.0	1.56	-28.0	0.6706
2.0	0.71	-3.5	0.2137

eter increases because of the more rapid decrease of the pressure loss with the increase of the diameter.

The crystallization induced through the flow in the convergent region of the entrance in the capillary appears at all studied diameters. The extruded polymer through capillaries with different diameters will have changes in the molecular and supramolecular structure. These changes are presented using the dates of thermal analysis, Figure 3.245.

For the diameters used in the studies, the range of the glass transition of the extruded polymer starts from lower temperatures than at the initial polymer because fracture of the interchain links takes place, Table 3.88.

In the case of the small diameter, 0.5mm, when the induced crystallization in the convergent region from the entrance in the capillary, affects seriously, some of the interchain links are remade and the extruded polymer has a smaller decrease of the glass transition temperature, 76.7 °C, compared to 73.5 °C at the 2 mm diameter.

In the case of diameters of 1 and 2 mm, the increase of T_{CC} is a clue for the decrease of the orientation degree in the flow direction and for the increase of global crystallinity.

The increase of the value of H_{CC} in the extruded polymer shows that a process of disorientation of the macromolecules takes place, compared to the state from the initial polymer, [768].

This loss in orientation is smaller in the case of the capillary with the diameter of 0.5 mm, in which the induced crystallization through the flow in the convergent entrance in capillary appears.

The decrease of the melting point of the extrudates is the result of the macromolecular chains degradation during the flow of the molten PET through capillaries with different diameters. The obtained chain fragments, more mobile will orient more easily in

Table 3.88. The influence of the capillary diameter on the thermal characteristics of the extruded PET [167]

Capillary diameter (mm)	T_g (^{0}C)	T_{cc} (^{0}C)	ΔH_{cc} (J/g)	T_m (^{0}C)	ΔH_m (J/g)
initial polymer	81.9	132.5	29.6	261.1	43.2
0.5	80.8	132.2	30.3	260.0	48.0
1.0	79.6	139.4	31.1	260.1	47.9
2.0	79.6	137.0	32.2	260.0	48.6

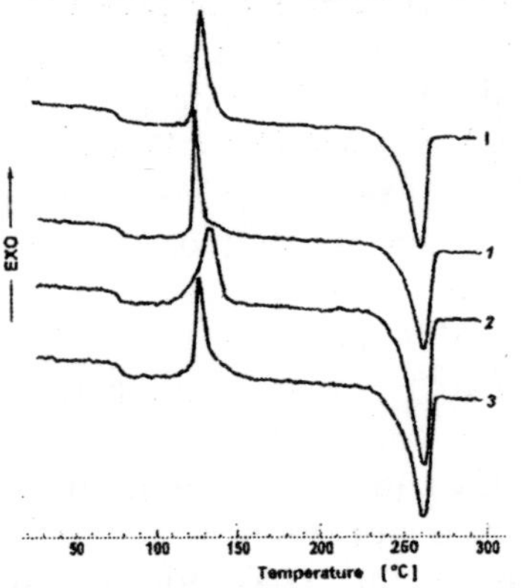

Figure 3.245. The DSC thermograms of the PET extruded through capillaries with different diameters; I) initial polymer; 1) d_c = 0.5 mm; 2) 1 mm; 3) 2 mm; (T_e = 280 °C) DSC- Mettler TC 11; sample of 10 mg; s.r. = 5 °C/min.

the conditions of the flow, in this way influencing the growth of the crystalinity degree in the extrudates.

The melt enthalpy of the polymer in extrudates had raised, as a proof of the increase of crystallinity and of the decrease of molecular weight.

(c) Flow regime

The study of the influence of the flow regime on the development of polymer processing in the fluid state can be made in two directions: the first evaluates the rheological characteristics of the polymer and the second one reveals the mechanochemical behavior of the polymer in the fluid state.

There are many possibilities to study the rheologic comportment, but the most important is that whose takes into account the effect of the time because depending on the duration of the flow process can intervene the thermo- and the mechanical degradation phenomenona.

The time dependence of the rheological behavior of the fluid state permits an interpretation concerning the macromolecular structure of the polymeric body, without make pertinent considerations on the chemical structure of the macromolecular entities during the flow.

The supramolecular order for a part of the polymers was discussed specially at the extrusion of the polyethylene in melted state. Rudin and Schreiber [769] had been found that the output rate of the extrudate modified in time, even if the mechanical stress remains constant.

The same effects have also been recorder for other polymers [770,771].

Generally, this behavior can be presented in the form of the curve in Figure 3.246, where the output rate is correlated with the piston position during extrusion.

Figure 3.247 shows graphs for the molten PET, representing the evolution of the extrusion force depending on the piston position.

The curve of the extrusion force has a peak followed by a decrease to a stationary level. The peak is evident especially at high rates of extrusion. The stationary regime is attained after the crossing of a transition zone whose length increases in proportion as the velocity of the piston increases.

The shear stress dependence of time contains two ranges, Figure 3.248.

The first, corresponding to the interval between the beginning of the solicitation and the position **h** of the piston, is attributed to an unstable state of deformation. In this period takes place the destruction of the supramolecular structure, corresponding to the static regime and a new structure, specifically for the dynamic regime, forms because of the decrease the mechanical shear which decreasing the density of the entanglements [772].

The shear stress shows a sudden increase to a maximum, similar to the extrusion force.

In the domain of the stabilized flow, the shear stress remains

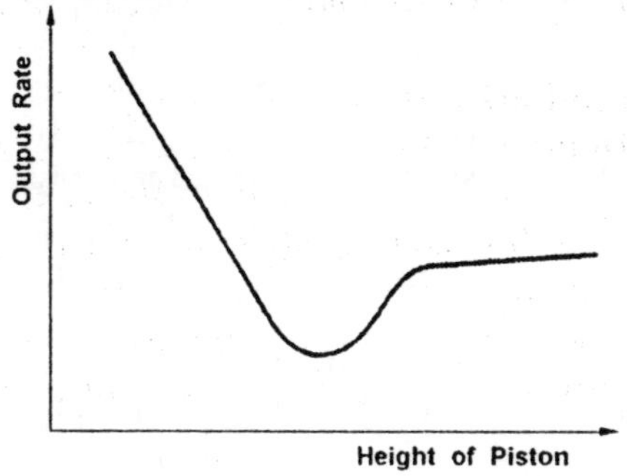

Figure 3.246. Correlation between the output rate and the piston position during extrusion.

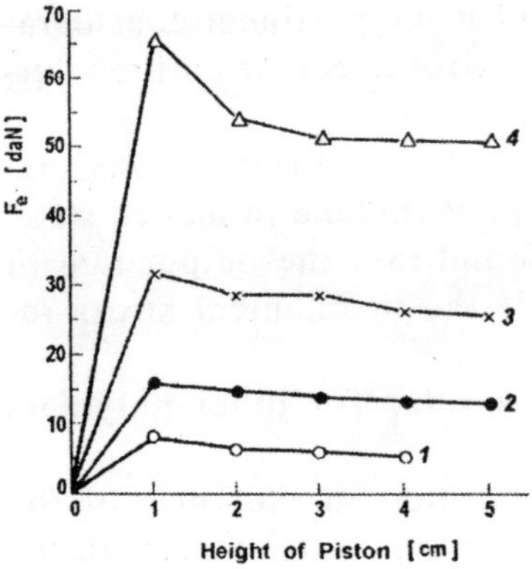

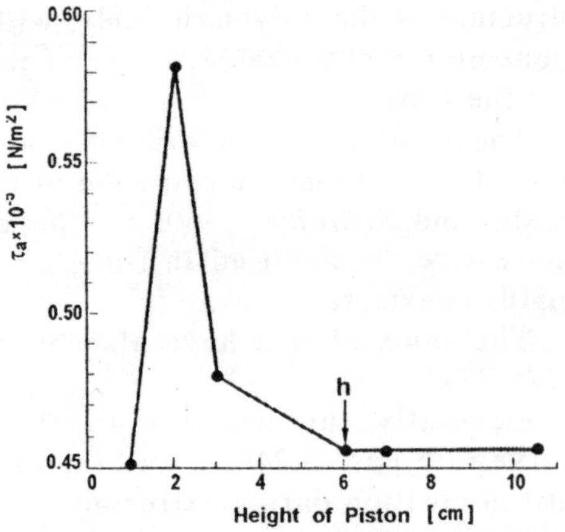

Figure 3.247. Modification of the extrusion force as a function of the piston position during extrusion of molten PET trough capillary; L_c = 40 mm; 1) r_e = 2 mm/min; 2) 5 mm/min; 3) 10 mm/min; 4) 20 mm/min (T_e = 280 °C; Reoscope 1000).

Figure 3.248 (right). Apparent shear rate at different positions of the piston during extrusion of molten PET trough capillary (L_c = 40 mm; r_e = 20 mm/min; T_e = 280 °C; Reoscope 1000).

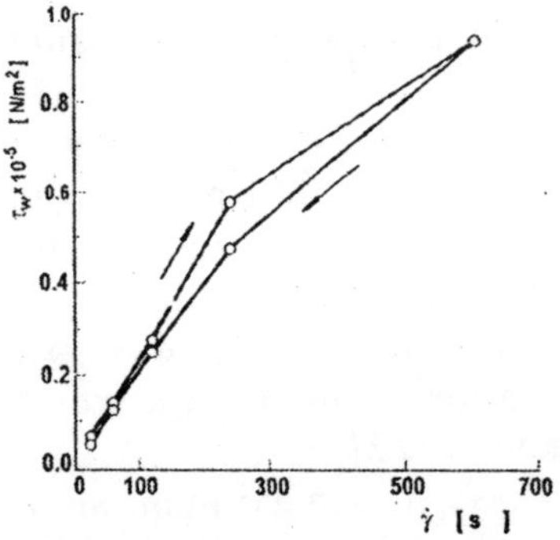

Figure 3.249. The flow curve of molten PET at 280 °C; (L_c = 40 mm; D_c = 1 mm; Rheoscope 1000).

constant owing to the realization of a dynamic equilibrium between the processes of destructuration and structuration of the polyester melt.

The rheological behaviour of the fluid polymer is indicated by the flow curves $\tau = f(\dot{\gamma})$, Figure 3.249. Figure 3.249, plotted for the increasing and decreasing shear stresses, show the appearance of two distinct curves resulting from the existence of a tixotropy phenomenon. The structure of the melt changes under the action of extrusion forces.

The rate of destructuration at the increasing of the shear rate is higher than the rate of structuration following in the decreasing period of the shear rate.

higher than the rate of structuration which corresponds to decreasing shear rate.

The slow reformation of the polymer structure observed with the decrease of the shear rate is expressed by a curve which describes in the same way (the low branch) this trend with repeating cycles. This behaviour of the PET melt suggests a pseudoplastic fluid, the tixotropic behaviour is evident only for the first cycle of stressing during the shearing of the PET melt through the capillary (Figure 3.249).

The pseudoplastic behaviour also appears at the length of the capillary of 20 mm and 40 mm.

The rheologic curves show that after a cycle of stressing at the shear stress the molten PET undergoes modifications at the supramolecular level of its structure.

Measuring the molecular weight of the polymer in the fluid state before and after the effect of shear stress, Kiang and Cuculo [757] found that the shear viscosity of molten PET at 280 °C decreases with time in accordance with curves like those in Figure 3.250; after a period of stability there is a period of decrease of viscosity, depending on the initial molecular weight. The authors attribute this change to a purely mechanical effect. The parameters used clearly indicate the possibility of degradation of the molecular structure. In this way it was possible to distinguish between the purely mechanical and chemical answer to the applied stress.

Generally, the decrease of the viscosity is a measure of the decrease of molecular weight and an argument in the favour of polymer degradation assessment. In consequence, we may note the problem of the quantification of the contributions of the thermo- and mechanical phenomena to degradation.

Using the same principle of observation, i.e. the influence of the solicitation time on the rheological behavior of polymer melts, Lenk showed that in the first period of stressing, the apparent viscosity of the viscoelastic fluid decreases until an equilibrium state is established, and in the second period the viscosity remains constant [773]. The authors' interpretation o fthe observed behaviour also relies on mechanical arguments.

In addition to the PET, a series of polymers have been investigated under shear conditions, at constant temperature, and the occurrence of degradation was found in all of them.

From these examples it has been ascertained that, during the flow, the polymers in the fluid state have a stationary behaviour for a while, either at the beginning of the process or at its end, which

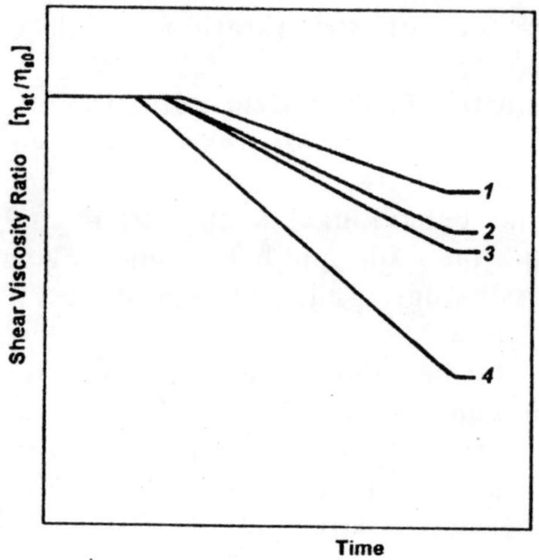

Figure 3.250. The decrease of the shear viscosity during the extrusion trough capillary ($M_1 > M_2 > M_3 > M_4$; T_e = 280 °C).

proves the existence of a stable equilibrium structure.

The rigorous interpretation of the loading time on the rheological response of the polymers in the fluid state is the subject matter of several papers [714,743,774].

In this sense, the time dependence of the behaviour of the polymeric fluid system is defined as a function of the value of the shear rate used during the experiment. At high shear rates the typical time dependence of the shear stress has the form depicted in Figure 3.251 [714].

The first domain corresponds to an unstable state of deformation. In this period the supermolecular structure is destroyed and a new specific structure for the hydrodynamic state is formed.

The shear stress suddenly increases, reaching a maximum. Above the critical value of the stress parameters, corresponding to the maximum, the modification of the supermolecular and molecular structures takes place during stress relaxation and a new state of equilibrium. This state is stable at the given flow conditions [775].

To determine the level to which the action of shear stress is evident in the molten polymer, the intrinsic viscosity of the extrudates obtained at different shear rates was examined.

For the polymer melts it is easy to follow the degradation as a function of the shear rate even though the correlations concerning the shear stresses are essential ones [769].

For the entire range of shear rates, the intrinsic viscosity of the

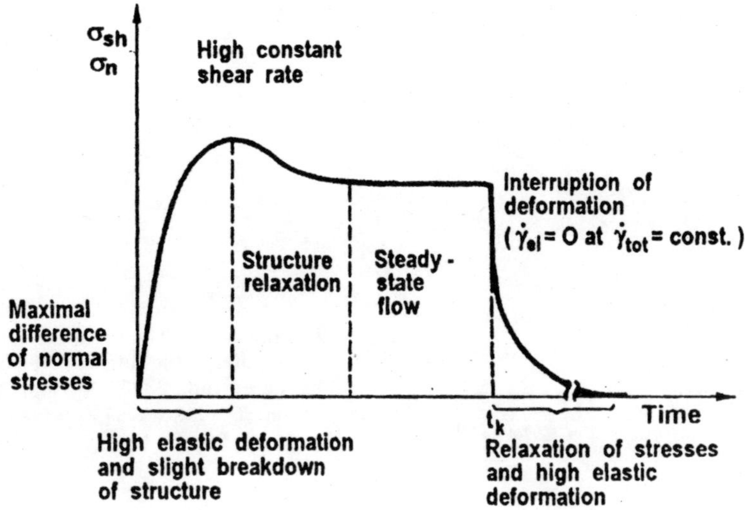

Figure 3.251. Modification of the normal and shear stresses as a function of the time of extrusion at high shear rate.

extrudate diminishes by comparison with the initial polymer. When the shear rate increase from 45 s^{-1} to 1500 s^{-1}, the intrinsic viscosity of the extrudate also increases, Figure 3.252.

An increase of the shear rate also increases the shear stress. The mechanical degradation also intensifies in proportion to the shear stress [776]. However, it was found that the degradation of the polymer diminishes when the shear rate increase.

Initially, thanks to the selective character of the mechanochemical degradation, break the longer macromolecules, those that deviate more from the average molecular weight of the polymer. In consequence, the length of the chains becomes uniform and the polydispersity narrows. So, the principal defect of the structure which causes the mechanical scission of the chains disappears. On the other hand, the number of the intermolecular junctions grows contributing to the intensification of viscous heating. The longer the residence time in the thermal field, the more intense the overall degradation of the polymer will be.

At low shear rates, the residence time of the polymer melt in the capillary is higher and, consequently, the degradation of the extrudate will be higher than for higher shear rates.

The appearance of the carboxyl groups in PET is a consequence of the thermo, mechano- and hydro-degradation reactions [777–779].

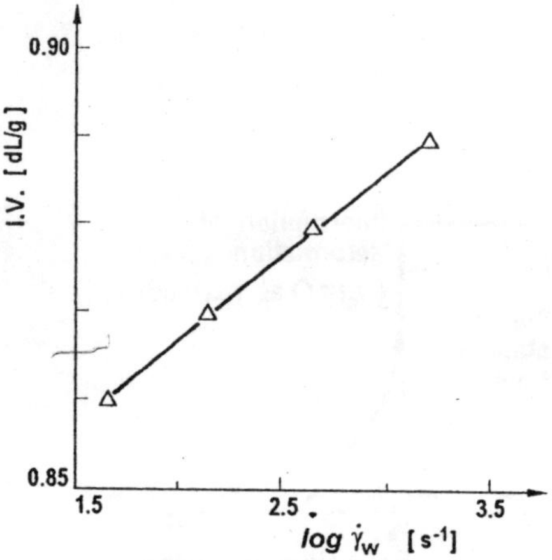

Figure 3.252. The influence of the shear rate on the intrinsic viscosity of the extrudate (L_c = 60 mm; D_c = 1 mm; T_r = 280 °C; IV_o = 0.91 dL/g; ICR Model 3212).

When the shear rate increases, the concentration of the carboxyl groups in the extrudate diminishes Figure 3.253, the values obtained being higher than in the initial polymer.

The evolution of the carboxyl groups concentration is determined by the development of a thermal degradation, the mechanical factor having, in essence, the role to decrease its activation energy.

For this reason, at small shear rates and high residence times, the extension of the degradation in all melted polymer us favoured and the extrudate is characterised by a higher concentration in carboxyl groups than then one obtained at high shear rates.

(d) Temperature

A series of factors contribute to the lowering of the effect of the mechanical factor in the processing of polymer melts. So, the temperature has a direct influence on the mechanical stress value from the polymeric system, during the flow.

The shear stress at the wall obtained during extrusion through a capillary of post-polycondensated PET granules is shown in Figure 3.254. The capillary entrance angle was 90^0 and the shear rate 45 s^{-1}.

Increasing the extrusion temperature from 270 to 310°C results in a decrease of the shear stress at the capillary wall.

If in the range 270–290 °C the shear stress value decreases by only 0.1×10^6 dyne/cm^2, in the range 290–310 °C the decrease of the shear stress is greater, exceeding 0.2×10^6 dyne/cm^2 [780].

Generally, during the extrusion of the polymer melts the shear stress at the capillary wall decreases linearly with an increase of

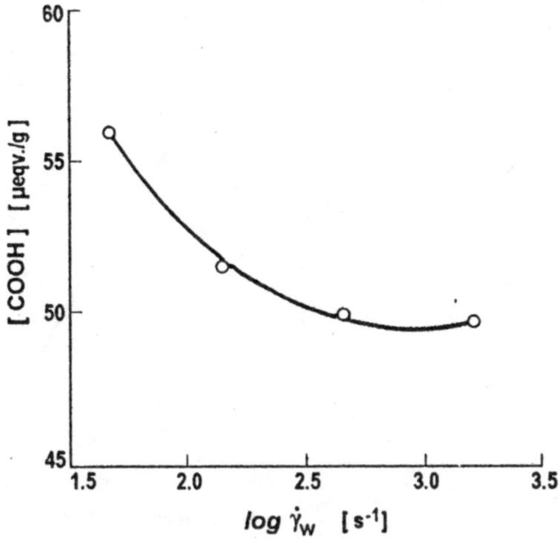

Figure 3.253. The effect of the shear rate on the concentration of the –COOH groups in extruded PET (L_c = 60 mm; D_c = 1 mm; T_e = 280 °C; ICR Model 3212).

temperature in the temperature range of 50–60 degrees.

Modification of the rheological behavior of PET melts in the range 290–310°C, is explained by transformations caused in the polymer supermolecular structure, and even in the polymer molecular structure.

Temperatures higher than 290 °C result in thermal degradation of PET with the generation of micromolecular compounds that act like lubricants in melted polymers, decreasing the shear stress.

The shear stress attained in the range 270–290°C is lower than 1×10^6 dyne/cm², i.e. values of the shear stress for which mechanochemical degradation of polymer melts was reported [282,781].

The evolution of the shear stress with increasing temperature follows the same laws like those observed for many other polymers in the fluid-viscous state [781,782,783].

Because the shear stress remains low, the process of degradation can evolve in accordance the thermal scission rules. The characterization of the produced extrudates at a certain temperature shows the proportion of main transformations produced in the polyester structure by the effect of stress.

The viscosities of the extrudates in the solution obtained at different temperatures are presented in Chapter 3.5.2., Figure 3.65. Under the dynamic conditions the increase of the extrusion temperature reduces the intrinsic viscosity of the extrudate.

It is important to show that two type of product were obtained after extrusion.

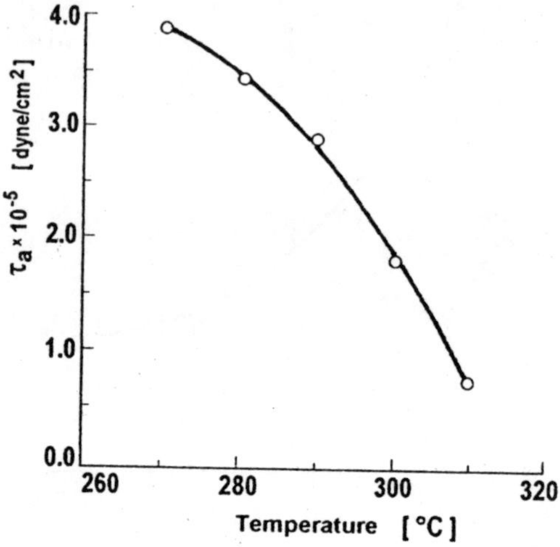

Figure 3.254. Influence of the extrusion temperature on the shear stress at the capillary wall (L_c = 60 mm; D_c = 1 mm; shear rate = 45 s^{-1}; ICR Model 3212).

The extrudates obtained in the temperature range 270–290°C, have an intrinsic viscosity higher than those obtained at temperatures of 290–310°C.

In comparison with the initial value of the intrinsic viscosity of PET, in the first temperature range, a decrease of viscosity of about 0.09 dL/g is recorded, while in the second range the decrease of viscosity amounts to 0.271 dL/g [780,784].

The decrease of the intrinsic viscosity of the extrudate is a direct indication of the development of degradation during extrusion.

In the conditions of a constant and small value of the mechanical factor, the degradation is attributed to the thermal factor.

Jabarin [785] showed a similar behavior for the PET melts. He mentioned that the thermal degradation of PET in static conditions is amplified four times in the interval 290–310°C, by comparison with the interval 270–290°C.

(e) Chemical factor

The chemical factor under the form of the micromolecular products (initial humidity, inhibitors of final groups, reactive gases) which are present in the polymer submitted to processing, and in the processing medium, can diminish the effect of the mechanical factor during the mechano-chemical transformation process.

The shear stress recorded in extrusion through capillaries at 290°C of the PET melts with different initial humidities is given in Table 3.89.

At a constant shear rate the shear stress decreases proportionally with the initial humidity of PET. Beginning with the shear rate of 450 s^{-1} the shear stress exceeds the value of 10^6 dyne/cm² from which are possible mechanical scission of macromolecules in the fluid state [786].

The lower values obtained for the shear stress at certain initial humidities are the consequence of the lubricating effect of water molecules. The water content influences the rheological properties of PET melts [787]. In the initial humidity range 0.015–0.025% a decrease of viscosity of 1.3% was recorded for a 0.001% increase in the initial humidity, at 284°C and low shear rate, <10^3 s^{-1} [788].

The content of humidity from PET melt is important also in connection with the hydrolysis of polyester macromolecules, because hydrolysis determines the decrease of the degree of polymerization.

The extrudate obtained at two shear rates for different initial humidities was characterized by viscosity in solution. On the basis of the values of the intrinsic viscosities the number of scissions per macromolecule was calculated, Table 3.90.

An increase in the initial humidity from the PET submitted to extrusion increases the number of scissions per macromolecule in the extrudate, although the shear stress does not attain the required values for the mechanical scission of macromolecules.

The concentration of carboxylic end groups in polyester is an important factor in correlation with the tendency at hydrolysis, because these final groups have a catalytic effect in the degradation process, [789].

In a series of determinations, the evolution of the carboxylic

Table 3.89. Influence of the initial humidity of the PET melt on the stress developed during extrusion through capillary

Initial humidity (%)	Shear stress (dyne/cm²)		
	45 s^{-1}	150 s^{-1}	450 s^{-1}
0.0010	319800	889200	2558400
0.0033	312400	886300	2357000
0.0063	302600	881050	2116400
0.0089	292500	877500	1852500

L_c = 60mm ; D_c = 1mm; entrance angle = 90°; ICR-3212 for extrusion.

Table 3.90. Influence of the initial humidity of PET on the number of scissions on macromolecule [187]

Initial humidity (%)	Scissions on macromolecule	
	45 s^{-1}	150 s^{-1}
0,0010	0,062	0,050
0,0033	0,088	0,081
0,0063	0,126	0,123
0,0089	0,167	0,169

T_e= 290°C; L_c= 60mm; D_c = 1mm; entrance angle = 90°; ICR-3213 for extrusion.

groups concentration in extrusion of PET melts with different shear rates was investigated.

The concentration of the final carboxylic groups in extrudates obtained from a polyester with an initial humidity of 10 ppm, after a shearing at 280°C, are presented in Table 3.91.

Even for a low concentration of water in the polyester submit to extrusion and at a temperature where the degradation occur with a low velocity, the concentration of the carboxylic groups in extrudates, increases. The presence of carboxylic groups in the polyester melt intensifies the hydrolysis of the macromolecular chains just by autocatalytic process.

For blocking the final carboxylic groups from the melted PET and for the inhibition of their action during the products utilization, some epoxydes can be used [790,791].

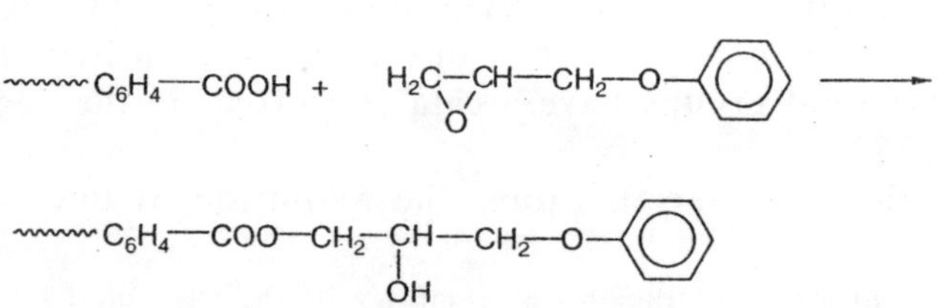

In a series of runs the influence of the glycidyls' nature on the evolution of carboxylic groups in extrudates was studied, Table 3.92.

Table 3.91. Modification of the final carboxylic groups concentration after PET extrusion as a function of shear rate [10]

Shear rate (s^{-1})	0	45	150	450	1500
COOH (μequiv/g)	16,7	17,4	17,9	18,2	19,2

L_c= 60mm ; D_c = 1mm; entrance angle = 90°; ICR 3212 for extrusion

When the epoxydes are introduced on the surface of PET chips, before the extrusion, the concentration of –COOH groups in the extrudates was lower than in the untreated polymer.

Depending on the glycidyl nature, different reductions of the carboxylic groups concentration in extrudates were obtained.

The concentration of the carboxylic groups is influenced by the nature of glycidyl and also by the quantity of inhibitor, when it was used. To explain this aspect, several extrusions of PET granules, treated on the surfaces with different quantities of PhGE, were made, Table 3.93.

With the increase of the quantity of glycidyl introduced in the polymer, the concentration of the COOH groups from the extrudates decreases to values, lower than 10 μequiv/g.

The decrease of the –COOH group concentration in the extrudate is not proportional with the weight of the glycidyl disposed on the surface of the granules.

Increasing the quantity of PhGE from 0.2 to 0.74 %, the –COOH groups concentration diminishes at a half from the initial concentration.

For comparison, doubling of the glycidyl concentration from 0.74 to 1.46 % doesn't modify the –COOH groups concentration in the extrudate.

It means that there is a limit to which the –COOH groups concentration from the extruded PET can be reduced, depending on the concrete processing conditions. The liquid inhibitor has a diffusion rate limited in the solid granules so, after a period of mixing, the inhibitor used did not penetrate entirely into the bulk polymer.

Even if there is enough inhibitor, the –COOH groups are not entirely neutralized. The molecules of the inhibitor do not diffuse in all the volume of the melt during extrusion to react with the carboxylic groups.

Table 3.92. Influence of the nature of glycidyl on the concentration of the –COOH end groups in extrudates [193]

Sample of polymer	-COOH (μequiv/g)	
	at 45 s^{-1}	at 150 s^{-1}
without glycidyl	29	28
with PhGE	22	20
with N-GPhI	11	9

L_c= 60mm ; D_c = 1mm; entrance angle = 90°; T_c = 280°C; ICR-3212 for extrusion

3.8.2.3.3.2. Correlation between the factors which affect mechanochemical transformations
(a) Residence Time and Extrusion Temperature

The magnitude of the shear stress acting on the molten polymer also determines different values of the shearing rate as well as of the residence time in the extruder which then in turn will influence the intensity of polymer degradation.

At the same time, the decrease of its molecular weight is influenced by temperature and extrusion rate.

The influence of the temperature on inherent viscosity (I.V.) at different residence times is shown in Figure 3.255; a continuous decrease in time of this property may be observed, this decrease being greater as temperature increases.

Thus, at temperatures higher than 290 °C, the macromolecule significant splitting occurs pointed out by the I.V. pronounced decrease.

Starting from Jellinek's equation for statistic degradation of a polymer and applying it to the PET melts, the following relation is arrived at:

$$\frac{1}{(IV)_n^{1,47}} - \frac{1}{(IV)_0^{1,47}} = 3{,}92 \times 10^4 \, k_{iv} t \tag{3.135}$$

where: k_{IV} is the I.V. reduction rate constant [89].

The values of the k_{IV} constant for different temperatures were determined from experimental data using equation (3.135); Figure 3.256 gives the linear plots that were obtained.

It may be seen that the degradation is more pronounced at temperatures higher than 290 °C and the degradation rate constant changes as function of temperature according to an Arrhenius law. Using this equation, the value of the activation energy of PET melt

Table 3.93. Influence of the concentration of the glycidyl on the –COOH groups concentration in extrudates [194]

Concentration of PhGE (%) in granules	-COOH in extrudate (μequiv/g)
0	19
0,2	16
0,74	8,5
1,46	9,5

$\dot{\gamma}$ l= 45 s^{-1}; T_e= 280 °C; $[COOH]_{in.t}$= 16,7 μequiv/g; ICR-3212 for extrusion

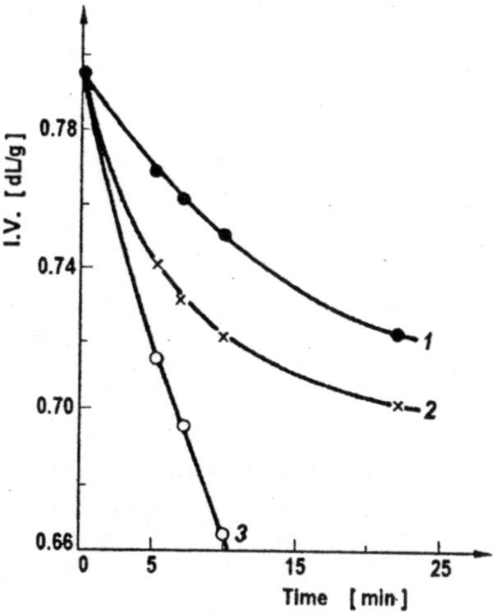

Figure 3.255. Influence of the residence time in the extruder on IV of the extrdate (L_c = 60 mm; D_c = 1 mm; T_e = 1 – 270 °C; 2 – 290 °C; 3 – 310 °C; ICR Model 3212).

degradation was obtained (E_a = 84 kJ/mole). The melt was forced to pass through the capillaries.

Consequently, the equation describing the reduction of PET inherent viscosity in time has the following form:

$$\frac{1}{(\text{I.V.})_n^{1.47}} = \frac{1}{(\text{I.V.})_0^{1.47}} + 3.92 \times 10^4 \,(e^{2.782 \frac{84}{RT}})\, t \qquad (3.136)$$

The equation allows the calculation of I.V. for the extruded PET as a function of the stationary time inside the extruder at different temperatures; in its turn, this includes both the preheating and the proper extrusion time.

Compared to the thermal degradation, the PET degradation under the combined action of the temperature as well as mechanical stresses needs a lower activation energy (84 kJ/mole as compared to 119 kJ/mole in stationary conditions). During all the experiments the shearing stress was maintained at high values (0.15 to 0.5 MPa) so that the mechanical splitting of the chemical bonds may develop predominantly. It was expected that the shearing mechano-degradation should intensify as the shearing stress increase and the residence time inside the extruder decrease. As shown in Figure 3.256, the bond splitting was more intense at long residence times. This fact as well as the temperature influence on the polymer degradation suggests that temperature has the main contribution to the reduction of PET inherent viscosity during melt spinning.

Although the mechanical stress is not the main cause, it has nev-

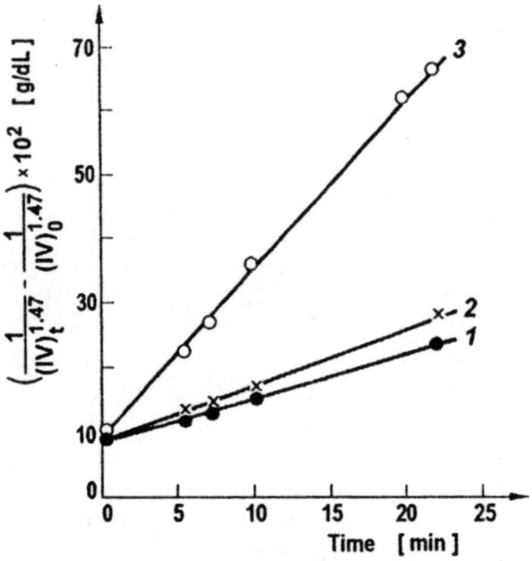

Figure 3.256. Plotting of Equation (134) for different remperatures: 1) 270 °C; 2) 290 °C; 3) 310 °C.

ertheless an important role in reducing sensibly (with over 25%) the value of the destruction activation energy.

The two destructive processes complete each other. Both of them act concomittantly and occur according to the same mechanism of free radicals formations, by homolytical splitting of the macromolecular chains.

(b) Shear Rate and the Chemical Factor

In the PET granules introduced at extrusion there always exists a small amount of water which cannot be eliminated using industrial processes of drying. The inhibitor of carboxylic groups always acts in the presence of such water traces.

The influence of water traces upon the activity of the inhibitor of carboxylic groups at different shear rate is presented in Figure 3.257. In the presence of 0.5% PhGE, and only 10 ppm water, the PET extrudates' intrinsic viscosity is close to that of the untreated polymer. At high shear rates, the inhibitor's presence contributes to reducing degradation and to obtaining a higher viscosity in the extrudate [279,786].

On maintaining the same amount of inhibitor, but in the presence of 114 ppm. water, extrudates with lower intrinsic viscosities than in the case of 10 ppm. water are obtained, over the whole domain of shear rates taken into study. The water present in high amounts in the initial polyester contributes to the chains' hydrolytic scission so that the extruded polymer will have shorter chains and a lower intrinsic viscosity.

At low shear stresses, water's action is stronger; with the in-

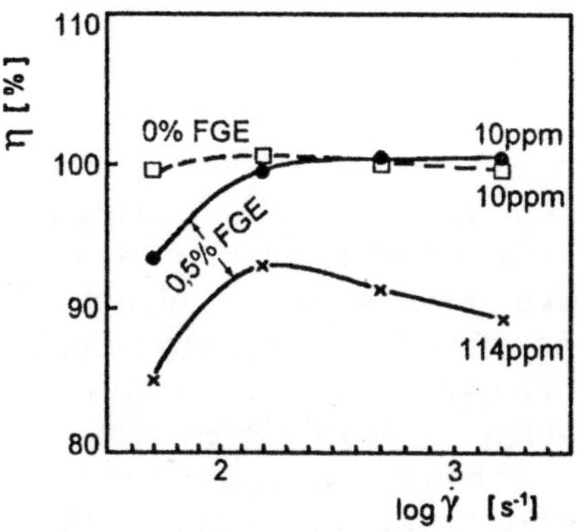

Figure 3.257. Influence of water traces on the activity of the inhibitor of COOH groups: 1) 10 ppm water and without PhGE; 2) 10 ppm water and with 0.5 % PhGE; 3) 114 ppm water and with 0.5 % PhGE; ICR Model 3212.

crease of the shear rate, the residence time in the shear field decrease, while water's action on the chains is reduced.

At high shear rates, the combined action of mechanical shearing and hydrolytical degradation increase the number of scissions, the extrudates having lower intrinsic viscosities.

At higher concentrations of water in the polyester, the inhibitor cannot block the final carboxylic groups formed as a result of degradation, Figure 3.258.

On increasing the shear rate, concentration of the –COOH groups in the extrudate obtained from the polymer with only 10 ppm water, increases.

On treating the initial polymer with 0.5% PhGE, concentration of the carboxylic groups decreases by the increase of the shear rate, in the case of PET containing 10 ppm. water, and increases with the shear rate at 114 ppm. water.

In the extrudates obtained from PET treated with 0.5% PhGE, the concentration of the carboxylic groups is lower than in the extrudates resulted from untreated PET, almost over the whole domains of shear stresses taken into study. Treatment of the PET containing 10 ppm. water, with 0.5% PhGE causes blocking of the carboxylic groups as early as during the polymer's residence in the rheometer's reservoir.

Their catalytic action no longer occurrs and, instead of increasing the carboxylic group concentration, their decrease takes place with an increase of the shear rate [786].

In the presence of larger amounts of water, the inhibitor neutralises part of the –COOH groups, but cannot longer block the groups appearing from the combined action of water and of mechanical shearing.

Consequently, on increasing the shear rate, concentration of the –COOH groups from the extrudate is increased.

The effect of initial humidity put into evidence in this series of determinations agrees with the results obtained by Wampler [792] on poly(1,4-cyclohexylenedimethylen-terephthalate) when, with increases of humidity over 20 ppm, concentration of the –COOH groups in the extrudate suddenly increases.

The inhibitor contributes to increasing mobility and to the macromolecules' alignment on the direction of the field of forces. In this way, intermolecular frictions are diminished, which is confirmed by the values recorded for the extrusion forces, Table 3.94.

In order to act as an inhibitor of carboxylic groups, the glycidyl should be introduced into polyester with water content below 20 ppm.

During extrusion, an increase of the shear rate in the presence of water intensifies degradation, which is reflected in the increase of the carboxylic groups concentration and the decrease of the viscosity of the final product.

Increase of the inhibitor concentration in the PET melt induces a decrease of the shear stresses at all shear rates. At constant concentrations of inhibitor in the melt, the shear stress increases with

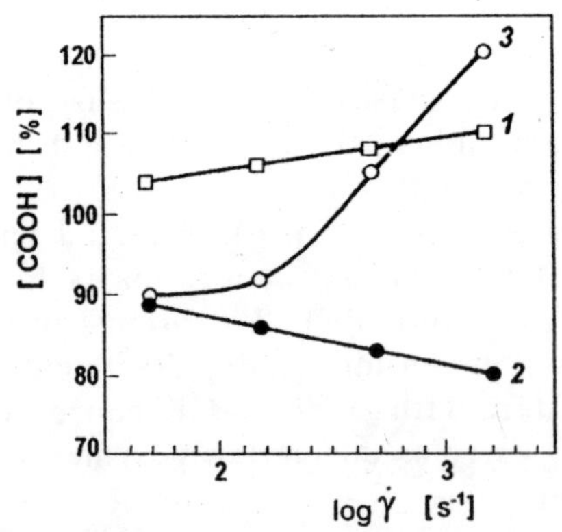

Figure 3.258. Influence of the shear rate and initial water content on the concentration of COOH groups at extrusion of PET treated with PhGE. 1) 10 ppm water and without PhGE; 2) 10 ppm water and with 0.5 % PhGE; 3) 114 ppm water and with 0.5 % PhGE; ICR Model 3212.

the increase of the shear rate. Introduction of glycidyl does not modify the rheological behaviour of the PET melt, once again it acts only on the level attained by the shear stress, as a function of concentration and of the shear rate.

Modification of the shear stress will produce different transformations in the structure of the extruded polymer. In order to elucidate this aspect, concentration of the –COOH groups was determined in the extrudates obtained at various shear rates and various concentrations of the inhibitor; the results obtained are plotted in Figure 3.259 [786].

When PhGE was used as an inhibitor, the concentration of the carboxylic groups in the extrudates decreases for all shear rates. When the inhibitor concentration is between 0.2–0.75%, the reduction in the concentration of the carboxylic groups from the extrudate is higher than the reduction recorded at initial concentrations of the inhibitor between 0.7–1.5%.

At higher concentrations of glycidyl in the initial polymer a part of the inhibitor remains in the melt, even if it had not been involved in the reaction of blocking all carboxylic groups. Such a situation may cause lowering of the extrudate intrinsic viscosity with the increase of the inhibitor's concentration in the initial polymer, Figure 3.260.

In Figures 3.259 and 3.260, the intrinsic viscosity and concentration of the extrudate's COOH groups are compared to the values of the same parameters determined for the initial polymer. For concentrations of the inhibitor between 0.2–0.5%, the extrudate intrinsic viscosity is improved.

In the same inhibitor concentration range, an appreciable reduction in the concentration of the –COOH groups in the extrudates is also obtained.

Even if for the initial concentration of the inhibitor of 1–3%,

Table 3.94. Values of the extrusion forces as a function of the shear rate and of the inhibitor's concentration [10].

PhGE concentration	Shear rate (sec^{-1})			
(%)	45	150	450	1500
		Extrusion	force (N)	
0	106	338	800	3000
0.739	98	308	690	1453
1.467	89	298	666	1300

L_c= 60mm ; D_c= 1mm; T_{ext}= 280°C; ICR-3212 for extrusion.

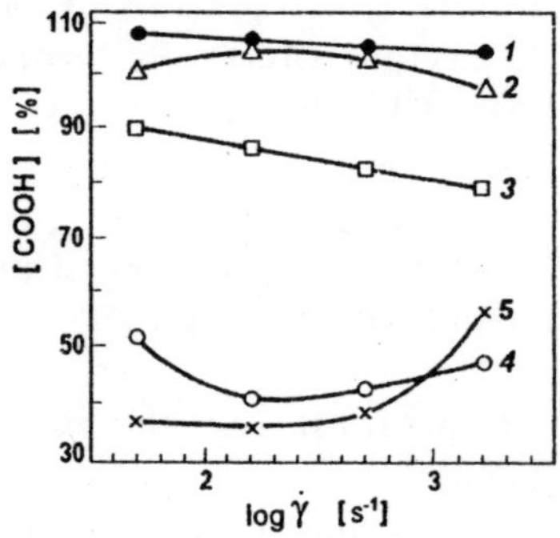

Figure 3.259. Influence of the shear rate and inhibitor concentration on the concentration of COOH groups in extruded PET. 1) without PhGE; with: 2) 0.2%; 3) 0.5%; 4) 1%; 5) 3% PhGE (L_c = 60 mm; D_c = 1 mm; T_e = 280 °C; ICR Model 3212).

spectacular reductions of the carboxylic groups occur, the extrudate intrinsic viscosity greatly decreases as compared with that of the initial polymer. This effect is related to the unreacted PhGE remaining in the final material which produces an opposite effect to the expected one.

(c) Shear Rate, Processing Temperature and Concentration of the Carboxylic Group Inhibitor

Influence of the three factors upon the characteristics of the extruded polymer and of the occurrence of mechano-chemical transformations was studied with a program similar to the factorial, centred, rotable, composed one.

The variation range of each factor is given in Table 3.95.

The extruded polymer was characterised by its solution viscosimetry and concentration in carboxylic groups.

Based on the values of the extrusion force measured under thethe conditions given in the experimental plan, the shear stress at the capillary wall was calculated for each combination of factors.

Processing of the experimental data was based on the specific methodology characteristic to the multifactorial, non-linear, regression analysis, a case in which a mathematical expression of the variation law of the parameter measured is a function of the variable factors, [793,794].

The dependence of shear stress at the capillary wall on the variable factors has the fowm:

$$\tau_s 1 = -37469.6 + 28589.7X_1 + 1045.4\, X_2 - 10390X_3 - 5399.3X^2_1 - -315.7X_1X_2 + 3684X_1X_3 - 279X^2_1X^2_3 - 0.1X_2X_3 + +2172.7X^2_3 \quad (3.137)$$

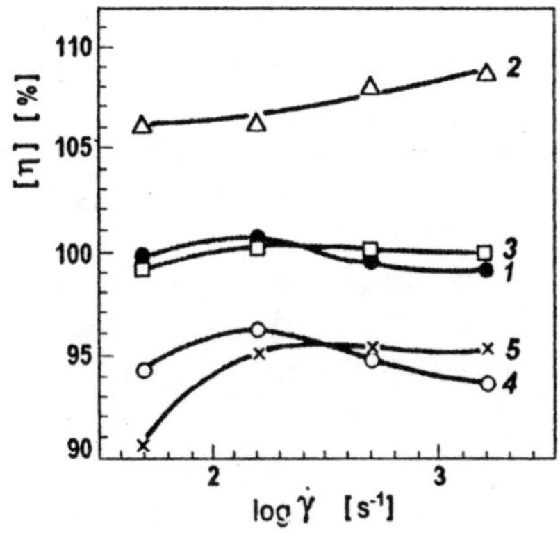

Figure 3.260. Change in intrinsic viscosity of extrudate as a function of shear rate and concentration of inhibitor. 1) without PhGE; with: 2) 0.2%; 3) 0.5%; 4) 1%; 5) 3% PhGE (L_c = 60 mm; D_c = 1 mm; T_e = 280 °C; ICR Model 3212).

Considering model's accuracy, shear stress has been calculated for different values of temperature, concentration in the inhibitor and shear stresses [795].

A comparison between the rheological curves of the PET curves and the curves provided by the mathematical model, for the polymer with and without inhibitor of carboxylic groups, was performed by monofactorial graphical processing.

At concentrations of the carboxylic group inhibitor of 0 ÷ 0.4%, the curves $\tau = f(\dot{\gamma}\ 1)$ show that the melt has a newtonian behaviours over the 270–300°C temperature range.

The shear stress increases with the increase of the shear rate both in the melt without and with inhibitor, Figure 3.261.

The tridimensional representation $\tau = f(T, c)$, at constant shear rate, evidences deviations from the rheological behaviour of the

Table 3.95. Codes and limits employed in the variation of factors.

codes	factor	limits employed		coefficient scalar
		minimum	maximum	
x_1	Temperature (°C)	270	300	10^{-2}
x_2	Shear rate (s^{-1})	15	1500	10^{-2}
x_3	N-GPhI concentration (%)	0	2	10

L_c= 6cm; D_c= 1mm; capillary entrance angle = 90°; IRC-3212 for extrusion

PET melt without inhibitor, Figure 3.262.

As such a deviation is not a monotonous one, there results that, besides the flow, some other effects are also manifested which should be considered in the explanation of the PET melt's behaviour stabilised with *N*-GPhI during flowing through the capillary, (crosslinking, degradation, etc.).

When increasing the shear rate, the shear stress at the capillary wall also increases. Relations $\tau_w = f(\dot{\gamma} 2)$ are linear, but the slope is smaller than 1 which confirms the pseudoplastic character of the PET melt's behaviour (Figure 3.263).

At low extrusion temperatures (270°C), the shear stress at the capillary wall decreases with the increase of the inhibitor's concentration; a minimum is recorded around a value of 1.5%, Figure 3.264.

At temperatures of 300°C, the shear stress evidences a maximum around the inhibitor concentration of 1% in the PET melt (Figure 3.265)

At low extrusion temperatures, the inhibitor does not react with the polyester, acting as a micromolecular product and contributing to lowering of the shear stress. When the inhibitor's concentration increases, branchings begin to appear, inducing a slight increase of the shear stress in the capillary.

At higher temperatures, the inhibitor reacts with the polyester, thus forming branched structures causing the stress stress to increase. The excess of inhibitor acts as a diluting compound, reducing shear stress.

Comparison of the behaviour of the polyester treated with inhibitor at the boundary temperatures of the investigated range leads to the conclusion that the branching reaction is initiated from a certain temperature, the value of which depends on the extrusion rate.

Consequently, at low extrusion rates, 15 s^{-1}, the curves plotted in Figure 3.266 were obtained.

At 270°C, the shear stress at the capillary wall decreases with the increase of the inhibitor concentration, a minimum being attained at 1.5%, followed by a slight increase induced by an increase of concentration towards 2%.

Between 277 and 285°C, the rheological behaviour of the polyester treated with inhibitor changes, a maximum appears on the curve of the shear stress which proves the existence of some synergetic effects of the external factors upon PET melt rheology.

Equally important in the utilisation of the PET treated with inhibitor is the extrudate's molecular weight, indirectly characterised

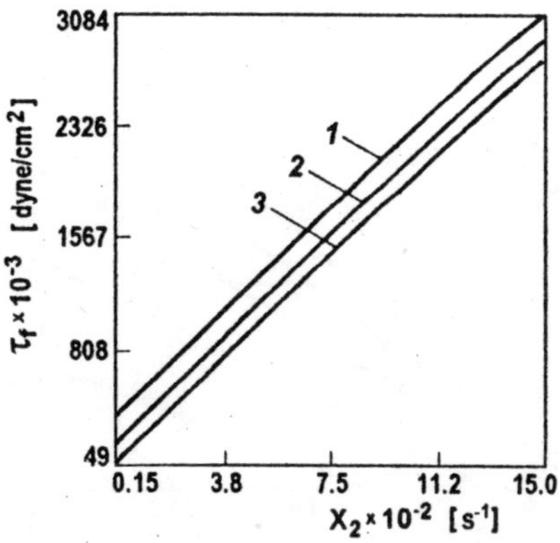

Figure 3.261. Relations shear stress–shear rate obtined from the mathematical model. 1) without inhibitor; 2) 0.5% N-GphI; 3) 1% N-GphI (T_e = 270 °C).

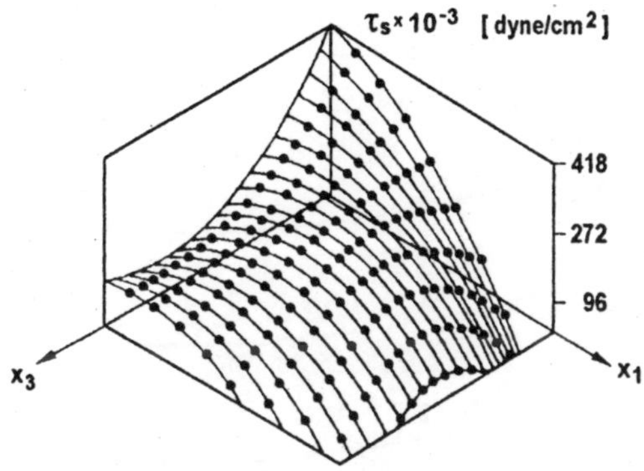

Figure 3.262. Shear stress as a function of extrusion temperature (X_1) and inhibitor concentration (X_3). (shear rate = 45 s^{-1}).

through intrinsic viscosity.

The influence of the external factors upon the extrudates' intrinsic viscosity was calculated with the mathematical model based on the results of the experimental program drawn in Table 3.95.

The response of the experimental program was the extrudates' intrinsic viscosity.

Processing of data evidenced the mathematical model for the dependence of [η] on the experimental factors [796].

$$[\eta] = -1261{,}6 + 1541X_1 + 42{,}1X_2 - 203{,}6X_3 - 306{,}3\,X_1^2 - 14{,}1\,X_1X_2 + 40{,}3X_1X_3 - 0{,}24X_2^2 + 5{,}96\,X_2X_3 + 8{,}47X_3^2 \quad (3.138)$$

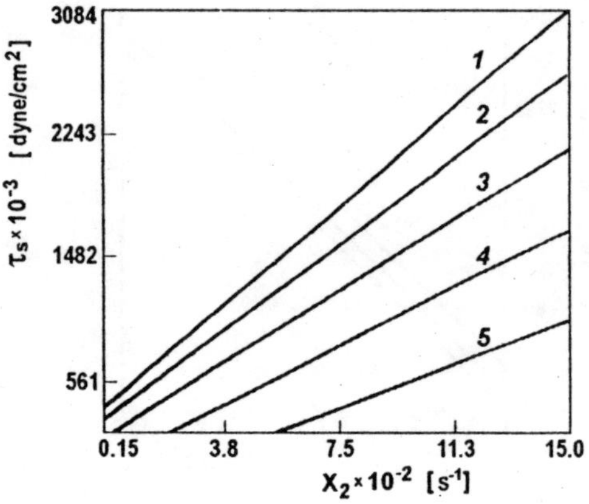

Figure 3.263. Shear stress as a function of the shear rate, at different extrusion temperatures, based on the mathematical model; 1) 270 °C; 2) 278 °C; 3) 285 °C; 4) 293 °C; 5) 300 °C.

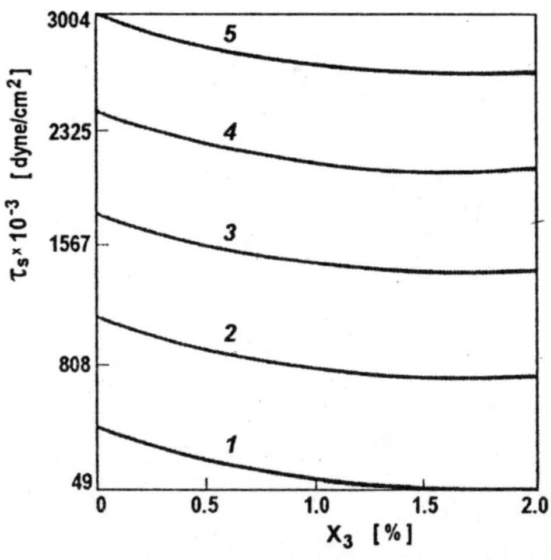

Figure 3.264. Relation shear stress–inhibitor concentration at 270 °C : 1) 15 s^{-1}; 2) 386 s^{-1}; 3) 757 s^{-1}; 4) 1128 s^{-1}; 5) 1500 s^{-1}.

The comparison between the values of intrinsic viscosity provided by the mathematical model for the polymer with and without inhibitor of carboxylic groups was based on mono-, bi-, and tridimensional graphical representations.

Mechano-degradation was checked in all the 20 experiments, the values of the extrudate intrinsic viscosity being lower than those of the initial polymer.

In the case of PET untreated with inhibitor, the extrudates' in-

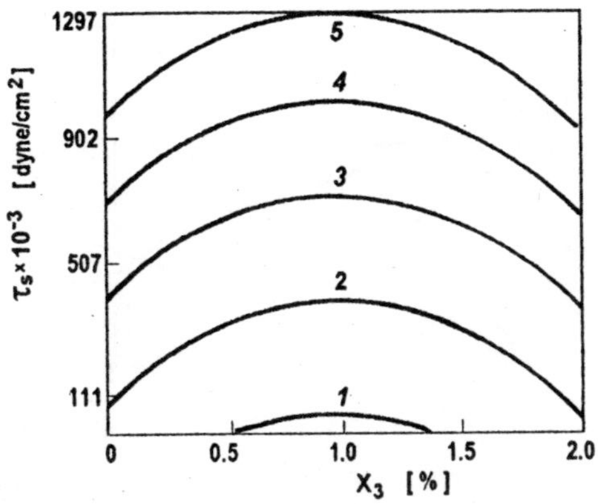

Figure 3.265. Relation shear stress-inhibitor concentration at 300 grade C: 1) 15 s^{-1}; 2) 386 s^{-1}; 3) 757 s^{-1}; 4) 1128 s^{-1}; 5) 1500 s^{-1}.

trinsic viscosity values recorded at various shear rates are presented in Figure 3.267.

The function $[\eta] = f(\dot{\gamma}3)$ has a maximum which is shifted towards low shear rates with an increase of the extrusion temperature.

The occurrence of the maximum may be explained by the intervention of at least two external factors.

The graphical representations given in Figure 3.268 confirm the effect of the extrusion temperatures upon the extrudates' intrinsic viscosity.

Thermal degradation acts up to the point of maximum. The lower is the extrusion rate, the longer is the residence time in the thermal field, while the extrudate's intrinsic viscosity is lower. Increase of the shear rate reduces the stationary time and thermal degradation.

Exceeding a certain shear rate stimulates mechanical splitting whose intensity increases with increasing shear rate.

Increase of the extrusion temperature contributes to increasing thermal degradation. In the partially degraded polymer, intervention of the mechanical factor is much easier; therefore it is manifested at low shear rates. The action of the mechanical factor is complementary to the thermal ones. Thus, the point of maximum occurs even under conditions of stress, at low rates.

In the situation in which only the thermal factor acts, the relations given in Figure 3.268 would be obtained. The extudates' intrinsic viscosity decreases with the increase of the inhibitor's concentration and temperature.

The thermal degradation occurring at high residence times and

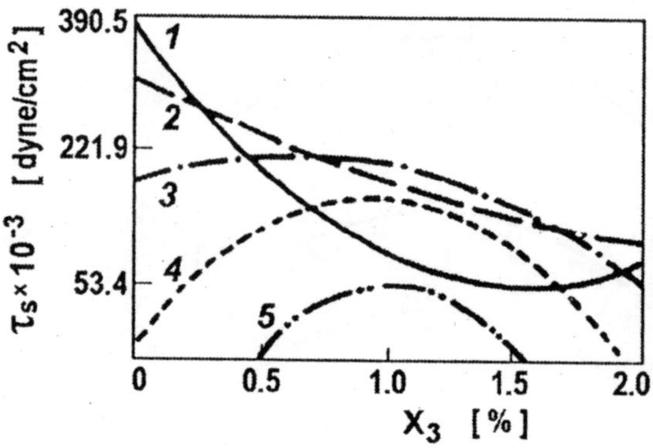

Figure 3.266. Shear stress as a function of inhibitor concentration at different extrusion temperatures. 1) 270 °C; 2) 278 °C; 3) 285 °C; 4) 293 °C; 5) 300 °C. (Shear rate = 15 s^{-1}).

the inhibitor's plasticization effects contribute to reducing the extrudates' intrinsic viscosity.

Intervention of the mechanical factor modifies the result of the global degradation process. The extrudates' intrinsic viscosities are plotted – as a function of the inhibitor's concentration – in Figure 3.269 and 3.270. At temperatures of 270 and 300 °C, the inhibitor's presence induces reversing of the intrinsic viscosity's evolution as a function of concentration.

At 270°C, this reversion occurs at low inhibitor concentrations (0,2%) and high extrusion rates (1100–1500 s^{-1}).

At a high temperature (300 °C), changing of the intrinsic viscosity's evolution direction takes place even in the absence of inhibitor, starting from low shear rates.

At low concentrations of inhibitor and high shear rates mechanical splitting occurs, an extrudate with lower viscosity than that those of the thermally degraded samples. Actually, at 270 °C, thermal degradation is quite low, its value increasing with the residence time.

Intervention of the mechanical factor amplifies chains splitting and extrudates with lower viscosity form even if the residence times are low.

Increase of the inhibitor concentration reduces the shear stress and also mechanical splitting, according to the relation given in Figure 3.264. As Figure 3.270 shows, extrudates with higher intrinsic viscosities form in this case.

At 300°C, the action of mechanical stress is exercised on an already thermally degraded polymer, $[\eta]$=0.604 *dL/g*, as compared

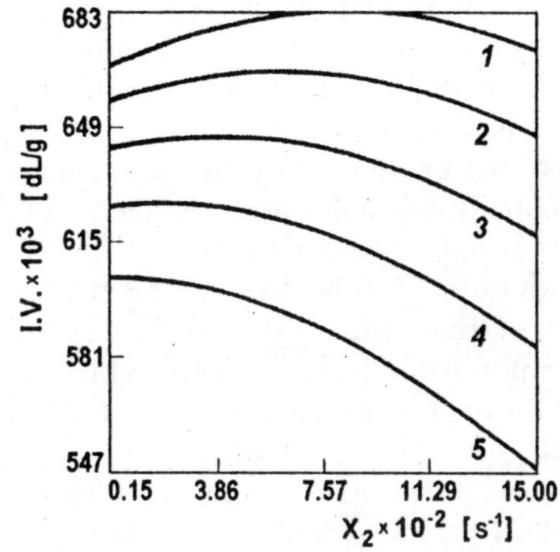

Figure 3.267. Intrinsic viscosity of the extrudates obtained from unstabilised PET, as a function of the shear rate and extrusion temperature. 1) 270 °C; 2) 278 °C; 3) 285 °C; 4) 293 °C; 5) 300 °C.

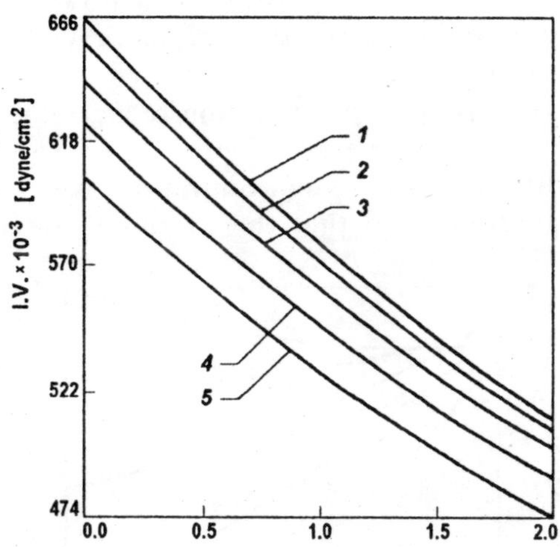

Figure 3.268. Change in intrinsic viscosity of the extruded PET as a function of inhibitor concentration, and extrusion temperature, at small value of mechanical factor. 1) 270 °C; 2) 278 °C; 3) 285 °C; 4) 293 °C; 5) 300 °C. (Shear rate = 15 s^{-1}).

with $[\eta]$= 0.695 *dL/g*, for the polymer extruded at 270°C.

At 300°C, in the absence of inhibitor, the extrudate intrinsic viscosity decreases with the increase of total – both thermal and mechanical – degradation, (Figure 3.270).

The conditions causing reduction of total degradation contribute to an increased intrinsic viscosity of the extrudate; at lower concentrations in the inhibitor, total degradation is amplified with the increase of the extrusion rate, the order to characterise the 0–1% domain being obtained. When the inhibitor's concentration attains a certain value, mechanical degradation is reduced or even cancelled, being followed by thermal degradation.

The higher the amount of inhibitor in the initial PET and the shorter the residence time, the higher is the extrudate intrinsic viscosity. More than that, the plasticizing effect of the micromolecular inhibitor appears, inducing reduction of the shear stress, Figure 3.270.

At a temperature of 300°C and maximum shear rate (1500 s^{-1}), in the absence of inhibitor, degradation is the highest, Figure 3.270.

The introduction of the inhibitor causes a reduction of thermal degradation during stationing in the rheometer reservoir. Even if the mechanical factor evidences increased shear stresses, their level is not sufficiently high to cause splitting, so that the effect of thermal degradation is predominant; finally, polymers with a lower degree of degradation are produced.

Increase of the inhibitor concentration reduces the initial thermal degradation, and, even if the shear stress increases, the extrudate intrinsic viscosity increases. A concentration over 1% also reduces mechanical degradation, so that the extrudate intrinsic viscosity

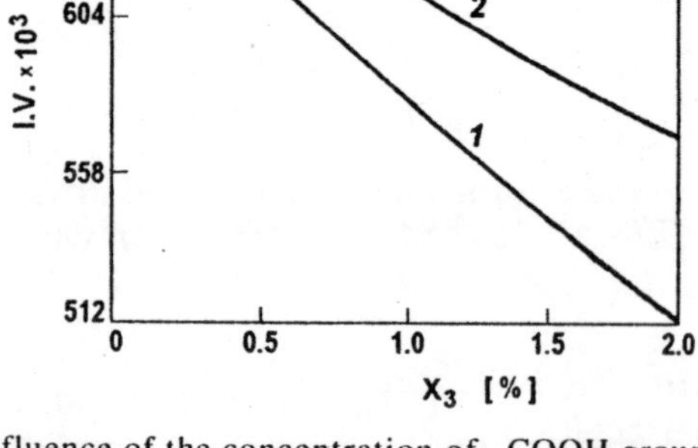

Figure 3.269. Influence of the concentration of –COOH groups' inhibitor and shear rate on the viscosity of extruded PET at 270 °C : 1) 15 s^{-1}; 2) 386 s^{-1}; 3) 757 s^{-1}; 4) 1128 s^{-1}; 5) 1500 s^{-1}.

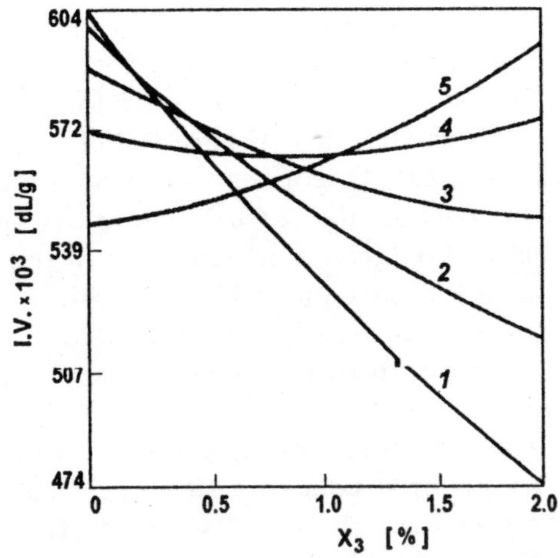

Figure 3.270. Influence of the concentration of –COOH groups'inhibitor and shear rate on the viscosity of extruded PET at 300 grade C: 1) 15 s^{-1}; 2) 386 s^{-1}; 3) 757 s^{-1};4) 1128 s^{-1};5) 1500 s^{-1}.

increases more quickly than in the first concentration range. At a shear rate of 1120 s^{-1}, the residence time is longer, thermal degradation is manifested and intrinsic viscosity is reduced, as compared with the values recorded at a rate of 1500 s^{-1}

Based on the mathematical model of the extrudate intrinsic viscosity, the maximum values of viscosities have been calculated over the whole range of factor variation, and the extreme conditions have been determined. The obtained values are listed in Table 3.96. As shown in the Table, the maximum values of the extrudate viscosity may be attained at low temperatures (270–275°C), and high extrusion rates (1500 s^{-1}), in the presence of a carboxylic group inhibitor.

Processing of the data obtained on the concentration of the extrudate carboxylic groups using the experimental program, the following mathematical model was obtained, [796]:

$$[COOH] = 256{,}9 - 275{,}0X_1 + 1{,}34X_2 + 306{,}6X_3 + 65{,}3X_1^2 + 1{,}42X_1X_2 - 108{,}9X_1X_3 - 4{,}5E - 02X_2^2 - 5{,}11X_2X_3 + 5{,}14X_3^2 \quad (3.139)$$

In the case of untreated PET, the dependence of the COOH groups concentration on the shear rate evidences two types of evolution, as a function of the extrusion temperature. Up to temperatures of 285°C, the COOH concentration shows a maximum at low shear stresses. On increasing the extrusion rate the –COOH concentration is reduced, passing through a minimum, followed by another increase, at high shear rates, Figure 3.271.

At temperatures over 285°C, concentration of the COOH groups increases with the increase of the extrusion rate. At very low shear, concentration of the COOH groups increases with temperature. The more degraded is the PET introduced in the capillary, the higher is the COOH groups' concentration in the extrudate.

Attainment of a temperature of 285°C induces changes in the rheological behaviour of the PET melt, as indicated in the relations given in Figure 3.266. Starting from a slightly thermally degraded polymer, the COOH groups' concentration increases with the increase of the shear rate (the already discussed effect of the shear rate on stress action).

For the extrudates treated with 1% *N*-GPhI the evolution of –COOH group concentration is shown Figure 3.272.

The maximum –COOH group concentration is attained at high stationary times and low extrusion temperatures. Viscous heating generates heat, which is propagated within the whole mass, thus amplifying chains' degradation, a phenomenon which cannot be restricted by the amount of inhibitor present in the melt mass. At 15 s^{-1}, the shear stresses decrease with increasing temperature, Figure 3.273.

To the same initial thermal degradation, the degradation produced during extrusion should be added.

At 270°C, in the PET treated with 2% *N*-GPhI, the shear stress is higher than at 300°C.

The inhibitor cannot compensate the viscous heating, occurring in the capillary at 270°C; a more degraded extrudate thus results.

Viscous heating affects the extrudate's external part, and even if the concentration of the –COOH groups increases, in the extrudate's centre there remain non-degraded molecules, which induce a high intrinsic viscosity.

Table 3.96. Maximum values of the extrudate's intrinsic viscosity.

Factor	Real value of the factor				
x_1 (°C)	270	277,5	285	292	300
x_2 (s^{-1})	1500	1500	1500	15	15
x_3 (%)	2	2	2	0	0
$[\eta]_{MAX}$(dl/g)	0.695	0.675	0.652	0.625	0.604

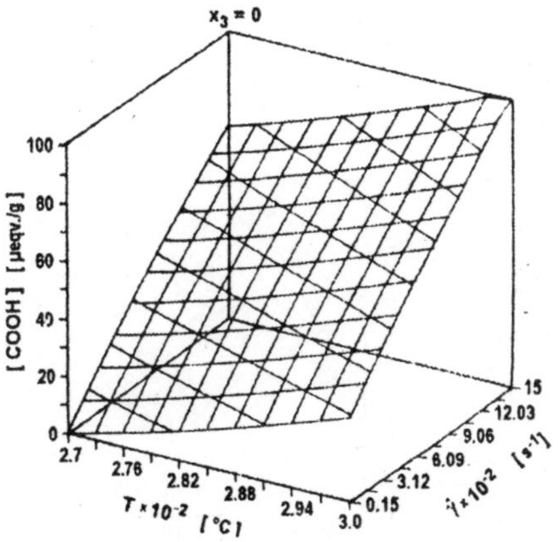

Figure 3.271. Concentration of –COOH groups in extrudates obtained from untreated PET, as a function of extrusion temperature and shear rate.

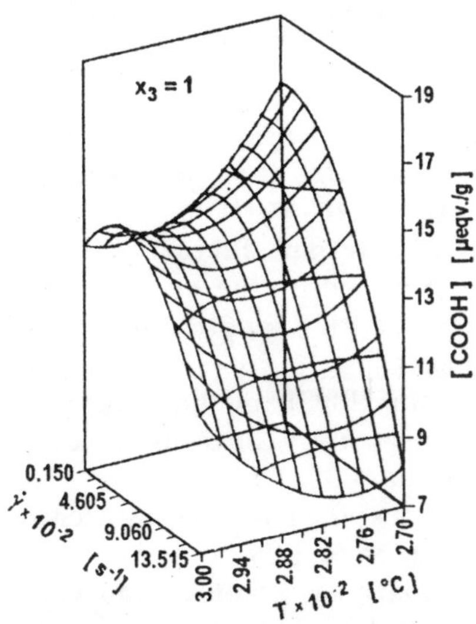

Figure 3.272. Concentration of COOH groups in extrudates obtained from PET treated with 1% NGPhI, as a function of extrusion temperature and shear rate.

At 300°C, the whole melt mass is thermally degraded prior to its entering in the capillary. During passing through the capillary, degradation is lower, an extrudate with less –COOH groups thus resulting.

A similar behaviour was also noticed when employing PhGE as an inhibitor of carboxylic groups, with increasing inhibitor concentration to 1.5%, extrudates in which the –COOH group concentration decreases with the increase of the shear rate are obtained, Figure 3.274.

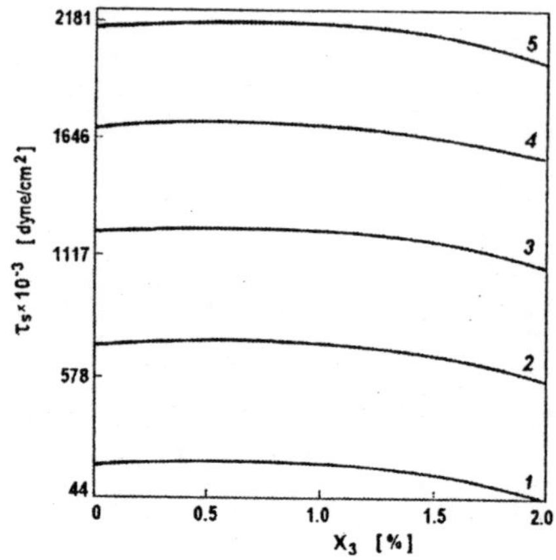

Figure 3.273. Influence of the inhibitor concentration and shear rate on the shear stress: 1) 15 s^{-1}; 2) 386 s^{-1}; 3) 757 s^{-1}; 4) 1128 s^{-1}; 5) 1500 s^{-1}. (L_c = 60 mm; D_c = 1 mm; T_e = 280 °C; ICR Model 3212).

(d) Reduction of Thermomechanical Degradation during Industrial Processing of PET Treated with PhGE

The mechanisms of action of the external factors on the characteristics of PET extrudates was verified by the utilisation of PhGE as an inhibitor to obtain a sort of PET, applicable in the fabrication of technical polyesteric yarns.

Preparation of PET granules containing PhGE in the bulk melt, was made in an industrial plant. The viscosity values obtained for PET mixed with PhGE are listed in Table 3.97.

The obtained granules and yarns have been characterized according to the methods usually applied in industrial processing of polyesteric fibres. For the sake of comparison, granules and yarns from current production have been utilised.

The viscosity values of wet granules obtained from PET with PhGE range within the limits prescribed by PET technology for technical yarns, i.e. 0.870 for sample N in Table 3.97.

The viscosity values of granules obtained in the second stage of granulation are lower when employing PhGE, possibly because of the inhibitor's uneven distribution along the whole reactor's height, and consequently its remaining in the upper half of the melt layer.

The value of the softening points, yellow index and brightness degree, listed in Table 3.98, range within the pre-established limits, being higher for PET granules synthesised with PhGE than for granules of current production.

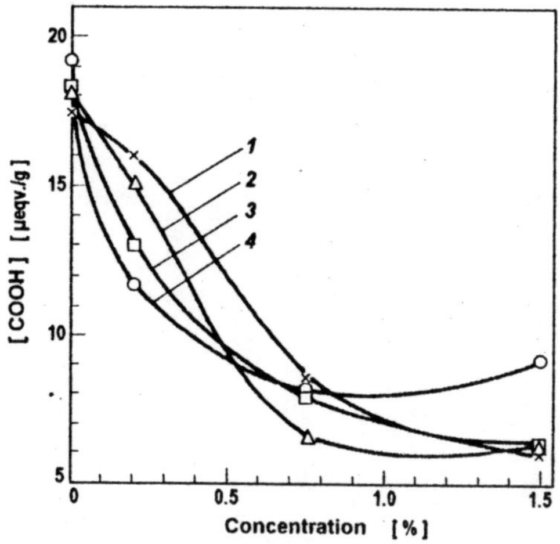

Figure 3.274. Influence of PhGE concentration in initial polymer and of the shear rate, on the concentration of the COOH groups in extrudate: 1) 45 s^{-1}; 2) 150 s^{-1}; 3) 450 s^{-1}; 4) 1500 s^{-1}. (L_c = 60 mm; D_c = 1 mm; T_e = 280 °C; ICR Model 3212).

Introduction of PhGE into the PET melt was aimed at reducing the content of –COOH groups in the granulated polymer.

This assumption was checked by determining the concentration of –COOH groups from the PET granules; the values obtained are listed in Table 3.99.

When employing PhGE, a slight reduction in the concentration of the –COOH groups in the polymer was recorded, which was quite expected, as due to the low amount of the inhibitor used.

The granules obtained both with and without COOH groups inhibitor have been processed in spinning process, for the fabrication of technical yarns. The extent to which the presence of PhGE is manifested in the quality of the obtained technical yarns was followed, Table 3.100.

For comparative purposes, yarns from the current production as well as yarns made of PhGE treated PET, obtained in the same drawing lines, have ben employed. Batch 1 contained granules with PhGE. The values of title and tenacity for the drawn yarn are within prescribed technological limits, i.e. 1100 + 5%, being better for the yarn obtained with PhGE. Elongation evidences slightly higher values for yarn obtained from PhGE treated PET. Finally, adherence to rubber recorded higher values when PhGE containing granules was used, i.e. 18.3 compared with 16.6 for the yarn not treated with the inhibitor.

Table 3.97. Specific viscosities of PET solutions in phenol/tetrachlorethane (3/2, g/g) [199]

Conc. of PhGE (%)	Charge nr.	Specific viscosity × 10^3			Difference between viscosities × 10^3		
		I	N	F	I-N	N-F	I-F
0,3	1	881	873	863	8	10	18
	2	896	883	869	13	14	27
	3	878	876	858	2	18	20
	4	875	870	867	5	3	8
	6	877	871	857	6	14	20
	7	886	884	857	2	27	29
	8	884	875	843	9	22	31
mean		**882**	**877**	**859**	**5**	**18**	**23**
0	10	920	891	889	16	14	30
	11	904	887	890	5	6	11
	12	890	878	878	4	9	13
	13	877	897	882	18	5	23
	14	872	879	871	1	12	13
	15	866	874	867	11	4	15
mean		**892**	**883**	**875**	**9**	**8**	**17**

I = initiation of granulation ; N = 15 min. after initiation of granulation ; F = end of granulation

Table 3.98. Influence of the COOH inhibitor on the characteristics of PET granules [199]

PhGE conc. (%)	Charge nr.	Softening point (°C)	Yellow index	Brightness degree
Prescribed values		259-262	Max. 35	min 11
0.3	1	259.5	11,9	22.5
	2	260	2,8	19.2
	3	259.5	15,8	24.3
	4	259	20.5	22.15
	5	260	10,3	21.2
	6	259,5	10,0	18.9
	7	259	13,5	20.0
	8	259	5,4	18.2
Mean value		**259.4**	**11,27**	**20.8**
0	10	259	11,9	20.5
	11	260	10,9	20.2
	12	259	19,7	21.4
	13	259	14,1	20.7
	14	259	9,3	21.0
	15	259	15,4	23.3
Mean value		**259.1**	**13,55**	**21.1**

Table 3.99. Influence of PhGE on the concentration of COOH groups from he granules [199]

PhGE concentration (%)	Charge nr.	COOH (equiv/ t polymer)
0.3	2	18.7
	3	17.8
	4	18.3
	6	16.9
Mean value		**17.7**
0	12	19.2
	13	20.5
	14	17.4
	15	18.3
Mean value		**18.8**

3.8.2.4. Mechano-chemistry of polymer processing in solution

The systems constituted of a macromolecular and, respectively, a micromolecular compound are equally important both from a theoretical and a practical point of view.

For theoretical investigations, diluted solutions are preferred, a case in which the interactions between macromolecules are considerably diminished, up to their cancellation.

Interpretation of the results obtained in studies devoted to diluted solutions allows determination of the structural pecularities of the macromolecules (such as shape, dimension, molecular weight), along with the thermodynamic characteristics of the solutions.

In practice, of special importance are the concentrated solutions, from which both artificial and synthetic fibres may be obtained through wet or dry spinning.

The same solutions play an important role in producing adhesives and lacquers as well as in polymer swelling, fractionation and plasticization.

The macromolecules in solution may be subjected to the action of some mechanical stresses resulted from several types of stresses applied, such as: ultrasonation [797,798], cryolysis [799,800], laminar or turbulent flow through thin tubes [801,802], ultrahigh speed stirring [803–806], flow between two concentric cylinders which rotate in opposite directions [210], forced flow through the capillaries [798–814].

Table 3.100. The quality of technical yarns obtained from PET treated with PhGE [199]

Batch	Title of drawn fiber(dtex)	Tenacity (gf/dtex)	Elongation (%)	Adherence (Kgf/cm)
		Experiment A		
1	1140	7.0 (6.6-7.5)	14.4 (12.8-16.4)	18.3
1	1120	7.0 (6.7-7.5)	14.4 (13.6-15.6)	15.9
2	1150	6.9 (6.5-7.7)	14.0 (12.4-16)	13.3; 15.2
2	1140	6.9 (5.8-7.6)	14.0 (12-15.6)	-
3	1140	6.9 (6.7-7.2)	13.6 (12.8-14.4)	14.4; 15.3
3	1130	7.0 (6.8-7.4)	13.2 (12.8-14.4)	15.3
		Experiment B		
1	1140	6.8 (6.6-7.2)	14.8 (14.4-15.6)	18.3
1	1130	7.1 (6.7-7.7)	14.4 (12.8-15.6)	16.6
4	1160	7.0 (6.7-7.7)	14.8 (13.2-16.4)	14.4
4	1140	7.0 (6.5-7.3)	14.4 (13.6-16)	14.3; 17.3

Regardless of the type of stress, intense shear stresses appear in solution, concentrated on the individual macromolecules' points of structural defects. When a critical value is attained, they cause homolytic splitting of the weaker covalent bonds, usually placed in the macromolecule's central region, thus generating two macroradicals. They are responsible for the initiation of an assembly of chemical reactions, the final result of which is either the polymer's mechano-degradation or its chemical modification, in the presence of suitable reactants.

3.8.2.4.1. Mechano-chemical reactions during spinning of polymer solutions

The polymeric solutions forced to flow through capillaries or low diameter tubes often suffer a reduction of the average molecular weight as a result of splitting of some chemical bonds from its main chain.

A typical process of flow through capillaries of a polymeric solution – having significant practical applications – is represented by spinning from solution. Analysis of a polymer's molecular weight, both before and after spinning, under industrial conditions, evidences its decrease, sometimes up to 25% (in the case of acrylonitrile based copolymers) [88].

Flowing through thin tubes and, respectively, through capillaries, of the polymeric solutions, is characterised by a parabolic distribution, induced by its non-Newtonian character; the highest speed was recorded in the capillary's axis. Once admitting that the ends of a macromolecule or of a sufficiently long chain segment are found in zones with different flow rate, on the mentioned macromolecule and, respectively, the chain segment, a shear stress will be manifested. At critical values of the rate gradient, the chemical bonds on which the shear stress was concentrated are homolitically split. For example, for a macromolecule with a value of the 10^3 order, breaking of a chemical link occurs at a shear rate of 10^5 s^{-1}.

According to Bueche [278], it is only the coiling between the macromolecular chains with extremities situated in distinct flow regions that are capable of producing, due to the rate gradient, sufficiently high stresses for causing breaking of the –C–C– bonds. The theory is supported by several results, attesting that mechano-degradation's efficiency increases with solution's concentration, as a result of the even increased number of "junctions" occurring between the macromolecular chains.

Intense phenomena of mechano-chemical degradation can be observed in polymers with a stress threshold, even in diluted solutions. Consequently, for flowing of PEO or PAA solutions, in a turbulent regime, through thin tubes, a drastic reduction of the polymer molecular weight was observed [815–817].

High molecular weight PAA is a very important polymer, with applications in oil's extraction.

Its injection under pressure, in the form of some aqueous solutions, through porous media, represents a typical process of forced passing through the capillaries.

For the process characterisation, parameter $\varepsilon \cdot L_D^{1/3} / \Phi^m$ was proposed:

$$\varepsilon = (flux) \cdot \Phi \cdot \Delta p \tag{3.140}$$

$$\Delta p = \frac{1-\varphi}{\varphi} \cdot \sqrt{150 - l_p / \varphi} \tag{3.141}$$

where: k_p – permeability; ϕ – porosity; L_D – pore length; Δp – pressure drop; m – the average diameter of the pore; m – an empirical constant.

For high values of ε comparable to relaxation ones, and especially for very high values, mechano-degradation occurs. The process intensity is very well correlated with ε, as well as with permeability, porosity and flow geometry.

The molecular weight distribution curve, Figure 3.275, obtained by GPC, indicates reduction of the macromolecule dimension, along with narrowing of the polydispersity domain towards low values [811].

Systematic studies devoted to this type of mechano-degradation, discussing also kinetic aspects, have been performed on PDMA in tetrahydrofurane, at concentrations ranging between 0.22–8.9% [819–821].

The process kinetics may be described with equation:

$$-dc_i / dt = k_i c_i^n \tag{3.142}$$

where i is the type of high molecular weight from the polymer's component.

For all concentrations taken into study, a first order kinetics of the degradation reaction was established. The constant of the reaction rate, of the 10^6 order, was found to be proportional to the hydrodynamic volume of the polymer's molecule, expressed by means of terms $[\eta]_i$ and $\bar{M}_i$. In a regime of turbulent flow a linear dependence is maintained between k_i and $M_i^{1.79}$, while in the regime of laminar flow, k_i varies directly proportional with $\bar{M}_i$, [821].

The constant of the mechano-chemical degradation rate depends on the solution concentration:

$$k_i = (K + b\,c)^{-1} \tag{3.143}$$

where: K and b represent constant parameters for each polymer's fraction, regardless of the value of its molecular weight, Figure 3.276.

Although spinning in solution of polymers constitutes a typical process of forced flow through capillaries, with well-known effects upon the polymer's molecular weight, it is quite surprising that the

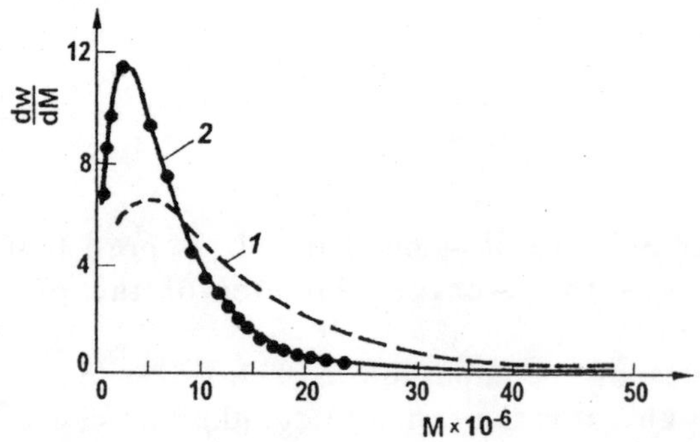

Figure 3.275. Molecular weights distribution of the initial PAA (1) and after forced passing through porous media (2).

mechano-chemical processes involved here have been less studied.

The first to mention a sensible reduction of the PAN 's molecular weight as a result of spinning in solution was Fichman [822]; he also stated that the capillary's length and especially the solution jet's flow rate through it may considerably influence the process.

Mashudanov [823] analysed the nature and part played by the forces acting during dry spinning from solution of the cellulose acetate in the aerodynamic and hydrodynamic field presents around the new-formed yarn.

The duration of action of such forces is 0.07–0.1 sec, a time interval over which yarn's formation usually occurs.

The forces' balance is the following:

$$F_i + F_g = F_{is} + F_{ii} + F_a + F_r + F \tag{3.144}$$

where F_i – force of yarn stretching at the end of the spinning capillary (the drawing force at the spinning nozzle); F_g – yarn's weight, $F_g = g\int_0^L s\rho dL$; g – gravitational acceleration; s – cross surface of a monofilament; ρ – density of the polymer's solution; F_{is} – stretching force on the yarn's surface; F_{ii} – inertial force, $F_{ii} = m{\cdot}a$; m – weight of a monofilament; a – acceleration of the yarn's movement; F_a – aerodynamic force during the yarn's

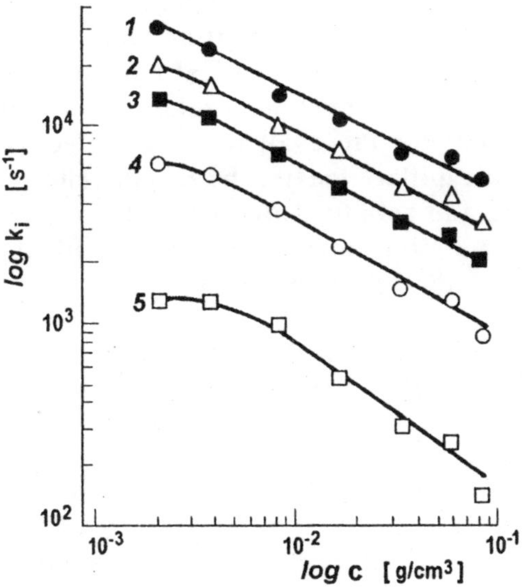

Figure 3.276. Dependence of the rate constant of the degradation reaction of PDMA in THF (k_i) on the solution concentration, for fractions of various molecular weight values. 1) $M = 17.5{\cdot}10^6$; 2) $13.5{\cdot}10^6$; 3) $11{\cdot}10^6$; 4) $7.5{\cdot}10^6$; 5) $3.2{\cdot}10^6$.

movement in the medium; F_r – rheological force, $F_r = \lambda \frac{du}{dl}$; λ – elongational viscosity of the solution jet passing through the capillary; du/dl – gradient of the yarn's motion rate; F – friction force of the solution jet with the capillary wall.

The values of the rheological force being determined by the rate gradient, it was concludent that it plays a special part in mechano-destruction processes.

Increase of the rate gradient means intensification of the shear stresses to the capillary's wall and, consequently, of mechano-chemical degradation.

3.8.2.4.1.1. Factors of influence

The dependence between mechano-destruction's intensity during forced passing of the polymer solution through capillaries, and some parameters characterising the process, have been studied even in the first papers in this field.

Thus, Grohn and Krause [812, 813] considered for example, that the determining factor of the mechanism of destruction is solution's concentration, while other authors draw the attention upon the capillary's length and, respectively, upon the number of forced passing of the solution through the capillaries [802, 803].

A systematic study on the influence of certain factors upon mechano-destruction, during forced passing through the capillaries of some polymeric solutions has been performed with a ternary acrylonitrile-based copolymer [830]. In view of getting really interesting results for industrial applications, mathematical modelling of the process was developed, aimed at establishing a relation between intrinsic viscosity (as a criterion for the evaluation of molecular phenomena) and its parameters: capillary length, the length/diameter ratio, solution's concentration, and spinning pressure [814].

The regression equation correlating intrinsic viscosity with the above-mentioned variables takes the following form:

$$\begin{aligned}[\eta 4] = {} & 1.427 + 0.007\times L + 0.002\times L/D + 0.021\times C - \\ & - 0.005\times P + 0.0001\times L^2 - 0.001(L/D)^2 + 0.0002\times L\times C + \\ & + 0.0002\times L\times C - 0.0001\times L\times(L/D) - 0.0002\times(L/D)\times C - \\ & - 0.0007\times(L/D)\times P - 0.0005\times C^2 + \\ & + 0.0007\times C\times P + 0.0026\times P^2 \end{aligned} \qquad (3.145)$$

where: L – capillary length; D – capillary diameter; C – solution concentration; *P* – spinning pressure.

(a) Capillary length

Increase of the capillary length results in a continuous decrease of the polymer's intrinsic viscosity. Once accepting that mechano-degradation is induced by the shear stresses generated by the parabolic profile of the flowing rate in the capillary section, it is evicent that the most intense shear stresses are manifested at the capillary wall. Increase of the capillary's length, assumes, for a constant spinning pressure, increase of the duration of action of the shear stresses upon the macromolecules in solution. Consequently, the polymer's intrinsic viscosity decreases continuously with the increase of the capillary's length (or of the spinnaret's thickness), Figure 3.277.

The effect is also confirmed by other experiments [802, 803]. Thus, during flow of solutions of PIB in toluene, through capillaries of varying lengths, the mechanodegradation process is intensified with increase of the capillary length. The effect is more intense when flowing occurs in a turbulent regime, when intensification of stresses is more amplified in the whole solution.

In a laminar regime, mechano-destruction is less intense since it is not related to the shear stresses acting on the capillary wall but it is related to the entrance effects in the capillary.

(b) Length/Diameter Ratio

This ratio represents an important geometrical characteristic of the capillary, the value of which may increase either with its length or with reduction of the diameter. The effect of increasing the capillary length has been already discussed.

The dependence between the polymer's intrinsic viscosity and the *L*/*D* ratio is plotted in Figure 3. 278.

The *L*/*D* ratio may increase by reducing the capillary diameter, which requires knowledge of this parameter's influence, as well.

(c) Capillary Diameter

This parameter influences considerably the flow's profile. Therefore, for a given pressure, the flowing rate of the solution polymer through the capillary will be higher for capillaries with reduced diameter. As a result, in the capillaries with a small diameter, more intense shear stresses will be generated to the wall, which should increase the intensity of mechano-destruction.

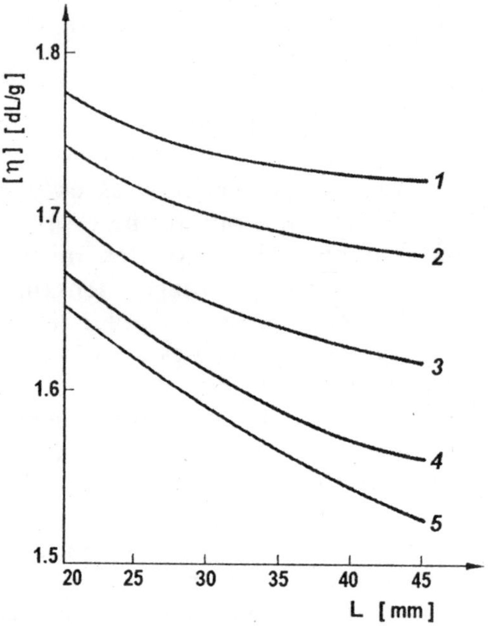

Figure 3.277. Variation of [η] with the capillary length (L) at different values of the L/D ratio: 1) L/D = 51; 2) 61.12; 3) 75.7; 4) 90.28; 5) 100 (mm/mm). (P = 10 at; c = 13.5%).

The experimental results confirm the effect, Figure 3.279.

The influence exercised by the mentioned such factors may the therefore explained by the parabolic distribution of the polymeric solution's flow rate, a consequence of their non-newtonian behaviour. As a result, only a small fraction of the macromolecules are present for a certain period of time in the vicinity of the wall where the shear stresses are high.

The pronounced mechano-destructive effect that occurs can not be solely explained through splitting of the macromolecules situated close to the capillary wall.

Responsible for its production are the primary mechano-radicals formed in the regions of a high gradient of flow rate, and also the shear stresses formed in the regions with a lower gradient of rate located towards the capillary centre.

An effective time is therefore defined representing the reciprocal of the flowing times corresponding to the rates appearing in the volume of solution from the capillary. This experimental parameter measures the mean time (the mean being calculated for a non-

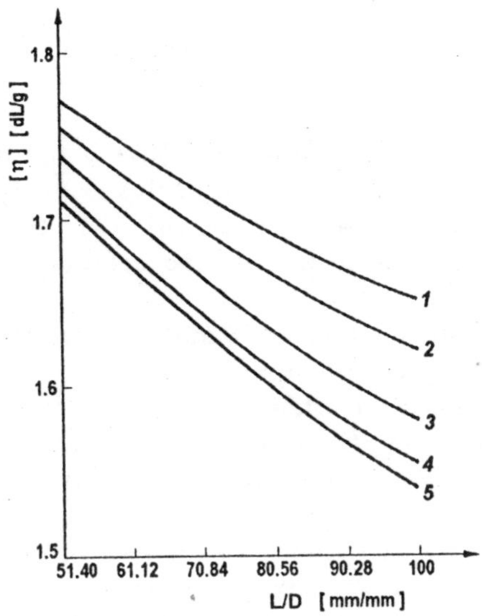

Figure 3.278. Variation of [η] with L/D for different values of the capillary length. 1) L = 20 mm; 2) 25 mm; 3) 32 mm; 4) 40 mm; 5) 45 mm. (P = 10 at; c = 13.5%).

newtonian distribution of rate), in which a macromolecule stands in the capillary. In the case of laminar flow, the parameter will be proportional to the mean time of the effect of stress on the chains by the shear forces.

(d) Concentration

The theoretical considerations of Bueche [278] predict that the macromolecule splitting in solution is the exclusive result of mechanical stresses related to the entanglements between chains.

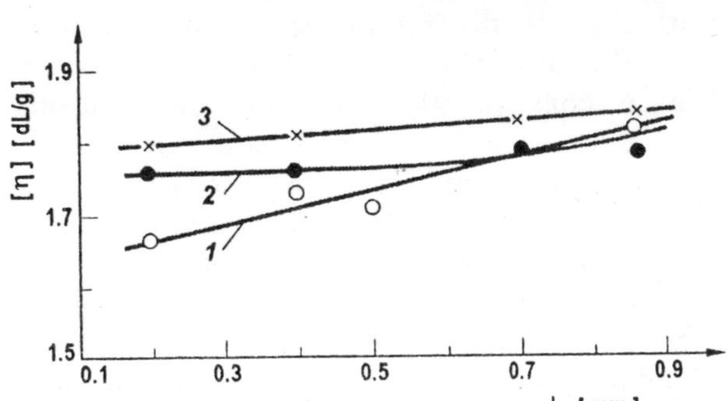

Figure 3.279. Influence of the capillary diameter upon [η]. 1) c = 10%; 2) 16%; 3) 22% (P = 7.5 at).

These entanglements are known to be able of producing sufficiently high stresses for splitting the C–C bonds. The more numerous are the interactions – i.e., the "junctions" – between the macromolecular coils, the higher should be mechano-destruction's efficiency and, respectively, the degradation rate. However, increase of the mutual interactions between macromolecules is only possible with the increase in concentration, which suggests the existence of some more intense mechano-destructive effects in concentrated solutions.

The results given in literature are quite contradictory as to the influence of the above-mentioned parameter.

Other scientists paid their attention to the existence of more intense mechano-destructive effects in diluted solutions [824, 825].

In the case of a ternary acrylonitrile-based copolymer, subjected to spinning from a solution, a more pronounced reduction of its intrinsic viscosity is observed, indeed, when working with more diluted solutions (having, nevertheless, a minimum concentration of 10%), Figure 3.280.

As a matter of fact, increase of the solution concentration intensifies mechano-degradation primary effects, and, consequently, the concentration of the primary macroradicals formed through homolysis is higher.

The fact has been confirmed in several experiments developed during ultrasonation in solutions of the same polymer, when the influence of concentration was also examined [289].

In the case of polystyrene subjected to mechano-destruction, in the presence of DPPH, increase of the solution's concentration causes lower and lower values of intrinsic viscosity [826].

(e) Pressure

It is an important mechanical parameter, which determines the rate of polymeric solution's flow through the capillary, respectively flow's turbulence; consequently, it will influence the intensity of the shear stresses generated in solution and thus the efficiency of mechano-destruction. Figure 3.281 presents the results obtained during spinning of ternary acrylonitrile-based copolymer.

The family of curves found shows a minimum value of intrinsic viscosity around the value of 10 at. The descending branch of curves (p = 5–10 at) evidences a decrease of [η] with pressure, the effect being induced, obviously, by the intensification of the shearing stress at the capillary wall.

At pressures over 10 at, the number of split chemical bonds in-

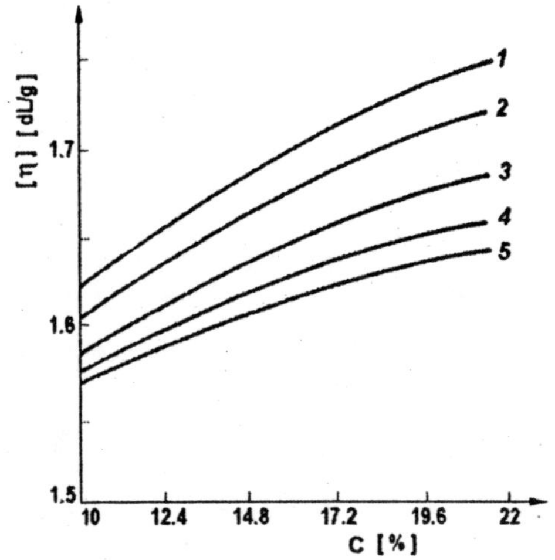

Figure 3.280. Variation of [η] with solution concentration at different value of the capillary length. 1) L = 20 mm; 2) 25 mm; 3) 32.5 mm; 4) 40 mm; 5) 45 mm. (P = 10 at; L/D = 75.7 mm/min).

creases, consequently, concentration in the primary macroradicals increases. Besides their reciprocal stabilisation through recombination and disproportionation, the chain transfer is more and more actively manifested, as facilitated by the reduction of the "free volume" of macromolecules, caused by pressure's increasing [827,828].

Recombination of the secondary macroradicals thus formed produces crosslinking effects, intensified with increase of pressure. In this way, the increase of intrinsic viscosity is explained.

All data agree with the observation that the polymer's solubility in DMF, after spinning, at pressures higher than 10 at, is quite low, comparatively with that of the initial one.

Performing of spinning at pressures over 10 at, with solutions in which a radicalic acceptor (DPPH) had been dissolved, evidences further decrease of [η].

It is beyond any doubt that, under such conditions, the primary macroradicals formed through mechano-degradation are stabilised by the reaction with the radicalic acceptor, so that the chain transfer reactions are prevented.

An experiment consisting in spinning of solution at pressures over 10 at, in the presence of benzidine, has also confirmed the mentioned hypotheses, Figure 3.282.

It was observed that, with the increase of pressure, to values even higher than 10 at, ever increasing amounts of diamine react, suggesting an ever increased number of primary macroradicals, concomitantly with the decrease of the polymer's intrinsic viscosity,

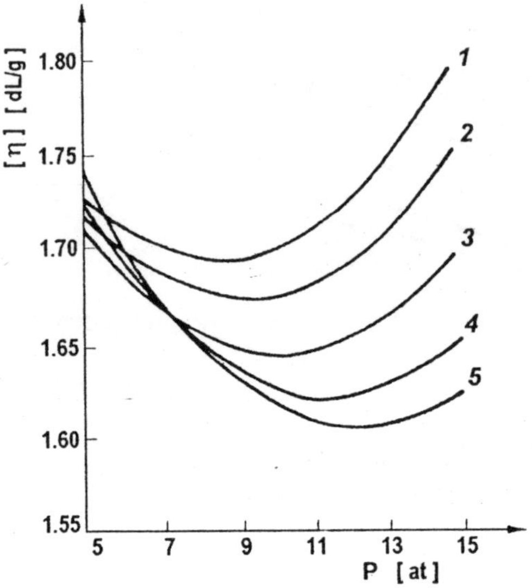

Figure 3.281. Variation of [η] with pressure for various length of the capillary: 1) L = 20 mm; 2) 25 mm; 3) 32.5 mm; 4) 40 mm; 5) 45 mm. (c = 13.5%; L/D = 75.7 mm/min).

once the reactions of macroradical stabilisation are hindered through chain transfer. Therefore, it becomes clear that the increase of [h], observed at pressures over 10 at, is induced by the polymer's slight crosslinking, and not by the reduction in the number of broken covalent links from the main chain.

(f) Number of Forced Passes of the Solution Through the Capillary

For any chemical reaction, the duration constitutes an important parameter; in processes of spinning in a solution, the influence of duration may be studied either on very long capillaries or by varying the number of passes of the solution through the capillaries. Figure 3.283 plots the influence of such a parameter on spinning in solution of a ternary copolymer of acrylonitrile.

The spinning process seems to occur according to some typical mechanochemical laws, being intensified with the number of passings through the capillary, up to the attainment of a limit minimum value of [η] [300] .

Similar results have also been obtained in the case of shearing through capillaries of solutions of PS in toluene, Figure 3.284 [812].

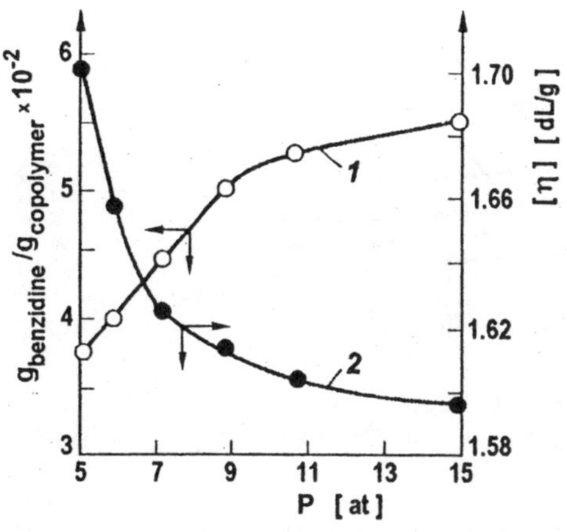

Figure 3.282. Variation of the amount of chemically reacted benzidine (1) and of the intrinsic viscosity [η] (2), with the spinning pressure (c = 10%; L/D = 100 mm/mm).

In principle, the process of mechano-destruction in conditions simulating drawing in solution may be influenced, too, by other factors, such as temperature, the polymer's initial molecular weight and polydispersity, the capillary's geometry, the entrance angle in the capillary.

From the practical point of view, the last ones are especially important in spinning process, although their influence has been studied only for the process of spinning extrusion from melts.

Many other studies have also been devoted to other types of mechanical stressing of polymers in the solution (such as ultrasonation, ultrahigh speed stirring, etc.) In all cases, the general rules of mechanodegradation were respected.

3.8.2.4.1.2. Chemical modifications

The experimental results obtained have shown that, under conditions

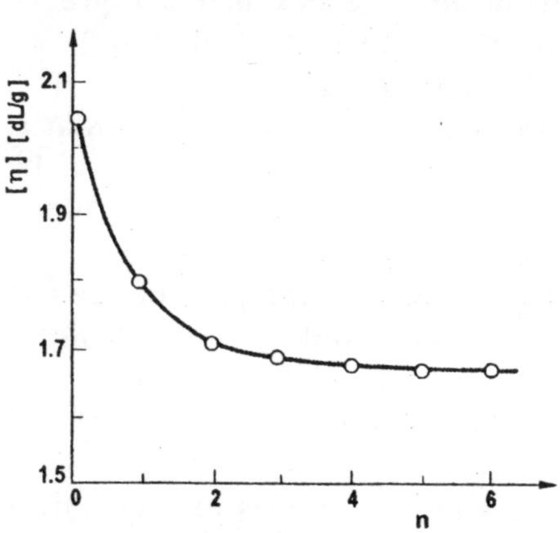

Figure 3.283. Influence of the number of forced passings through the capillary (n) upon intrinsic viscosity (D = 0.2 mm; c =10%; P = 5 at; L/D = 100 mm/mm).

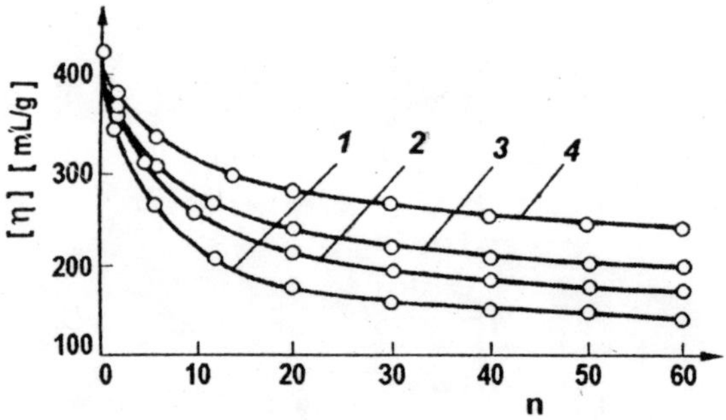

Figure 3.284. Dependence of mechanochemical destruction, by shearing in capillaries, on the capillary' s length and of the number of passings (P = 20 kgf/cm^2): 1) L = 8 cm; 2) 4 cm; 3) 2 cm; 4) 0.25 cm.

simulating spinning in solution, the polymer's degradation occurs, as manifested by lowering of intrinsic viscosity. The cause is the overstressing of the chemical bonds, mainly of the central ones, situated in the vicinity of the macromolecule's center of gravity or near those from the 'junctions' between the macromolecules. Such an effect intensity increases with the square of the chains' length and, on attaining a critical rate gradient, it causes their homolytical splitting, accompanied by the formation of macroradicals.

The radicalic mechanism of such a process of destruction activated by the polymeric solutions' forced passing through the capillaries has been demonstrated by the introduction of some radicalic acceptors, either in the monofilaments' coagulation bath or in the solution of polymer subjected to spinning.

Coagulation of monofilaments of ternary, acrylonitrile-based copolymer in a methanol solution of DPPH produced their violet colouring, seen as stable on repeated extractions in methanol, even after the monofilament's dissolution and the polymer's reprecipitation. Obviously, it is induced by the DPPH, that had stabilised, through recombination, the primary macroradicals.

Similar effects have been obtained when utilising aromatic diamines (such as benzidine, *o*-PhDA) as radicalic acceptors. In this case, too, the obtained monofilaments are coloured, the colour being also maintained in time.

An indirect proof of diamine's binding is also the possible development of a reaction of diazotization and coupling with phenolic compounds such as β-naphtol and resorcine. The diazotization reaction is accompanied by the formation of new colours of the treated fibres.

All these experimental data support the hypothesis of a chemical radicalic mechanism, initiated by the primary macroradicals generated through homolytic splitting of some chemical bonds from the basic chain, on which the shear stress is concentrated:

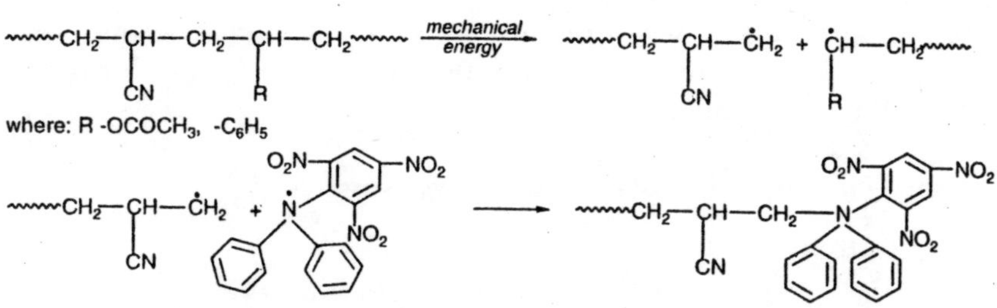

The macroradicals thus formed may react with the micromolecular substances present in the medium, either through recombination (DPPH) or through chain transfer, followed by recombination (aromatic diamines).

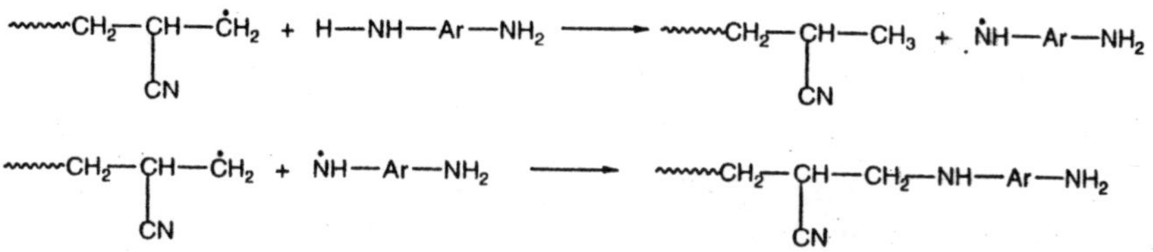

In the case of diamines with high reactivity (*p*-PhDA, benzidine), structures of the following type are obtained:

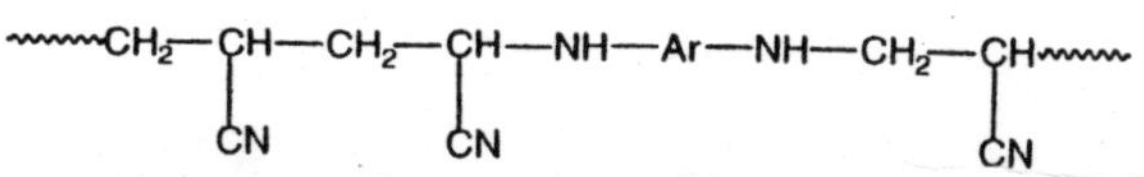

It is therefore explained why the polymer spun in the presence of *p*-PhDA has a higher intrinsic viscosity than the one spun in the presence of DPPH: because its molecular weight increases.

Such a reaction occurs, as well, in other conditions of mechanical stress applied to the polymers, in the presence of aromatic diamines, regardless of their physical state or phase. These reactions are termed 'mechano-chemical polycondensation' [85].

In the absence of radicalic acceptors, interruption occurs through disproportionation of radicals, the final product showing a lower molecular weight:

Increase of the polymer's intrinsic viscosity in conditions of high pressure spinning or when employing concentrated solutions, may be explained by the chain transfer reactions to the polymer, followed by recombination:

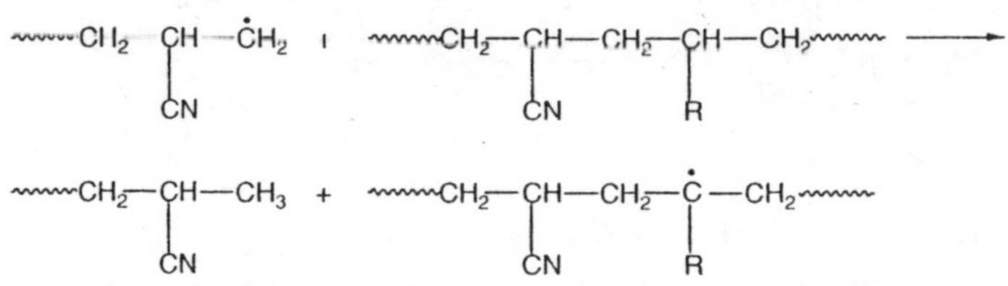

Branched,

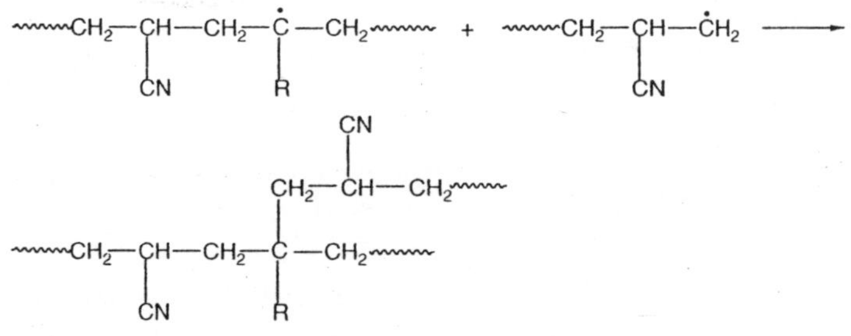

or reticulated structures:

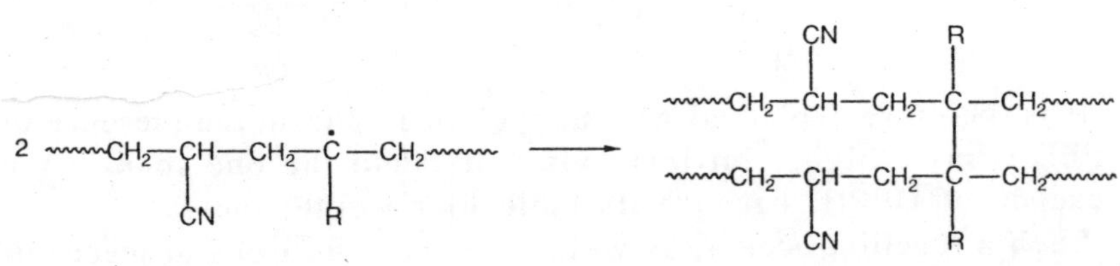

are thus obtained.

Of special practical interest in spinning from solution of some polymers may become the macroradicals' reactions with aromatic diamines. In this way, chemically modified polymers -containing free aromatic aminic-groups may be obtained. In their turn, they may be either diazotized or coupled with phenols or tertiary mixtures of amines, azoic structures being thus formed at the ends of the macromolecular chains. Depending on the chemical nature of the coupling agent employed, different colorations may results:

Consequently, the inevitable and undesired effects of polymers' mechanodegradation induced by spinning may contribute to improvement of some of their deficitary properties. Acrylic fibres are known for their low tinctorial capacity, so that the above - described reactions had been applied only for the obtainment of chemically dyed monofilaments, in a large range of colours.

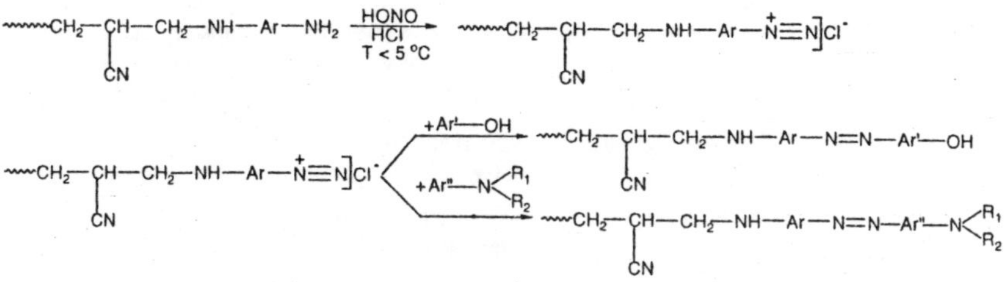

The tests performed at pilot scale confirmed the idea of chemical dyeing of acrylic monofilaments, coagulated in alcoholic solutions of aromatic diamines.

In some investigations, diamine has been introduced even in the solution subjected to spinning. Some of such experiments' results are listed in Table 3.101.

Under the stated conditions, the highest reliability is manifested for a 15% concentrated solution, which does not obstruct the spinning nozzle, coagulates rapidly at an average immersion depth, and generates monofilaments with good mechanical strength. An inferior behaviour is manifested by the 12% concentrated solution, from which monofilaments lacking mechanical strength are obtained; therefore, a higher immersion depth is required, which nevertheless complicates the technology. The spinning capacity is, also, different for the three solutions and, in conditions of coagulating the monofilaments in ethanolic solutions of diamines, the tinctorial properties vary, too, from one sample to another.

Most efficient best dyeing is performed on monofilaments obtained from 15% concentrated solutions which are uniform and stable at repeated extractions. Obviously, these monofilaments contain higher amounts of chemically linked diamine.

Introduction of diamine in the solution subjected to spinning intensifies the colours of the obtained monofilaments, yet dyeing is non-uniformly distributed and less resistant in time.

The amount of chemically-linked diamine is, of course, higher, while the chemically-non-linked one remains included in the monofilaments, which affects their mechanical properties.

Table 3.101. The behaviour of the acrylic type monofilaments under spinning conditions on a micropilot installation, in the presence of aromatic diamines as radical acceptors

No.	Solution concentration (%)	Composition of the coagulation bath	Observations	
			on spinning with acid colorants	on dyeing with acid colorants
1	12	ethanol	- reduced spinning capability reduced mechanical strength of the monofilament	- reduced tinctorial capacity the dyeing - does not resist to repeated extractions
2	15	ethanol	- good spinning capacity	
3	20	ethanol	- satisfactory spinning capability	
4	12	ethanolic solution of benzidine 1(%)	- reduced mechanical strength of mono-filament needs small speed of spinning	- satisfactory dyeing, various shade ranges
5	12	ethanolic solution of *p*-PhDA 1(%)		- good dyeing
6	15	ethanolic solution of benzidine 1(%)	- very good spinning capability good mechanical strength of monofilaments	- very good dyeing
7	15	idem	idem	- very good dyeing witl a less loss of shade at repeated extractions
8	idem	ethanol	idem	- good dyeing with a slight loss of shade at repeated extractions
9	20	ethanolic solution of benzidine (1%)	- small depth for dipping	- good but nonuniform dyeing
10	20	ethanolic solution of *p*-PhDA (0.3%)	- quick clogging of the capillary	- weak and non-uniform dyeing

The best dyeing, comparable with the one performed on wool, was obtained with monofilaments coagulated in the presence of benzidine or *p*-PhDA, and employing an acid dye.

Inferior results were attained when utilising *p*-PhDA as a radical acceptor: the free aminic group from the *o*-position cannot react with the voluminous molecule of reactant, as due to the steric hindrances.

All results indicate that good knowledge of the reaction mechanism laws, as well as of the kinetics, and of the factors influencing the mechano–chemical processes, permits one to intervene in polymers' processing and modify them structurally and, implicitly, from the new point of their properties, or to prevent the undesired processes of mechano–destruction, with all their unfavourable consequences.

3.8.3. Mechano–chemical reactions during polymers exploitation

3.8.3.1. Polymers fracture by fatigue

A continuously increased attention was paid in the last half of century on the fatigue phenomenon of polymeric materials that is reflected in the great number of studies, published articles and books dealing with this topic.

3.8.3.1.1. Polymers fatigue by solicitation in constant field of forces

The static fatigue is the phenomenon that occurs after a certain period of constant stress application. The stress acts for a long period of time on the sample and it is usually lower than the stress required for breaking under the conditions of monotonous loading. E.H. Andrews observed that it is hard to make a clear distinction between the static fatigue and the material stressing with a monotonous–increasing effort, since, in the last case, the effort is continuously applied and a finite period of time always appears between the stress applying and sample fracture [840]. The aggressive media diminish the sample durability [841].

Mechanical properties of a material are quantitatively appreciated by the term of mechanical resistance that expresses the maximal value of the stress, corresponding to a given type of solicitation, at which the macroscopic fracture occurs. For instance, the polymers tensile strength depends, among many other factors, of macromolecules conformation and dimensions. Thus, in the case of linear polymers, having the widest technical utilisation, its value, σ_r, varies in the range from 50 to 1000 Kgf/cm^2, in function of polymerisation degree, Figure 3.285. Following the evolution of this property, for a polymer–homologous series, it can be seen that until a polymerisation degree of about 100, σ_r increases with the increase of $\overline{DP}$, the increase being lesser evident one above this limit. For $\overline{DP}$ < 500, σ_r tends to a limit value. The explanation of this behaviour consists of the fact that, in the case of low

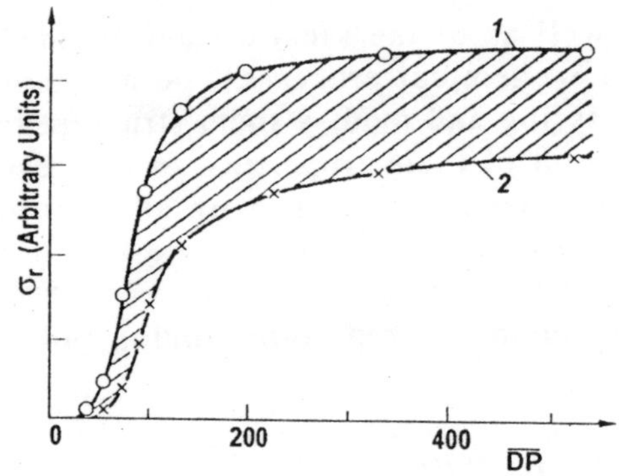

Figure 3.285. Variation of σ_r with DP: 1) polymers with high mechanical strength; 2) polymers with low mechanical strength.

molecular weight polymers, i.e. with $\overline{DP}$ < 100, at the same time with the increase of chain length the physical interactions between the chains intensify, diminishing, in this way, the possibility of their relative displacement (flowing). Apart from them, in the case of polymers with $\overline{DP}$ > 100, the deformation and fracture mechanism is changed. J. Sohma and co–workers, comparing the fracture processes of a low molecular weight compound, i.e. ethane at very low temperature, with a polymer constituted by methylene units (polyethylene), concluded that for $\overline{DP}$ > 83 the fracture is related to the mechano–chemical act, consisting in the scission of covalent bonds from the main chain. By empirically estimating the fracture activation energy as 1/2.45 of the vaporisation energy of the low molecular compound, they concluded that polymer fracture occurs at a value E_a = 1.01 Kcal/mole. If a compound with n structural units of methylene–type is considered for calculations, it is evident that for n > 83 the sum of the interactions between two neighbour macromolecules overtakes the value of 83 Kcal/mole, corresponding to the activation energy of chemical bond, –C–C–, splitting from the main chain. Consequently, the fracture process will further evolve by homolytical splitting of the chemical bonds and not of physical ones. Above this limit, the evolution of the considered mechanical property with $\overline{DP}$, changes the shape and tends to a limiting value. This interpretation can be extended form polyethylene to all other carbo– and heterocatenary polymers, with the mention that the last ones contain –C–X– bonds weaker than –C–C– bonds.

It is clear that the mechanical resistance of the polymers must decisively depend on the main chain chemical bonds strength.

However, the experimental studies proved that, in reality, the mechanical resistance of a polymer body is several times lower than the calculated value from the energy of interatomical bonds. This discrepancy appears and is manifested due to the existent faults at the all levels of polymer structural organisation. These faults act as factors of the stress concentration. In this way, it is possible as in the case of applied stress, which is significantly inferior to that causing macroscopic fracture, concentrated the local stress on a structural fault and to reach a high value overtaking the energy of the chemical bond and, therefore, causing its scission [1]. Therefore, the mechano–chemical step, being present both during deformation and fracture processes and consisting in homolytical splitting of the main chain covalent bonds, is the determining step of fracture. Polymer chain scission is followed by the appearance of free macroradicals, which determines all the subsequent polymer transformations, according to the already described mechanism.

Systematic studies dealing with the static fatigue have been carried out both on the semicrystalline and amorphous polymers. The obtained results decisively contributed to the improvement of the technological processing of these polymers into final products as well as to the forecast of their long term exploitation under specific conditions.

Thus, the stretching under constant stress (creep), variable stress, or at constant deformation (relaxation) constitute the most frequent solicitations of the textile fibres during exploitation. This is way, the knowledge of polymer behaviour presents an peculiar theoretical and practical interest. From the practical point of view, the obtained experimental results allows the correct and optimal determination of a given product exploitation regime, with the aim of increasing its durability. From theoretical point of view such results contributes to the elaboration and improvement of the theories concerning the fracture under load of the polymers materials and objects.

The study of polymer fracture processes is based on the generalised relation time–temperature, given by S.N. Jurkov, which was expressed by G.N. Bartenev in the following form [843–846]:

$$\tau(\sigma, T) = \varphi(\sigma, T)\exp\left(\frac{U - \omega\beta_{(0)}\sigma}{KT}\right) + \frac{L - l_k}{\nu_k} \qquad (3.146)$$

where: $\varphi(\sigma, T)$ – the preexponential factor (the influence of

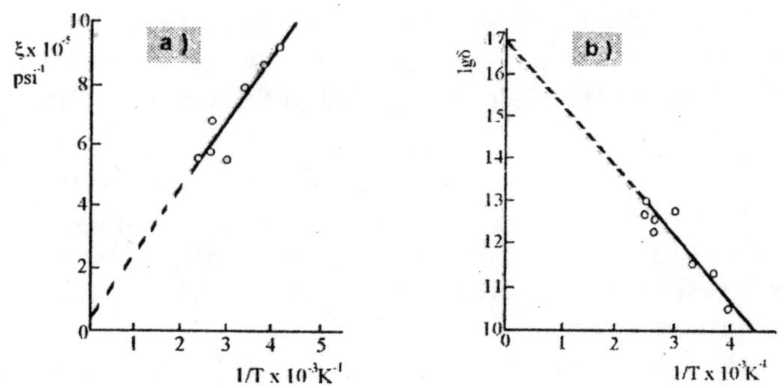

Figure 3.286. Effect of temperature on the activation constant, x, (a) and preexponential constant, δ (b) [69].

parameters σ and T on φ is relatively reduced as compared with the effect of the preexponential factor on the relation. Consequently, φ = constant, for the range of usually applied stresses; U – the activation energy of chemical bond splitting by thermal fluctuations; $\omega = \lambda\lambda_{\pi} \cdot \lambda_{m}$ – the microvolume within which the thermo–fluctuational splitting of the chemical bonds occur (λ – elementary distance covered by the crack tip, at each fluctuation that determines the fracture of chemical bonds; λ_{π} – elementary circumference of the fracture front, containing many chemical bonds; λ_{m} – chemical bond length just before its scission); $\beta_{(0)}$ – stress concentration coefficient; L – sample width; and λ_{k} – crack critical length corresponding to the passing from thermal–fluctuational mechanism of splitting to that athermal one, when the crack propagation rate is practically constant.

In equation (3.146) the first term reflects the contribution of from thermo–fluctuational stage of fracture and the second one represents the contribution of the athermal factor. The theory introduces the notions of limit stress, σ_0, for the crack generation, and critic stress, σ_k, which were practically confirmed [847, 848].

In the range of low stresses, for the brittle state characterised by small deformations, a deviation from linearity of the function lg τ(σ) was found and the durability increases, tending to the value of σ_0. At high values of applied stress ($\sigma \geq \sigma_k$) the curve passes through a minimal value, $\tau_k \cong 10^{-6}$ s, for λ = 1 cm, which was related to the occurrence with a maximal rate of the crack growing, v_k. This assumption was confirmed by graphical representation of lg τ(σ) vs. time and argues the hypotheses of passing from thermal–fluctuational mechanism to the athermal one, Figure 3.286. On

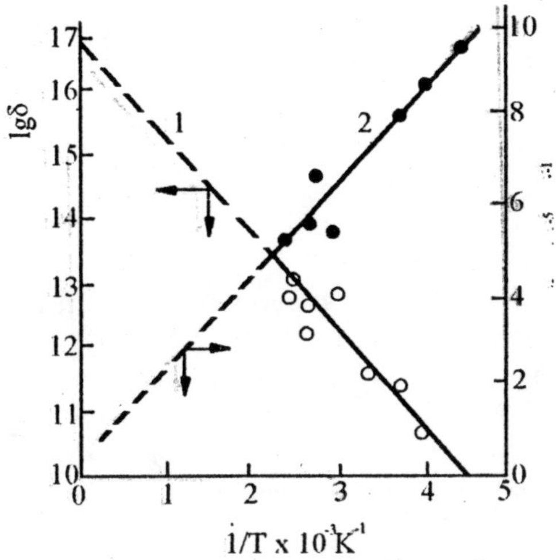

Figure 3.287. Effect of temperature on the preexponential constant, δ, (1) and activation constant, *x*, (2) in the case of tensioned nylon-6 fibers [69].

the basis of the general equation, Bartenev developed the theory of 'complete isotherm' of polymer fibre durability which was verified on many concrete examples.

At the molecular level, the following equation was derived between the applied stress and free radical concentration::

$$[R\cdot] = \delta \exp \xi\sigma \qquad (3.147)$$

where: $\delta = \alpha \exp(-\gamma / KT)$ and $\xi = \beta' KT$; α, β', and γ are polymer constants, rigorously depending on temperature, Figure 275. In the case of polyamide 6 based fibres this dependency is depicted in Figure 3.287.

The simplest element of the supramolecular structure of the fibres is the microfibril, with ϕ = 100–200 Å, that is composed by parallel macromolecular chains, which are characterised by amorphous–crystalline inhomogeneity. During the fibre uniaxially stretching, the amorphous zones are firstly stressed, some discontinuities occur and the nucleus of a submicroscopical crack arises. Initially it has a disc-like shape that is perpendicular to the stretching direction and which determines the material brittle fracture. In this case, polymer fracture is located just in the crack proximity, where the overstress σ^* that activates the covalent bond splitting is higher that the average stress corresponding to rest of sample volume. According to the theory of thermo-fluctuational fracture, the following relation describes the increase of the crack formation rate:

$$v(l,\sigma,T)=2\lambda\upsilon_0\exp\left(\frac{U-\omega\sigma_0^*}{KT}\right)\sin h\left[\frac{\omega}{KT}\left(\sigma^*-\sigma_0^*\right)\right] \quad (3.148)$$

where: υ_0 – thermal oscillations frequency, which causes chemical bonds splitting and rebounding ($\upsilon_0 \approx 10^{13}$ s^{-1}); σ_0 and σ^* – the limit stress and overstress, respectively, in the volume $\omega(\upsilon_0 = \alpha\lambda/\lambda_m)$.

For a disk-shaped (penny-shaped) axial-asymmetrical crack in a sample axially stretched with a constant stress, we have:

$$\sigma^*=\beta_{(r_0)}\sigma\sqrt{r_t/r_0} \quad (3.149)$$

where: r_t – crack radius at the moment t; r_0 – crack initial radius; σ – coefficient of stress concentration in the crack proximity.

The critical stress is given by:

$$\sigma_k=\frac{U}{\omega\beta_{(0)}} \quad (3.150)$$

Finally, the following expression is obtained:

$$v(r,\sigma,T)=\begin{cases}2\lambda\upsilon_0\exp\left(-\frac{U}{KT}+\alpha\sigma_0\right)\sin h\left[\sigma\sqrt{r_t/r_0}-\sigma\right] & for \quad \sigma_0\le\sigma\le\sigma_k\\ v_k & for \quad \sigma\ge\sigma_k\end{cases} \quad (3.151)$$

The durability of a fibre having the radius, r, and stressed with an stretching effort, σ, in the range form σ_0 to σ_k is expressed as:

$$\tau=\int_{r_0}^{r_k}\frac{dr}{v(r,\sigma,T)}+\frac{R-r_k}{v_k} \quad (3.152)$$

where the second term, represents the contribution of the athermal factor.

By integration, the following expression is obtained:

$$\tau = \tau_{\varphi} + \tau_k = \varphi(\sigma, T)\exp\left(\frac{U - \omega\beta_{(0)}\sigma}{KT}\right) + \frac{R - r_k}{v_k} \quad (3.153)$$

$$\varphi(\sigma, T) = \begin{cases} \dfrac{2r_0 f_0}{\lambda \upsilon_0 \alpha\sigma} & \text{for} \quad \sigma_0 \le \sigma \le \sigma_k \\ 0 & \text{for} \quad \sigma_0 \ge \sigma_k \end{cases} \quad (3.154)$$

$$f_0 = \sum_{n=0}^{\infty} \frac{1}{2n+1} \exp\left[-2n\alpha(\sigma - \sigma_0)\right] \quad (3.155)$$

By decreasing the stress, when $\sigma \to \sigma_0$ and $f_0 \to \infty$, the curve lg $\tau(\sigma)$ tends to the axis $\sigma = \sigma_0$. In the range of stress $\sigma > \sigma_0$, $f_0 \approx 1$ and the influence of σ and T on the preexponential factor becomes negligible as compared with those of the exponential one. Practically, the function $\tau(\sigma)$ is linear one in this range of tensions, which corresponds to the usual range of durability, ranging from 10^{-3} to 10^{-8} s. For this range, it can be written:

$$\tau = \tau_{\varphi} = \frac{2r_0}{\lambda v_0 \alpha\sigma} \exp\left(\frac{U - \omega\beta_{(0)}\sigma}{KT}\right) \quad (3.156)$$

which is similar to Jurkov's equation.

Whenever $\sigma(\sigma_{\varphi} \to \sigma_k)$ the second term of the sum from the equations (3.152) and (3.153) plays the essential role and the critical durability becomes:

$$\tau_k = \frac{R - r_k}{v_k} = \frac{R - 4\lambda\beta_{(0)}}{v_k} \quad (3.157)$$

It results that the function lg $\tau(\sigma)$ is the strength line in this range of stresses.

Because the fracture activation energy, U, depends on temperature, the following two situations are expected, namely – for $\sigma_0 \le \sigma \le \sigma_k$:

$$\tau = \frac{2r_0 f_0 \exp(-q/K)}{\lambda \upsilon_0 \alpha\sigma} \exp\left(\frac{U_0 - \omega\beta_{(0)}\sigma}{KT}\right) + \frac{R - r_k}{v_k} \quad (3.158)$$

and for $\sigma \ge \sigma_k$:

$$\tau_k = \frac{R - 4\lambda\left(\beta_{(0)}\right)^2}{\nu_k} \tag{3.159}$$

Equation (3.158) adequately describes many experimental data, especially those regarding to the dependency of fibre durability with temperature on the stress range between σ_0 and σ_k (when $f_0 = 1$). Equation (3.156) also proved its validity in practice, the molecular significance of some parameters included in it being the following:

$$\tau_0 = \varphi(\sigma, T)\exp(-q/K); \quad \gamma = \omega\beta_{(0)} \tag{3.160}$$

The validity of the durability isotherm was experimentally proved, on the oriented fibres obtained from many polymers such as: polyamides [849], polypropylene [850], polyethylene [851], and cellulose triacetate [852, 853], at different temperatures of solicitation.

In the case of oriented polyamide 6, the theoretically calculated distance between the neighbouring chains, is equal to the intermolecular distance, 4 Å, when all the chains are oriented alongside of the fibre axis. Therefore, $\lambda = \lambda_0 = 4$ Å and $\lambda_k = \lambda_0 = 4$ Å, respectively. It results that the thermo-fluctuate volume, ω, is equal to $2.4 \cdot 10^{-23}$ cm^3..

The crack initial radius may be calculated with the relation $r_0 = 4\lambda_0\beta_{(0)}$, where is experimentally determined, the fibre diameter was equal to 2 μm and the crack growth rate $v_k = 800$ m/s, which finally allows the calculation of critical and limit stresses. The types of radicals generated by the cracking of highly oriented semicrystalline polymers are presented in Table 3.102.

A similar shape of the durability isotherm has been found in the case of semicrystalline polymers. Four distinct domains have been studied, namely:

I) $\sigma \leq \sigma_0$, the range before the brittle fracture;

II) $\sigma_0 \leq \sigma \leq \sigma_\phi$, the range of fracture by the thermal-fluctuate mechanism;

III) $\sigma_j \leq \sigma \leq \sigma_k$, the transitional range, with the athermal mechanism prevailing with the increase of σ; and

IV) $\sigma_k \geq \sigma$, the range of athermal fracture.

The range of stress from σ_0 to σ_ϕ delimits the exploitation range of polymer fibres that is determined by their durability. By increasing the temperature, this range becomes narrower, but the domain comprised between σ_ϕ and σ_k, where the athermal mechanism prevails, becomes broader.

Table 3.102. The type and concentration of free radicals tat appear during the fracture of some highly oriented polymers

Polymer	Type of free radicals, $R\cdot$	Concentration, $[R\cdot]$ (spin/g)
Polyethylene	$...-CH_2-\dot{C}H_2$	5×10^{16}
Polypropylene	$...-CH_2-(CH_3)\dot{C}H$	5×10^{16}
Nylon 6	$...-CH_2-CO-NH-\dot{C}H_2$	10^{17} - 10^{18}
	$...-CO-NH-(CH_2)_3-\dot{C}H_2$	
Nylon 6,6	$...-CO-NH-(CH_2)_4-\dot{C}H_2$	10^{18}
	$...-CH_2-CO-NH-\dot{C}H_2$	
Poly(ethylene terephthalate)	$...-C_6H_4-COO-CH_2-\dot{C}H_2$	1.5×10^{17}
	$...-OCO-C_6H_4-CO\dot{O}$	
Poly[*p*-(2-hyroxyethoxy) benzoic acid]	$...-C_6H_4-CO\dot{O}$	4×10^{16}
	$...-C_6H_4-O-CH_2-\dot{C}H_2$	

It was proved that the knowledge of the macromolecule orientation that influences the parameters from the durability equation is of the primary importance for efficient exploitation of the final products [851]. Thus, the coefficient γ is related to the polymer orientation degree, its change being reflected in the modification of fibre rigidity. Taking into account that $\omega = 2.4\cdot10^{-20}$ mm^3, $\beta = 7.6$, the value that is close to that characteristic for the polyamide fibres ($\beta_{(0)} = 9$ for the non-oriented polyamide fibres): $\omega\beta_{(0)} = 1.3\times10^{-18}$ mm^3, therefore, $\beta_{(0)} = 9.3$ [854], which is close to the value of $\beta_{(0)} = 9.3$ for the amorphous polyamide. This means that at the same time with the polymer tensile strength also increases with the increase of the macromolecules orientation degree, this fact being related to the decrease of γ which, in turn, is caused by the decay of $\beta_{(0)}$.

The theory of polymer fibres durability, derived by Jurkov's thermal–fluctuational theory assumes the origin of deformation and fracture processes in the atomic–molecular and supramolecular mechanism. Its peculiar importance resides in the fact that this theory explains the role played by the chains orientation during the occurrence of deformation and fracture. On can conclude that the increase of fibres resistance and, therefore, durability can be achieved by decreasing the values of parameter γ.

Under the exploitation conditions, the polymers are supposed both to the action of continuous efforts (creep conditions) and cyclical ones and the fatigue phenomenon accompanies these processes. The polymer fatigue is manifested not only to the high values but also to the low ones, acting for prolonged periods, of the applied stress. In the last case the polymer durability is expressed as [855]:

$$\int_0^{t_i} \frac{dt}{\tau[\sigma(t)]} = 1 \tag{3.161}$$

where the denominator is replaced with the known expression from Jurkov's equation, the parameters τ_0, ν_0, U and γ being experimentally determined. A particularity of cyclical fatigue is related to the fact that after any load/unload cycle an amount of elastic energy is dissipated in the polymer, being converted in heat, which causes local overheating that, in turn, determines the structural changes in the zone of fracture propagation. With other words, due to the parameter γ is continuously changed during cyclical solicitation. This is way, the durability equation must be modified, by introducing in it both the real temperature, measured at the place of fracture development and the corresponding value of the parameter γ. The study of the microcracks growing process, based on the measurement of the accompanying thermal and luminous effects, proved that during any cycle of polymer stressing its rate jumps to higher values, Figure 3.288; the magnitude of the crack growth rate modification is essentially influenced by the stress rate and medium temperature [858].

The molecular mechanism of the fatigue process was investigated for a series of spinning polymers, such as polyamide 6 and polycarbonates [875–860], poly(ethylene terephthalate) and polypropylene [861], polyacrylonitrile and viscose-type cellulose [830, 862–864]. The majority of published papers are focused on the aspects concerning the other structural levels, i.e. supramolecular and morphological one.

Thus, the study of mechanical properties of semicrystalline fibres (PET and nylon 6,6) allowed the establishing of a distinct mechanism of fracture during fatigue process [865]. The crack is initially initiated at the fibre surface and propagates alongside of it, with a small angle with respect to the axis Ox, by 'wedge effect'. The fibre cross section gradually decreases and the fracture occur due to the creep process. As Figure 3.289 shows, A.R.

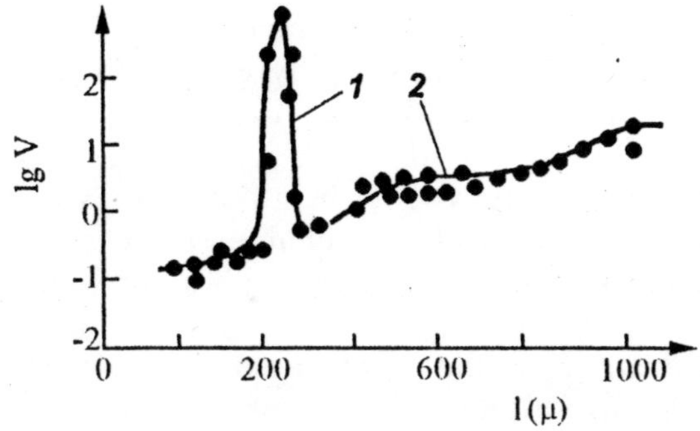

Figure 3.288. Dependence of crack growth rate by crack length for cellulose hydrate [856]. Solicitation speed: 1) 240 mm/s: 2) 0.7 mm/s.

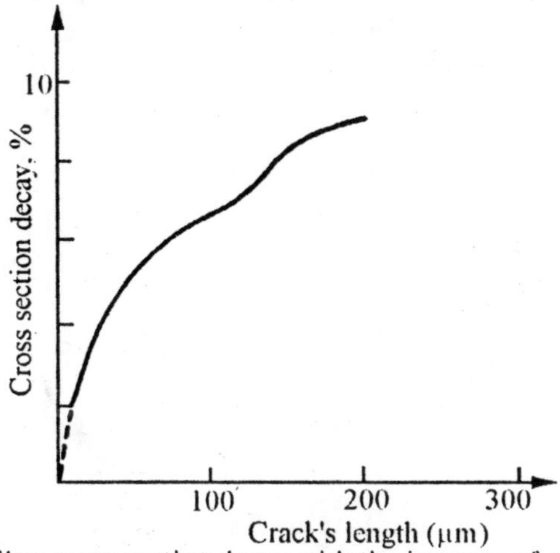

Figure 3.289. Fibre cross section decay with the increase of crack's length [866].

Bunsell and Ch. Ondet quantitatively determined the decrease of the fibre cross section, caused by the 'wedge effect' [866]. Using IR spectroscopy and X-ray diffraction, decay of crystallinity was found during fatigue of poly(ethylene terephthalate) yarns. The yarns were composed from cylindrical microfibrils with the diameter ranging from 15 to 20 μm. Concomitantly with the decrease of crystallinity, the formation of some extended amorphous zones in the fibres was evidenced by electron microscopy. The fracture trajectory passes just through the amorphous region, generating voids. The formation of the amorphous region is the result of crystalline phase destruction and of amorphous domains coalescence, which was observed in the proximity of the crack tip.

On the other side, the fibres present a peripheric-axial anisotropy, on which basis a "core-shell" morphology is characteristic for this type of fibres. The analysis, by electronic microscopy, evidenced that for the most different types of fibres the external surface is more oriented but presents many irregularities, such as voids or thickness faults [868]. In all the investigated cases, the fibres present a porosity that is characterised by pores, cavities, exfoliates, and vacuole whose presence was proved using different techniques

In the first stage of the mechanical solicitation, the resultant deformation greatly affects the amorphous zones, which under the action of developed forces temporary "ameliorate" their organisation, especially by macromolecules parallel orientation. Concomitantly, the alignment of the preexistent voids take place, by their migration and regrouping in bigger voids and cavities, thus generating cracks on the fibre's mantle. Under the conditions of a prolonged or repetitive solicitation, new microcavities and cracks are generated that cumulated whit the existent ones, bring the system in the state of 'destructibility' and fracture by fatigue. In these conditions the existent faults on the morphological level attain the maximal dimensions, which correspond to the superior threshold of the applied stress, σ_k, [869–871]:

$$\sigma_k \geq \sqrt{\frac{2ES}{a\pi}} \tag{3.162}$$

where: σ_k – critical uniaxial nominal stress; E – Young's modulus of elasticity; S – superficial energy; and a – elliptical cavity half of the long axis. Therefore, under the conditions of polymers fatigue, the characteristic barrier of potential is modified during the stress applying in the favour of fracture occurrence.

Based on studies by J. Murray and D. Hull, the changes of the existent faults in the fibres in the shape and dimensions are illustrated in Figure 3.290 [872]. Due to stress concentration to the level of any kind of fault, the splitting of physical and chemical bonds occurs with the generation of free radicals.

On the macroscopic level, the fatigue process is stimulated by some possibilities of faults evolution at this level: 1) preexistence of the superficial cracks; 2) cavities preexistence or their generation; 3) broken of microfibrils or "linking" macromolecules inside of cracks and release of the "active cavities". The catastrophic fracture will occur with different speeds, in function of the type and

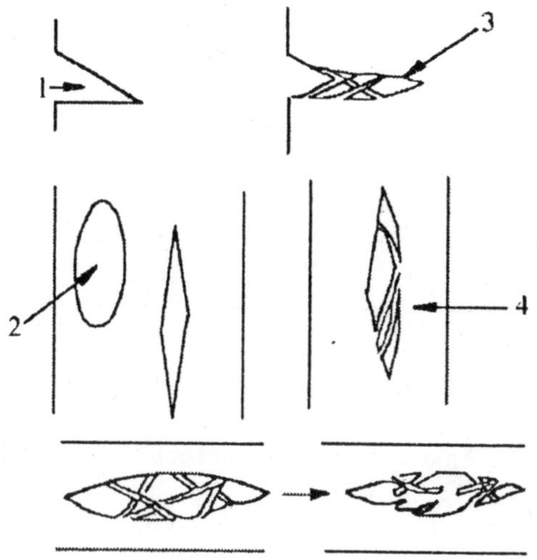

Figure 3.290. Different types of faults in structure [872]: 1) crack; 2) cavity; 3) open cavity with connection fibrils; 4) close cavity with fibrillar straps.

origin of fault that caused the fracture. Thus, it was found that the crack preexistence: 1) favours the rapid fracture propagation whether the applied stress overtakes the critical value σ_k; cavities formation 2) can occur in any region of the fibre, and their growth, depends by the position of crack or of the already existent cavities; when the dissipated energy, by internal friction, assures the scission of much more bonds inside of the cracks, accelerating the catastrophic fracture 3).

The presence of cracks or cavities rapidly lead to the catastrophic fracture, in function of time this one should appear after few cycles or after a period, t_{i_1}, when the probability of the mechanism 1 occurrence is maximal, Figure 3.291.

The crack or cavity progressive growth, developed by the regrouping of several small faults, leads after a certain period, t_{i_2} to a fatigue that occurs by mechanism 2.

If the first two mechanisms do not occur up to a relatively high period of solicitation, t_{i_3}, or after a large number of cycles the first two mechanisms do not occur but the coalescence of a large number of cavities and the splitting of the linking molecules does take place, a general fatigue state is installed in the system. The stress concentration in these zones overtakes the critical value, and, therefore, the macroscopic fracture is accelerated. In this case the third mechanism is responsive by totality of events that take place.

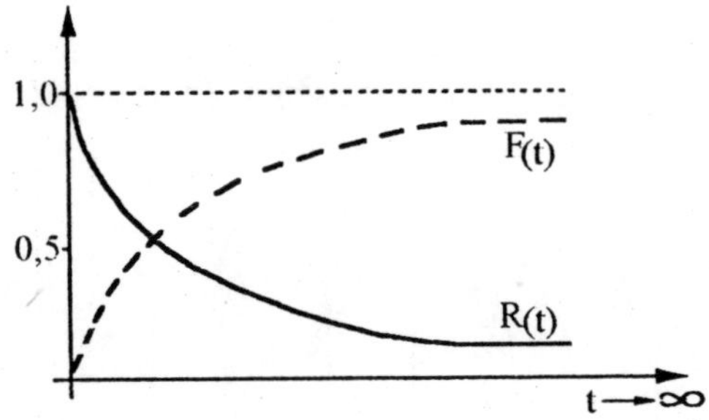

Figure 3.291. Graphical representation of the three types of mechanisms [871].

In reality, in the case of textile materials, regarded an assemblies of fibres, threads, and fabrics, the three mechanisms can simultaneously acts, but with differentiate evolutions in time, Figure 3.292.

From statistical point of view, the density function of probability of instantaneous degree of material loosing properties can be expressed as:

$$h(t)=\frac{R(t)}{R(t)} \tag{3.163}$$

where: $f(t)$ – "non-endurance" or desertion function; and $R(t)$ – "survive" or reliability function.

The graphical representation of this equation leads to the curve represented in Figure 3.293. Between the two curves, a complementary relation of reciprocal exclusion and non-simultaneous occurrence exists:

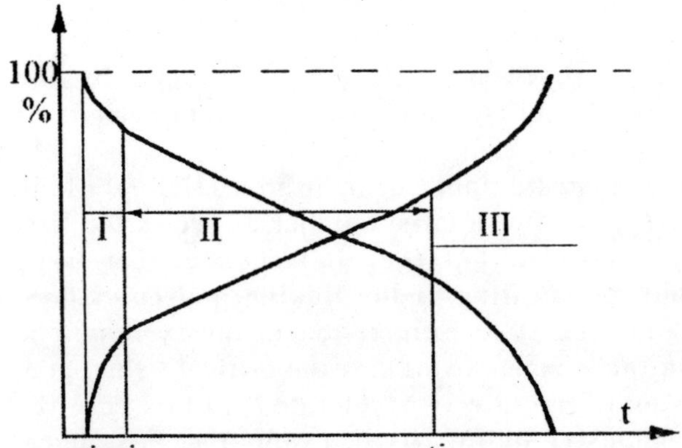

Figure 3.292. Representation of the combined action of the three mechanisms [871].

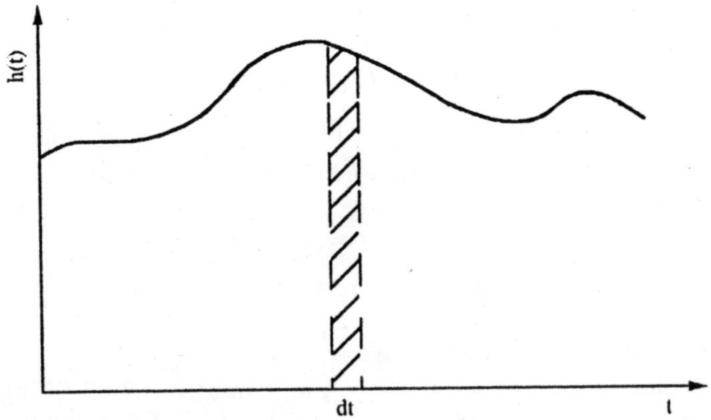

Figure 3.293. Graphical representation of the function $h(t) = f(t)/R(t)$ [871].

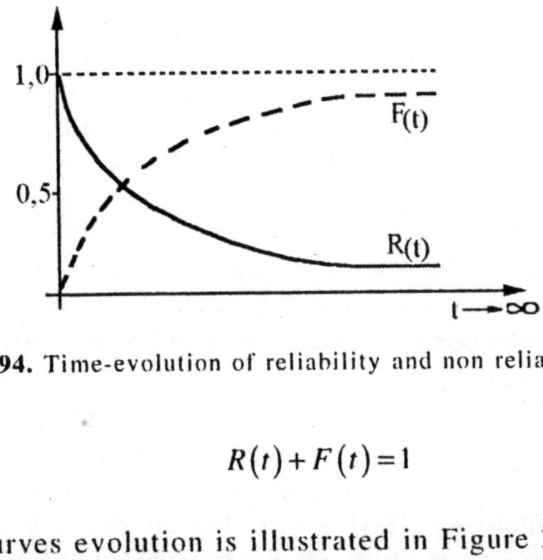

Figure 3.294. Time-evolution of reliability and non reliability [871].

$$R(t)+F(t)=1 \tag{3.164}$$

The two curves evolution is illustrated in Figure 3.294. Initially, at t = 0, the reliability is maximal, $F(t)$ = 1. When the non-reliability reach its maximal value, the product is non-functional.

From probabilistic point of view the three mechanisms are independent each to other. The overall probability of "survive" (reliability), $R_g(t)$, is given by the product of the survival probabilities of the three individual mechanisms:

$$R_g(t)=R_1(t)R_2(t)R_3(t) \tag{3.166}$$

In a similar manner the overall fatigue processes (non-reliability), F_g, may be written:

$$F_g(t)=1-R_g(t)=F_1(t)+F_2(t)+F_3(t)-F_1(t)F_2(t)-$$
$$F_2(t)F_3(t)-F_1(t)F_3(t)+F_1(t)F_2(t)F_3(t) \quad (3.166)$$

and the overall density function of probability, $h(t)$, will be:

$$h_g(t)=h_1(t)+h_2(t)+h_3(t) \quad (3.167)$$

As Figure 3.294 shows, each fatigue mechanism is characterised by a different shape of the "non-reliability" curve, namely:

- Mechanism 1, corresponding to the reliability $R_1(t)$ is characterised by a descending function, $h_1(t)$ in time t, for $t \geq t_{i_1}$ (t_{i_1} – the time for the first mechanism releasing);
- Mechanism 2 corresponds to the reliability $R_2(t)$ being characterised by a function $h_2(t)$ that is practically independent of time, for $t \geq t_{i_2}$ (t_{i_2} – the required time for the second mechanism initiation);
- Mechanism 3, which corresponds to the reliability $R_2(t)$ and is characterised by a sharply increasing function $h_3(t)$, for $t \geq t_{i_3}$ (t_{i_3} – the time after which the third mechanism becomes effective). In this case, the "instantaneous" fatigue degree strongly increases in time, after t_{i_3}.

One can be noticed that the second mechanism of fatigue describes the essential phenomenon, which occurs during the whole period of product exploitation. The effects caused by the first mechanism are the consequence of the initial faults existence, enough of significant in order to compromise the material, under the severe conditions of exploitation, for instance, during the textile materials manufacturing. The initially weakened zones by the faults presence must be eliminated by specific procedures, for instance, fibres spooling. In this way, the appearance of the second mechanism, which is responsive by the catastrophic fracture, is avoided or at list delayed.

Static fatigue (creep)

The static fatigue of textile fibres determines internal changes in the fibre structure, with negative implications on their quality long time before as the macroscopic fracture to occur. This is the reason why the quantification of the fatigue degree is absolutely required and usually it is achieved using indirect methods, i.e. periodical measurement of some structural characteristics or following the variation in time of some properties, especially mechanical ones [875, 876].

A systematic study of the creep fatigue has been realised on the

Table 3.103. Conditions of mechanical solicitations of the monofilamentary thread SOLVRON-SX and the modification of some of its characteristics [877]

Solicitation		Residual recovering at $t_2 = 15$ min				Birefringence
(min)	Dt_{rez}[a] (%)	ε_1[b] (%)	ε_2[c] (%)	ε_3[d] (%)	Ge_{rez}[e] (%)	$\Delta_n \cdot 10^3$
1	1.25	52.00	40.00	8.00	92.00	14.21
5	2.80	47.85	39.28	12.87	87.13	14.59
10	3.00	48.33	35.00	16.67	83.33	15.38
15	3.12	49.00	35.00	16.00	84.00	16.25
1	4.50	50.00	38.88	11.12	88.88	14.86
5	6.50	49.15	37.69	16.15	86.84	15.45
10	7.00	49.98	35.00	15.71	84.28	16.40
15	7.75	51.61	33.54	14.83	85.18	17.02
1	8.50	54.86	34.85	10.29	89.71	16.25
5	10.00	51.00	33.50	15.50	84.50	17.02
10	10.12	55.61	29.56	14.81	85.17	17.71
15	11.00	58.20	28.36	13.36	86.56	18.53

[a] – total residual deformation, recorded in the moment of removing the load;
[b] – the instantaneous elastic residual deformation ;
[c] – the retarded elastic residual deformation;
[d] – the remained residual deformation;
[e] – residual elasticity degree

soluble poly(vinyl alcohol) fibres, SOLVRON – SX (100/20 den), Japan [877]. The solicitation of filamentary thread was done in the similar conditions with the creep experiment. After the imposed relaxation, the measurement was repeated. It was found that the total residual deformation, Δt_{rez} is lower than the total deformation, recorded for the initial creep. Its value presents a tendency of increase both with the applied stress and stressing duration. Conclusive and very important are the modifications that occur during the relaxation period, which are apparently in contradiction as compared to those obtained for the initial creep, performed on the non-stressed fibres. Thus, the components of elastic residual recovering (ε_1, %) and elastic retarded recovering (ε_1, %) are characterised by higher values than the unstressed fibres, which increase with increasing of applied stress and duration. As Table 3.102 shows, in the case of the fatigued fibres, the residual elasticity degree (Ge, %) also presents superior values.

Following the evolution of the difference between the elasticity degree of fatigued and initial fibres, the results depicted in Figure 3.295 are obtained.

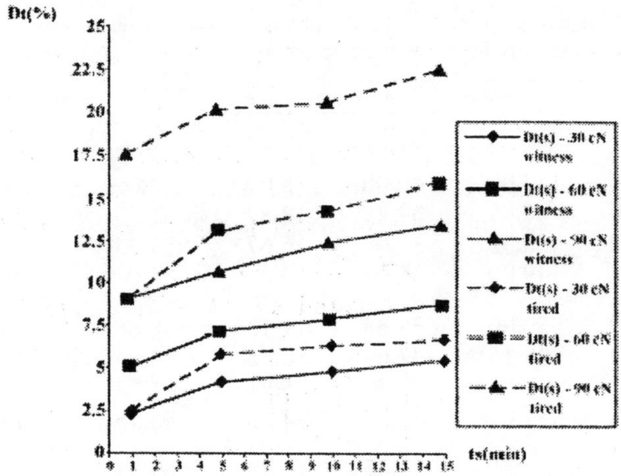

Figure 3.295. Variation in time of the differences between the elasticity degrees of fatigued and non fatigued samples, respectively [877].

The correlation of the obtained results lead to the conclusion that the soluble PAV thread, after a certain fatigue, suffers important changes not only in the amorphous regions, which suffered already a previous orientation, but also to the level of crystalline structure. In certain conditions, by accentuate change of the cohesion forces of some chemical bonds, the energetic value of the system and the thermo-dynamical equilibrium are so much modified that during relaxation the sample returns with a supplementary force to the new state of thermodynamic equilibrium. In this stage, some chain segments adopt a disorganized state of a coil. This behaviour could be assimilated to a strongly tensioned arch, which after tension removing is squeezed even more, having after relaxation a lower length, l_1, as compared to its length in tensioned state, l_0. This effect becomes more accentuated with the increase of stressing force and its period of action. These changes of the energetic potential of the fatigued samples induce difficulties during the textile mechanical processing and negative effects on the sample quality.

The filamentary thread fatigue under creep conditions, for different periods, determines significant changes both for the control samples, previously unstressed, and especially for the already fatigued ones, Figures 3.296 and 3.297.

As expected, in both cases the deformations increase with the increase of force and solicitation period. The most important increase was found in the case of the fatigued sample, under the

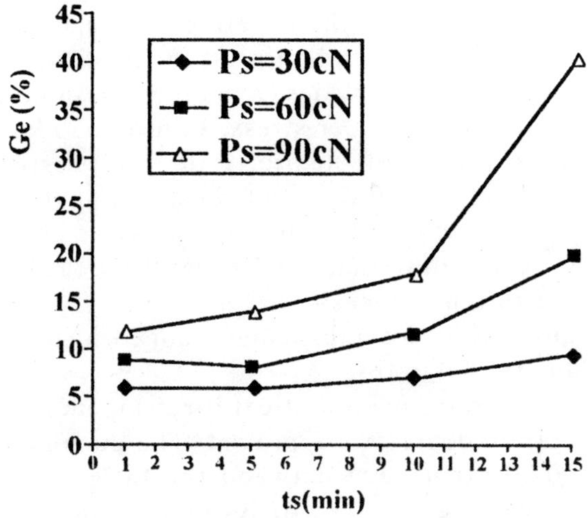

Figure 3.296. Variation of total deformation under load with the solicitation period [877].

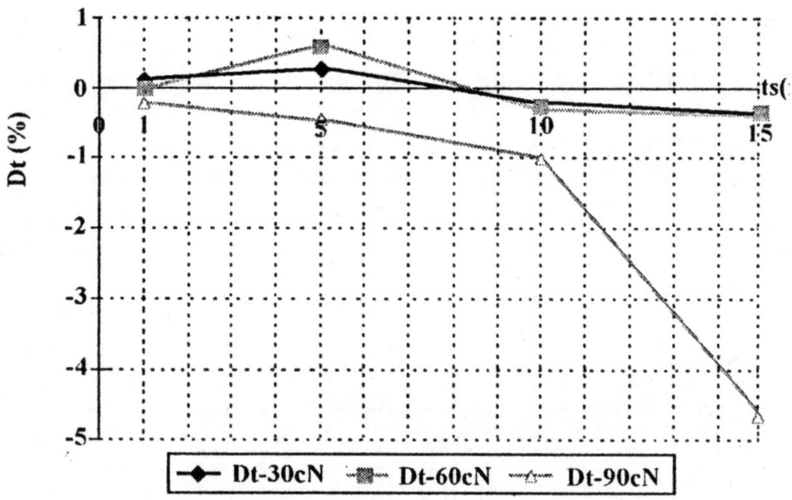

Figure 3.297. Variation of deformation differences between fatigued and control samples during de recovering step in function of soliuitation period [877].

maximal experimental stress, 90 cN, Figure 3.295. On the other side in the case of the fatigued sample, for 15 min of tensioning, one can note the occurrence of the negative recovering of deformations which are lower than those corresponding to the control samples. This behaviour is manifested even for the lowest applied force and shorter maintaining periods of 10 and 15 min. Corresponding to the maximal force, the negative deformations are more intense and start to appear after 1 min of solicitation, after which the "contractile" behaviour amplifies, reaching the value of –4.60% after 15 min of maintaining under stress, Figure 3.297.

The observed "increase of elasticity degree" residues in the above mentioned effect, which does not represent anything else but a supplementary relaxation, caused by the energetic changes of the fibres structural formations that tend to occupy the most stable positions after the fatigue process.

The most important structural changes concern the changes of the crystalline/amorphous ratio.

As Figure 3.298 shows, using optical birefringence method, the changes of orientation degree were proved. The comparative analysis of the optical birefringence between the fatigued sample and control one evidenced a positive difference, which is higher at small values of the applied stress and solicitation duration. With other words, additional orientations occur in the fatigued sample, but these ones diminish with the increase of the stress and solicitation duration.

After 15 min of stressing, irrespective of the applied force value, the difference in birefringence becomes negative; this means that the macromolecules orientation decay in the fatigued samples as compared with the control ones. Maintaining the sample under

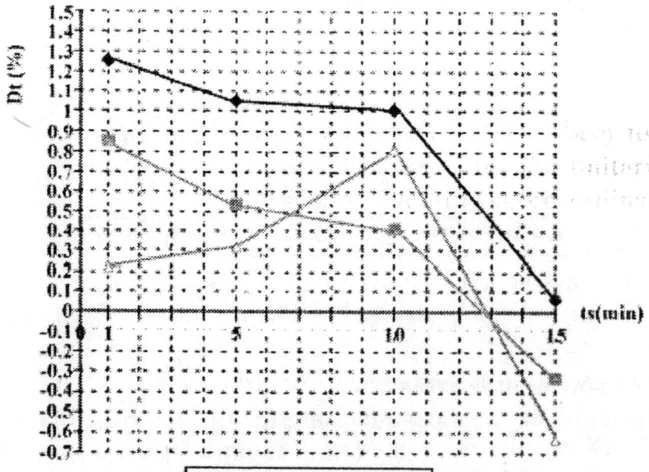

Figure 3.298. Variation of optical birefringence of the fatigued and control samples with the applied stress and solicitation period [877].

stress for a prolonged period and increasing the applied stress induces the increase of the disorganisation degree. At the maximal stressing force, 90 cN, for 15 min the negative difference of birefringence between the two samples is highest. In this case, the decrease of orientation degree is the effect of a disorganisation process caused by the fatigue process. Otherwise, the evolution of internal disorganisations occur non-regularly in time. Thus, after a short solicitation periods, 5 min, when a increase of difference is observed until 10 min of solicitation, after that the decay of this difference is very pronounced.

The structural changes at the level of crystalline and amorphous phases have been confirmed by the variation of the mechanical properties, determined for the stressed samples in the conditions previously described. As Table 3.104 shows, the effects of fatigue process are clearly evidenced both by the evolution of specific stress, i.e. elasticity, yield, and fracture limit, and the corresponding strains.

It can be noted that in the case of the specific forces for the elastic range, 30 and 60 cN, the efforts difference in not so high, but are a little higher as compared to those corresponding for the control sample. The most important changes were recorded at the maximal solicitation force, 90 cN, which characterise the visco-elastic domain. The elastic limit diminishes up to the value of control sample, and the yield and fracture limits are slightly higher. The strain corresponding to the fracture is lower than of the control sample.

Important changes were found in the variation of elasticity modulus, E, which decrease with the increase of the applied stress bellow the value corresponding to the control sample with the exception of the sample stressed at 30 cN, where a slightly increase was recorded. The changes occurring both in the elastic and post-elastic range of the stress-strain curves prove the important role-played

Table 3.104. The mechanical characteristics of the fibres fatigued by creep

P_S	E	Elasticity limit			Flow limit			Fracture limit		
(cN/thread)	(cN/tex)	P_e (cN/thread)	σ_e (cN/tex)	ε_e (%)	P_c (cN/thread)	σ_c (cN/tex)	ε_c (%)	P_r (cN/thread)	σ_r (cN/tex)	ε_r (%)
C	566.37	80.0	7.08	1.2	100	8.84	5.70	295	26.10	27.3
30	575.20	88.3	7.81	1.6	110	9.73	5.86	300	26.55	25.6
60	442.40	85.0	7.52	2.4	125	11.06	7.20	290	25.66	26.2
90	398.20	80.0	7.08	2.4	140	12.39	6.00	300	26.55	22.0
Maximal solicitation without relaxation										
90	353.9	55	4.87	1.2	125	11.06	4.00	350	30.97	14.4

by the amorphous phase in the static solicitations at which the samples were previously supposed. At loads higher than this range, on the final part of the stress–strain curve, the stress seems to be reestablished and the strain considerably diminishes for the maximal applied stress during the fatigue period. Comparative results, for the recorded curves in relaxed and non-relaxed states are presented in Figure 3.299. A series of characteristics, such as the elasticity modulus, elastic stress, and fracture strain are significantly changed.

In the conditions of fibres – representing highly oriented structures – mechanical stretching solicitation, the appearance, evolution and microcracks development, by free radicals mechanism, sequentially occur, by a similar way with the ductile fracture, Figure 3.300 and 301 [878 and 879]. This fact was evidenced using SEM technique [880].

The explicit evidence of the macroradicals presence, under the conditions of cyclical fatigue by stretching, was demonstrated on the example of monofilaments based on poly(acrylonitrile-*co*-vinyl acetate-*co*-α-methylstyrene) copolymer. For this purpose, a special technique was applied, namely the filaments were impregnated with a solution of a stable radical 1,1-diphenyl-2-picrylhydrazyl (DPPH) and supposed to cyclic fatigue. The generated macroradicals during the solicitation period reacted with a part of DPPH. Using ESR method, the difference between the initial concentration of the stable radicals and their final concentration allowed the calculation of

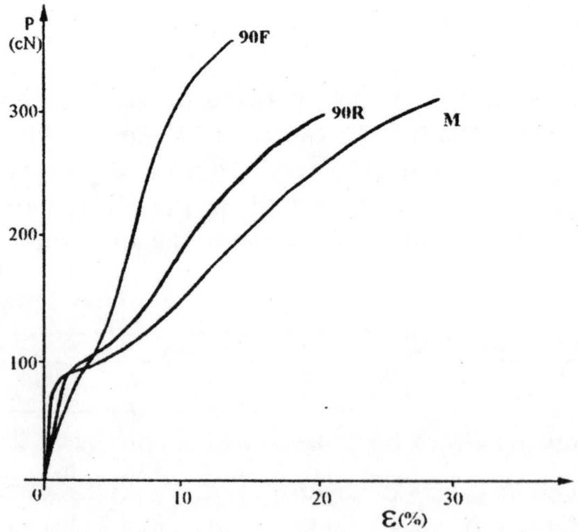

Figure 3.299. Stress-strain curves for [877]: 1) control sample, *M*; 2) relaxed sample, *R*; 3) without relaxation, *F*.

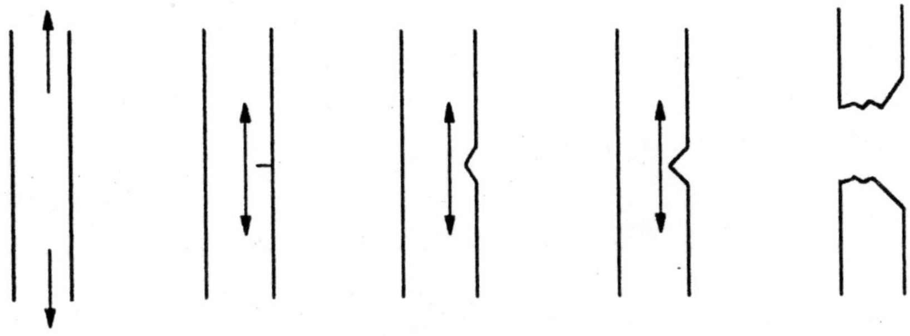

Figure 3.300. Changes in fiber structure under the conditions of mechanical solicitation, by stretching, until the failure [878]: a) before solicitation; b) apparition of a superficial crack; c) crack opening; d) *V*-shape growing of the opened crack; e) final fracture.

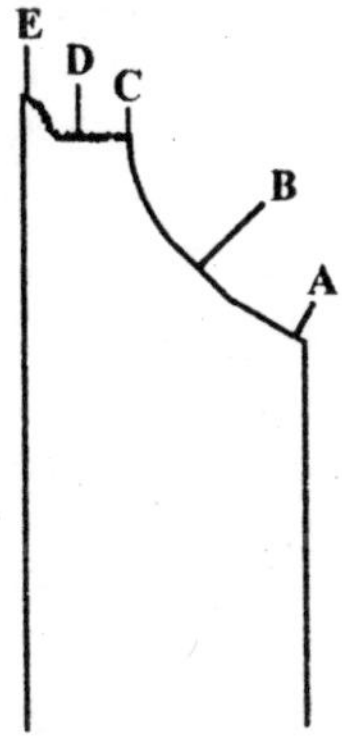

Figure 3.301. Fracture evolution [878]: A) apparition of the first crack; B) crack opening; D) rigid sliding and fracture growth; E) fracture finalisation.

the macroradicals relative concentration, Figure 3.302 [90].

ESR spectra of the samples impregnated with DPPH and mechanically solicited consist by singularly lines, with the span varying from 0.92 to 16 s. The absence of hyperfine structure denotes severe interactions of exchange. Since DPPH acts as a radical acceptor, at the end of stress period it remains in a small amount being strongly coupled by exchange reactions. Some supplementary results concerning the effects of mechanical solicitation?? are given in Table 3.105.

The increase of deformation in the range of 50 to 75% of total deformation to fracture determines the decrease of DPPH concentration due to the increase of the mechanoradical concentration. One can note that these radicals appear in high concentrations even at

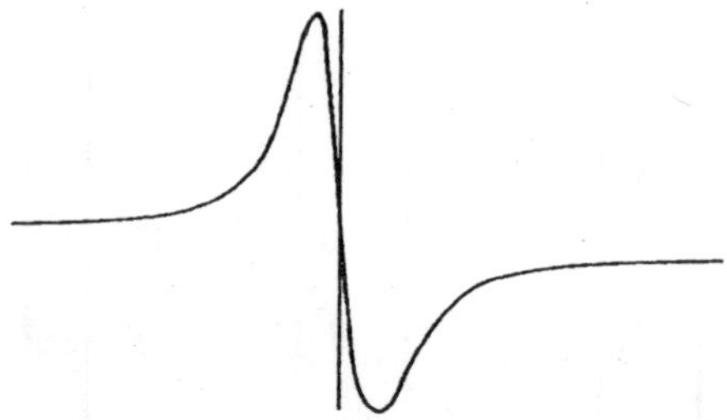

Figure 3.302. ESR spectrum of the mechanically stressed monofilaments based on ternary copolymer poly(acrylonitrile-*co*-vinyl acetate-*co*-α-methylstyrene) with fineness of 2.5 den, impregnated with DPPH (deformation, 50% from the fracture elongation; solicitation rate, 5 mm/min) [90].

deformations well below the fracture deformation, which is in good accordance with Peterlin's mechanism concerning the fracture of high-oriented semicrystalline polymers [882, 883].

The 'linking' macromolecules, having more or less strength conformations, traverse the inter-crystalline domains. Under the deformation conditions, these ones gradually adopt stretched conformations, are tensioned, and suffer the homolytical splitting. Therefore, under the creep fatigue conditions, the fracture process is directly influenced not by the applied stress but the induced deformation. An important effect is played by the increase of solicitation rate, due to the fact that the stress duration becomes comparable with the relaxation time. The overtensioned macromolecules are not able

Table 3.105. Relative concentration of free radicals and the content of grafted vinyl chloride by mechanical solicitation of MELANA type monofilaments (samples impregnated with DPPH) [90]

No.	Mechanical parametres		Width of resonance line, H_{pp} (Gs)	DPPH conc. (relative units) (%)	Free radicals conc. (relative units) (%)[b]	Content of grafted vinyl chloride (%)
	Deformation (%)[a]	Deformation rate (mm/min)				
1*	-	-	0	0	0	0
2**	-	-	0.9±0.04	100.0	0	1.68
3	50	5.0	0.96	72.0	18.0	2.07
4	75	5.0	0.92	22.0	78.0	2.85
5	100	5.0	1.00	21.0	79.0	2.92
6	100	2.4	0.96	48.5	51.5	2.36
7	100	8.1	0.92	12.0	88.0	2.87
8	100	13.5	0.92	8.5	91.5	3.02

[a] - with respect to the deformation to fracture;
[b] - with respect to the amount of stressed PAN;
* - non impregnated with DPPH, unstressed;
** - impregnated with DPPH, unstressed.

to rearrange under load, consequently suffering the scission of chemical bonds.

It was found that the macroradicals recorded by ESR spectroscopy usually have the unpaired electron located to a tertiary carbon atom. It means that the chemical bonds splitting preferentially occur to the chemical bonds containing a totally substituted carbon atom, i.e. at the α-methylstyrene structural units, Figure 3.303.

Since the stress is distributed on a small region of sample, the splitting of chains tend to accumulate into a limited volume. Additional to these primary scissions, generated by the applied stress, other scissions arise, which are induced by the free radicals and in this way a chained mechanism is activated:

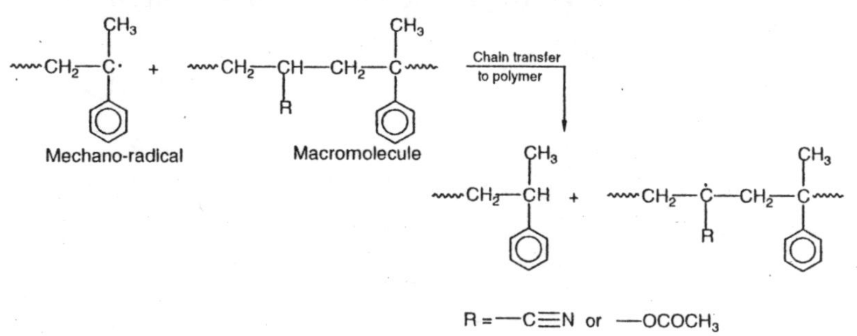

The detachment of carbon atom from the chain decreases the dissociation energy of adjacent covalent bond to 1/3 from the dissociation energy of a – C – C – bond. Consequently, the macromolecular chains easily split near the radical, in such way propagating the fracture kinetic chain:

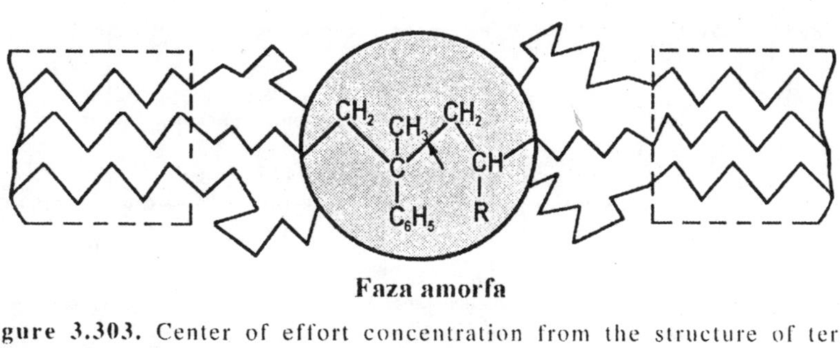

Faza amorfa

Figure 3.303. Center of effort concentration from the structure of ternary copolymer [830].

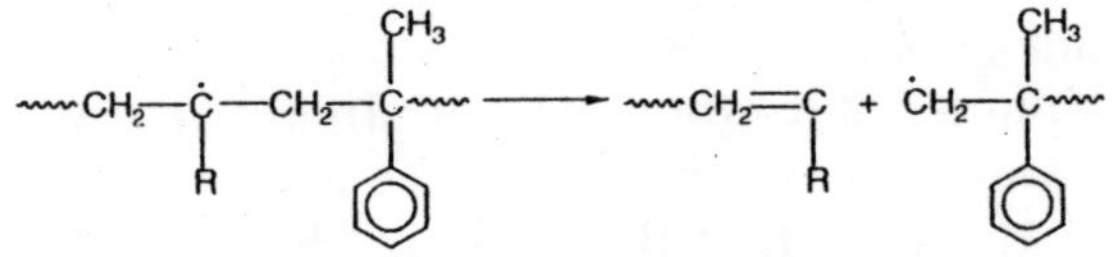

The formed macro-radical reinitiates a sequence of reactions that continues until the stressed macromolecule will generate a new central radical. According to the data from literature, the number of split bonds due to a pair of mechano-radicals can reach the value of 3000 [830].

The fracture molecular mechanism as well as the mechano-radicals existence were also proved by grafting reactions with unsaturated monomers, at the fracture surfaces. The reactions with the sable radical DPPH, or grafting reactions occur as follow:

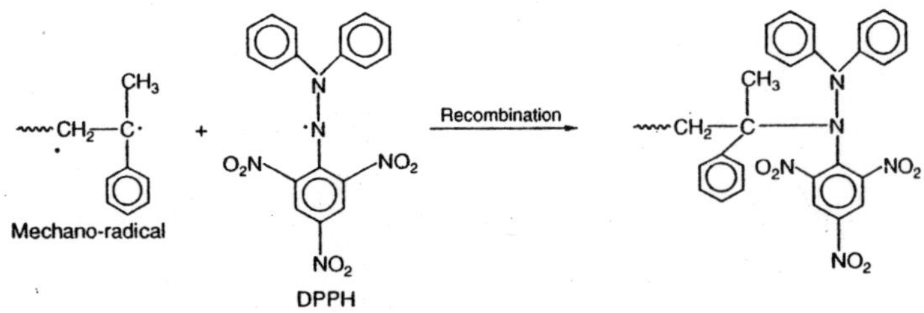

The grafting reactions with vinyl chloride at the fracture surface:

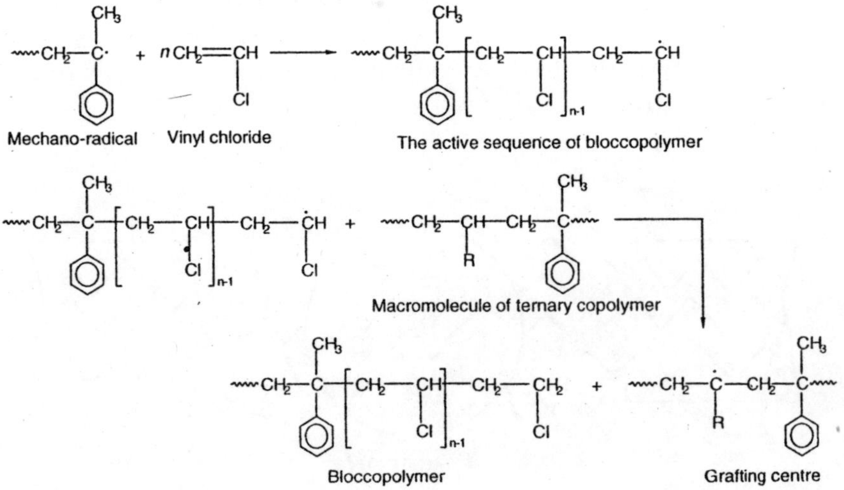

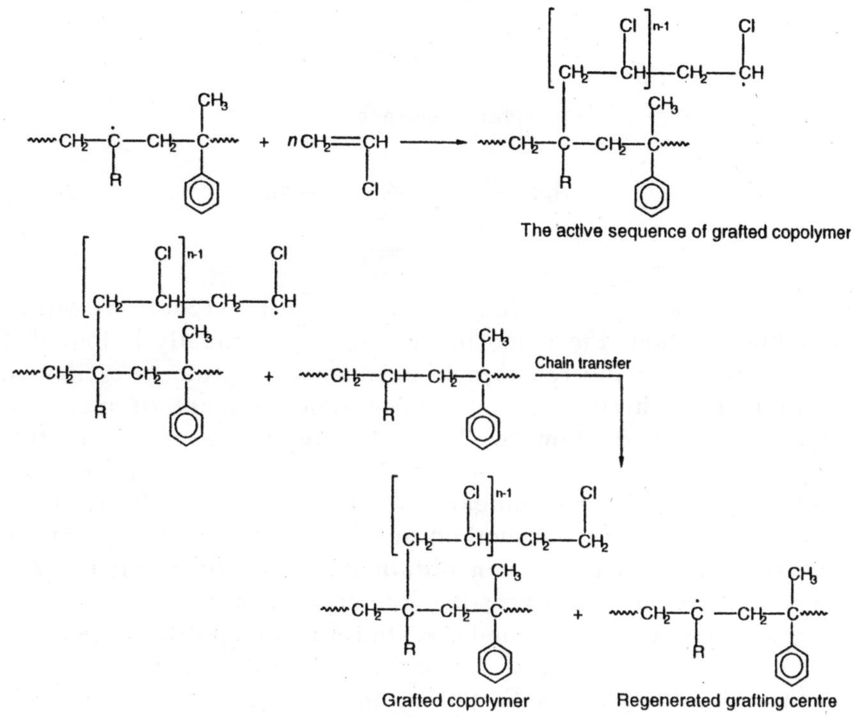

Diamines have also been utilised as radical acceptors, being able to react with the mechano-radicals, when the latter ones are generated in their presence [830]:

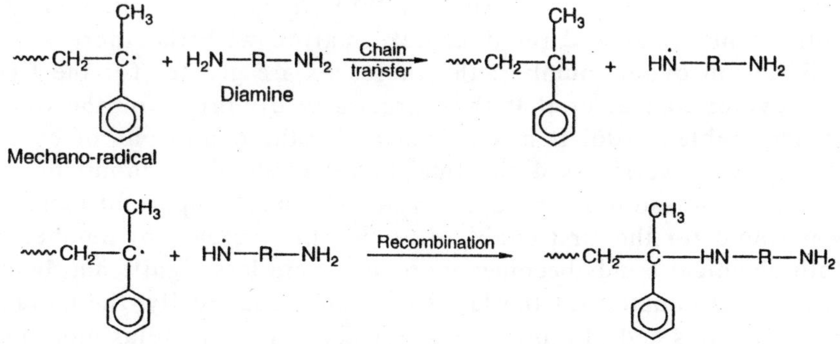

Whenever R is an aromatic residue such as:

p-Phenylenediamine

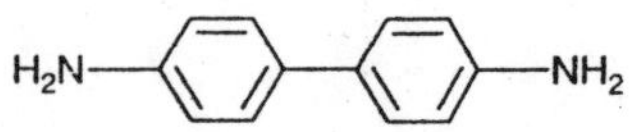

Benzidine

is possible as by diazotization and coupling reactions with a suitable colorant, the monofilament to be chemically coloured. It was found, that the reaction is extremely sensible, the filament coloration occurring even for very small amount of mechano-chemically fixed diamine. This reaction has a great practical importance. Thus, it can be used either for mechano-radicals identification of for chemical coloration of the synthetic fibres, replacing the traditional method of coloration in bulk. The above-described results have been obtained on a pilot plant, C.F.S. – Savinesti, Romania. Interesting results concerning the cyclical fatigue have been obtained by studying the polyurethane-type polymers. The crosslinked polyurethane Moldotan D was supposed to the axial stretching solicitation, under the creep conditions. Six stretching-recovering cycles were carried out for each investigated sample [207, 885]. The cycle duration was 2000 hrs, according with ASTM-D, 2990-71 [884]. As Figure 3.304 shows, the sample solicitation with a constant load, 14.3 MPa, during the whole period of the six cycles did not determine sample's macroscopic fracture. The calculation of characteristic deformations proved the increase both of maximal and residual deformation with the increase of solicitation cycles number; the increases are greater for the first two cycles and after that they gradually diminish for the other cycles, Table 3.106. It means that under the conditions of cyclic fatigue, the overstressed chemical bonds gradually accumulate and suffer the scission during each cycle. The most important increase was found for the first cycle, after that the increase of number of split chemical bonds becomes more and more less significant, being strongly attenuated for the last two cycles. Practically, the number of split bonds and of those, which remade during the relaxation time is constant. For a new progress of the deformation process, the increase of unitary stress applied to the sample is required. For a certain value of the constant stress applied, the samples are able to support and undetermined number of stressing-recovering

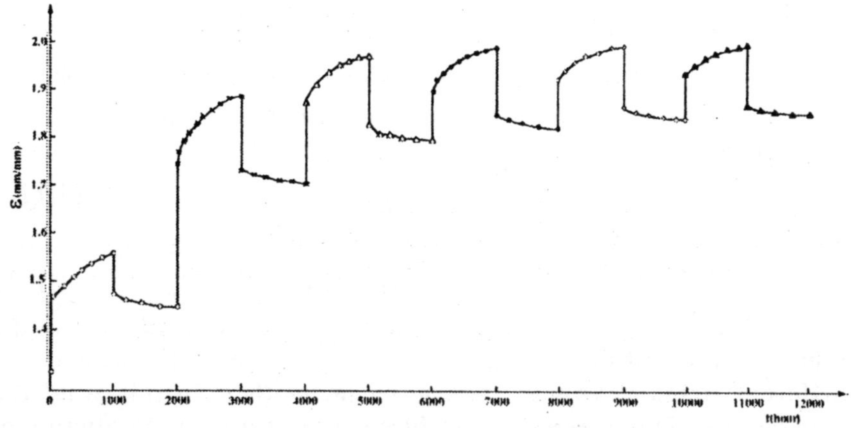

Figure 3.304. Creep curves, ε – τ, for the cyclic solicitation under creep condition of the polyurethane elastomers Moldotan *D* (Moldotan *D* is a crosslinked polyurethane based on 4,4'-dibenzyldiisocianate, 98.05%-NCO groups, poly(ethylene adipate), $\overline{M} = 2000$, and ethylene glycol) [886].

Table 3.106. The values of characteristic deformations in the case of cyclical solicitation in creep conditions (σ_c = 14.3 MPa, $\overline{M}_c$ = 6000 g/mole) [207]

Cycle no. → Deformation ↓	1	2	3	4	5	6
Instantaneous elastic deformation	1.200	1.740	1.860	1.900	10920	1.940
Retarded elastic deformation and viscous flowing	0.360	0.150	0.115	0.090	0.075	0.060
Maximal deformation	1.570	1.890	1.975	1.990	1.995	2.000
Instantaneous elastic recovering	0.080	0.170	0.145	0.140	0.195	0.100
Retarded elastic recovering	0.025	0.020	0.035	0.030	0.020	0.040
Residual deformation	1.455	1.705	1.795	1.845	1.845	1.860

cycles.

The linearisation of the creep curves, by representing in double logarithmic co-ordinates the variation of deformation with time can be done applying the method developed by B.W. Chery and D.F. Kindles [887], i.e. of overlapping of the stress vs. time, for the prediction of the polymers behaviour to creep for long periods of solicitation. In the case of the vitreous polymers, the creep rate can depend, in certain conditions, only by the applied stress and instantaneous deformation.

$$\dot{\varepsilon} = a(\sigma)/b(\varepsilon) \tag{3.168}$$

the above equation can by rearranged in the manner:

$$b(\varepsilon)d\varepsilon = a(\sigma)dt \qquad (3.169)$$

and for a constant stress after integration the following expression is obtained:

$$b(\varepsilon) = a(\sigma)t \qquad (3.170)$$

or

$$\varepsilon = c\left[\log a(\sigma) + \log t\right] \qquad (3.171)$$

where: log $a(\sigma)$ is the displacement function, which allows as the representations of the variation of ε with log t in such a manner for different stress values to superpose each other. In this way, the equation (3.171) grounds a valuable method for creep prediction at long solicitation periods, using the experimental data, obtained at short periods. By drawing of each creep curve, alongside of a convenient time-axis, a 'master' curve is obtained in such way that the creep surface to be defined by the relationship:

$$\varepsilon = f\left[S(\sigma)\cdot T(t)\right] \qquad (3.172)$$

where: S and T are functions depending of σ and t, respectively.

Comparing the equations (3.171) and (3.172), it can be observed that the representations of ε with log $T(t)$, for different values of stress, overlap whenever the curves are translated alongside of the $T(t)$ axis with the value defined by the displacement factor $S(\sigma)$.

When the material behaves as a linear visco-elastic solid at small applied stressed, the equation (3.172) is reduced to:

$$\varepsilon = \sigma \cdot J(t) \qquad (3.173)$$

where: J – linear compliance to creep of the investigated material and which for the particular case of uniaxial stretching solicitation can be noted as $D(t)$.

Applying this method to the cyclical fatigue of polyurethane Moldatan-D, the master curve was drawn ε vs. log $D(t)$, Figure 3.305. It should be mentioned that, in this case, the master curve was obtained without the translation of the corresponding curves of each solicitation cycle and a perfect superposition of the experimental data was obtained only by their simple graphical representation. The representation of the master curve for static fatigue under creep conditions permits determination of the behaviour of the polyurethane elastomer for a practically unlimited number of cycles, the data obtained in short-term measurements being easily extrapolated for long-time solicitation, by master curve prolongation which in this case is a strength line.

The polyurethane-based elastomers are characterised by exceptional mechanical properties; additionally, they have a good resistance to oxygen, ozone, and dried air, up to a temperature of 70 °C. Their utilisation is limited especially by the poor resistance to the photooxidative degradation, hydrolysis, and micro-organisms action [888–902].

In the case of some semicrystalline highly-oriented polymers as polyamide 6 fibres, the uniaxial stretching, concomitantly carried out with UV irradiation, accelerates appearance of the free radicals and microcracks thus decreasing the durability with about four orders of magnitude [900–902].

In the case of polyurethane based on 4,4'-dibenzyl diisocyanate, uniaxial stretched with an effort of 9.7 MPa, concomitantly with a accelerated ageing, by UV irradiation (λ = 355 nm), at constant temperature 50 °C for 200 hrs, the results are presented in Figures 3.306–3.309. It was found that the mechanically fatigued and concomitantly aged sample is broken after 50 hrs of solicitation (curve 3 in Figure 3.306). Apart from it, the sample only fatigued (curve 1 in Figure 3.306) and that one previously accelerated aged for 200 hrs and subsequently fatigued (curve 2 in Figure 3.306) suffer the fracture only after 1000 hrs of solicitation.

The calculation of the characteristic deformations of visco-elastic behaviour, Table 3.107, reveals their dependence to the solicitation conditions. The maximal difference was recorded for the stretched and concomitantly accelerated aged sample. The sample firstly accelerated aged presents a maximal value and its residual

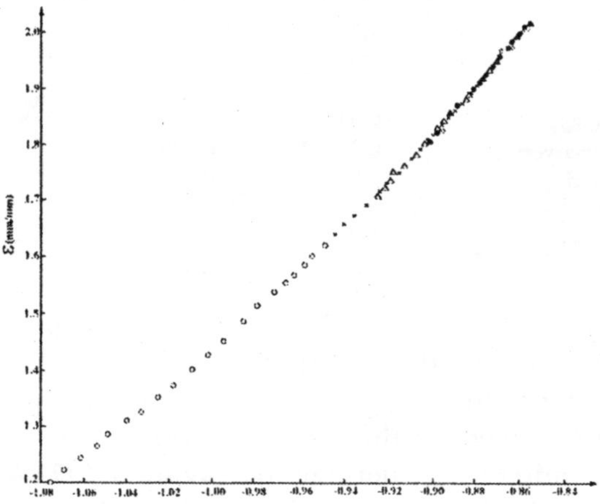

Figure 3.305. Master curve, in co-ordinates ε–log $D(t)$, in the case of cyclic solicitation under creep conditions (σ_c = 14.3 MPa) [207].

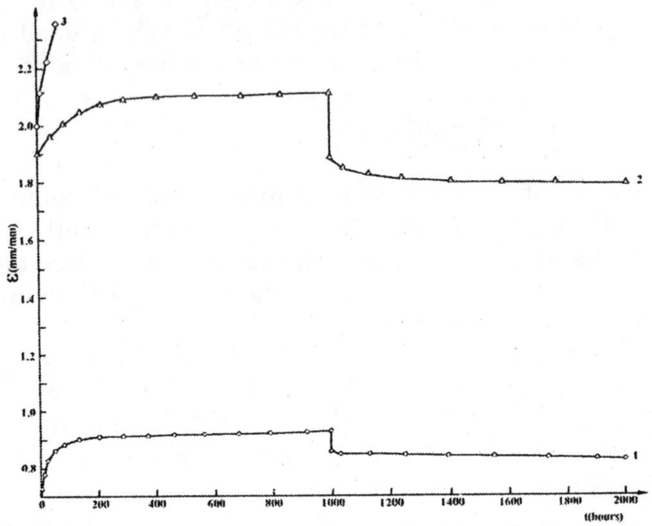

Figure 3.306. Creep curves, ε - *t*, for the cyclic solicitation under creep conditions (σ_c = 9.7 MPa) and accelerated ageing (M_c = 6000 g/mol) [207 h]: 1) stretching only stressed sample; 2) aged sample for 200 hr and subsequent solicited by traction; 3) concomitantly aged and mechanically stressed sample.

Table 3.107. Characteristic deformations in the conditions concomitant of creep fatigue and accelerated ageing [207]

Solicitation conditions → Deformation, ε↓	9.7 MPa	9.7 MPa after accelerated ageing	9.7 MPa concomitantly with accelerated ageing*
Instantaneous elastic deformation, ε_0	0.725	1.850	2.00
Retarded elastic deformation and viscous flowing	0.21=5	0.265	0.35
Maximal deformation, ε_{max}	0.940	2.115	2.35
Instantaneous elastic recovering	0.075	0.220	-
Retarded elastic recovering	0.015	0.080	-
Residual deformation	0.850	1.815	-

* - the sample was broken after 50 hr of solicitation

deformations are of 2.25 and 2.13 times higher, respectively, than the sample only mechanically solicited???.

The effects occurring by cyclical fatigue of polyurethane elastomer are influenced by the unitary stress values and accentuated by the concomitant polymer ageing, Figure 3.307. As expected, the deformations, Table 3.108, and durability, Figure 3.308, are changed, too.

Under the conditions of static fatigue and concomitant acceler-

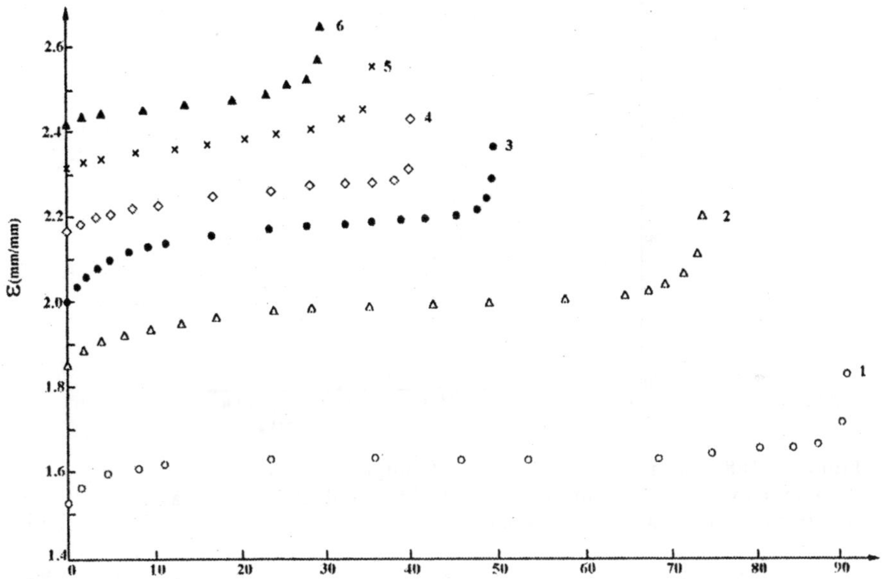

Figure 3.307. Creep curves, ε – *t*, under the conditions of concomitant static fatigue and accelerated ageing (M_c = 6000 g/mol) [207]: 1) σ_c = 7.0; 2) 8.1; 3) 9.7; 4) 11.2; 5) 12.6; 6) 14.3 MPa.

Table 3.108. Characteristic deformations of the crosslinked polyurethane (M_c = 6000 g/mole) concomitantly stressed at different unitary efforts and accelerated aged [207]

Effort, Mpa → Deformation, ε ↓	7.0	8.1	9.7	11.2	12.6	14.3
Instantaneous elastic deformation	1.53	1.86	2.00	2.17	2.32	2.43
Retarded elastic deformation and viscous flowing	0.28	0.35	0.35	0.25	0.23	0.21
Maximal deformation	1.81	2.18	2.35	2.55	2.55	2.64

ated ageing, the durability of polyurethane samples can be predicted by drawing the master curve. The creep compliance calculation, $D(t)$, representation of the experimental data in co-ordinates ε vs. log $D(t)$ for different applied stress, the linaerization of the obtained creep curves, their translation parallel to Ox axis led to a perfect overlapping, Figure 3.309.

The fatigue effects are also influenced by the certain structural polymer characteristics, namely the chemical structure of the macromolecular chain, in the case of linearly polyurethane or of the

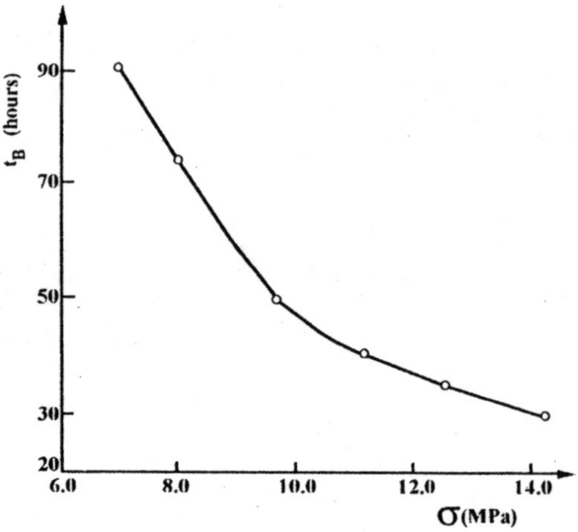

Figure 3.308. Variation of durability of polyurethane elastomer samples under the conditions of concomitant static fatigue and accelerated ageing with the solicitation effort (M_c = 6000 g/mol) [207].

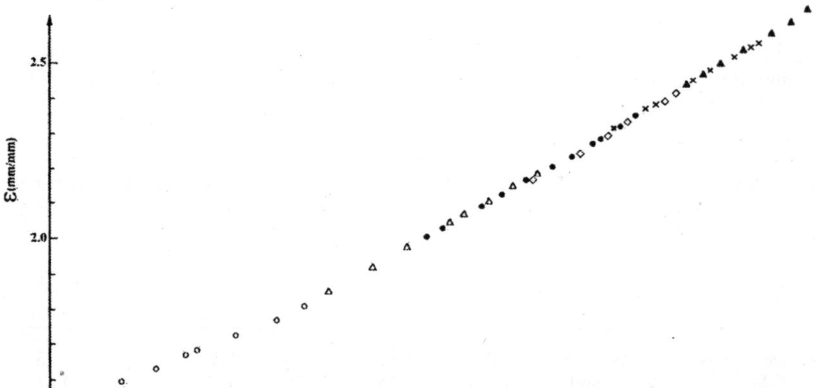

Figure 3.309. Master curve obtained by plotting the experimental data in coordinates å – log $D(t)$ (σ_R = 7,0 MPa) [207].

network, in the case of crosslinked polyurethanes. Thus, in the case of polyurethane based on 4,4'-dibenzyl diisocyanate, the durability decrease with increasing the crosslinking degree, i.e. with the decrease of $\bar{M}_c$, Figure 3.310. This behaviour is related to the decay of the number of intercatenary secondary bonds, hydrogen and van der Waals bonds, which is reflected both in the shape of creep curves, Figure 3.311, and the deformation values, Table 3.109 [207].

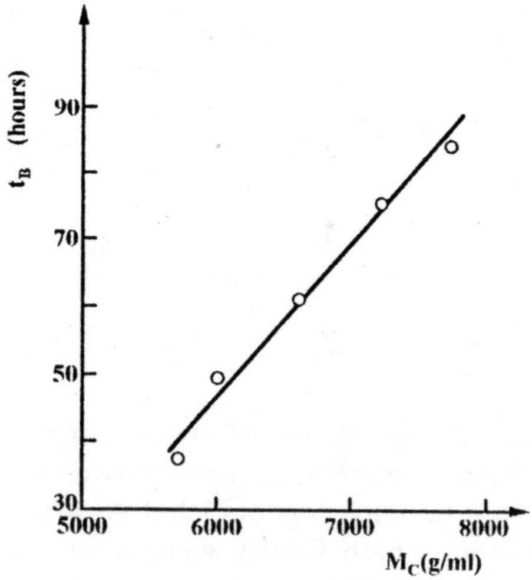

Figure 3.310. Variation of durability of polyurethane elastomer Moldotan *D* with the crosslinking density (σ_c = 9.7 MPa) [207].

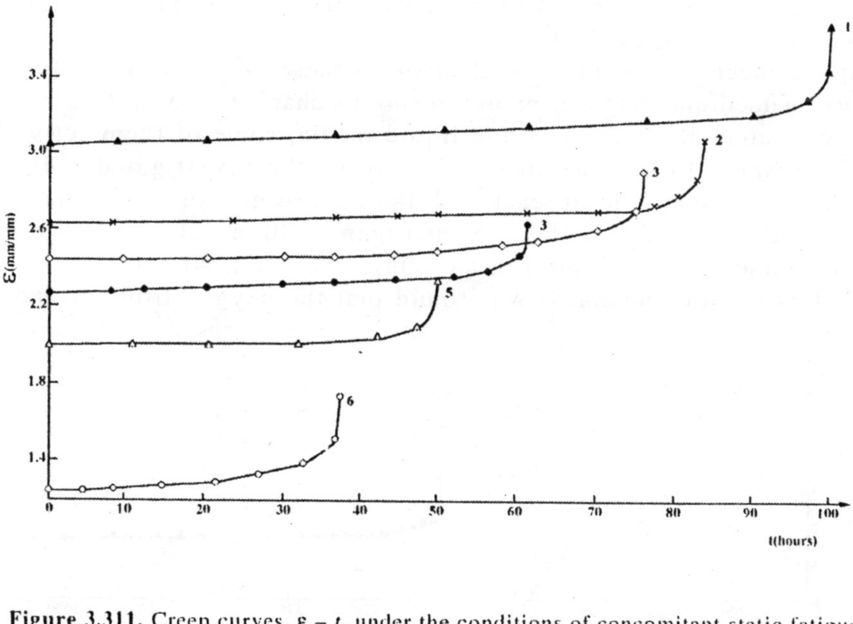

Figure 3.311. Creep curves, ε – *t*, under the conditions of concomitant static fatigue and UV irradiation of polyurethane elastomer samples of various crosslinking density (σ_c = 9.7 MPa) [207]: 1) M_c = 8; 2) 7700; 3) – 7134; 4) 6613; 5) 6000; 6) 5690 g/mol.

Table 3.109. Characteristic deformations of the crosslinked polyurethane with different crosslinking degrees, concomitantly supposed to static fatigue (σ_c = 9.7 MPa) and accelerated ageing [207]

$\overline{M}_c$, g/mole → Deformation, ε ↓	∞	7 700	7 134	6 616	6 000	5 690
Instantaneous elastic deformation	3.05	2.63	2.44	2.26	1.95	1.24
Retarded elastic deformation and viscous flowing	0.62	0.46	0.48	0.37	0.40	0.51
Maximal deformation	3.67	3.09	2.92	2.63	2.35	1.75

The solicitation in conditions of static fatigue of a linear polyurethane, type Estane, gave the curve presented in Figure 3.312 [207]. The concomitant ageing also led to a sharply increase of samples deformation, which finally were fractured only after 84 hr of solicitation, Figure 3.313. In these conditions, as Table 3.109 shows, a important increase of maximal deformation, of about 1.57 times, was found.

The linear thermoplastic segmented polyurethanes of $-(AB)-_n$ – type, are widely used on the industrial scale, due to their physico-mechanical properties and of good stability in atmospheric conditions or in hydrocarbons. In this sense and for understanding of fatigue mechano-chemical mechanism of these polymers, the influence of medium chemical nature during mechanical solicitation was investigated. Both gaseous and liquid media, some of them active with respect to the chemical structure of the investigated polyurethanes, being able to react with the formed macroradicals, have been selected. Thus, the air, nitrogen (100 mmHg), nitrogen monooxide, NO, (130 mmHg), and vinyl chloride (140 mmHg) were used as gaseous media. It was found that the oxygen from air and

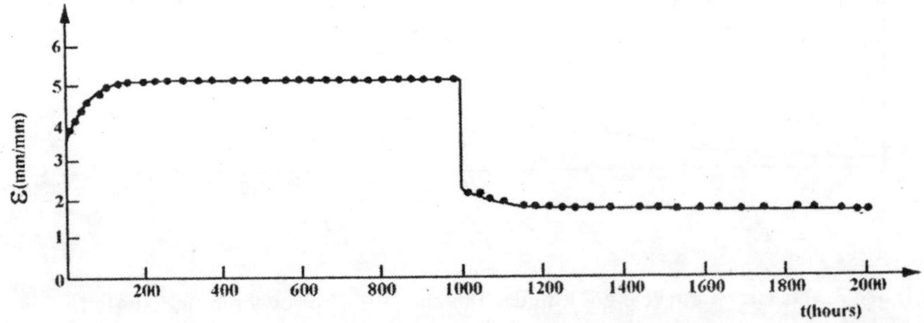

Figure 3.312. Creep curves, ε – *t*, of the linear polyurethane ETANE-type (σ_c = 42,2 MPa) [207].

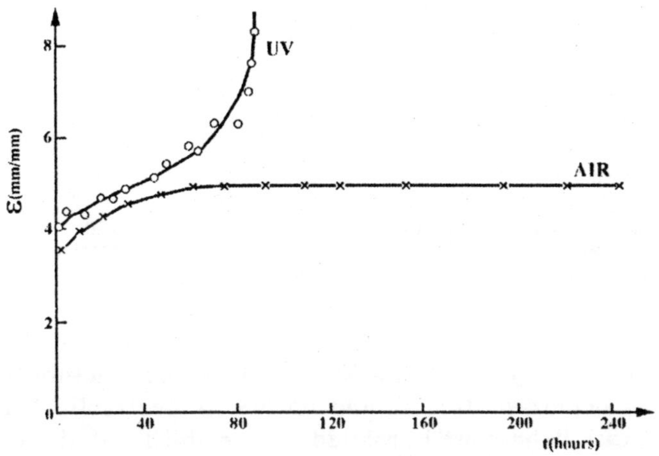

Figure 3.313. Creep curves, ε–*t*, of the linear polyurethane ETANE-type steressed under the conditions of concomitant static fatigue and accelerated ageing (σ_c = 42.2 MPa) [207].

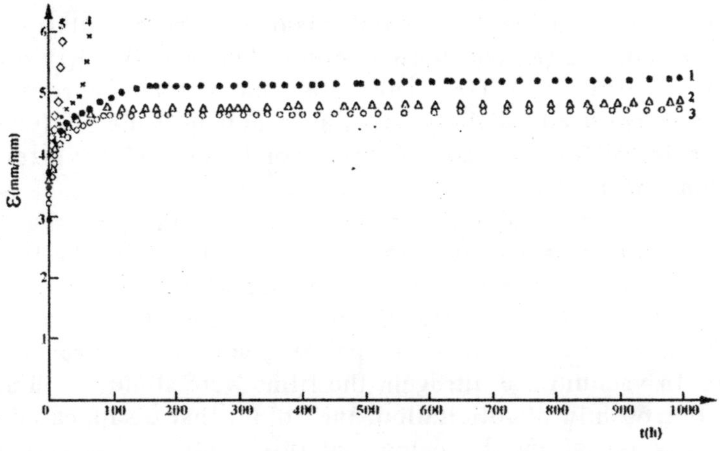

Figure 3.314. Creep curves, ε–*t*, in various gaseous media [207]: 1) air; 2) nitrogen; 3) vacuum; 4) vinyl chloride; 5) NO [207].

NO behaved as radical acceptors and vinyl chloride, unsaturated monomer, proved its ability to polymerise under the action of polyurethane macroradicals. The creep curves, recorded in these conditions, Figure 3.314, allowed the calculation of the elastical deformations, i.e. instantaneous, retarded, and overall one, respectively, Table 3.110. In the active media, able to accept electrons, the total deformation is higher than in inert ones, vacuum and nitrogen, for the same applied stress, σ_c = 42.20 MPa. In the presence of

Table 3.110. The deformation values for σ_c = 42.20 MPa, in different gaseous media [207]

Type of deformation	Medium				
	Air	Vacuum	N_2	NO	Vinyl chloride
Instantaneous elastic deformation, mm/mm	3.25	3.10	3.00	3.50	3.45
Retarded elastic deformation and viscous flowing, mm/mm	1.84	1.60	1.55	2.50	2.15
Maximal deformation, mm/mm	5.09	4.70	4.55	5.95	5.65

nitrogen oxide and vinyl chloride the macroscopic fracturation of stressed films occurred after 22 and 65 hrs, respectively. The explanation of this behaviour is related to the ability of these substances to sustain the propagation of chained reactions, by their fixation to the active radicalic centres and impeding the stabilisation of free radicals, by interruption reactions. Once again, the representation of the acquired date in semilogarithmic or logarithmic coordinates al allowed the curves linearisation. The calculation of creep compliance, $D(t)$, and the representation of ε vs. $D(t)$ and ε vs. log $D(t)$, giving a perfect superposition of the experimental data, which were used for the prediction of polyurethane elastomers behaviour in different media in the conditions of long-terms solicitation [207].

The mechano-chemical reactions that occurred during mechanical solicitation, in different media, have been visualised, by the investigation of the films aspect. In air, an intense whitening "silvering" was observed on the film surface, due to the formation of a great number of microcracks, which are partially closed after removing the stress. In vacuum and nitrogen, the films were slightly coloured in blue after 60 min of solicitation, the colour that disappeared after removing the stress. It seems that this colour belongs to the polyurethane mechano-radicals and its disappearance is related to the radicals termination reactions. In nitrogen oxide, the films were coloured in intense red-brownish and in the presence of vinyl chloride the observed colour was yellowish. In the last two cases, the colour was maintained after the stress removing, which suggests the formation of stable chemical bonds between elastomer and the chemical reagents used as environmental medium in creep experiments.

The above mentioned chemical transformations are also reflected in IR spectra, Figure 3.315, which show a intensification of the

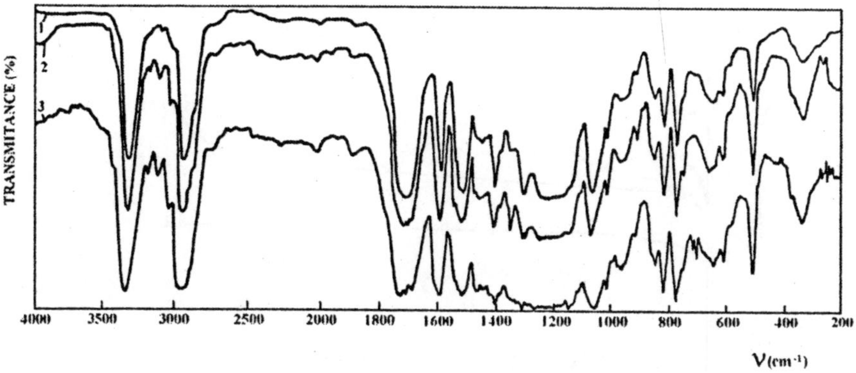

Figure 3.315. IR-spectra of the ESTANE-type polyurethane [207]: 1) unstressed; 2) stressed in NO; 3) stressed in vinyl cloride.

band from 1360 cm^{-1}, characterising the $\gtrdot$C—NO group, and of those from 1600 and 620 cm^{-1}, specific for $\gtrdot$C—ONO group, in the case of nitrogen oxide, indicating the bonding of NO to the formed radicals by splitting of urethane bonds from the main chain. The presence of the peak located at 700 cm^{-1}, curve 3 in Figure 3.315, proves the chemical fixation of the vinyl chloride structural unit at the same active centres.

As liquid media were used the following: ethyl alcohol, alcoholic solution of 0.5% DPPH, and monomers (acrylonitrile and isoprene). The typical creep curves obtained in this case are given in Figure 3.316. In ethyl alcohol the samples did not break, the deformation remaining at constant values after 20 hrs of solicitation. In the presence of active media, i.e. alcohol solution of DPPH, acrylonitrile and isoprene, fracture took place after 8, 48, and 140 hrs, respectively.

The deformation speed in these media is lower then in gaseous ones, the liquids favouring the relative sliding of macromolecules, at the beginning of solicitation, and increasing polymer's durability [207,903].

The stressing of polyurethane film in alcoholic solution of DPPH, determined their coloration in red-purple, colour which is maintained after the stress removing, which demonstrates the chemical bounding of the radical acceptor to the polymer mechanoradicals. The presence of DPPH structural units on the main chain were proved by UV spectrometry, Figure 3.317, and of acrylonitrile

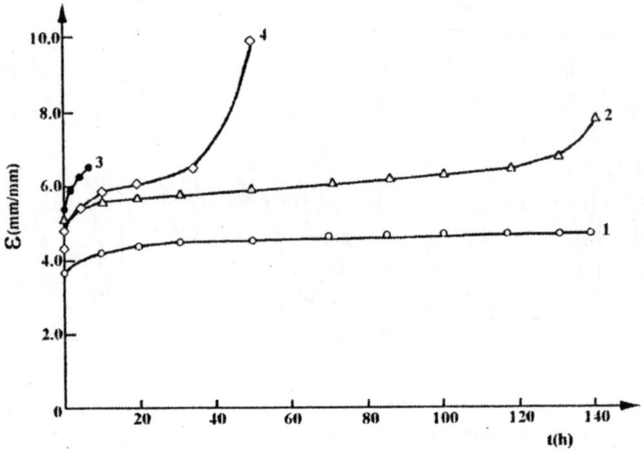

Figure 3.316. Creep curves, ε–*t*, in various liquid media [207]: 1) ethyl alcohol; 2) isoprene; 3) DPPH; 4) acrylonitrile.

ones, by IR spectrometry, by the appearance of the characteristic band of cyano groups, –CN, Figure 3.318.

The study of period of maintenance of polymer films in the liquid monomer, after the macroscopic fracturation, evidenced the increase of the bounded monomer with the post reaction duration. This effect determined in increase of the band specific to the cyano group in the IR spectrum.

The correlation of the obtained results allowed the elucidation of the mechano-chemical mechanism of static fatigue, in cyclical regime, of polyurethanes.

When a solid polymer is deformed, its macromolecules slide each to other, tending to adopt an uncoiled conformation, concomitantly with the scission of physical bonds. If their relative displacement is impeded by some of polymer structural elements, the macromolecular splitting occurs, by chemical bonds scission. Thus, into a semicrystalline polymer, the crystalline domains can act as anchoring points, which restrict the chains relative displacement. The same role is played by the crossbridges in the case of polymer networks as well as the physical knots in the case of amorphous polymers, especially at high values of deformation [306]. This means that by tensioning the polyurethane elastomer with a constant effort in time, initially a sliding of macromolecular chains occur, due to the scission of the hydrogen bonds. Along with the increase of deformation, this process is also accompanied by the some chemical bonds overstressing, which at a certain value of deforma-

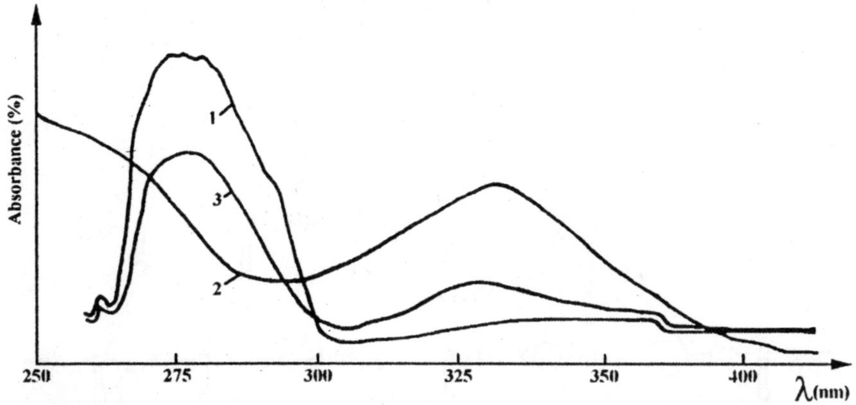

Figure 3.317. UV-spectra of the polyurethane elastomer films stressed in different liquid media [207]: 1) ethyl alcohol; 2) alcoholic solution of DPPGH; 3) alcoholic solution of DPPH.

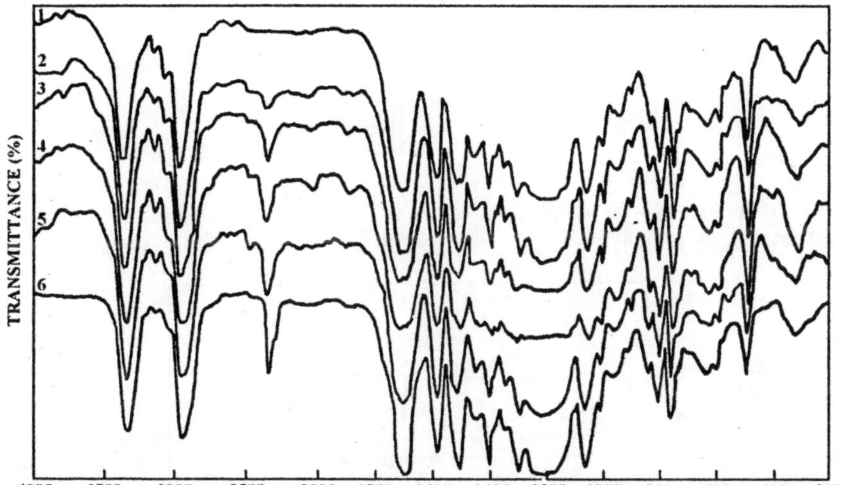

Figure 3.318. IR-spectra of the polyurethane elastomer films stressed in presence of acrylonitrile for different times of post-reaction [207]: 1) unstressed; 2) 24 hr; 3) 48 hr; 4) 72 hr; 5) 96 hr; 6) 120 hr.

tion start splitting, generating macroradicals that will react with the active reagents when these ones are present in the environmental medium of the stressed polymer.

The homolytical splitting occurs to the energetically weakest chemical bond, –C–X (X = heteroatom), in the concrete case

$-CO-\overset{\downarrow}{O}-CH_2-$, the fact evidenced by identification in IR spectrum of the stressed samples in the presence of NO of the characteristic bands both of –C–NO (1360 cm^{-1}) and –C–ONO (1600 cm^{-1}) groups.

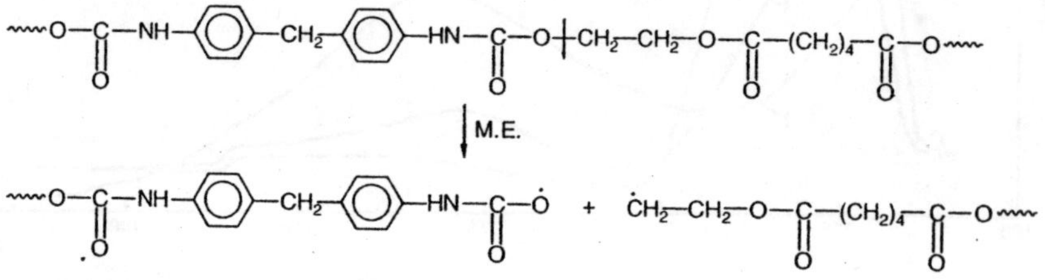

In the presence of radical acceptors (DPPH, NO, benzidine, etc.), the interruption reactions are impeded and the mechanoradicals are stabilised by reaction with acceptors:

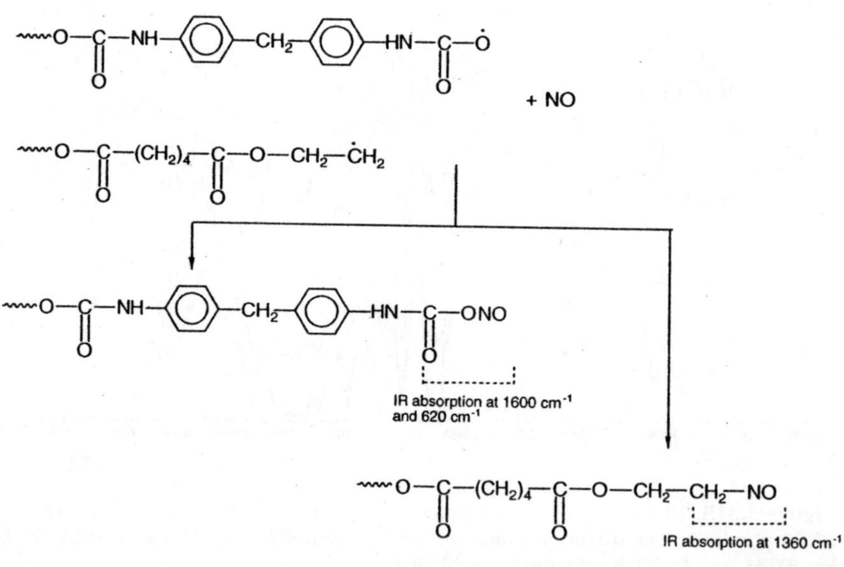

In the presence of unsaturated monomers, for instance of acrylonitrile, the bloccopolymerisation and grafting reactions are favoured; either the two mechanoradicals being capable of splitting the monomer double bond, acting as a polymerisation initiator:

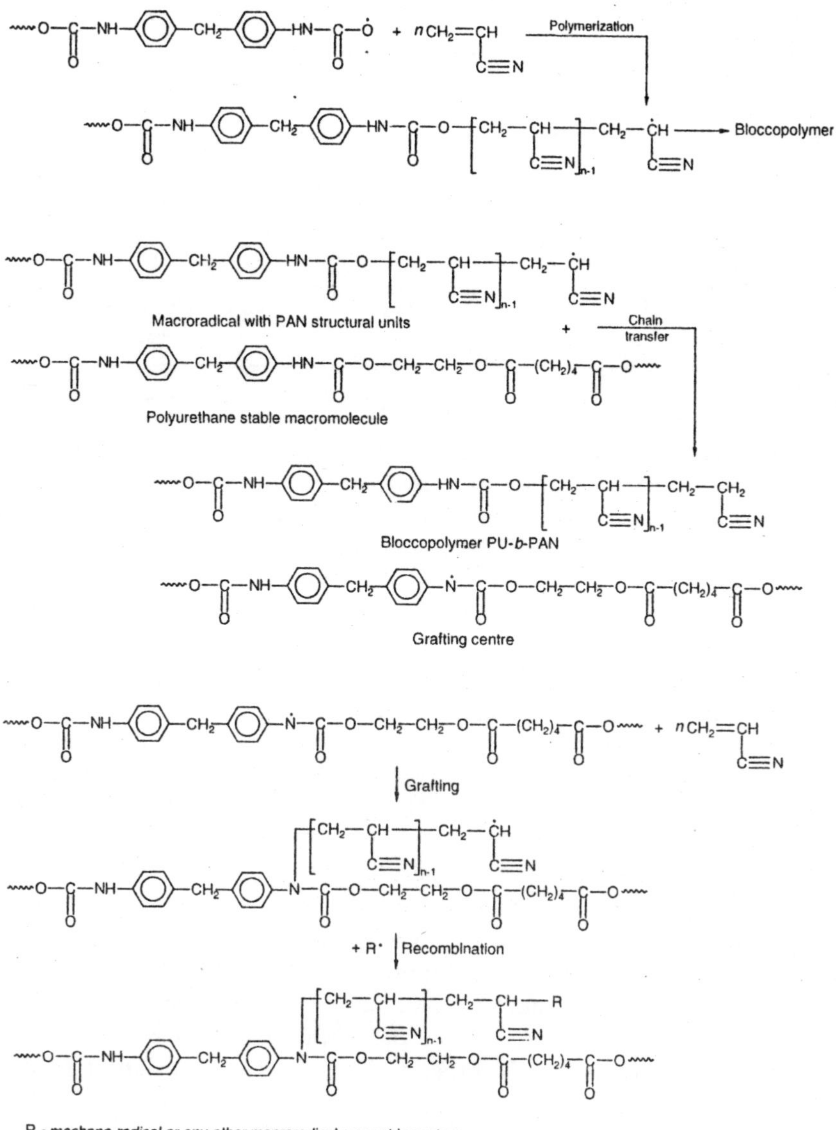

R - *mechano-radical or any other macroradical present in system*

Using the experimental data obtained for the thermoplastic polyurethanic elastomer based on 4,4'-diphenyl diisocyanate in the temperature range from 323 to 238 K the values of U_0 and γ have been calculated using Jurkov's equation, Table 3.111. As seen in this Table, in the range of positive temperatures, the fracture of the polyurethane films takes place at the relatively low stresses,

Table 3.111. The values U_0 and γ of the polyurethane elastomer based on 4,4'-diphenyl diisocyanate, at different temperatures

Temperature (K)	γ (10^{-6} $m^3 mole^{-1}$)	U_0 (kJ mole^{-1})
323	5050.526	94.746
313	5844.803	92.892
303	6921.266	92.245
293	8493.337	92.847
273	10678.201	93.304
268	12636.473	95.239
238	14079.390	96.608
233	18430.856	100.825

between 10 and 40 MPa, which is preceded by a pronounced deformation. In this range of temperature the fracture presents a ductile character. By decreasing the temperature, near the polymer T_g (–37 °C), the macromolecules loose almost complete their motility and the fracture has a brittle character.

The value of γ, obtained from the slope of the strength lines drown in co-ordinates τ–σ_0, constantly increases from 5844.803·10^{-6} m^3mol^{-1}, at T = 313 K, to 18430.856·10^{-6} m^3 mol^{-1}, at T = 233 K. The values of U_0 are smaller than those specific to the semicrystalline polymers stressed in similar conditions, i.e. low density polyethylene (LDPE), polypropylene (PP) [904], oriented polyamide 6 [905], which have U_0 equal to 117, 125, and 184 kJ/mole, respectively.

These data show that the dissociation energy of the urethane-type bonds is lower than that of –C–C– bonds from polyethylene or polypropylene. It is also two times lower than the dissociation energy of the bond from the polyamide chain. In this last case, the energetic barrier that must be overtaken in order as a elementary crack to appear is higher due to the supplementary orientation of the amorphous domains of fibres, realised by forced orientation during stretching.

The data from Table 3.111 also reveals that concomitantly with the decrease of temperature bellow 0 °C, the energetic barrier of mechano-chemical reactions release increases due the decrease of chains mobility and the corresponding increase of mechanical resistance [906]. In addition, a slightly increase of U_0 even for temperatures higher then 40 °C was observed. This finding was ascribed to an effect of thermal plasticization which firstly occurs by the splitting of the secondary bonds, phenomenon that imposes higher applied stresses for passing the energetic barrier required for chemical bonds splitting.

Along with the indirect chemical methods (radical acceptors and monomers), the mechano-chemical mechanism is sustained by the data obtained using electronic paramagnetic resonance spectroscopy (EPR) for the characterisation of free radicals [907]. The identification of the new end-functional groups, obtained by radicals stabilisation has been performed by FTIR spectroscopy [908]. Complementary results were obtained using mass spectroscopy, MS, and small angles X-ray diffraction techniques.

The fracture surfaces have been analysed by transmission electron microscopy (TEM) and so-called electronic fractography is a widely used method of fracture investigation, evidencing the changes that occurs at the supramolecular level. This method allows making the difference between the ductile or fragile character of materials fracture.

Applying this method, the fracture surfaces of statically fatigued polyurethanes in different conditions have been characterised and the results were compared with those for unstressed and non-irradiated reference samples. Reference sample fracture was achieved after its immersion in liquid nitrogen, Figure 3.319 [207]. The replica of fracture surfaces evidences their rugged character, consequence of the brittle character of the samples brought into a

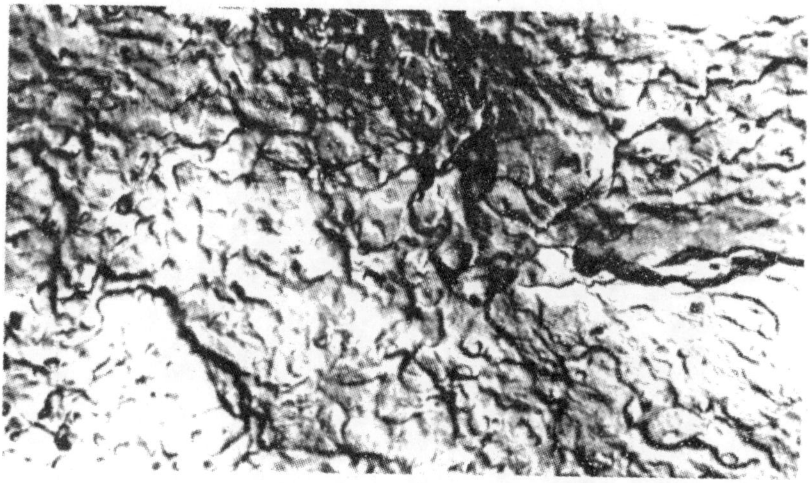

Figure 3.319. Electron micrograph of the reference sample fractured after the introduction of liquid nitrogen, ×18,000 [207].

state of high rigidity. On can also see the globular formations, which are specific ones to the amorphous polymer in the vitreous state.

In the case of uniaxial stressed samples, under a constant effort in time and concomitantly exposed to accelerated ageing, the micrographs allowed the argumentation of the fact that under the action of mechanical energy and UV radiation, the fist scission of macromolecules occur in the sample cross section.

Based both on the selective character of mechano-cracking and polymer polydispersity, it results that the first scissions occur at the level of longer macromolecules. The scissions accumulation in great number into a small volume will contribute to the appearance of the nascent microcracks. In Figure 3.320 the micrography of a sample just at the beginning of the solicitation period is depicted. The relative smooth aspect of its surface reveals the ductile character of fracturation. At a higher magnitude, on the relative smooth surface, the existence of some incipient microcraters and of globular formations, which are characteristic for the investigated polyurethane, can be observed, Figure 3.321. This implies that already on the first stages of solicitation, the molecular scissions occur, which subsequently will nucleate the nascent microcracks.

Along with the prolongation of the solicitation time, the above-mentioned small "craters" increase their diameter and become more and more numerous, Figure 3.322, being, in fact, the result of nascent microcracks coalescence, which, in turn, will pass in bigger formations, evolving toward the catastrophic fracture. On this stage,

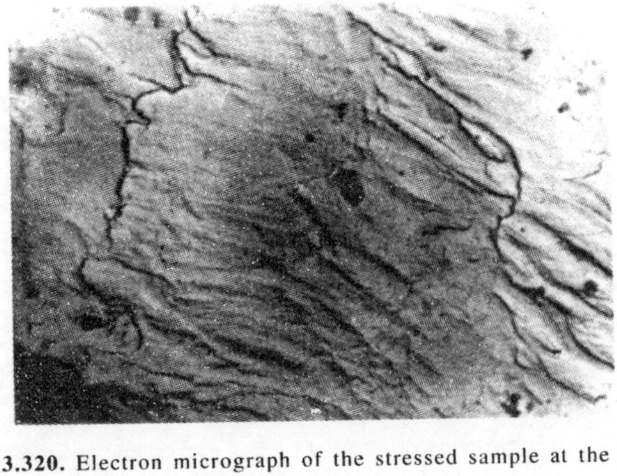

Figure 3.320. Electron micrograph of the stressed sample at the beginning of solicitation period, ×11,300 [207].

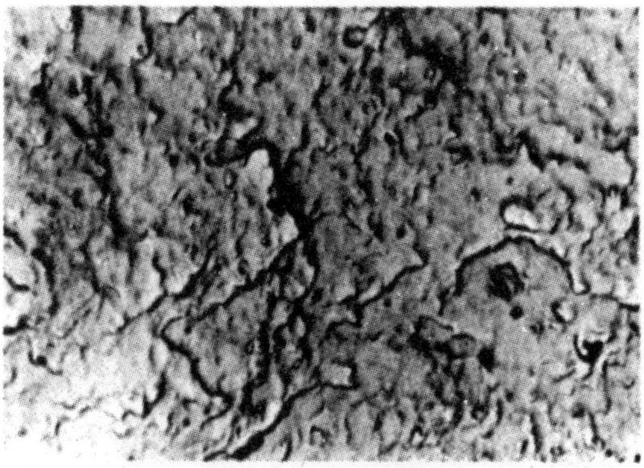

Figure 3.321. Electron micrograph of the stressed sample at the beginning of solicitation period, ×18,000 [207].

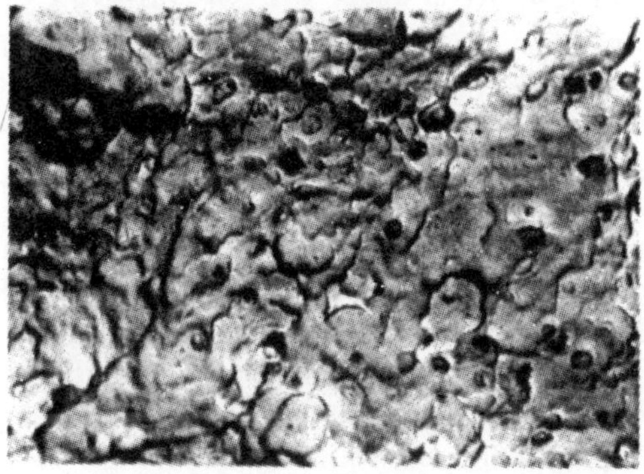

Figure 3.322. Electron micrograph of the stressed sample after a certain period from the beginning of solicitation, ×18,000 [207].

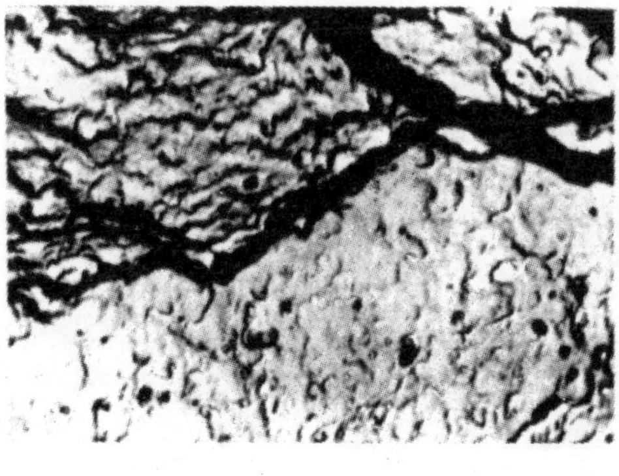

Figure 3.323. Electron micrography of the fracture surface evidencing the magistral crack, ×18 000 [207].

the unitary effort did not reach the value that determines the macroscopic fracture; it acts lent but constantly, having the comparable duration to that of relaxation times of highly elastic material. The ductile fracturation is favoured, because the macromolecules have enough time to regain their initial conformation, rearranging in such manner that the material still presents a relatively smooth surface.

A growing crack of high dimensions diminishes the samples cross area so much that a certain moment the unitary effort, even initially lower than that required for macroscopic fracture, becomes higher and causes the samples fracturation. The propagation rate of a crack, which attained the dimensions of the magistral crack, considerably increases and causes its catastrophic propagation through the sample. The relaxation time of polymer chains being now higher then the crack propagation rate, the fracturation takes a fragile character, generating a surface with various relief, Figure 3.323. It can be seen a clear demarcation between the smooth and rugged zones, representing the front of macroscopic crack in the moment of catastrophic fracture release.

Figure 3.324 presents a detail of the rugged zone, where both the individual microcracks and those ones than by coalescence give rise to bigger cracks are clearly evidenced. The crack coalescence process is stopped in the moment of macroscopic failure.

The adjacent zone of the deformation one is characterised by fibrilar formations, appeared by the breaking of those globular, representing bundles of macromolecules, Figure 3.325, which do not evolves to crystalline formations. On this image one can observe the transversal cracks to the solicitation axis, which appeared be-

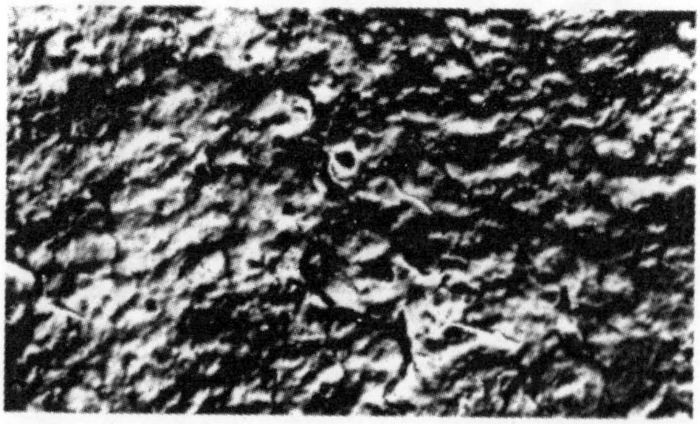

Figure 3.324. Details of fracture surface showing the "rough" zone in the moment of macroscopic fracture release, ×18,000 [207].

Figure 3.325. Electron micrography of the deformed zone in the proximity of fracture zone, ×11,300 [207].

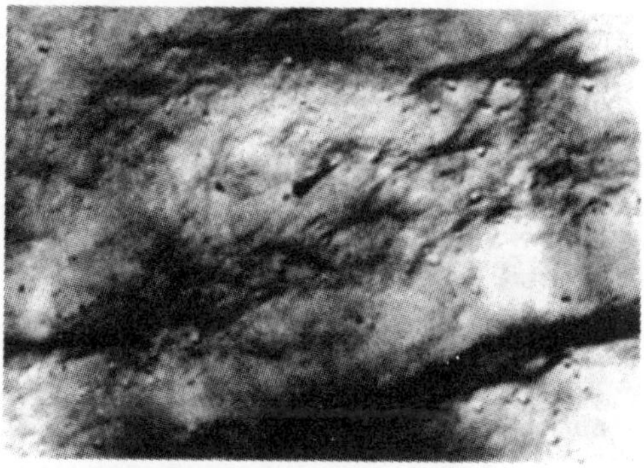

Figure 3.326. Electron micrography of the zone adjacent to fracture zone for a unstressed sample fractured in liquid nitrogen, 11,300 [207].

fore the magistal crack occurrence. The micrography of the adjacent zone to those of fracture for the unstressed sample and fractured in liquid nitrogen is presented in Figure 3.326.

The structural changes occurring under the conditions of cyclical fatigue and accelerated ageing affect a series of fundamental properties of polymer, such as: the swelling degree at equilibrium, thermal and chemical stability. Both the evolution of kinetic curves and of swelling rate constant, established for the crosslinked polyurethane elastomer, type Moldotan-D, demonstrate the splitting of a great number of transversal chemical bonds. Consequently, decrease of macromolecules packing degree occurs and the liquid access is favoured, Figure 3.327 a and b.

The combined solicitation regime also determines the modification of some important exploitation properties of polymers, for in-

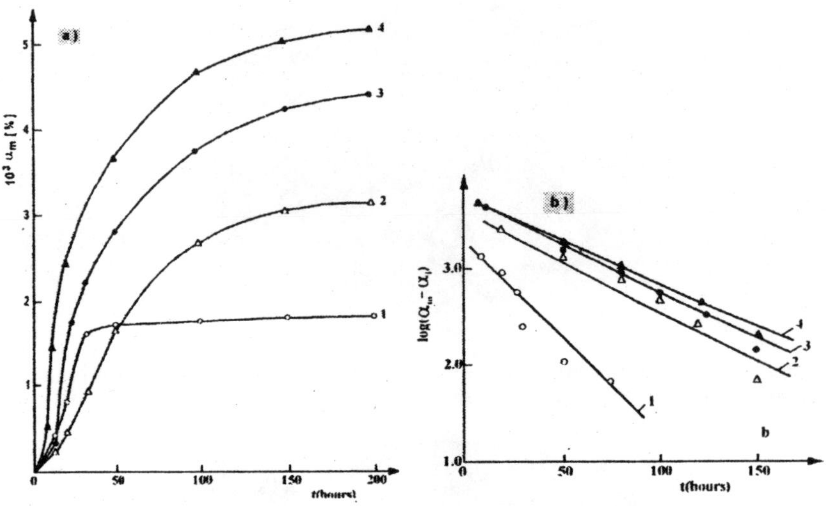

Figure 3.327. (a) Kinetic curves of swelling for the polyurethane elastomer supposed to static fatigue and accelerated fatigue [207]: 1) unstressed; 2) σ_c = 9.7 MPa; 3) σ_c = 9.7 MPa and subsequently accelerated ageing; 4) static fatigue at σ_c = 9.7 MPa and concomitant accelerated ageing; and (b) Graphical determination of swelling constant [207]: 1) σ_c = 0; 2) 7.0; 3) 9.7; 4) 12.6; 5) 14.3 MPa.

stance the chemical stability. In the case of O-rings, fittings, and amortisation elements based on polyurethane, the correlation of the fatigue and ageing regimes with the chemical stability, in different solvents, such as: water, sulphuric acid of different concentrations, alkali, or mineral oils, is of the first importance.

The variation of chemical stability, evaluated by the variation of sample's weight in time, for the crosslinked elastomer based on 4,4'-diphenyl diisocyanate is presented in Figures 3.328 and 3.329. This one show a good stability, the smallest values, approximately 2% weight looses, characterises the samples stressed in the most severe conditions of the cyclical fatigue and concomitant accelerated ageing. As expected, the highest stability was obtained for the sample unstressed previously, 1.6% weight loses in mineral oil and 1.3% in water, respectively, after 200 hr of tensioning.

The variation of chemical stability in sulphuric acid, 10% aqueous solution, in different solicitation conditions, is presented in Figure 3.330 a and b, and in sodium hydroxide in Figure 3.331.

In accordance with the applied regime of mechanical solicitation, the thermal characteristics are changed, too. Thermogravimetric analysis (TGA) proved that the samples stressed under creep con-

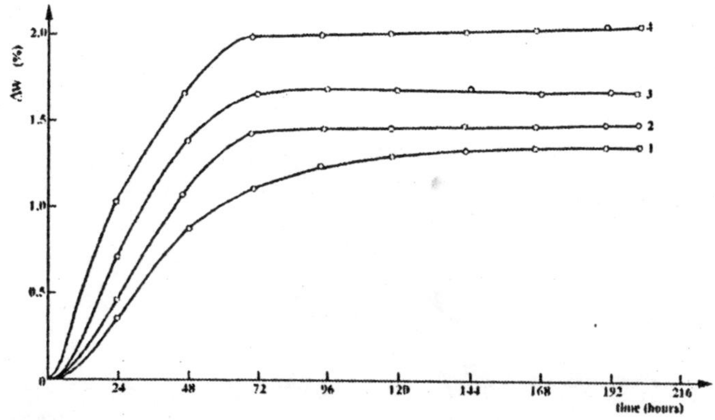

Figure 3.328. Variation in time of chemical stability of polyurethane in water [909]: 1) unstressed; 2) stressed at 9.7 MPa; 3) aged for 200 hr and subsequently stressed at 9.7 MPa; 4) stressed at 9,7 MPa with concomitant ageing.

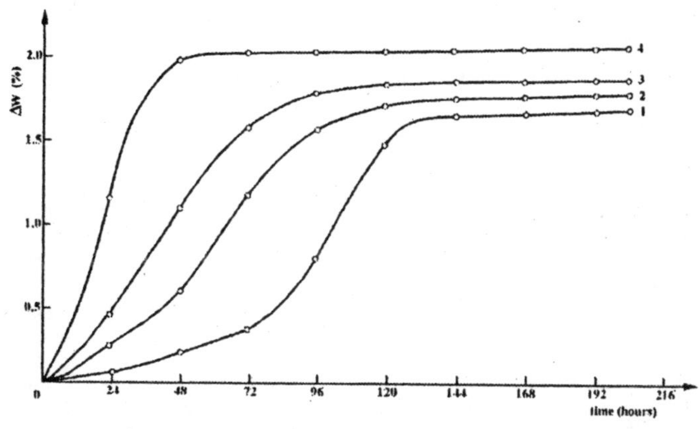

Figure 3.329. Variation in time of chemical stability of polyurethane in mineral oil [909]: 1) unstressed; 2) stressed at 9.7 MPa; 3) aged for 200 hr and subsequently stressed at 9.7 MPa.

ditions have a lower chemical stability than the control sample, the weight loses increasing with solicitation intensity, Figure 3.332, [909].

The differential curves (DTG) indicate the existence of three processes of thermal decomposition, in accordance with the type of split bonds. Thus, the secondary van der Waals bonds, physical hydrogen bonds, and the homolytic splitting of chemical bonds occur at 280, 320, and 360 °C, respectively. The intensity of these processes varies with the magnitude of the applied stress. For instance, the characteristic process of valence bonds splitting is more

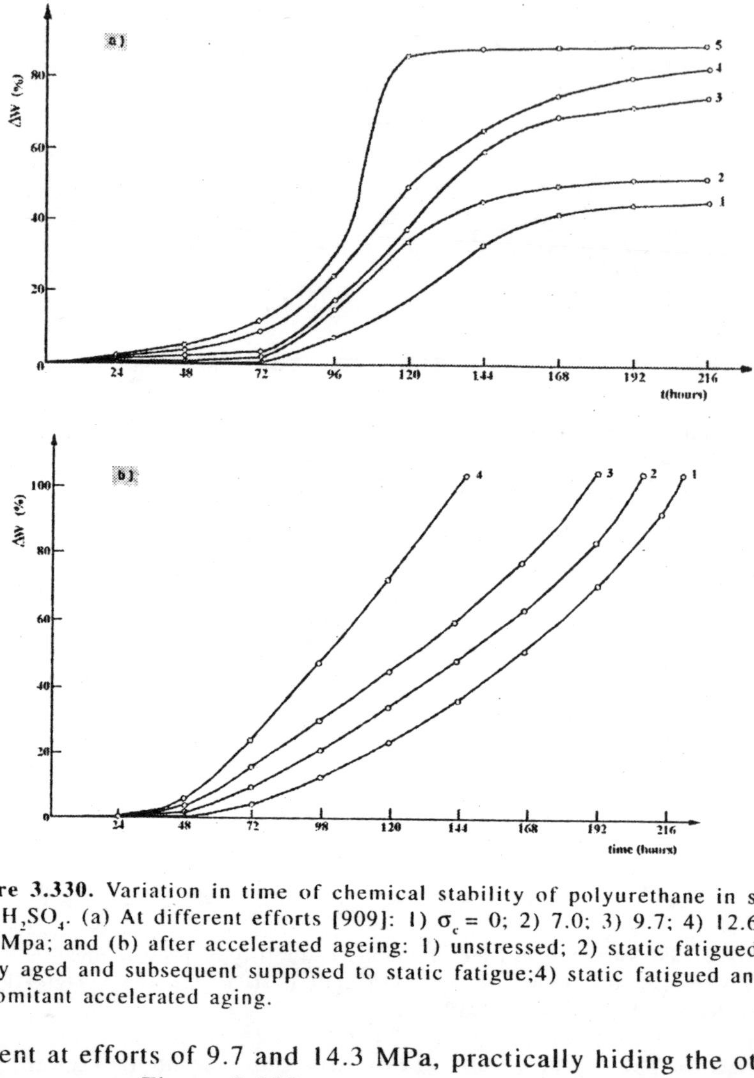

Figure 3.330. Variation in time of chemical stability of polyurethane in soln. 10% H_2SO_4. (a) At different efforts [909]: 1) $\sigma_c = 0$; 2) 7.0; 3) 9.7; 4) 12.6; 5) 14.3 Mpa; and (b) after accelerated ageing: 1) unstressed; 2) static fatigued; 3) firstly aged and subsequent supposed to static fatigue;4) static fatigued and to concomitant accelerated aging.

evident at efforts of 9.7 and 14.3 MPa, practically hiding the other two processes, Figure 3.333.

The activation energy, corresponding to the thermal decomposition, is characterised by negative values, which are specific for the endothermal transformations, being of about 1.8 times higher for the stressed sample at 14.3 MPa as compared to the unstressed one, Table 3.112. In all cases to reaction order is about equal to 1.

Under the combined conditions of cyclical solicitation and accelerated ageing, the ATG data shown, for the Moldotan-D polyurethane the maximal weight loss of 98% at 550 °C, the control

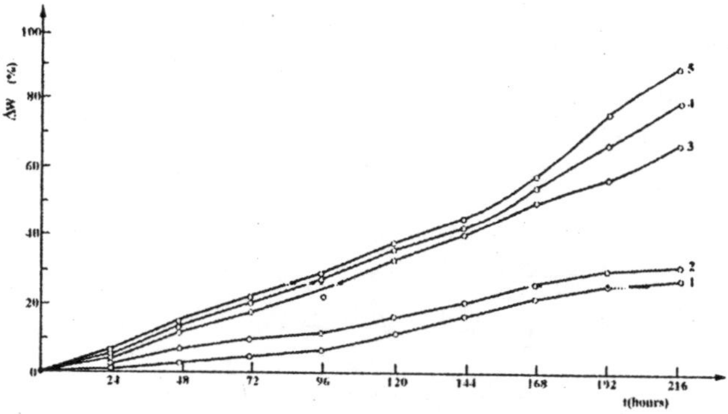

Figure 3.331. Variation in time of chemical stability in soln. 10% H_2SO_4 for stressed polymer [207]: 1) σ_c = 0; 2) 7.0; 3) 9.7; 4) 12.6; 5) 14.3 MPa. [207].

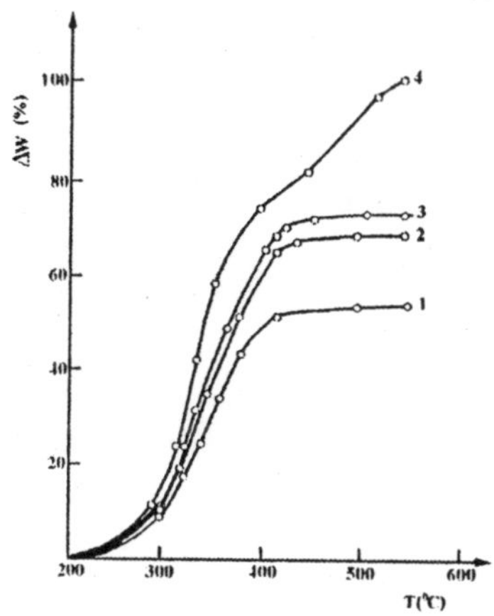

Figure 3. 332. ATG curves of the polyurethane elastomer supposed to static fatigue and accelerated ageing [910]: 1) control; 2) 9.7 MPa; 3) aged 200 hr at 9.7 MPa: 4) static fatigue at 9.7 MPa and concomitant accelerated ageing.

sample being the most stable, Figure 3.334, [910].

DTG curves, Figure 3.335, shows that the main process of thermal degradation occurred at 360 °C and it is most intense in the conditions of combined cyclical solicitation and accelerated ageing. The high number hydrogen bonds in polyurethane-based elastomers explains the intensity of corresponding process that occur in the temperature range from 320 to 330 °C. When the cyclical fatigue and accelerated ageing occur concomitantly this effect is superposed to that of valence bonds splitting, of higher intensity. The same effect refers to the weakest bonds, of van der Waals type. Once again, the activation energies of the thermal decomposition present mini-

Table 3.112. Activation energy and reaction order of polyurethane elastomer thermal decomposition [969]

Unitary effort applied, MPa	0	7.0	9.7	14.3
Activation energy·10^5, kcal/mole	- 0.16428	- 0.20503	- 0.26253	- 0.29824
Reaction order	1.0	1.1	0.6	1.0

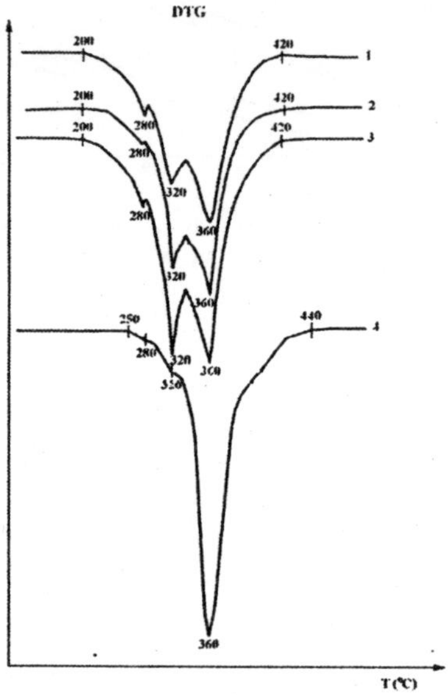

Figure 3.333. DTG curves of polyurethane elastomer after solicitation at: 1) σ_c = 0; 2) 7.0; 3) 9.7; 4) 14.3 MPa.

mal and negative values for the tensioned and accelerated aged sample and maximal for the control one (Table 3.113).

The data concerning the thermal stability prove that the lowest stable samples are those tensioned under the conditions of static fatigue with the highest applied stress and those ones tensioned and concomitantly supposed to accelerated ageing.

3.8.3.1.2. Polymers fatigue by solicitation in variable field of forces

In the case of mechanical solicitation into variable field of forces, the rates of transmission and elastic energy absorption, respectively, are well lower than the rate of mechano-chemical reaction, which is initiated by accumulation in polymer structure of this energy. Usually these solicitations are periodical and occur with frequencies located in the range of mechanical losses characteristics to the stressed material.

Table 3.113. Activation energy and the reaction order of thermal decomposition of polyurethane elastomers MOLDOTAN-D, concomitantly supposed to static fatigue and accelerated ageing [910]

Testing conditions	σ_c = 0 MPa	9.7 MPa	9.7 MPa after accelerated ageing, 2000 hr	9.7 MPa concomitant with accelerated ageing
Activation energy·10^5 (kcal/mole)	- 0.16428	- 0.16856	- 0.17011	- 0.26033
Reaction order	1.0	1.0	1.1	2.5

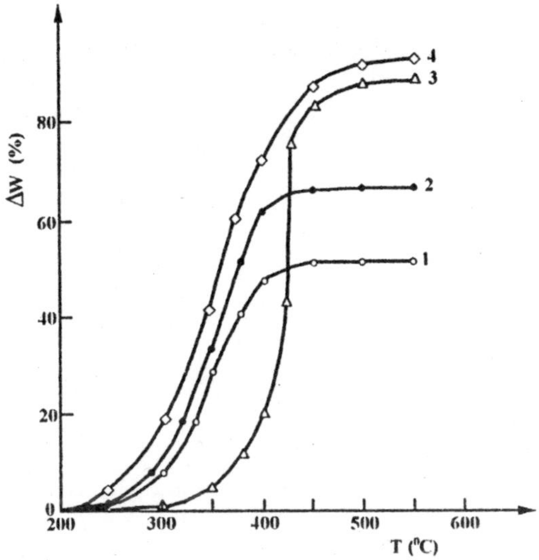

Figure 3.334. ATG curves of polyurethane elastomer after solicitation at: 1) σ_c = 0; 2) 7.0; 3) 9.7; 4) 14.3 MPa.

The most important types of solicitation in variable field of forces are the dynamical fatigue, vibratory milling, and ultrasonic treatment.

Dynamic fatigue

Dynamic fatigue represents the material fracturation or destruction phenomenon under the action of a repeated or oscillating mechanical solicitation [912]. The importance of dynamical fatigue consists in the fact that under the action of fluctuating efforts, the materials are damaged at the efforts values well lower than those supported by material during an usual mechanical solicitation, under the action of a monotonous increasing effort. On the other side, for a given amplitude of effort, the durability of the polymers supposed

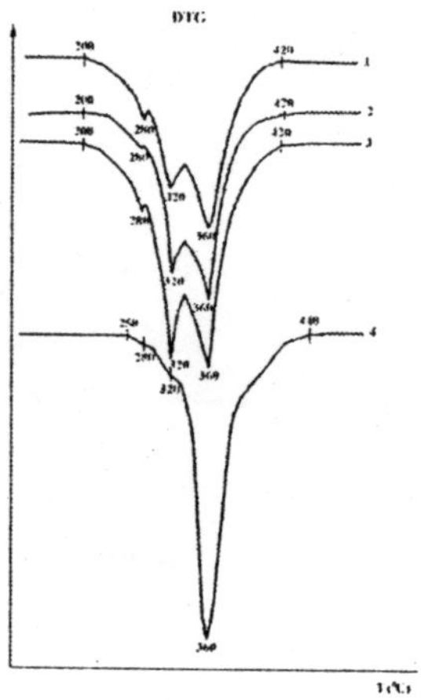

Figure 3.335. DTG curves of the polyurethane elastomer supposed to static fatigue and accelerated ageing [910]: 1) control sample; 2) stressed under creep conditions; 3) firstly aged and subsequently supposed to static fatigue; 4) concomitant static fatigue and accelerated ageing.

to the dynamical fatigue will be lower than that corresponding to the static fatigue, when a constant effort is applied in time. With the increase of the applied stress value the number of solicitation cycles (*N*) until fracture decreases.

The accelerating effect of the stress-relaxation cycles on the decay of polymers resistance originates in the two characteristic proprieties of this class of materials: 1) the non elastic structural response that leads to a dissipation of mechanical energy as heat during each cycle; and 2) the activation of molecular rearrangements, having as effects the knots unbinding, sliding of chains segments, local reorientation, and/or generation of some cracks/discontinuities in polymer structure [888].

Three processes can contribute the polymer fracture, by fatigue, namely - 1) thermal softening; 2) flowing or accelerated creep; and 3) initiation and propagation of fatigue's cracks [834]. The parameters of dynamical fatigue have been defined by E.H. Andrews, [831], and are the following:

- the effort periodical fluctuation, which is characterised by the corresponding amplitude, $\Delta\sigma$;
- the periodical amplitude, corresponding to the deformation variation, $\Delta\varepsilon$;
- the stress average value, σ_m;
- the deformation average value, ε_m;

- the solicitation frequency, ν (ω = 2πν);
- a characteristic wave shape (sinusoidal, squared, etc) both for stress and deformation;
- medium temperature and that developed in sample during solicitation; and
- samples' geometry, including the cracks and faults presence.

In practice, five types of fatigue tests are most often used, namely: 1) periodical stressing at traction or compression, between the preestablished limits of stress; 2) periodical stressing at traction or compression, between the deformation preestablished limits; 3) bending reversible tensioning into one direction; 4) reversible bending tensioning in two directions; and 5) reversible shear tensioning.

On the other side, the dynamical fatigue is influenced both by the polymer intrinsic parameters, (i.e. molecular structure, supramolecular morphological structure) and of solicitation, such as: applied stress and deformation, frequency, cycles number.

In the case of crosslinked polyurethane based on 4,4'-dyphenil diisocyanate, characterised by different degrees of crosslinking, expressed by the value of $\overline{M_c}$, depending on the number of solicitation cycles, N, and the effort amplitude, σ, for a constant value of frequency, ν, the results collected in Table 3.114 have been obtained. Taking as criterion of the fatigue resistance the number of solicitation cycles, NF, these ones diminish with the increase of crosslinking degree, respectively with the decay of $\overline{M_c}$. This finding was explained by the decrease of polymer cohesion energy due to the presence of supplementary transversal bridges of alophantate type. Consequently, the specific intercatenary interactions, of van der Waals type and especially hydrogen bonds, characterised by energy close to that of covalent bond, are manifested only at short distance, thus decreasing the elastomer resistance to dynamical fatigue. In addition, concomitantly with the increase of solicitation stress, or of deformation, respectively, diminishes the number of cycles until fracturation.

During the dynamical fatigue, the variation of material resistance with the stress or deformation amplitude is usually determined by drawing of Wöhler curves (variation of σ vs. N_F), Figure 3.336, [914]. On these curves, three distinct regions can be delimited:

- region I – where the cracks generation start, during the first quarter of cycle (Figure 3.336 a). The cracks are visible even without magnification. Samples fracture occurs by the initially lent increase of a crack, followed by its catastrophic propagation. The number of microcracks essentially depends

Table 3.114. Variation of the number of cycles until fracture, N_F, with the crosslinking degree and unitary effort, in the case of casting polyurethane elastomer (ν = 4 Hz) [207]

$\overline{M_c}$, g/mole → σ, MPa ↓	7 700	7 134	6 613	6 000	5 690
0.810	30 850	23 250	22 200	20 400	16 810
0.820	23 820	19 510	18 250	13 830	12 850
0.850	20 510	16 790	14 340	9 710	9 070
0.905	10 250	9 310	8 770	8 120	7 510
1.000	9 520	8 920	7 930	7 410	7 200
1.100	7 410	6 830	6 240	6 590	5 010
1.200	6 310	5 700	4 910	4 150	3 810

on the stress magnitude in indirectly by the cycles number;

- region II – is characterised by a period of crack nucleation that slowly occurs and the catastrophic propagation. In this region the dependence σ – N_F is well diminished due to both the retardation of the microcracking process initiation and low crack growing rate for well lower values of the stress;
- region III – delimits the borders of material resistance during fatigue. In this region the dependence σ – N_F is practically equal to zero. The initiation of magistral crack requires a long induction period.

The fatigue process that characterises the regions II and III of Wöhler curves is composed by at list two steps, namely – the crack initiation step due to the homogeneous response of polymer and a second one "heterogeneous" that evolves through the appearance of some discontinuities of material, which continuously evolve.

The typical Wöhler curve of the linear polyurethane, Figure 3.336 b, is similar to that one corresponding to the crosslinked polyurethane, with the mention that the values of N_F are of several times higher in the first case. This fact demonstrates the peculiar mechanical resistance of this polymer to dynamical fatigue, in the given conditions of stressing. Its utilisation to the artificial skin manufacturing, particularly for shoes when the cyclical solicitation are very frequent, is based just on this property. During the dynamic solicitation, very sharply sounds, which accompanies the deformation process, can be heard even by the human ear. The acoustic emission, called "Kaiser effect", is the result of cracks generation and growing [915]. Immediately after the generation of acoustical emission, the microcracks appearance is observed. These

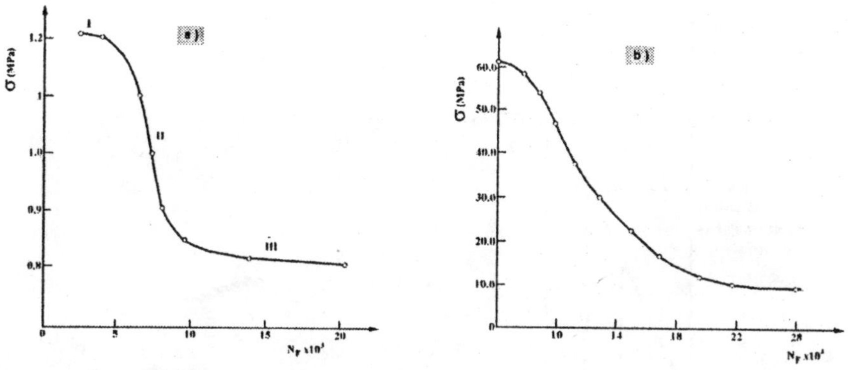

Figure 3.336. Wöhler curves of: (a) crosslinked polyurethane ($\bar{M}$ = 6000 g/mole; ν = 4 Hz); (b) linear polyurethane, Estane, (ν = 4 Hz) [208].

ones, on the subsequent stages of mechano-chemical process generate the magistral crack and cause the macroscopic fracture. A rigorous dependence exists between the mechano-chemical phenomena and acoustic emission. On this basis, the special devices for acoustic emission detection, which placed on any dynamometer, record and convert the acoustic signal in electrical impulses. By the correlation of acoustic diagram with the stress-strain curve, the moment of catastrophic crack appearance and propagation can be exactly established, Figure 3.337, [830, 916].

The decrease of material resistance to fatigue also occurs by the increase of the cycles frequency, for instance from 1 to 5 Hz for the crosslinked polyurethane, Table 3.115. Generally, for a given material and a fixed frequency the material durability diminishes in accordance with the increase of unitary effort.

The investigation of the influence of the main parameters of dynamical fatigue is not so easy to be made because of the long period of experiments. From this reason, frequently some simplified programs are used, when certain parameters are maintained at constant values. By automatic processing of the recorded data, the constant level curves are obtained, Figure 3.338, which evidence that by increasing of solicitation frequency, ν, and sample deformation, ε, the number of cycles to fracture, N_F, decays, with other words, the samples durability to dynamical fatigue decreases.

Practically, at the same time with de decrease of experiences number, by the establishing of regression equation and dispersion analysis, the material mechanical resistance, under the condition of dynamical fatigue, can be foreseen, if the most important param-

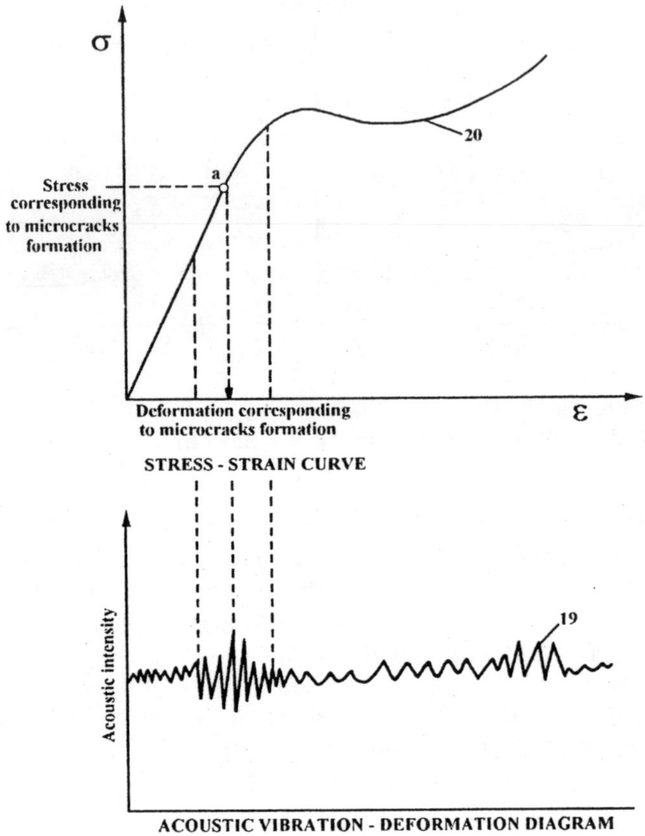

Figure 3.337. Acoustic diagram for establishing the moment of apparition and propagation of magistral crack in polymeric materials [830, 916].

eters of solicitation, i.e. frequency and maximal deformation, are known.

For the same crosslinking degree and effort amplitude, the samples similarly characterised, but in conditions of accelerated ageing present a decrease of durability with about 30–40%, due to the UV radiation and temperature contribution to the diminish of energetic barrier required for thermo-fluctuating scission of overstressed chemical bonds. These results once again confirm the common mechanism, the radicalic chained one, which governs all the destructive processes, Table 3.116 [917]. In the case of thermoplastic elastomer, Estane, in the same conditions, the durability decay

Table 3.115. Influence of the solicitation frequency on the durability of the crosslinked polyurethane elastomer samples ($\bar{M}$ =5690 g/mole) [208]

Frequency, Hz → σ, MPa ↓	1.0	2.0	3.0	4.0	5.0
0.810	138 000	85 200	52 350	16 810	9 020
0.820	124 000	80 510	48 050	12 850	7 160
0.850	107 010	77 280	41 750	9 070	5 880
0.905	92 720	73 050	37 110	7 510	4 710
1.000	87 250	68 290	35 050	7 200	3 930
1.100	74 110	57 270	26 800	5 010	2 820
1.200	69 080	52 050	24 180	3 810	1 950

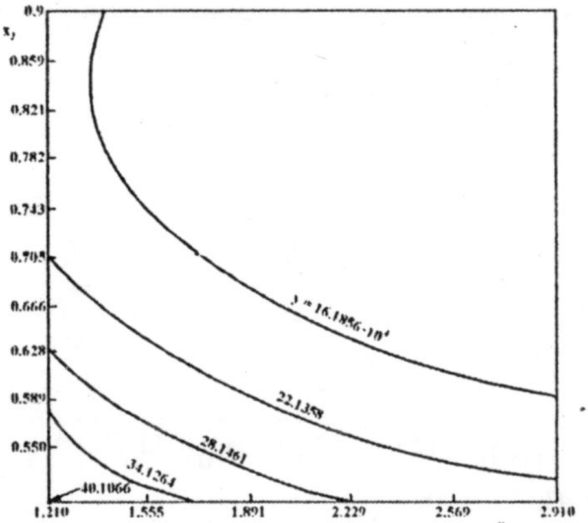

Figure 3.338. Curves of constant level in the experimental plane ν-ε obtained under the conditions of dynamical fatigue of a thermoplastic elastomer [207]. Regression equation is: N_f = 13,409853 – 3,619595ε – 5,632082ε + 1,554812ε² + 1,580?ε + 3,671674ε² [142].

is of about 18–20%, Table 3.117. By increasing temperature to 35 °C, and concomitant irradiation (λ = 355 nm) the number of cycles to fracture slightly increases (5–15%) due to thermal plasticization, which intensifies the relative displacements of macromolecules, by splitting of a number of physical bonds. Subsequently the increase of temperature at 60 °C intensifies the mechano-chemical process, causing the decrease of polymer durability.

The photo-chemical component of polymer fracture can be limited using some additives that acts as photo-stabilisers [918]. The understanding of main function of additives allows their use at the

Table 3.116. Variation of the number of cycles until fracture, N_F, of a crosslinked polyurethane elastomer, suppose concomitantly to dynamical fatigue and accelerated ageing (ν = 4 Hz, T = 60 °C) [917]

$\overline{M_c}$, g/mole → σ, MPa ↓	7 700	7 134	6 613	6 000	5 690
0.810	21 750	18 900	17 750	15 500	13 900
0.820	19 730	16 820	13 510	11 540	10 020
0.850	14 670	13 250	9 810	7 470	6 050
0.905	8 140	7 850	5 920	5 100	4 570
1.000	7 640	6 950	5 420	4 220	3 140
1.100	5 950	4 200	3 650	3 550	3 050
1.200	4 530	3 820	3 030	2 640	2 040

Table 3.117. Variation of number of cycles until fracture with frequency in condition of accelerated ageing [917]

Conditions of solicitation → Frequency ↓	without UV 20 °C	UV 20 °C	UV 35 °C	UV 60 °C
1.5	184 100	152 000	164 200	123 700
2.5	113 200	91 100	91 100	79 100
3.0	101 400	82 700	82 700	75 400
4.0	85 100	60 200	60 200	51 500

complementary reactivity levels and thus the prevention of concomitant effect of the destructive factors.

In the case of polyurethane elastomers, very interesting results have also been obtained in the case of other types of mechanical solicitation than traction, namely compression, bending, sonic, etc [207, 919].

In the field of textile fibres, the study of dynamical behaviour contributes to the clarification of the relation between the exerted forces and suffered deformations during the mechanical solicitation. This relation is the basis of their visco-elastic state and is essential for the understanding of their mechanical behaviour and of amortisation properties [877–881, 920–926].

The mechanical behaviour of the soluble poly(vinyl alcohol) fibres, of SOLVRON-SX type, supposed to repeated cyclical fatigue, based on a scheduled regime, on a apparatus STATIMAT-M that allowed both the forces and number of cycles variation in a large range, is described in Figure 3.339 [877].

The thread testing has been performed by its solicitation between two limits of oscillating stress, for each level a maximal force F_1

and minimal one F_2, respectively, and for a number of cycles varying between 0 and 500, for 7 levels of solicitations, with a constant maximal deformation and recover until to constant minimal deformation, Figure 3.340. Following the evolution of relaxation variation with the number of cycles, three type of relaxation have been found: 1) simple (normal) relaxation that occurs when to a constant deformation, the force decays in time (Figure 3.340 2) inverse relaxation, appears during the force increase in time (Figure 3.340 b); 3) mixed relaxation, when firstly an increase of force with time occurs and after that the force decays (Figure 3.340 c).

The occurrence of inverse and mixed relaxation was explained by visco-elastic manifested during the previous dynamical solicitations and evidences the influence of thread "history"/"memory" [922–924, 927]. Cyclical solicitation represented in Figure 3.340 adapted for a narrow range of forces between F_1 and F_2, starting with the

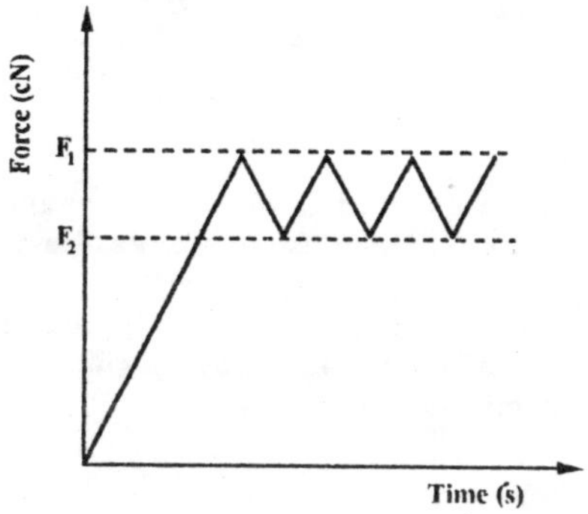

Figure 3.339. Dynamical solicitation between two levels of applied effort (F_1 ?si F_2) [877].

minimal level of solicitation, *A*, and finishing with the maximal one, 6, Tables 3.118 and 3.119.

From typical diagrams of solicitation the following characteristics were determined: the dynamical strain, ε (%), in function of solicitation steps and number of cycles; the force at which the relaxation is released, F_s, bellow 3 min; the residual force, F_r, for any step of solicitation in the range from 0 to 3 min; E_d (cN/tex) for each step of solicitation in function on number of cycles.

The relaxation, expressed by the residual force F_r was calculated with the relationship:

$$F_r(\%) = \frac{\text{Force measured during the relaxation test}}{\text{Force atwhich the relaxation measuring starts}} \cdot 100 \qquad (3.174)$$

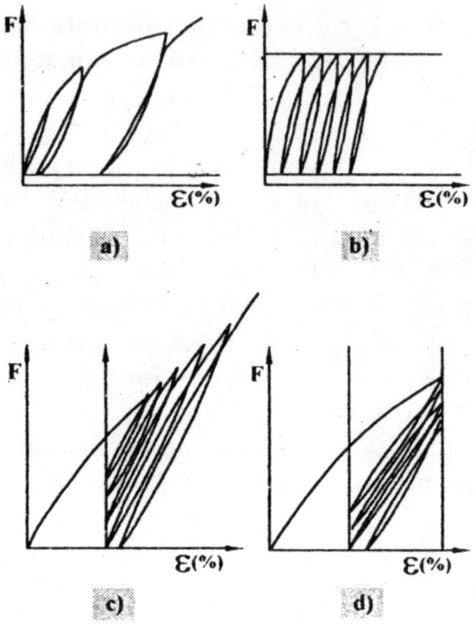

Figure 3.340. Illustrative representation of fibres dynamical solicitation in various conditions [921]: (a) maximal force increasing and reverting to an imposed and constant minimal value; (b) maximal force constant and reverting up to a minimal and constant level; (c) maximal increasing deformation and reverting to a constant minimal deformation; d) maximal deformation constant and reverting to a constant minimal deformation.

and the dynamic modulus:

$$E_d = \frac{F_2 - F_1}{\varepsilon_2 - \varepsilon_1} \tag{3.175}$$

where: F_2 – the superior limit of oscillating stress (cN/tex); F_1 - the inferior limit of oscillating stress (cN/tex); ε_1 – strain superior limit; ε_2 is the strain inferior limit.

Dynamic strain was calculated as:

$$\varepsilon_d = \frac{\varepsilon_2 + \varepsilon_1}{2} \tag{3.176}$$

The data from Tables 3.118 and 3.119 prove that for a relatively small number of cycles, 50, in the elastic range, reorient-ations of the macro-molecular chains from the amorphous phase firstly occur. After that, deformation freeze takes place. By increasing the solicitation forces, in the viscoelastic domain, the deformations increase again, as effect of chains segments displacement and/or of some highly organised formations, which proves that the splitting of both secondary bonds and strongly tensioned cov-alent ones occurred.

The data concerning the variation of dynamical strain with the number of cycles and the applied stress show that this one varies in a large range, from 0.85% (variant A) to 18.30% (variant 6), and its behaviour can be analysed for two distinct regions. In the first region, between 0 and 50 cycles, even if the deformation increases, the process is characterised by a slow rate.

It comes out that especially from the evalu-ation of the residual force, which at 3 min of relaxation varies from 67.52%, for the solicitation level A, up to 72.16%, for the maximum level of solicitation (level 6), passing through a maximal value, 81.26% for the level D, which is defined as a critical level of affecting the fibres quality. The calculation of the ratio between the values of residual forces from the first zone (force at 20 s) and the second one (force at 180 s), shows an increase of this ratio until the solicitation level D. In other words, the most important changes occur in the zone I and after that the changes are shifted in the region II.

On the whole, the obtained results evidence the influence of the solicitation range and of cycles number on the fibre structure, which in the conditions of cyclical fatigue enters into an unstable regime, showing labile formations that have been evidenced by electron microscopy, Figures 3.341 and 342.

Even from the first 10 min of cyclical solicitation on the filaments surface appear some modifications of texture, in sense of

Table 3.118. The behaviour of the filamentary thread SOLVRON during cyclical solicitation, with a constant maximal force and recovering to a constant minimal solicitation. Variation of dynamical elongation [877]

Samples code	Level of solicitation (cN/tex)	ε_{d0} (%)	ε_{d50} (%)	ε_{d500} (%)	Zone I $\varepsilon_{d50} - \varepsilon_{d0}$	Zone II $\varepsilon_{d500} - \varepsilon_{d50}$	$R = \frac{Zone\ I}{Zone\ II}$
A	1 – 3.5	0.20	0.50	0.85	0.35	0.35	0.85
B	3 – 5.5	0.60	2.00	3.80	1.40	1.80	0.77
C	5 – 7.5	2.80	5.20	6.80	2.40	1.60	1.50
D	7 – 8.5	4.20	6.20	7.20	2.00	1.00	2.00
E	8 – 10.5	6.20	8.00	9.50	1.80	1.50	1.20
F	10 – 12.5	7.80	10.20	12.30	2.40	2.10	1.14
G	14 – 16.5	10.80	14.25	18.30	2.45	4.05	1.65

some small dimensions inhomogenities proliferation, Figure 3.341. At 30 cycles, the asperities are more evident and the deepnesses are gradually changed in channels, Figure 3.342. The morphological changes are the result of deformations that accelerated occur, accompanied by the reorient-ations in the amorphous phase, which are evidenced by the increase of dynamical modulus, Table 3.120.

Enhancing the solicitation level, for instance at 50 cycles, Figure 3.343, or 100 cycles, Figure 3.344, on one side the number of faults increases and on the other side at 100 cycles a region of faults association, constituting the nucleation place of magistral crack, can be observed. At 200 cycles of solicitation, Figure 3.345, dark zones of intense deformation appear, suggesting the presence of a macrofault with transversal orientation, which will evolve until to the fracture of the structural elements.

Enhancing more the solicitation level, by increasing the number of cycles at 300 and 500, respectively, Figures 3.346 and 3.347, the dominant effect is that of micro and macrocracks association in inhomogeneous

Table 3.119. The behaviour of SOLVRON fibres during cyclical solicitation with the solicitation level. Variation of the residual force [877]

Samples code	Level of solicitation (cN/tex)	Force at the start of relaxation, F_0 (cN)	Force at 20 s of relaxation, F_{20} (cN)	Force at 180 s of relaxation, F_{180} (cN)	Residual force (%)	
					at 20 s	at 180 s
A	1 – 3.5	43.93	35.00	29.48	79.67	67.52
B	3 – 5.5	67.10	55.00	47.25	81.85	70.32
C	5 – 7.5	91.92	80.00	73.54	87.03	80.00
D	7 – 8.5	103.64	91.00	84.22	89.73	81.26
E	8 – 10.5	127.89	115.00	101.86	89.92	79.65
F	10 – 12.5	151.16	135.00	116.05	98.31	76.77
G	14 – 16.5	197.59	172.00	142.58	87.01	72.16

formations, giving rise to voids, deepnesses, and craters, preliminary to final failure. The progressive accumulation of faults determines the fracture of structural element, Figure 3.348.

The fracture by cyclical fatigue is the result of interdependence between the nature, magnitude, and intensity of deformation forces and the fibre morphological structure. In the case of SOLVRON

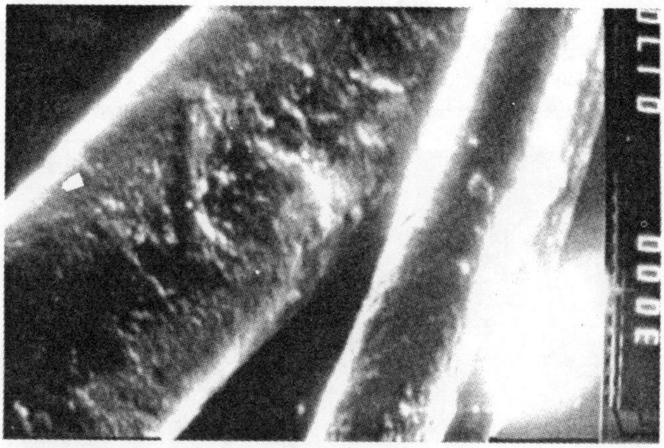

Figure 3.341. SEM of individual filament SOLVRON-SX after 10 cycles of solicitation [877].

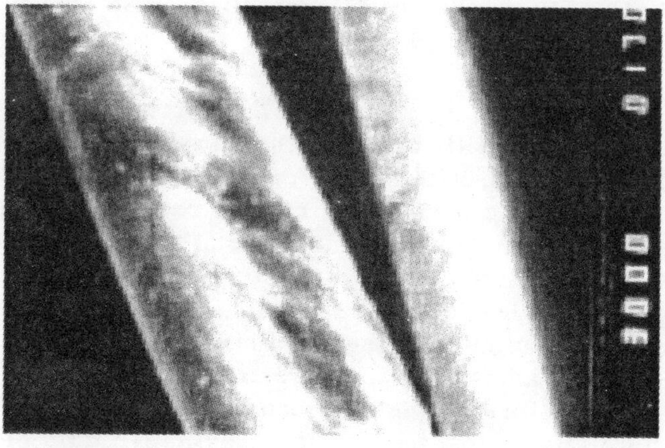

Figure 3.342. SEM of individual filament SOLVRON-SX after 30 cycles of solicitation [877].

Table 3.120. Variation of the dynamical modulus with the number of cycles for the seven levels of solicitation

Cycles no. → σ (cN/tex)↓	0	10	20	30	40	50	100	300	500
A	835	790	785	783	780	782	770	760	760
B	810	695	645	635	637	637	650	700	715
C	715	627	637	660	700	700	717	780	804
D	715	645	665	677	700	707	737	787	807
E	720	687	730	740	750	760	785	830	850
F	810	777	777	800	817	822	855	909	880
G	1040	930	930	980	1020	990	990	970	960

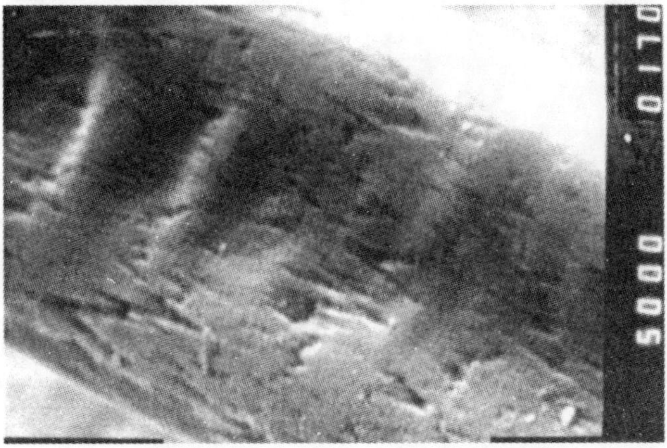

Figure 3.343. SEM of individual filament SOLVRON-SX after 50 cycles of solicitation [877].

fibres, characterised by a core-shell structure, the mechanical solicitation differentially modifies the morphology of surface texture. Thus, the shell, with a more enhanced orientation degree, is able initially to take over higher efforts, but representing a more rigid structural element with the increase of solicitation regime it is the fibre part that suffers the most profound and numerous changes. The core, with a less ordered structure, even if at repetitive and relatively small solicitations suffered many re-orientations, mainly of the structural inhomogeneities, presents a sufficiently high deformation capacity for limiting the fracture acceleration. In this case, the fracture rate is governed by the quantitative ratio of the core and shell structures. The knowledge of these particularities of the fatigue effect, strictly related to the fibre structure, furnishes data

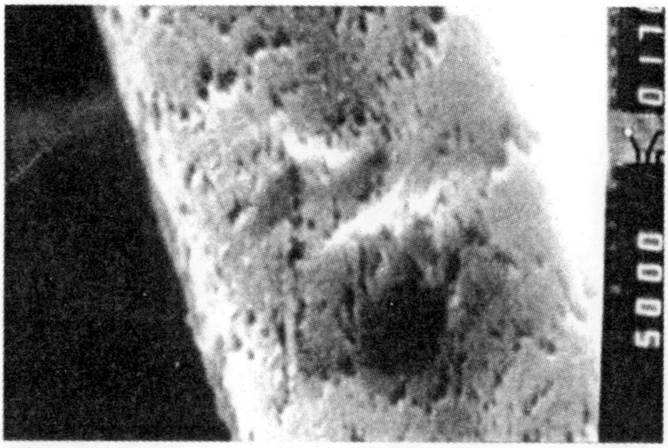

Figure 3.344. SEM of individual filament SOLVRON-SX after 100 cycles of solicitation [877].

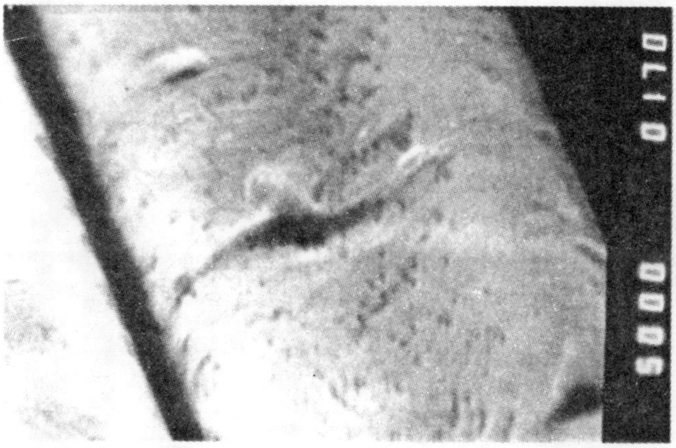

Figure 3.345. SEM of individual filament SOLVRON-SX after 200 cycles of solicitation [877].

that constitute the base of the mechanical regime optimisation program during the textile manufacturing process, thus allowing the improvement of end products quality in the exploitation conditions.

A similar study carried out on crosslinked PAV fibres, of KUKALON type, evidenced that in the case of cyclical axial solicitations, under the conditions of severe mechanical regime, the covalent bounds are broken both of the main chain and transversal bridges, accompanied by the free radicals appearance. Their main effect is the interaction with the neighbouring macromolecules, de-

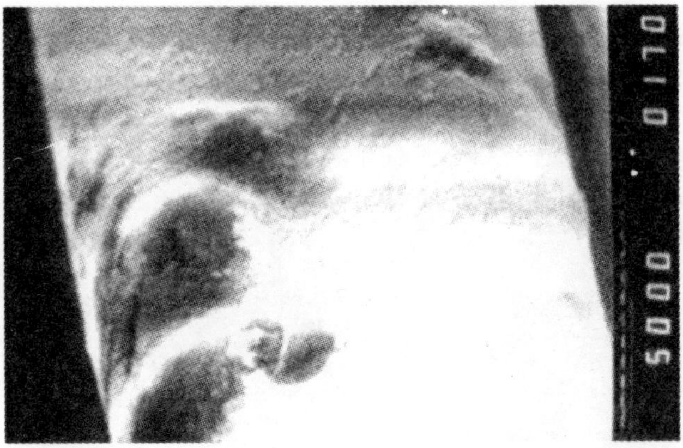

Figure 3.346. SEM of individual filament SOLVRON-SX after 300 cycles of solicitation [877].

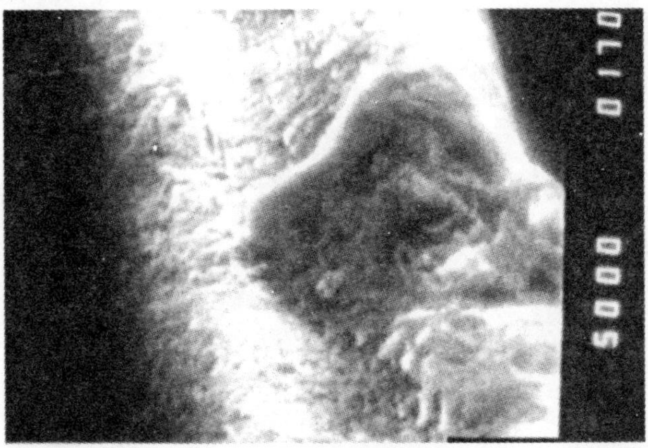

Figure 3.347. SEM of individual filament SOLVRON-SX after 500 cycles of solicitation [877].

termining "permanent deformation" [928]

The fracture by torsion of the filamentary thread, type SOLVRON-SX, solicited at 900 torsions 150cm, alternating the torsion sense, *S* and *Z*, respectively, was realised on a Zmeigler D 312 model apparatus, and is depicted in Figure 3.349. The fracture presents a "craggy" morphology, determined by the pulling out of some labile structural formations. The process occurred by the individual breaking of filaments composing the thread.

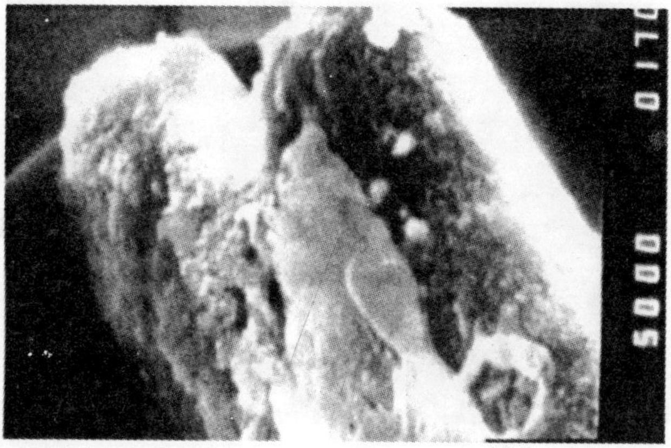

Figure 3.348. Fracture of the individual filament SOLVRON-SX due to the cyclic solicitation [877].

Sonic fatigue

The ultrasounds propagation, with a frequency higher than 20 kHz, constitutes an adequate method of evaluation not only of the polymer structural characteristics but also of the changes they suffer during mechanical solicitation. On the other side, the prolonged sonic irradiation of the macromolecular structures determines their irreversible change, by the accumulation of mechano-chemical transformations, having as final effect the installation of fatigue state.

W.W. Mosley [942] and R. Samuels [943] made the correlation between the sonic modulus, E_s, with the fibres structural parameters, proposing the following equation:

$$\frac{1}{E_s}=\frac{2}{3}\left[\frac{\beta(1-f_{cr})}{E^0_{t_{cr}}}+\frac{(1-\beta)(1-f_{am})}{E^0_{t_{am}}}\right] \qquad (3.176)$$

where: E^0_t - intricate sonic transversal modulus, for both the crystalline and amorphous phases ($E_{t_{cr}}$ = $3.68 \cdot 10^{10}$ dyne/cm^2 and $E_{t_{am}}$ = $1.82 \cdot 10^{10}$ dyne/cm^2); β - crystallinity.

The rate and sonic modulus directly influenced the fibre structure and can be utilised as criteria of evaluation of the changes that occur both in the orientation state of macromolecules and crystallinity, under the conditions of mechanical solicitation.

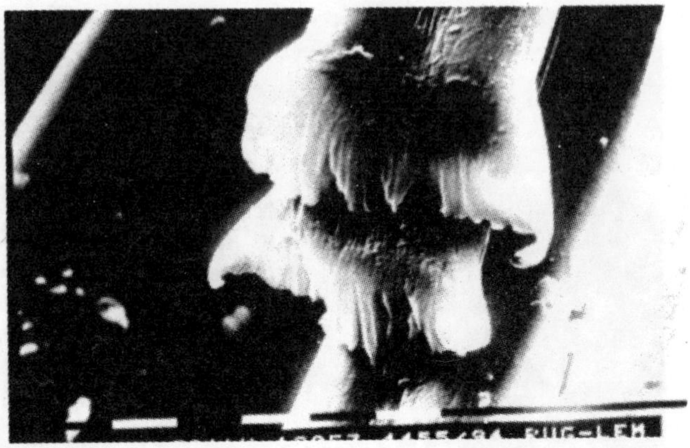

Figure 3.349. Fracture of the individual filament SOLVRON-SX by torsion [877].

In the case of based on poly(vinyl alcohol) soluble fibres, SOLVRON-SX, the sonic solicitation was realised using two ceramic piezoelectric transducers, placed at distance of 12.5 and 17.5 cm, respectively, in different conditions of applied force, P_s. During solicitation step, the experiments were carried out in standard environmental conditions and both in air and acetone (non solvent of PAV). As Table 3.121 shows the solicitation induced in the relaxation stage the corresponding modification of the ultrasonic rate. The dependence of ultrasonation modulus, E_U, and rate, v_u, with the developed force, P_s, in the solicitation and relaxation stage, respectively, is illustrated in Figures 350 and 351.

In the solicitation stage, in the force range from 30 to 70 cN, the both parameters tends to constant values after that a rapid increase of their valued is recorded with the increase of solicitation. This evolution demonstrates that in the first moments of solicitation (10–30 cN/thread) a molecular orientation in amorphous phase takes place, concretised in the increase both of dynamic modulus and ultrasonation rate. It follows a structural stabilisation, in the force range from 30 to 70 cN/thread, after which the orientation occurs predominantly in higher organised structural formations, i.e. crystalline regions and microfibrils.

In the relaxation stage, both in air and acetone the same tendency was found. In air the relaxation leads to lower values of ultrasonation rate and modulus than in the solicitation stage. It results that even if the relaxation time is short, some rearrangement

Table 3.121. The modification of the SOLVRON-SX fibres characteristics by sonic solicitation [877]

P_S (cN)	Characteristics	Solicitation in: air conditioned		Relaxation in: air conditioned		acetone	
		inst.	after 10"	inst.	after 10"	inst.	after 10"
10	v_u (m/s)	3918.14	3933.9	3412.9	3486.7	3998.8	3933.9
	E_u (cN/tex)	15.3538	15.4755	11.6478	12.1570	15.9744	15.4755
20	v_u (m/s)	3974.5	4000.0	3631.08	3714.7	4038.7	3961.9
	E_u (cN/tex)	15.7966	16.0000	13.1847	13.7989	16.3110	15.6966
30	v_u (m/s)	4035.5	4016.06	4012.8	3946.3	4084.9	4081.6
	E_u (cN/tex)	16.2852	16.1287	16.1025	15.5732	16.6864	16.6594
50	v_u (m/s)	4055.1	4025.7	4048.5	3990.4	4091.6	4098.3
	E_u (cN/tex)	16.4438	16.2062	16.3903	15.9232	16.7411	16.7960
70	v_u (m/s)	4152.4	4084.9	4098.3	4035.5	4336.5	4142.5
	E_u (cN/tex)	17.0189	16.6864	16.7960	16.2852	18.8052	17.1603
90	v_u (m/s)	4230.1	4145.9	4152.8	4108.4	4541.3	4504.5
	E_u (cN/tex)	17.9837	17.1884	17.2457	16.8789	20.6234	20.2905
100	v_u (m/s)	4233.7	4295.5	4208.7	4233.7	4646.8	4659.8
	E_u (cN/tex)	17.9242	18.4513	17.7131	17.9242	21.5927	21.7137
120	v_u (m/s)	4262.5	4321.5	4237.2	4280.8	4655.4	4960.3
	E_u (cN/tex)	18.1689	18.6753	17.9538	18.3252	21.6727	24.6045

of macromolecules occurs, these ones passing in other conformations, which correspond to the new level of thermodynamic equilibrium.

The ultrasonation of the MELANA type monofilaments, base on the ternary copolymer poly(acrylonitrile-*co*-α-methylstyrene-*co*-vinyl acetate), PAN, was carried out at a frequency of 2 MHz, using the water as medium of ultrasound propagation. The waves were generated by a piezoelectric crystal of quartz, for periods varying from 10 to 100 hrs. The filaments tensile strength, σ was used as a criterion of the consequences of the effect of ultrasound. For three types of monofimanents, differing by fineness degree, the results are presented in Figure 3.352. In all cases, σ diminished in time, tending to a limit of acoustic solicitation, above 100 hr. The evolution of this parameter in time reflects whole particularities of a mechano-chemical transformation, namely: 1) its dependency on the level of mechanical soliciation; and 2) the decrease until a certain limit of the parameter taken as a criterion of evaluation, tensile strength in this particular case.

For the mathematical modelling of the sonic fatigue process of PAN monofilaments, the following equation was considered:

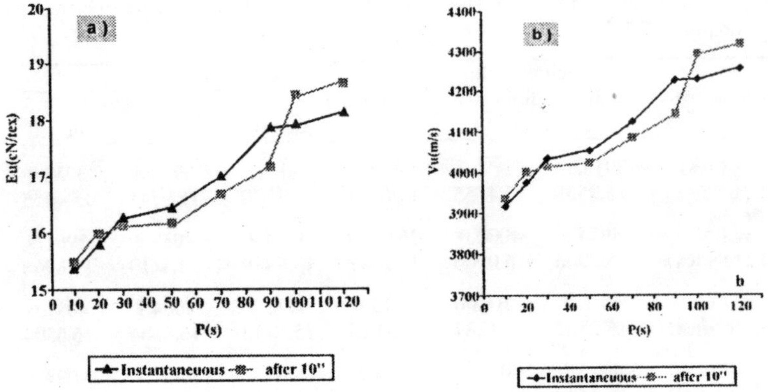

Figure 3.350. Dependence of the sonic modulus (a) and of ultrasonic rate (b) by the applied force during solicitation step [877].

$$\sigma(t)=\frac{\sigma_{(\infty)}}{1+\frac{\sigma_{(\infty)}}{\sigma_{(0)}}-e^{\sigma_{(\infty)}kt}} \tag{3.178}$$

where: $\sigma_{(t)}$ – the unitary stress at time t, (gf); $\sigma_{(\infty)}$ – limit value to which the fracture effort tends, (gf); $\sigma_{(0)}$ – the stress applied to the non-ultrasonaned monofilament, (gf);

$$k=\frac{\nu}{\sigma_{(\infty)}}\cdot\frac{\Sigma y_i N_i}{\Sigma N_i^2} \tag{3.179}$$

$$y_i=\frac{\sigma_{(\infty)}/\sigma_{(0)}-1}{\sigma_{(\infty)}/\sigma_{(r)}-1} \tag{3.180}$$

ν - frequency of the piezoelectric crystal (2.814 MHz); N_i – the number of fatigue cycles.

The validity of the proposed model is proved by the good accordance between the experimental and calculated curves form Figure 3.352.

The mechano-chemical nature of the MELANA fibres sonic fatigue was proved by measuring the average molecular weight, which diminished in time. The formed free macroradicals were reacted with an aromatic diamine, benzidine. By free primary amino groups diazotizing, the resorcinol chemical bounding was possible, which irreversible coloured the fibres in red.

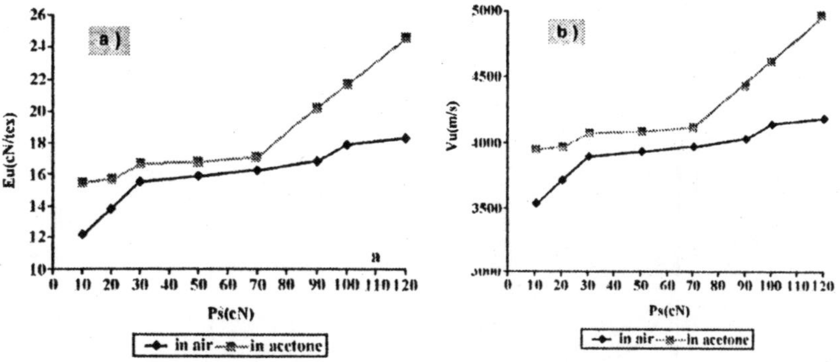

Figure 3.351. Dependence of the sonic modulus (a) and of ultrasonic rate (b) by the applied force during the relaxation step [877].

In conclusion, the mechano-chemical act of chemical bonds splitting on the ternary copolymer main chains stays on the basis of the mechanical resistance decay in the case of MELANA fibres supposed to sonic fatigue. The generated mechano-radicals initiate o chain of reactions that in time affects the supramolecular-morphological structure, reflected by the decrease of mechanical properties [830–951].

It was proved that the equation (3.178) adequately describes the sonic fatigue of other polymers [952, 953].

3.8.3.2. Polymer fracture by wearing

The tear is defined as the evolving process of loosing the material from the surface of a solid body, from mechanical reasons, by intimate contact and relative displacement of other body, which should be in solid, liquid, and gaseous phase, respectively. It is concretised by submicroscopic, microscopic, or macroscopic looses as tear particles as well as by transformations in substance and shape of the superficial layers, tribologically solicited [954].

The tribologic solicitation occurs to the contact surfaces by the overlapping of three components, namely – mechanical, thermal, and chemical one – being accompanied by the loosing of friction energy and by concomitant conversion of the kinetic energy (dissipation) in heat, especially in the contact points.

The mechanical component includes the transformation of all deformation, elastic, and plastic energies in thermal energy; chemical solicitation is manifested by mechanical and thermal activation of chemical reactions, at the contact surface, with the external or internal medium. Therefore, the friction processes through the

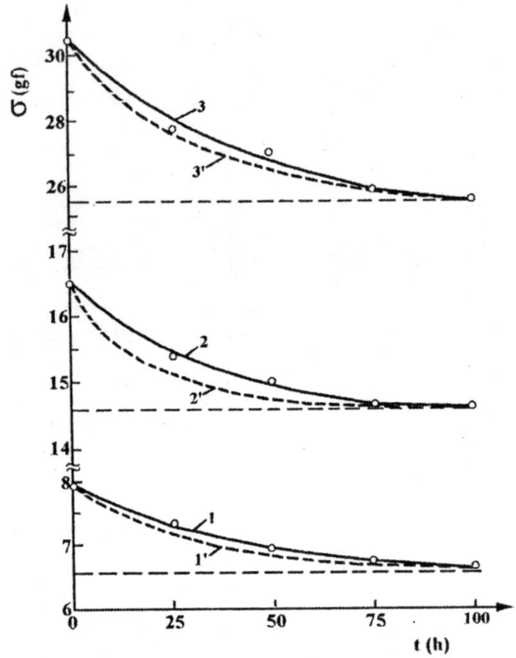

Figure 3.352. Influence of ultrasonic irradiation period on the tensile strength of Melana monofilamente of the finesses: 1) 2.5 den; 2) 6 den; 3) 15 den (with dotted line the theoretical curves are presented).

energetic transformations release a series of irreversible phenomena of mechano-chemical nature. Generally, in the technical field such effects are of peculiar importance, since under the exploitation conditions a diversity of polymers-based objects, mixed objects, constitutive parts of different devices and engines, and installations suffer intense tribologic solicitations.

The most important mechanisms of tear are depicted in Figure 3.353 [954]. On can conclude that, in fact, these ones refer to the different concrete cases of fracture, implying both the splitting of intercatenary physical and covalent chemical bonds. In the adhesion case, it is proved that the fracture implies not only the breaking of adhesive bounds, but also of those cohesive ones, accompanied by mechano-emission [955].

During exploitation, the polymer products suffer deformation and friction effects, and finally these ones are destroyed, as effect of gradual accumulation of some structural modifications that diminish their resistance to repeated deformations. Two are the most important reasons of finite objects deterioration: 1) polymer ageing;

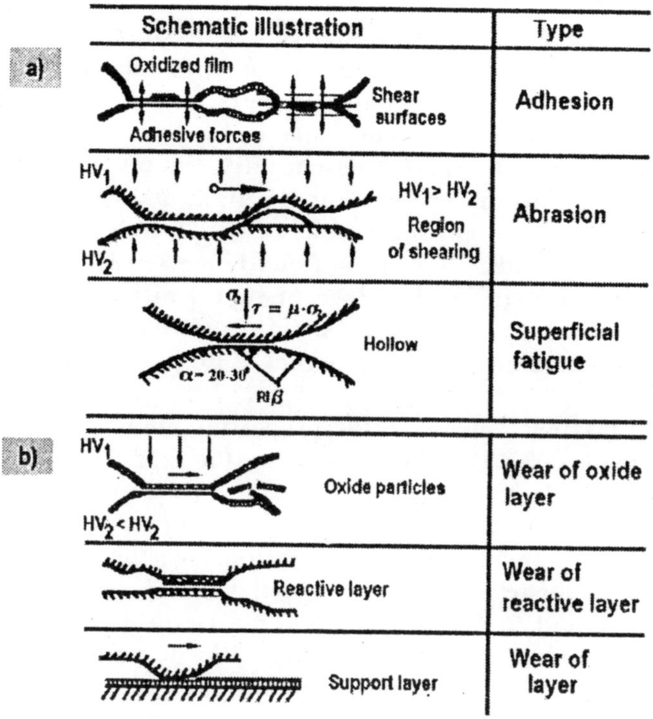

Figure 3.353 Mechanisms of wear: a) inside limit layer; b) outside limit layer [954].

and 2) action of mechanical deformation.

The first effect is caused by the modification in time of utilisation properties of material from which the object was made, under the action of the external medium factors (light, heat, oxygen, humidity, etc) that constitute component parts of the ageing process. The second factor is based on the appearance and accumulation of the material fatigue, under the action of mechanical energy.

Whenever the mechanical solicitations are accomplished during friction process, the changes in the superficial layer of material, which lead to its fracture, can be defined as wear. Frequently this process can be accelerated by different mechanical operations, such as surfaces polishing or mechanical processing of different mechanisms joints.

Practically, the ageing, tear, and fatigue processes concomitantly occur and are strongly intercorrelated each to other. The medium nature also plays an important role on these processes. In the case of polymer fracture, the medium effect may be concretised in macromolecular chains splitting, polymer crosslinking, or the absence of any effect [956–958]. For instance, in the ozone medium the double bonds from unsaturated rubbers are weakened and the splitting of the main chain occurs, mainly by decomposition of the intermediary-formed ozonides. Consequently, the polymer molecular weight decreases. In the presence of UV light and oxygen, the unsaturated rubbers are oxidised by photolysis, followed by network formation, finally a decrease of plasticity being recorded. Bellow of glassy temperature, in the presence of liquid nitrogen, the rubber suffers an intense cracking process, followed by the nucleation of the magistral crack and of its catastrophic propagation through the sample.

The reasons that determine the properties modification by tear and fatigue can be the following: 1) the appearance of free macroradicals under the deformation conditions; 2) the activation of some reactions with medium components; 3) chains reorientation and packing in supramolecular formations; 4) the appearance of adsorption and superficial-active properties of the mechano-destruction products; and 5) the ordering and plastic properties modification of the contact surfaces by friction with the polymer [595].

Any of these components of tear or fatigue can play a main role or a secondary one, in function of the polymer concrete structure and the ingredients it contains, the medium nature, mechanical regime, the characteristics of the surfaces in contact with the polymer during friction.

Under the friction conditions, the formed macroradicals are able to release the same subsequent transformations as any other mechanoradicals, namely: reaction with acceptor radicals, with other macroradicals, chained reactions of mechano-degradation and crosslinking, transfer reactions, grafting and block copolymerisation reactions, or other reactions with the medium components.

It was found that the metal tear to the dynamical contact with the polymers is diminished in the oxygen presence than in argon atmosphere, even if the peroxy macroradicals are active with respect to metallic surfaces. Probably in the competition of the chemical events occurring there, the oxygen succeeds to protect in a way the metal surface, by increasing its stability against to the polymer radicals.

It worth to be mentioned the tear of polymer objects, by friction, therefore the fracture of small particles from the contact surface is of mechano-chemical nature. The theoretical treatment of this process is complicated by a series of physical factors, such as the necessity of considering the required mechanical work for deformation that proceeds the fracture. However, in essence taking into account the interaction with the active components of the surrounding medium and the contribution of other chemical factors, a great analogy can be made between the wear process and fatigue one, which is just related to their common mechano-chemical component.

It was established that metal wear in the polymers presence increases with the increase of the energy required for polymer thermal degradation, under the action of heat released by friction. As Figure 3.354 shows, The number of macroradicals generated increases with the decay of friction heat and the metal surfaces are more affected [960–976]. On the other side, the metal tear increases with the increase of the polymer average molecular weight, Figure 3.355. The curve that correlates the tear rate with the molecular weight passes through a maximum, since at high polymer concentrations, to a certain value of mechanical forces, some "amortisation" effects may appear. The friction pressure also induces a double effect on the tear rate. Thus, at low pressures the polymer plays a role of lubricant while at high pressures the mechano-destruction intensifies and, consequently, the metal particles detachment is favoured. This finding must be taken into account in the case of metal-polymer couples, because otherwise the polymer will becomes a fracture focus.

The study of some radical acceptor influence (iodine, hydroquinone) proved the chained radicalic mechanism of tear, these ones being fixed with the same efficiency as in the case of polymer mechano-degradation. On the other side, it was established that the mechano-degraded polymer is partially fixed on the metallic surface,

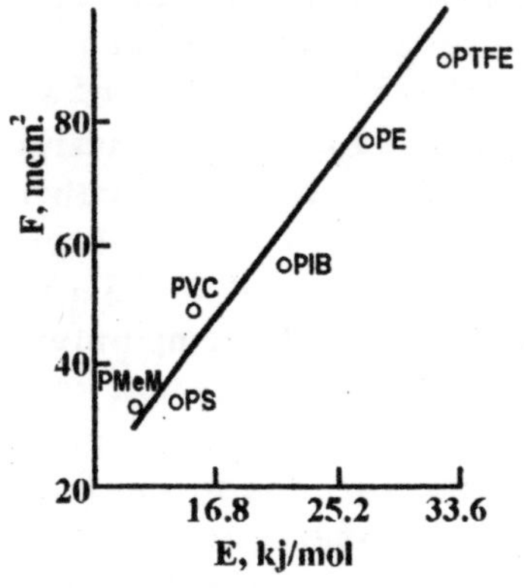

Figure 3.354. Variation of the iron grinding degree with the activation energy of polymer thermal decomposition [959].

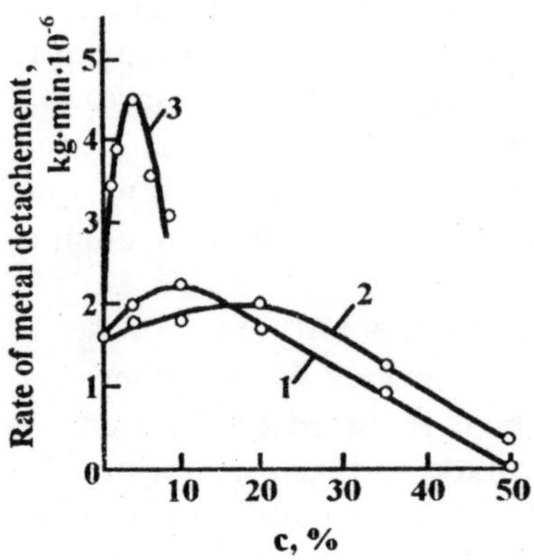

Figure 3.355. Dependence of the iron detachement rate on the content of polyisobutylene [959]: 1) $\bar{M}$ = 1.7×10^4; 2) $\bar{M}$ = 1.2×10^4; 3) $\bar{M}$ = 2×10^5.

penetrating in its superficial voids and acting as plasticizer it delays the subsequent tear. Exactly similar phenomena occur in the case of polymers mechano-degradation, i.e. polymer binding on the surfaces of metal particles of the colloidal size, detached from the reaction walls or milling bodies. As it was discussed in the paragraph 3.8.1.1.4, the binding is realised by co-ordinative bonds metal-polymer. This fact was postulated and new reactions of mechano-chemical complexation were achieved. The presence of some electrolytes complicates the wear mechanism, the electrochemical processes overlapping in this case, especially the formation of galvanic pairs.

In the case of polymer-polymer friction, the transfer from the lower polar polymer to the more polar one takes place [971], the kinetic of this process depending both on the molecular and supramolecular structure of the two polymers and the friction regime (speed, pressure, orientation direction). Finally, the all events are controlled by the free radicals formation, grafting to surfaces, and subsequent transformations. On its turn, the formation of free radicals is governed by temperature [975]. Thus, the friction of an abrasive polymer at –196 °C releases a radicals concentration of about $6.24\cdot10^{17}$ radicals/Kg, with respect to the volume, or $4.42\cdot10^{17}$ radicals/m^2, with respect to the surface, while at 20 °C the corresponding values are $1.08\cdot10^{19}$ radicals/Kg and $0.155\cdot10^{17}$ radicals/m^2, respectively. Therefore, with increasing the temperature the radicals mobility increases that favours their migration into polymer volume.

The fatigue by wear is a characteristic phenomenon of the friction of elastomers with a counter body with sharpened excrescences, when the detachment of polymer particles occurs due to of multiple deformations of its surface, leading to their fatigue by tear.

For a quantitative description of tear process, the Jurkov's equation was used, within which the relaxation time, τ, was replaced with the tear rate, I, τ_0 with I_0 – the value of unitary tear, and the constants γ and σ with λ' and μ, respectively, where: λ' – constant; P – specific friction pressure; and μ – friction coefficient in the friction zone. In this case, the durability equation can be written as:

$$I = I_0 \exp\left(-\frac{U_0 - \lambda'\mu P}{RT}\right) \tag{3.181}$$

accepting that μ is slightly affected by the friction conditions and $\lambda = \lambda'\mu$, the equation (3.181) becomes:

$$I = I_0 \exp\left(-\frac{U_0 - \lambda P}{RT}\right) \tag{3.182}$$

It is known that for abrasive tear $P = P_r$. To the level of contact surface between the bodies in abrasive friction this value is enough high, so that the particles detachment take place at each act of mechanical solicitation for which $\lambda P_r > U_0$. In the case of plasticized poly(methyl methacrylate) with E = 700 MPa and when $\lambda P_r >$ 400 kJ/mole, which was well higher than U_0 = 200 kJ/mole, the values $P_r >$ 27 MPa and λ = 0.3 mole/kg were obtained.

When $\lambda P_r > U_0$ the particles fracture is practically non-activated "without the energetic barrier"; it represent a tearing process, occurring into a single step, whose intensity is determined only by pressure and that is independently by temperature.

In the case of fatigue occurring by thermal-activated tear, the tear intensity dependence with temperature $\ln I = f(1/T)$ gives a straight-line [977]. As Figure 3.356 shows, the straight lines family tends toward a pole. The temperature corresponding to this pole is just the temperature at which the solicited material completely looses its specific properties, which coincides with the thermoplastic polymers melting temperature, or with crosslinked ones thermal decomposition, Table 3.122. Figures 3.357 and 3.358 illustrate the tear dependence by temperature, pressure and the type of counter body.

In the case of the mixed process of wear fatigue and abrasive wear the smaller is the activation energy the higher is the contribution of abrasive components.

In the case of somne rigid polymers worn by friction the influence of the nature of the counter-body on the mechano-degradation intensity is shown by the data collected in Table 3.123 [978,

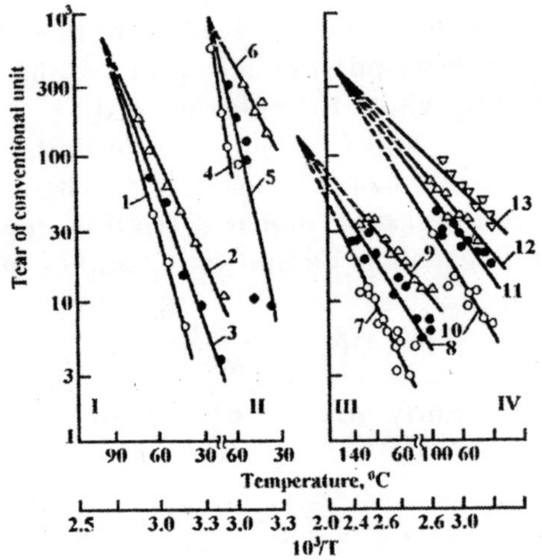

Figure 3.356. Dependence of wear of some plastic materials by temperature [959]: I) plasticized PVC; II) plasticized PMMA; III) pressing powder K-18-2 based on phenolformaldehyde resin; IV) epoxy resin of high-thermal stability. Applied load: 3, 9) 13 MPa; 8, 12) 0.13 MPa; 2, 11) 0.10 MPa; 7, 10) 0.05 MPa; 1) 0.05 MPa; 6) 0.003 MPa; 5) 0,015; 4) 0006 MPa.

Table 3.122. The values of critical temperatures [959]

Polymer	Temperature, °C		
	Corresponding to polymer	flowing	decomposition
Plasticized poly(methyl methacrilate)	75	70	-
Plasticized poly(methyl methacrilate)	105	90	-
Low density polyethylene	110	115	-
Polyurethane	160	-	160
Epoxy polymer	450	-	300
Phenoplaste K-18-2	520	-	400

Table 3.123. Influence of counterbody nature on the intensity of mechanochemical degradation by friction of certain rigid polymers [978]

Polymer	Characteristic viscosity			
	Initial	After friction with		
		Fillings	Abrasive paper	File's teeth
Poly(methyl methacrylate)	3.0	-	2.2	1.2
Polycarbonate	0.6	0.58	0.50	0.55
Polystyrene	0.8	-	3.0	0.5

979].

The fundamental modification of polymer surface chemical nature that suffered friction was experimentally established by NMR and IR spectroscopy. Thus, in the case of polyethylene, natural rub-

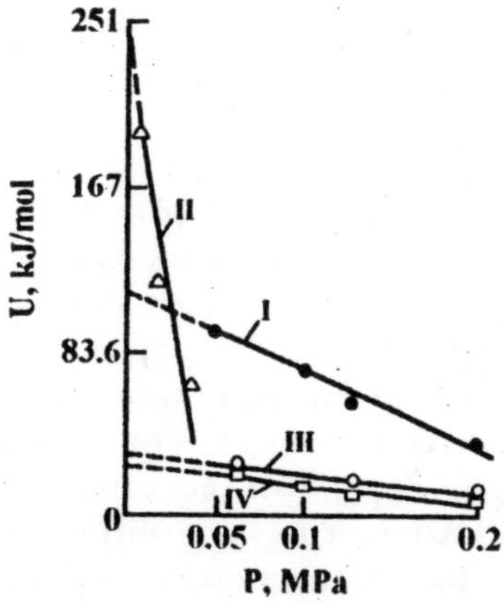

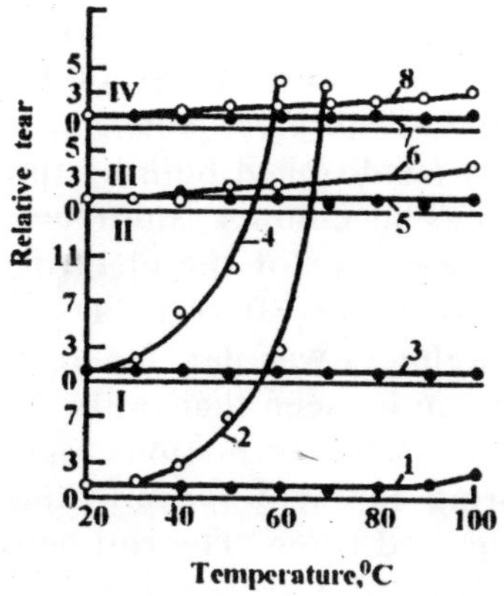

Figure 3.357. Influence of friction pressure on the energy values (conventional notations, see figure 3.356) [959].

Figure 3.358. Dependence of the value of the wear by friction by the nature of counter body and temperature (conventional names, see Figure 354) [959]: 1, 3, 5, 7) film; 2, 4, 6, 8) network.

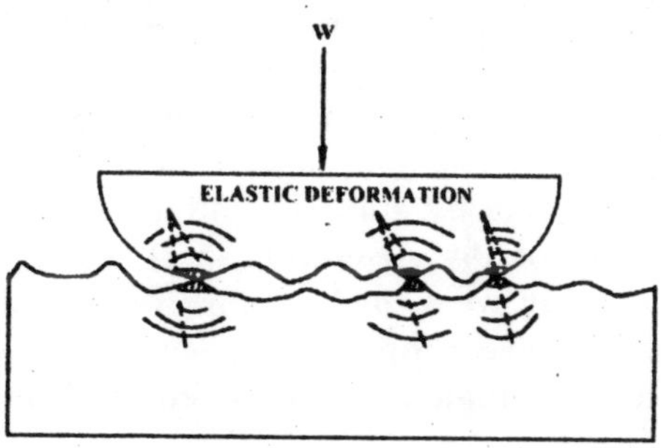

Figure 3.359. Model of deformations at the contact point of asperities [983].

ber, and butadiene-styrene rubber the formation of functional groups was proved. These groups characterise the oxidative destruction, such as –OH, and –COOH enhance the surface polarity. In addition, the decay of the average molecular weight and of chains orientation degree in the polymer superficial layers was found [980, 981].

The friction of the polymer surface induces not only changes in their superficial layer, but, as result of mechano-destruction, it is accompanied by the elimination of some low molecular weight compounds, such as: monomers, peroxy derivatives, etc. The particularities of this process are mainly related to the polymer and environmental medium nature [982].

Interesting data concerning the change of polymer structure at supramolecular – morphological level have been obtained by filamentary fibres, poly(vinyl alcohol), solicitation by friction with a standard abrasive surface. Starting from the idea that the friction is conditioned both by the mechanical action between the two surfaces in contact (their geometry, area, and exerted pressure) and the nature of the electrostatic charge, developed by surfaces friction, a suggestive model of deformations in the contact points of roughness was developed, Figure 3.359 [983].

In can be seen that in the junction points, at the same time with the specific deformations and/or elasto-viscous ones, fractures by shearing can also appear. Fibres resistance depend to the value of the applied force. The following particularities of the tear fatigued SOLVRON-SX thread were evidenced by the electron micrography, Figure 3.360:

- the filament surface is strongly modified, showing transversal striations of different profiles, with deepnesses and even small craters. This relief is caused not only by material fatigue but also by weight looses;
- the number of irregularities and their size increase on the end stages of solicitation by abrasion;
- the fracture morphology is of lower complexity as compared to other cases of mechanical solicitation.

The mechanical fracture by abrasive solicitation is mainly caused by the higher rigidity of the structural shell, strongly oriented, of filaments; this one being weakened by abrasive frictions accelerates the scissions and fracture propagation, which are shifted toward filaments core with an enhanced speed even for lower stresses.

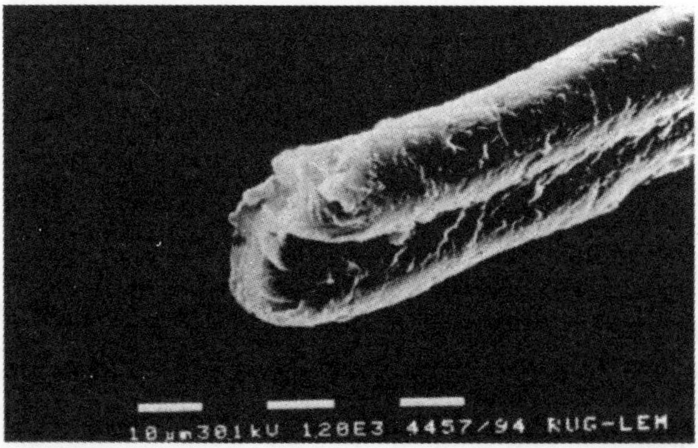

Figure 3.360. SEM of SOLVRON-SX filament fracture by abrasion [877, 984].

3.9. MECHANO-CHEMICAL REACTIONS OF POLYMERS IN DILUTED SOLUTIONS

3.9.1. Molecular asymmetry - the main reason of polymers mechano-degradation in diluted solutions

The systems consisting of a macromolecular component and others composed by low molecular weight molecules are of peculiar theoretical and practical importance. The diluted solutions are especially investigated, since the description of different properties of macromolecules could be performed only on this type of models. In diluted solutions the interactions between the macromolecules are practically cancelled. In this way, the determination of the structural particularities of polymer chains (shape, dimensions, and molecular weight) as well as of thermodynamic characteristics of polymer solutions became possible.

The behaviour of diluted solutions is related to the relation between the viscosity and the chain characteristics (structure, configuration, conformation, etc). Usually, the polymer solutions are treated as two-phase systems, consisting from mechanical elements, the macromolecules, immersed into a continuous media, the solvent. For long time, it was considered that the solvent acts to the polymer macromolecules in the same manner in which a fluid exerts forces about a small particle suspended in it. However, the extension of this model to the polymer solution is not adequate since, the ratio between the dimensions of macromolecules and those of solvent molecules essentially differs by that between the dimensions of a solid immersed particle and solvent molecules. On the other side, the flexible macromolecules, randomly coiled, can not be assimilated with the solid particles and therefore the typical relations applied to solid suspensions in liquids can not be used in this case.

One of the veridical quantitative evaluation about the viscosity of polymer solutions [985–988] belongs to P. Debye [989]. He enounced the hypothesis of the isolated macromolecules existence in a domain of simple shearing, Figure 3.361, and considered a range of speed that may be assimilated with the fluid speed $U = (\dot{\gamma} x_2 0, 0)$ with respect to macromolecule weight centre. By fixing the origin of co-ordinate axis in this centre, some simplified hypothesis can be considered, namely: 1) macromolecule structural units do not interact each another and do not influence the liquid movement. Such macromolecules have been called "free-draining coil"; 2) macromolecule's inertia is negligible; 3) when a structural unit is displaced with respect to the solvent, a delay appears, due to the friction that is proportional to the relative speed.

During the fluid movement the weight centre of macromolecule suffers a simple translation, while the flow in the far-off points exerts a moment on the macromolecular coil, which will determine its rotation. In these conditions, the value of angular rate, ω, is a half of shear rate:

$$\omega=\frac{1}{2}\dot{\gamma} \tag{3.183}$$

The relative speed of structural unit of co-ordinates (x_1, x_2) due to the rotation with respect to its weight centre may be expressed as:

$$U=\left(\omega x_{2.}-\omega x_{1.0}\right) \tag{3.184}$$

and with respect of the solvent as:

$$U_r=U-u=\frac{1}{2}\dot{\gamma}\left(x_{2.}-x_{1.0}\right) \tag{3.185}$$

The retarding force given by any structural unit is equal to:

$$F=\xi(U-u) \tag{3.186}$$

where: ξ - the friction coefficient of the structural units.

Using the relation that allows the calculation of dissipated energy on the whole macromolecule, its effective radius can be defined:

$$R_e^2=\frac{5}{2N}\sum_{i=1}^{N}\left(x_1^2-x_2^2\right)\cdot i \tag{3.187}$$

Analogous with the equation which describes the energy dissipation into a viscous homogeneous fluid during the simple shearing, the increase of viscosity due to the macromolecules presence can be calculated with the following relationship:

$$\eta_p=\phi/\dot{\gamma}^2 \tag{3.188}$$

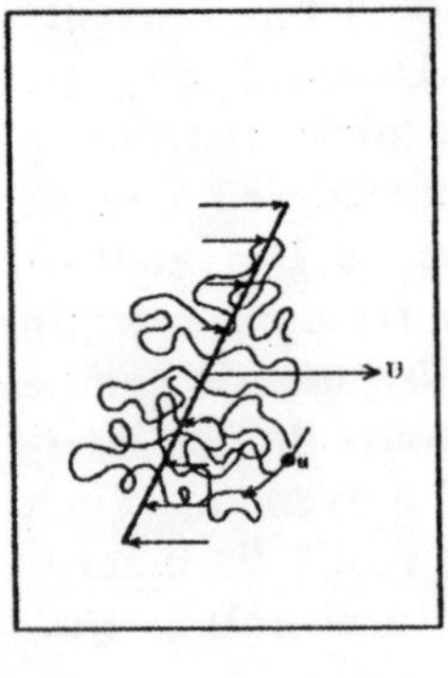

Figure 3.361. Rotation of a macromolecular chain around of its weight centre into a domain of simple shear [990].

where: $\phi = \frac{1}{4}\upsilon\xi\dot{\gamma}^2\sum_{i=1}^{N}\left(x_1^2 - x_2^2\right)\cdot i$, representing the volumetric speed of energy dissipation for n macromolecules per volume unit.

The solution viscosity is given by the sum of solvent viscosity, η_s, and the increment of polymer viscosity, η_p.

Starting from the relation describing the intrinsic viscosity:

$$[\eta] = \lim_{c \to 0} \frac{\eta - \eta_s}{c\eta_s} \qquad (3.189)$$

where c is the polymer concentration, and making all the replacements according to *P*. Debye's theory, equation (188) becomes:

$$[\eta] = N_0 \xi R_e^2 / 10\eta_s m \qquad (3.190)$$

where N_0 is Avogadro's number.

For a linear polymer, with enough long chains, R_e^2 can be considered proportional to the molecular weight, therefore:

$$[\eta] = k_m M \qquad (3.191)$$

In equation (3.191), k_m is related to the chemical nature of structural units, solvent viscosity, and friction coefficient, ξ. The chain configuration should influence the last parameter, in the measure within which it determines the values of ξ and R_e^2, respectively.

According to Bueche's theory, the mathematical function that correlates the viscosity, η_0, at the shear rate equal to 0, and the average molecular weight, $\overline{M_c}$, when the last one overtakes a certain critical weight presents a discontinuity. This relation was firstly established for polymer melts [989] and subsequently assimilated for the polymer solutions [991]. However, according to the data of Gupta and Forsman [992] and R.S. Porter and J.F.Johnson [993] this discontinuity can be neglected. For a certain critical value of the molecular weight, $\overline{M_c}$, the following relations may be written:

$$\eta_0 = k_1 \overline{M_w} \qquad \text{for} \qquad \overline{M_w} < \overline{M_c}$$

and

$$\eta_0 = k_2 \overline{M_w}^{3.4} \qquad \text{for} \qquad \overline{M_w} < \overline{M_c}$$

based on the above two equations, the following general equation is used:

$$\eta_0 = k_1 \overline{M_w} + k_2 \overline{M_w}^{3.4} \tag{3.192}$$

which describes the experimental data better than any of the two previously given previous equations. Starting from the experimental observation that for high molecular weights, the increase of viscosity occurs faster than with the power equal to 3.4, by using the probabilities theory, a more general equation is obtained:

$$\eta_0 = a_0 + a_1 \overline{M_w} + a_2 \overline{M_w}^{3} + a_3 \overline{M_w}^{5} \tag{3.193}$$

which can be reduced, with a sufficiently good approximation, to the three terms:

$$\eta_0 = a_1 \overline{M_w} + a_2 \overline{M_w}^{3} + a_3 \overline{M_w}^{5} \tag{3.194}$$

Excepting the case of very low concentrations, in solution, the interactions between the polymer macromolecules determine the variation of reduced viscosity η_{sp}/c with polymer concentration. The characteristic value of the isolated macromolecule in solution is obtained by the extrapolation of η_{sp}/c to zero polymer concentration. In this case the following equations are valuable:

$$\eta_{sp}/c = [\eta] + k'[\eta]^2 c \qquad \text{Huggins's eqn. (3.195)}$$

$$\left[\ln(\eta/\eta_0)\right]/c = [\eta] + \beta[\eta]^2 c \qquad \text{Mead and Foss's eqn. (3.196)}$$

$$\log\left(\frac{\eta_{sp}}{c}\right) = \log[\eta] + k'[\eta]^2 c \tag{3.197}$$

The graphical representation of η_{sp}/c *vs.* c, gives straight lines, whenever the polymer concentration is not high.

The macromolecules existence into a shear domain causes significant modifications of the diluted solutions viscosity. A detailed analysis concerning the dynamic behaviour of macromolecule into a shear domain was given by Bueche [995]. The author started from Debye's representation of a macromolecule that freely rotates into a domain of simple shear, Figure 3.362, with a frequency equal to a half of shear rate. In this case, the structural units are supposed to the stretching forces action, *AA′*, which tend to line up the chain; consequently this one will suffer a compression on the direction BB'. Therefore, during the rotation around the macromolecule weight centre, a chain segment suffers periodical stretchings and

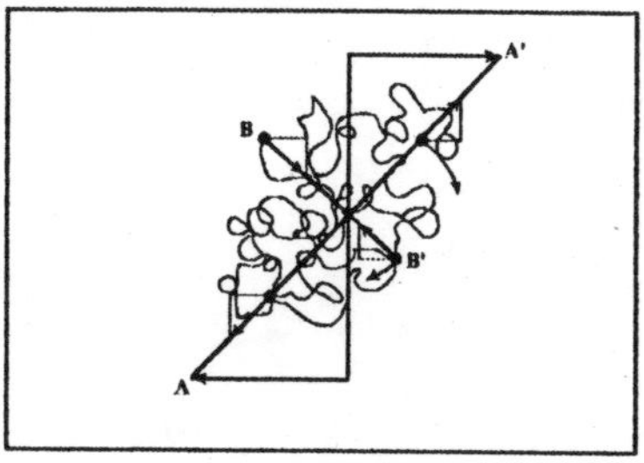

Figure 3.362. Action of shear gradient on a macromolecular chain in solution [995].

compressions on the directions AA' and BB', respectively.

It was proved that for molecular weights above 400,000 the shear effects becomes important even to the usual concentrations for the determination of intrinsic viscosity, [η], at the specific shear gradients of capillary viscometers. The shear effects are higher in the case of polyelectrolytes solutions in good solvents and also in solutions containing rigid macromolecules.

The viscosity measurement of a polymer solution into an increasing range of shear efforts firstly evidences its negligible variation with the shear rate (Zone I). At higher shear rates, the viscosity decays with the increase of shear rate (Zone II). In this domains the solution presents a non-newtonian behaviour. Finally, at very high shear forces, the viscosity becomes once again independent by the shear rate (Zone III), Figure 3.363, [996]. The third zone is currently called "the second Newtonian viscosity". The shear rate at which the transition from a zone to other occurs depends by the polymer molecular weight, its concentration, temperature, and polymer nature.

In stationary flow regime, the non-Newtonian behaviour of polymer solutions and melts is directly related to the two structural characteristics: 1) asymmetrical shape of macromolecules; and 2) the dimensions of flowing elements. Due to the very high length of macromolecules with respect to its diameter, during the stationary flowing their reorientation occurs. Thus, a rigid macromolecule, having its extremities in two zones with different flow rates, will be oriented with the long axis parallel to the flow direction without suffering any deformation. The complete orientation will be achieved whenever the shear rate becomes enough high in order to hide the effects of brownian movements. In this way the Newtonian behaviour is explained in the range of very small shear forces, when dominant is the effect of disordered movement that do not allow the

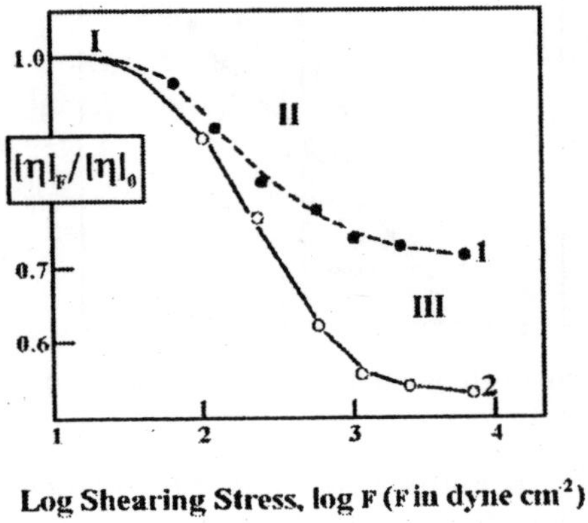

Figure 3.363. Variation of the ratio between the intrinsic viscosity at a shear stress noted F and the intrinsic viscosity at shear gradient equal to zero with the logarithm of shear stress in the case of polystyrene with cu $M_w = 282 \times 10^6$: 1) in tetraline; 2) toluene [996].

macromolecules orientation. At the other extreme, very high shear gradients, the Brownian movement is insignificant as compared to the orientation rate due to the shearing. In this case, the shear gradient does not influence the orientation degree, the behaviour being Newtonian one.

It can be observed that the two Newtonian viscosities differs each another. With other words, the orientation is a required but not sufficient condition for the achievement of non newtonian behaviour, typical for intermediate shear gradients. The mechanism of moment transfer from the fluid rapid elements toward those lent ones is changed. Consequently, the viscosity sensible diminishes, tending in some cases to those of solvent, constituted by small molecules.

The change of macromolecules conformation, under the action of shear forces, makes as the viscosity of solutions containing "infinite long" molecule to decrease with the increase of shear rate, passing through a minimum, and, finally to increase again. However, when the macromolecules are not enough long, only the decay of viscosity is recorded. The shear effect on the intrinsec viscosity is huge in the case of very high molecular weights. Thus, cellulose nitrate, $\overline{M_w} = 1 \cdot 10^6 - 2 \cdot 10^6$, in acetone, poly(methyl metacrylate), $\overline{M_w} = 16 \cdot 10^6$, in benzene, diminish at list at half their values of $[\eta]$ to the shear gradients varying in the range from 15,000 to 30,000 s^{-1}, Figure 3.364, [997]. Graphical representation η_{sp}/c vs. c, for different shear rates, proved that the reduced viscosity at $c \to 0$ diminishes with the increase of shear rate [998, 999].

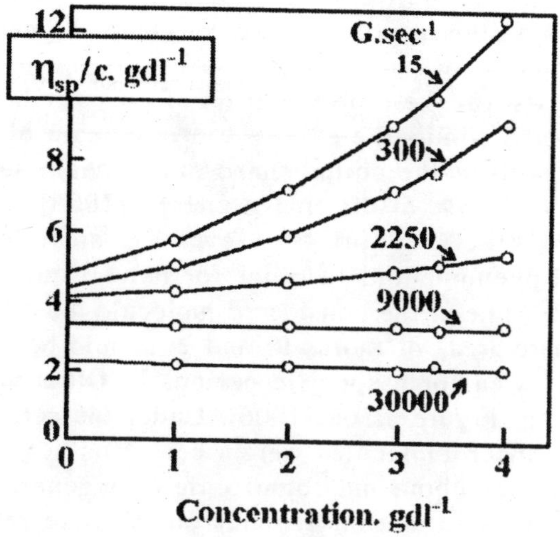

Figure 3.364. Variation of the reduced specific viscosity of cellulose nitrate ($\bar{M}$ = 2.6 × 10^6) in acetone, at 20 °C with polymer concentration for different shear rates [997].

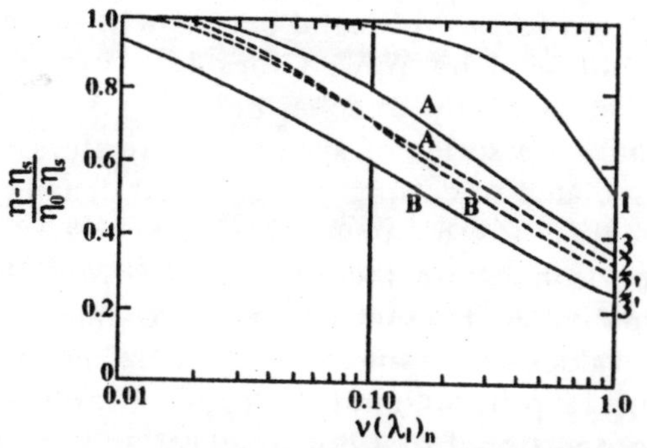

Figure 3.365. Verification of polydispersity theory on the polyisobutylene model: 1) according to F. Bueche; 2, 2') according to N. Yamaguki [998]; 3, 3') R.S. Porter [999].

The shear effects, depending in a great measure by polymer molecular weight, are inevitably influenced by polymer molecular polydispersity, Figure 3.365, [997, 998].

The increase of the polymer concentration in solution leads to the increase of macromolecule interpenetration. In this situation, polymer solution behaves to retain the polymer macromolecules in position by temporary bending.

At very high concentrations the macromolecule agglomeration occurs in clusters [1000] or temporary networks. This happens in both concentrated solutions or in melts [1000–1005]. The assessment concerning the macromolecules or chain segments rotation being few probable in concentrated solutions or in melts, another model was used for the representation of polymer behaviour in this case. The model admits the existence of a domain of non newtonian viscosity as well as the coiled shape of the chain segments having as result the increase of dissipated energy [1006].

From the kinetic point of view, the coiling process is a complicated phenomenon. In order for it to occur, two conditions are essential, namely: the considered molecule must be placed in the "coiling sphere area" of radius R; and 2) should be able to maintain itself in this area for a specific period, λ. Otherwise, the coiling does not occur, Figure 3.366 [1006]. Under the action of shear rate gradient, the macromolecules weight centre passes in the attraction sphere of the neighbour macromolecule thus generating knots. The duration of knots formation depends on the time required for covering the distance AB which decreases with the increase of the shear rate gradient. The increase of the shear rate gradient determines the decrease of the chain coiling density, since the chains that remain in the coiling area for a prolonged period will be less numerous.

The result obtained by W.W. Graessley, $\eta/\eta_0 = F(\lambda\dot{\gamma})$, is very similar with that given by Bueche, but the viscosity at zero shear rate will depend by a series of specific parameters such as: friction coefficient associated to coiling process, coil density at equilibrium, and the coil radius. The parameter λ that expresses the coiling duration is proportional with the value of λ_1 from Bueche's theory, the proportionality coefficient being approximately equal to unit. For very high values of parameter $\lambda\dot{\gamma}$, according to W.W. Graessley the ratio η/η_0 is proportional to $(\chi\dot{\gamma})^{-3/4}$ and according to F. Bueche it is proportional to $(\chi\dot{\gamma})^{-1}$.

The calculation of the elastic energy associated to the coiling process leads to the following relation of the normal tension coefficients, $\Psi_{1,2}$:

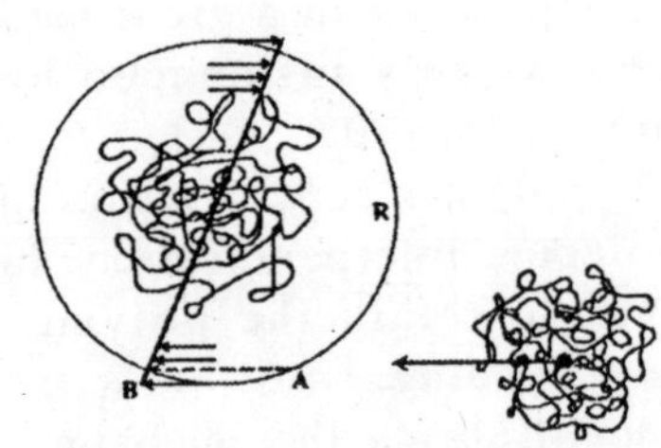

Figure 3.366. Representation of molecular coiling according to Groessley [1006].

$$\Psi_{1,2} = \Psi_{1,2}^{0} (\eta / \eta_0)^2 \tag{3.198}$$

which shows that at the same time with the increase of shear rate $\Psi_{1,2}$ decreases more quickly that η.

According to M.C. Williams, in concentrated solutions intermolecular interactions are developed, whose intensity derive from both the potential function and chain segments distribution. Without specifying the nature of intermolecular forces, the author started from the theory of the polymer solutions equilibrium, elaborated by M. Fixman [1008], and proposed the following relation:

$$(\eta - \eta_s)/(\eta_0 - \eta_s) = F(\lambda\dot{\gamma}) \tag{3.199}$$

where: λ – scalar parameter, having the time dimensions, which expresses the efficiency of shear rate gradient to the elongation/orientation of a macromolecule to the flow direction, i.e. it measures the "narrowing" of segments function of distribution. The relation $\eta_r = F(\lambda\dot{\gamma})$ is qualitatively similar those from other theories, but at very high shear gradients more rapidly decays.

Starting from the model proposed by M.C. Williams, R.A. Schultz expressed the dependence of η_0 and λ with polymer concentration and molecular weight [1009].

$$\eta_0 = \left(c \cdot M^{0.025}\right)^2 D_{(c)}^{1.15} \tag{3.200}$$

$$\lambda = (\eta_0 - \eta_s)/c^2 T F_{(c)} \tag{3.201}$$

Many concentrated solutions of polymers are characterised by a complex rheological behaviour, being classified in systems with instantaneous reversibility, or quasi-instantaneous, systems with retarded reversibility [1009, 1010]. The last ones present so-called "the memory" of the supported shear rate gradients; thus the structure of the hydrodynamic units is influenced by shear rate gradients, tending to a limit. The systems presenting a quasi-instantaneous reversibility are characterised by short periods of structurisation which are inferior to the measurement accuracy.

On this base, the phenomena that appear during laminated flow, Figure 3.367, [1009] were explained. By applying a very low concentration gradient, the initial agglomerates (a) are deformed, until the attainment of a threshold, the behaviour being non newtonian one (b); follows the detachment of big agglomerates/associates (c), and after that the (d) state whose corresponds the inflexion point, (γ_0, τ_0), of the rheologic curve, Figure 3.368. At high shear rate

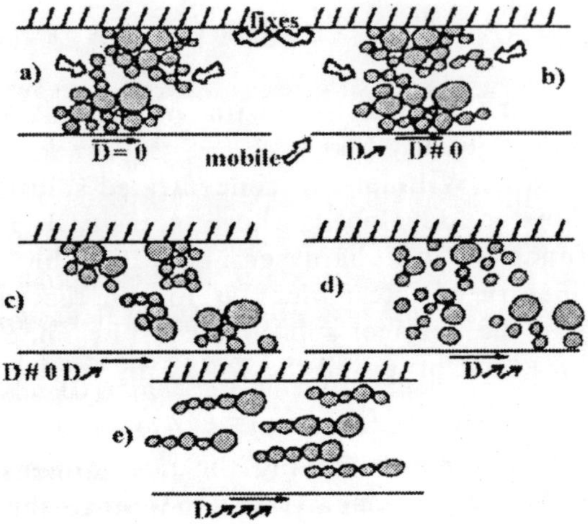

Figure 3.367. Schematic illustration of the behaviour of a concentrated solution (starch in water) supposed to the action of a increasing shear gradient [1009]: (a) repose; (b) intermediary state; (c) unbinding of big associates in smaller ones; (d) complete desassociation; e) chains alignment.

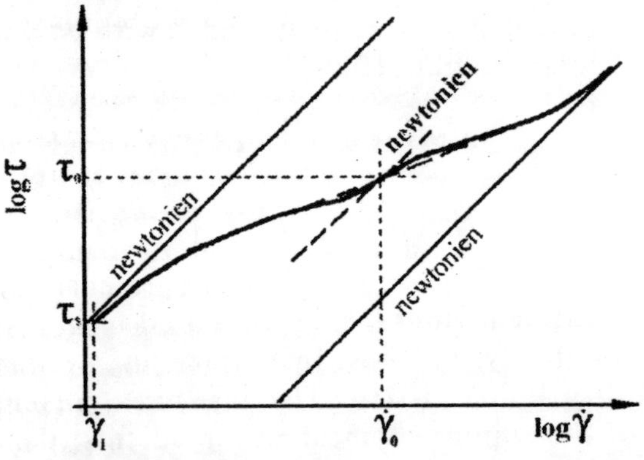

Figure 3.368. R.A. Schultz-rheogram [1009].

gradients (e) the hydrodynamic units line up and the behaviour becomes once again newtonian one.

Taking into account the thermodynamic and hydrodynamic interactions of the macromolecular chains in solution as well as the solvent displacement, the calculation of concentrated solutions viscosity, containing polymers with flexible chains, was performed as well as the calculation of "apparent energy" of flowing activation [1011–1014].

3.9.2. Hydrodynamic degradation by shearing through capillaries of polymer diluted solutions

The forced flow of polymer solutions or melts through capillaries or small orifices has as effect the appearance of shear forces, able to promote the scission of the main chain covalent bonds, causing the mechano-chemical degradation, which has as the first result the decrease of the polymer molecular weight.

H. Staudinger and H. Heuer were the first to observe that a significant decrease of viscosity takes place during the forced passing of a polystyrene solution through a platinum capillary. The result was ascribed to the scission of polymer chains under the shear conditions [1015]. The obtained results have also been verified on the cellulose nitrate in different solvents [1016].

J. Frenckel was the first scientist which tried to explain the phenomena that occur in the flowing conditions of polymer solutions [1017]. He assumed that the chemical bonds of the main chain, especially those located closely to the middle of chain, suffer stretching in the flow direction, while the rest of chain segments adopt a coiled shape. The magnitude of this effect increases with the chains length square, which at the reaching of a critical value of shear rate gradient are split. He also realised a direct correlation of the shear rate gradient with the degree of polymerisation of the stressed polymer. For instance, for splitting of a macromolecule with $\overline{DP} \sim 10^3$, the critical gradients must be of about 10^5 s^{-1}.

A lot of subsequent experimental works come to consider thoroughly, confirm, and develop these results. Thus, on one side J.W. Morish and R. Schnurman [1017 and 1018], and R. Kiyama and H. Kinoshita [1019] on the other side, increasing the value of shear forces succeeded to establish a clear correlation of the critical gradient with the polymerisation degree.

The researchers from U.S. National Bureau of Standards exploited this method, applying it to the investigation of poly(methyl methacrylate), polystyrene, and, especially, polyisobutene mechanodegradation. The shearing was realised using a consitometer having a capillary with d = 0.2 mm and the stainless steel disk thickness of 0.25 inch. The polymer solutions were forced the flow through capillary with variable pressures and passing numbers. It was found that the shear stress is a function of shear rate, temperature, initial polymer concentration and molecular weight [1020–1029]. The experimental results are correlated with some mechanical parameters, capillary geometry, especially the solution entering point being of first importance [1027]. Other works, of technical interest, deal

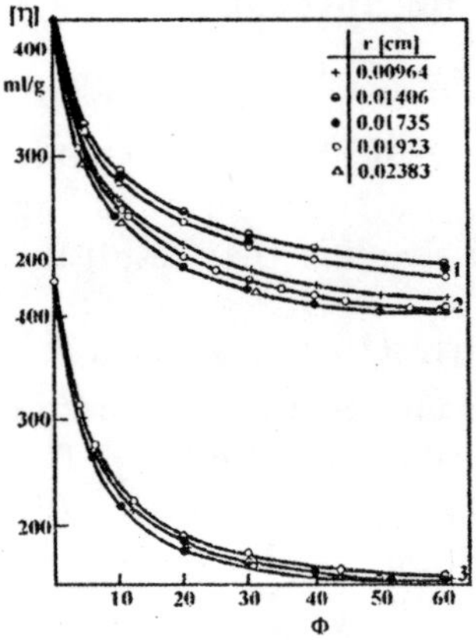

Figure 3.369. Influence of capillary geometry on the mechanochemical destruction by shearing of PMMA in toluene [1031]: 1) l = 4.0 cm; 2) 1 cm; 3) 0.25 cm; p = 20 kgf/cm^2.

with the polymer-based lubricants, supposed to shear solicitation, under the action of shear gradients of about 10^7 s^{-1} [1029, 1030].

A systematic study of the different factors influence, was initiated by H. Grohn and K. Krause; under the shear conditions through capillaries of poly(methyl methacrylate) solutions, they demonstrated the dependence of mechano-chemical degradation, evaluated by the number of split bonds, by the magnitude of mechanical forces applied (the pressure, number of passing, radius and length of capillary) and the solutions concentration, Figures 3.369 -3.372, [1031, 1032]. The results obtained by shearing polyamide solutions in *m*-cresol lead to the same conclusions concerning the effect of the mechanical regime on mechano-degradation [1033].

Mechanochemical degradation in solution of the dezoxiribo-nucleic acid, ADN, by capillaries shear, had as effect the transversal splitting of ADN macromolecule, with a double helix conformation, and not the scission of the secondary hydrogen bonds that assure the maintaining of this conformation in solution [1034–1037].

Both in the apparatus with capillary or in that with cylinder and piston the flow rates distribution is parabolic one, due to the non newtonian flowing. Under these conditions, only a small fraction of macromolecules will be at a given moment near the capillary wall where the shear rates are the smallest. For an accurate evaluation of the shearing of a polymer solution during its passing through a capillary or through the ring-shaped piston-cylinder space, the macromolecule average resident time in the apparatus was calculated and experimentally established. The calculations were made for a non-newtonian distribution of the flow rate. For the laminar flow this time is proportional to the average duration of

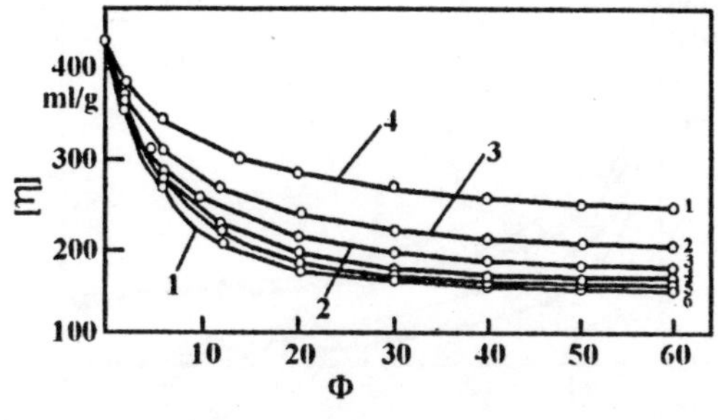

Figure 3.370. Influence of the capillary length and of number of passings through capillary on the mechanodegradation of PMMA toluenic solutions [1031]: 1) l = 8.0 cm; 2) 4.0 cm; 3) 2.0 cm; 4) 1.0; 5) 0.5 cm; 6 - 0.25 cm; p = 20 kgf/cm^2.

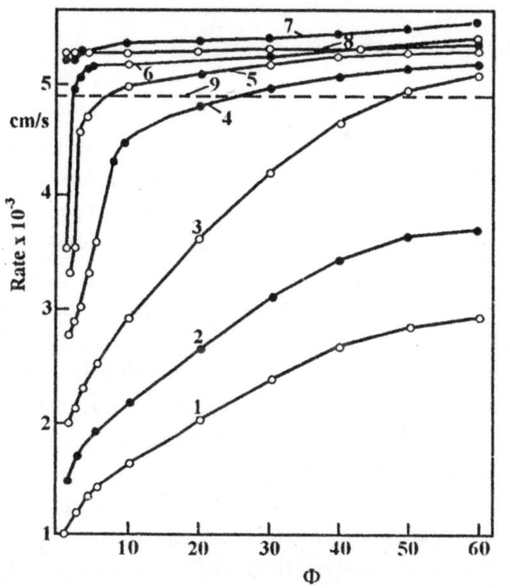

Figure 3.371. Modification of average flow speed with the number of passings at different concentrations [1032]: 1) 40 g/l; 2) 30; 3) 20; 4) 10; 5) 6; 6) 3.6; 7) 1.0: 8) 0.5; 9) 0.0.

macromolecules solicitation by shear.

Due to the parabolic distribution of the flowing rates, whenever the shear rates overtakes only slightly the critical value, the splitting points will be located in the proximity of the weight centre of the macromolecule and the destruction fragments should be almost perfect half parts of the initial chain. Such ideal behaviour was found to the shearing of T_2 ADN (dezoxiribonucleic acid from

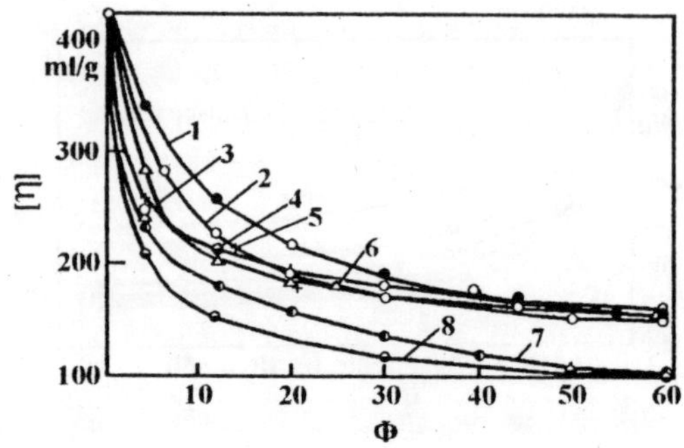

Figure 3.372. Dependence of intrinsic viscosity on the number of passes through a capillary at different concentrations [1032]: 1) 40 g/l; 2) 30; 3) 20; 4) 10; 5) 6; 6) 3.6; 7) 1.0; 8) 0.5 g/l.

bacteria T_2) solutions, in the conditions described by A.D. Hershey and E. Burgi [1038].

The analysis of mechano-destruction occurring in these conditions is complicated by two factors: 1) in a shear field, the macromolecule has not a perfectly stretched conformation, being randomly coiled [1039–1041]; and 2) for a given distribution of the average molecular weights the longer macromolecules will be dominantly split in the proximity of their centre, until reaching the limit dimensions, which can not be affected by the mechanical force. In this case, the longer macromolecules are split at the critical value of the developed forces, for the given mechanical regime, while other shorter ones are not affected by the applied forces. After destruction, the new molecular weight distribution is characterised by the presence of the longest macromolecules that are twice the length of the shortest ones. Evidently, the new distribution is narrower than the initial one which characterises the technical polymers.

In the case of mechanodestruction processes, under the conditions of polymer solutions flowing, the molecular weights distribution is an important characteristics and it constitutes one of the main reasons of the chemical bonds splitting.

If it is admitted that the shear force, expressed by $\eta\dot{\gamma}$ (where η - viscosity and $\dot{\gamma}$ - shear rate), is transmitted perpendicularly on the surface unit of the mechanically stressed solution and $\eta_0\dot{\gamma}$ (η_0 - solvent viscosity) represents a part of the exerted/transmitted force by the solvent, the difference $(\eta-\eta_0)\dot{\gamma}$ will express the force exerted by macromolecules passing through the surface unit. Knowing the average number of macromolecules that cross the surface unit, the exerted force by each molecule can be calculated.

Accepting that the macromolecules number crossing the surface unit is the same with the their number per volume unit, n, and noting by z the dimension of the macromolecular chain projection on the flow direction, the average shear force, called "average friction coefficient" can be calculated using the following relation [1035]:

$$\langle f \rangle_{med} = \frac{\dot{\gamma}(\eta - \eta_0)}{n \cdot z} \tag{3.202}$$

Taking into account that this force is not uniformly distributed on the chain, but after a parabolic function with its maximum on the chain middle, the calculation of average force should be made with the relationship:

$$\langle f \rangle_{med} = \frac{4\eta_{sp}\eta_0 M}{\pi r^3 cNf} \cdot \frac{dv}{dt} \tag{3.203}$$

where: M – molecular weight of the initial polymer; c – solution concentration; N – Avogadro's number. By replacing in this equation the geometrical dimensions of the cylinder-piston apparatus, we obtain:

$$\langle f \rangle_{med} = \frac{M \cdot \phi \cdot (b-a) \cdot \eta_{sp}}{2\pi \cdot L \cdot a^2 \cdot c \cdot N \cdot z \cdot \eta_{red}} \tag{3.204}$$

where: b – cylinder radius; a – piston radius; L – cylinder lenght; $\dot{\gamma} = \frac{v}{a} A$; and the developed force has the following value

$\phi = \frac{2\pi\eta L v a}{b-a}(A+1)$; $A = \frac{b^2 - a^2}{(b^2 + a^2)\ln\frac{b}{a} - (b^2 - a^2)}$; v and η are the piston speed and solution viscosity, respectively. Under these conditions, the sear rate becomes: $\dot{\gamma} = \frac{\phi(b-a)A}{1\pi\eta La^2(A+1)}$ and for a very small difference between a and b, $\frac{a}{b} \cong 1$ and $A \cong A+1$, the expression of shear rate is: $\dot{\gamma} = \frac{\phi \cdot b}{2\pi \cdot \eta \cdot L \cdot a^2}$.

It results that the chains splitting represent a chemical process

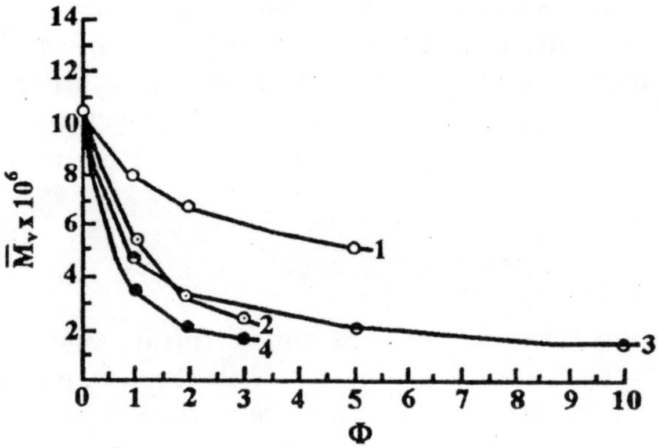

Figure 3.373. Decay of the average molecular weight with the number of passings through capillary into an apparatus with controlled shear [1034]: 1) polystyrene in toluene, capillary 200 psi (14 kgf/cm^2); 2) polystyrene in toluene, through perforated disk; 3) polystyrene in methylnaphtalene into a cylinder-piston apparatus, pushing force 320 psi (22.5 kgf/cm^2); 4) polystyrene in methylnaphtalene into a cylinder-piston apparatus (piston with a different shape), pushing force 320 psi (22.5 kgf/cm^2).

for whose activation the mechanical energy is spend (Figure 3.373).

The value of critical force can be deduced from equation (3.202), making the following replacements: n with $\left(c/M_f\right)N$ for the final value of the molecular weight, M_f, shear rate with the shear force, and the final value of viscosity η_f, it results that the critical force to the wall is [1042]:

$$\overline{f}_{cr} = \frac{\left(\eta_f - \eta_s\right)}{\eta_f} \cdot \tau\omega \cdot \frac{M_f}{cNz} \tag{3.205}$$

where: N – Avogadro's number; c – polymer concentration, g/cm^3; $\tau\omega$ - friction force to the wall, dyne/cm^2.

In physical sense, the critical force represents the minimal force required to decrease the molecular weight until to M_f, by mechano-degradation for a given regime of solicitation. Taking an average value for z, the equation (3.205) becomes:

$$\overline{f}_{cr} = k \cdot y \tag{3.206}$$

where: $y = \frac{\left(\eta_f - \eta_s\right)}{\eta_f \cdot c} \cdot \tau\omega \cdot M_f$.

For the same polymer and solvent $\overline{f}_{cr}$ is univocally determined; parameter y can be considered constant and for very diluted it is expressed as:

$$y=\left(\frac{\eta_{sp}}{c}\right)\cdot\left(\frac{\eta_{sp}}{\eta_f}\right)\cdot\bar{M}_v\cdot f\cdot\tau\omega\approx k_1\bar{M}_v f^{n+1}\tau\omega \tag{3.207}$$

where: $\bar{M}_v$ is the final viscosimetric average molecular weight; the intrinsic viscosity $[\eta]_f\cong\eta_{sp}/c\cong\frac{\eta_f-\eta_s}{\eta_s c}$; $[\eta]_f=k_1\bar{M}_v$; $\frac{\eta_f}{\eta_s}=\eta_n\to 1$ at very high dilutions.

The influence of the solution concentration on the mechano-degradation was studied for two domains, with extreme values of concentration, i.e. very low and very high, respectively [1031–1032].

In the case of concentrated solutions, the mechanical solicitation dominantly causes the detachment of intercatenary interactions; the decrease of average molecular weight is insignificant one. Contrary, in the case of very diluted solutions, the splitting of chemical bonds is the dominant process. For diluted solutions, the number of split bonds per macromolecule, for each passing through capillary is constant, while for concentrated solutions it decreases and tends to a limit by increasing the number of passings.

3.9.3. Mechano-chemical degradation of polymer solutions by high speed stirring

The stirring effect on the chemical reactions occurrence is well known, the hydrodynamic regime being decisive for many chemical processes, favouring the diffusion and mass transfer as well as thermal uniformity of the reaction systems. The mechanical energy released by stirring diminishes the energy of thermal activation of chemical reactions and whenever its value becomes comparable to the chemical bonds energy it can cause their splitting.

The mechano-chemical degradation of the polymer solutions under the intense stirring action (stirrer speed varying from 12,000 to 45,000 rot/min) determines the decrease of the average molecular weight by a chained radicalic mechanism [1042–1049]. In this case, the solution concentration and solvent quality, which governs the macromolecules conformation in solution, were found to be the fundamental factors that affect the process [1045–1050].

In poor solvents, the macromolecular chains are in the shape of compact coils, with a high packing degree, at limit tending to spherical shape. Thus formed macromolecular "clew" is individualised and does not allow intrahydrodynamic interactions. Increasing the solvent quality with respect to the polymer the coils are elongated, taking the ellipsoidal shape and braking the flow.

In very diluted solutions, the intermolecular interactions are cancelled and the macromolecules freely move. Under stirring conditions, with constant speed, the constant intensity rate gradients are developed. When the macromolecular coils are into a good solvent their interactions with solvent molecules are minimal and the shear forces will tend to adopt an entangled shape, the mechanical destruction becoming more efficient with the increase of solvent quality, Figure 3.374.

H. Grohn and G. Opitz have found that even into a good solvent but in the range of high polymer concentration, the destruction rate is strongly decreased while into a poor solvent it increases [1044]; similar results have been reported by K. Goto and H. Fujiwara [1049].

Under the conditions of hydrodynamic solicitation of concentrated solutions, in order to explain the polymer concentration effect on the chain splitting mechanism, Y. Minoura and co-workers stated from two hypotheses, namely: 1) the interaction between the polymer coil and solvent molecules; and 2) the interaction of coiling-uncoiling type are essential for mechano-degradation [1045 – 1049]. If the last one acts, the increase of concentration must favour the chains splitting, after an exponential law. This behaviou was not experimentally proved. K. Arai shown that the mechanochemical destruction is not caused by the coiling-uncoiling interac-

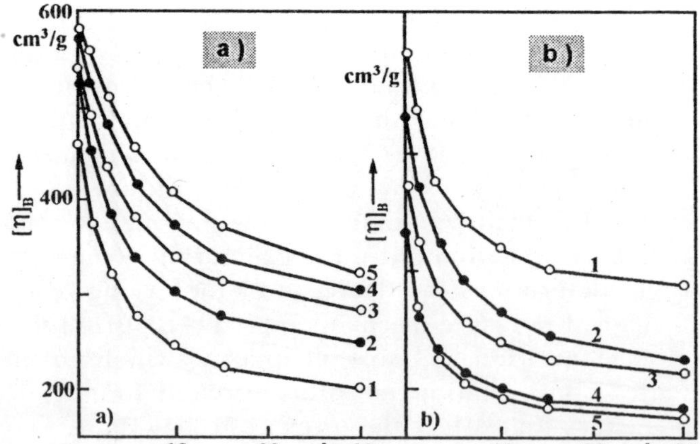

Figure 3.374. Influence of the polymer concentration on the mechano-chemical destruction of the PMMA solutions into a good solvent (a) and poor solvent (b) [1050]: 1) c = 1 g/l; 2) 10 g/l; 3) 20 g/l; 4) 35 g/l; 5) 50 g/l

tions, but the shear forces developed between the packed chains segments [1050]. The correlation of solution concentration with the viscosity decrease, as result of hydrodynamic shear, for a series of polymers, such as: poly(methyl methacrylate), polystyrene, polyisobutene, poly(ethylene oxide) is illustrated in Figure 3.375 and should be expressed as [1048]:

$$k = k_0 - A[\eta] \tag{3.208}$$

where: k_0 and A are constants that depend on the polymer nature.

One can see that according as the [η] diminishes, the degradation rate constant, k, tends to a constant value. The complete characterisation of the polymers behaviour during mechano-degradation by high speed stirring is possible by establishing the relation between k and [η] for the most different classes of polymers.

In the experimentally range of concentration investigated by Y. Minoura and co-workers, from 1 to 4% for poly(ethylene oxide), the interactions between the chains are enough strong. Whether under shear conditions, when the macromolecular coils suffer a stretching process, which supplementary intensifies the interactions between them, the fracture of chemical bonds should depend by these interactions and the rate constant is expected to increase. However, this result was not experimentally confirmed [1048, 1050]. On can conclude that even if the coiling-uncoiling type interactions exist, these ones are lesser important from the destruction and the factor that direct affect the process is the solvent quality. Under

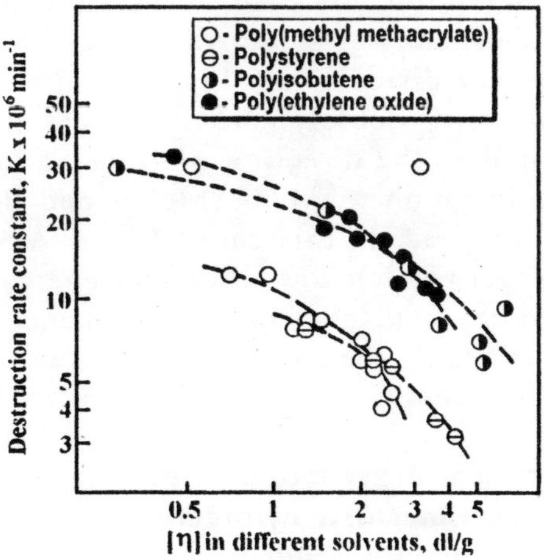

Figure 3.375. Dependence of the destruction rate constant by the intrinsic viscosity in various solvents for: 1) poly(methyl methacrylate); 2) polystyrene; 3) polyisobutylene; 4) poly(ethylene oxide) [1048].

shear conditions, in function of solvent nature the macromolecules adopt more or less coiled conformations. Under the action of developed hydrodynamic gradient, their stretching degree is very important for the mechano-degradation rate; the stretching degree determines the number of chain folds in the shear conditions. As is seen in Figure 3.376, the chain is lesser folded when it is stretched that than the coil is more packed. According to Minoura's model, the mechanical force developed under the conditions of high speed stirring more efficient acts on a strongly folded chain that on a stretched one [1048]. The mechano-degradation reaches maximal value whenever this force is maximal, therefore the force corresponding to the maximal shear rate gradient developed [1051].

According to R.E. Harrington, the mechano-degradation is initiated in the limit layers of the polymer solution, which soak the edge of stirrer blade and where under the action of the maximum gradient of the shear rate. A stretching force is developed and propagates along the macromolecular chains, causing the mechano-chemical destruction.

Starting from the R.E. Harrington representation and Minoura's model, the mechanism of hydrodynamic destruction can be interpreted for two distinct cases as follows: 1) whenever the macromolecular coil is not strongly compacted, therefore the polymer is into a good solvent, the maximal shear rate gradient will generate a force that meets a relatively small number of chains or chain segments with their axis disposed in the same direction to the direction of force sense. In this case, the biggest amount of mechanical energy is dissipated as sonic wave or as thermal effect, in inter- and intracatenary dislocations, Figure 3.376 a; and 2) when the macromolecular coil is densely packed, a great number of chain segments are oriented in the direction of force action and, consequently, the splitting number will be higher and the mechanical looses are well diminished. In this case, the conversion degree of mechanical energy conversion into chemical energy is improved, Figure 3.376 b. We can conclude that the mechanodegradation efficiency mainly depends by the solvent's characteristics that determines the more or less compacted shape of the macromolecular coil.

3.9.4. Mechano-chemical degradation of polymer solutions by laminar flow between concentric cylinders

The flow between concentric cylinders is a method that allowed the evaluation of polymers mechano-degradation, at low shear rates, under the conditions of strictly laminar flow, when a homogeneous

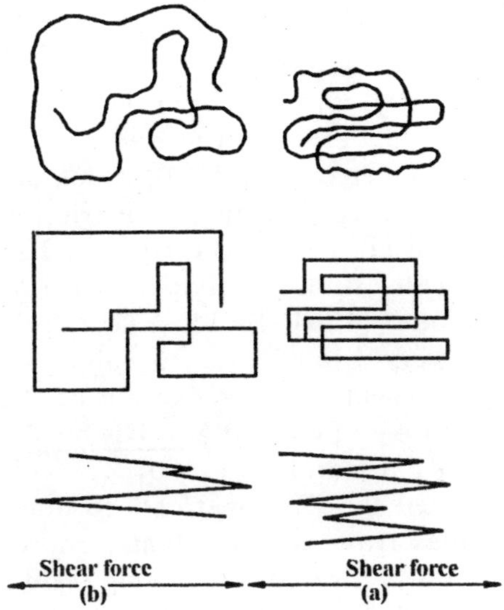

Figure 3.376. Models of coiled chains of the same length polymer chains: (a) stretched model; (b) compacted model [1048].

field of forces acts, the thermal and oxidative effects being avoided [262, 350, 1052, 1053].

R.S. Porter and co-workers supposed to shearing a series of polymers especially polyisobutene, which as a thin liquid film suffered a laminar flow among the coaxial cylinders, at low values of Reynolds criterion. The molecular weight decrease was correlated with the shear force, temperature, and polymer concentration. In addition, the change of the molecular weight distribution was proved by GPS measurements [262]. The polyisobutene laminar flow among the two concentric cylinders, the mobile one working at 1500 rpm. Once again it was proved that the mechano-degradation is a selective process, the longest macromolecules being preferentially split [1054–1058].

3.9.5. Shear by turbulent flow through pipes. Drag reduction

During the laminar or turbulent flow of very diluted polymer solutions, an unexpected phenomenon was observes, namely that of the drastic decrease of viscosity, sometimes lowed than solvent viscosity. On the explanation this phenomenon were involved researchers implied both polymer science and fluid flowing fields [1059–1079].

The first description of this effect belongs to K.J. Mysels and refers to the flow of aluminium salts of fatty acids in gasoline [1059]. In the case of polymers, B.A. Toms firstly found important decrease of he friction coefficient, about 50%, during the turbulent flow through pipes of the poly(methyl methacrylate) solutions in monochlorobenzene [1060]. Subsequently, this behaviour was found for other polymers and was called "Toms effect". The linear polymers with average molecular weight higher than 100,000, which are soluble in water or other solvents, decrease their turbulent friction coefficient until 70–80% from its initial value. The first studies referred to the investigation of the water soluble polymers, such as poly(acrylic acid) and poly(acrylamide) [1061–1069]. A.G. Tabula found that poly(ethylene oxide), as very diluted aqueous solutions, is the most efficient polymer with respect to the reduction of the friction coefficient during the flow through pipes [1066]. His results were subsequently confirmed by other researchers [1070].

Subsequently, in the effort to extent the "Toms effect" to other polymers, a series of polysaccharides from aquatic plants, microscopic algae, and bacteria were been investigated. Interesting results were found by J.W. Hoyt [1063] and P. Kenis [1071–1073].

Thus, J.W. Hoyt resorting to the critical value of the Reynolds number, specific to the turbulent flow inside of pipe:

$$\text{Reynolds number}_{\text{pipe flow}} = \frac{VD}{\nu} > 2.300$$

and for the turbulent external flow, for instance as over a ship hull:

$$\text{Reynolds number}_{\text{external}} = \frac{VD}{\nu} > 500.000$$

where: V – average flow velocity; D – pipe diameter; L – axial length; and v –viscosity of polymer solution.

It was found that the polymer amount required for the occurrence of a drastic decay of turbulent friction is very small, for instance, at the laboratory scale, about 1 ppm. The concentrations ranging from tenths and hundreds ppm were found to be optimal from the efficiency viewpoint.

Water soluble polymers such as poly(ethylene oxide) or poly(acrylamide) and oleo-soluble polyisobutene were proved to be extremely efficient compounds in the reduction of turbulent friction. The same effect was reported in the case of the polymer compound extracted from Cyamopsis and Xanthamonas, trademark guar.

For a aqueous solution of poly(ethylene oxide) with the average molecular weight of $4{\cdot}10^6$, flowing through a tube with a very small diameter, ϕ = 1.3 mm, the obtained results are depicted in Figure 3.377.

In this case, the Reynolds number was 14 000 and 1 ppm of

polymer produced drag reduction with 18%. The drag reduction was calculated with the following relation [1080]:

$$\% \text{ drag reduction} = \frac{\Delta P_s - \Delta P_p}{\Delta P_s} \times 100 \tag{3.209}$$

where: ΔP_s and ΔP_p – pressure drop of the solvent and polymer solution at the same flow rate, respectively.

Drag reduction is also obtained in the case of laminar flow of polymer solution, in the limits of usual concentrations, accompanied by a slightly increase of viscosity without to modify the linear equation shear-stress/shear-flow rate, Figure 3.378, the behaviour being typical newtonian one, Figure 3.379. Based on the last Figure a small increase a friction coefficient could be expected but not an important drag reduction as was found.

It was experimentally established that the higher are the polymer molecular weights the more drag reduction is achieved. The polymers with $\bar{M}_w < 100.000$ are totally inefficient. Thus, by increasing the poly(ethylene oxide) molecular weight from $2 \cdot 10^5$ to $5 \cdot 10^6$, a reduction with 70 % of the resistance to the rotating disk takes place, Figure 3.380. Very interesting are the results concerning the correlation between the molecular weight distribution and drug reduction [1081]. As Figure 3.381 shows, using the previously fractionated poly(acrylamide) samples, for a given polymer concentration and value of Reynolds number, the higher drag reduction was obtained for the highest molecular weights. It worth to be mentioned that for a mixture of samples with the extreme molecular weights, i.e. the lowest molecular weight of 1.43×10^6 and the highest one of 8×10^6, the drag reduction efficiency is higher as com-

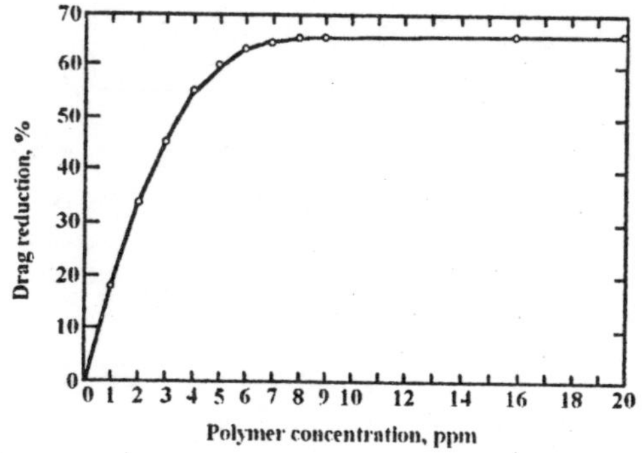

Figure 3.377. Drag reduction of poly(ethylene oxide) in water, at a Reynolds number of 14,000, in a small pipe [1080].

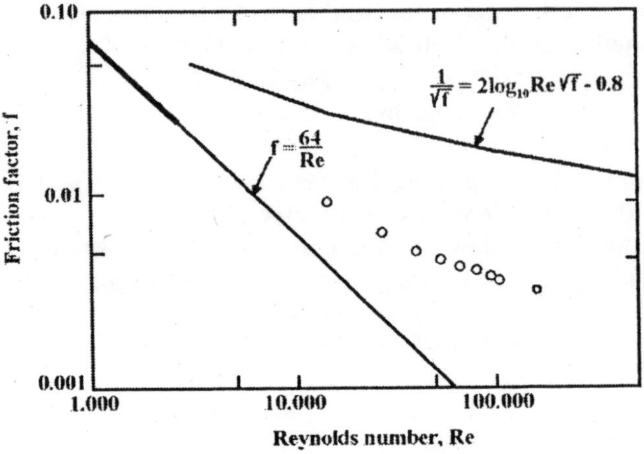

Figure 3.378. Typical data for drag-reducing polymer solutions falls between the turbulent solutions line for pipe flow, and laminar line, 64/Re, greatly extended beyond its usual limit of a Reynolds number of 2300, where f = pipe friction coefficient in engineering equal to pressure drop per length times the diameter, divided by $1/2\rho V^2$, and ρ = density, [1080].

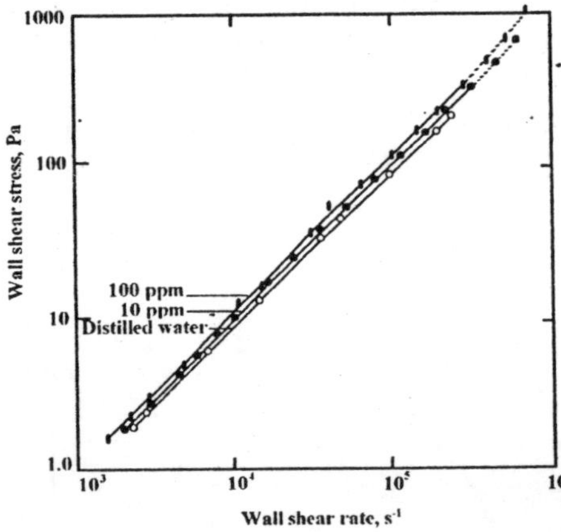

Figure 3.379. Rheogram for poly(ethylene oxide) at 25.0±0.1°C showing linear relationship over usual range of wall shear-rates, and increase in viscosity with increasing concentration; there is thus no evidence of non-Newtonian behaviour, [1080].

pared with the corresponding fractionated samples. This effect is characteristic for the small values of the Reynolds number, at its higher values the effect is diminished due to the polymer mechano-degradation, i.e. the scission of the longest chains and, on this basis, the accumulation of the shorter ones [1082].

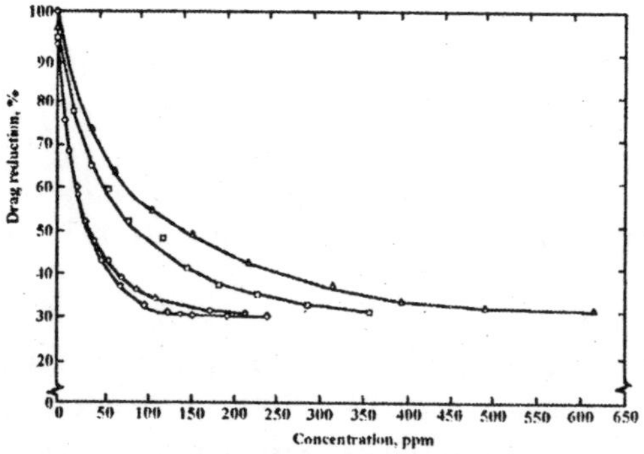

Figure 3.380. Effect of molecular weight on torque required to spin a 457-mm-dia disk at 240 rpm: (Δ), ca 2×10^5; (□), ca 6×10^5; (O), ca 4×10^6; (◇) $>5\times10^6$; water temperature, 14–15.5 °C. Increasing molecular weight reduces the torque at lower concentrations of poly(ethylene oxide). Disk torque ratio = percent of pure water torque, [1080].

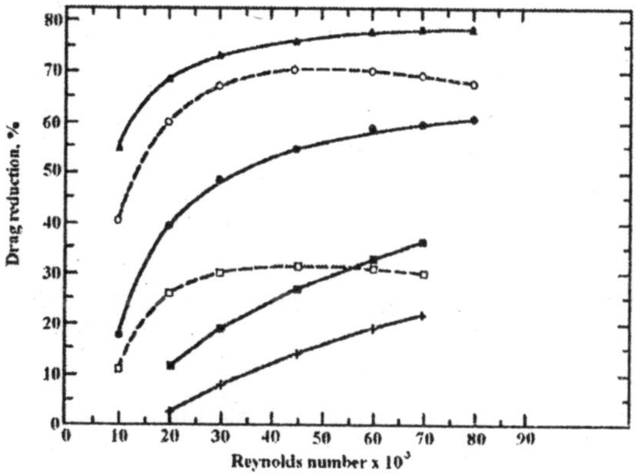

Figure 3.381. Drag reduction in a 6.1-mm-dia pipe as a function of polyacrylamide molecular weight and Reynolds number at a concentration of 50 ppm. $\bar{M}_w$; (▲), 8.0×10^6; (●) 4.26×10^6; (¦) 1.7×10^6; (+) 1.43×10^6; (O) 4.26×10^6 (mixture); (□) 1.7×10^6 (mixture), [1080].

The linear structure is an essential condition for the drag reduction efficiency, a maximal stretching of the macromolecular coil, for a given molecular weight, being required. Thus, the best results are obtained in the case of completely linear poly(ethylene oxide) and polyisobutene, followed by poly(acrylamide) [1080].

The same requirement of chains linearity was also established in the class of natural polymers such as polysaccharides and DNA. Polymers such as dextranes and gum arabic are inefficient [1082, 1083]. The other molecular characteristics, such as chain flexibility and configuration also play an important role [1080].

On the other side, extricates factors, such as solution concentration, are very important. One can note that in the case of polymer solutions, drag reduction occurs in the range of extremely small polymer concentrations, Figure 3.382. Thus in the case of poly(ethylene oxide) the optimal concentration is about 7500 ppm [1084], but not below 0.02 ppm [1085].

Therefore, in the case of turbulent flow, the drag reduction effect depends by the individual macromolecules stretching degree [1086]. By photographing J.W. Hoyt succeeded to visualise the elongated particles [1087]. The possible extrapolations and the limits of this effect have been defined by Hoyt starting from the friction velocity, u^*, caused by the friction to the pipe wall, which is expressed as:

$$u^* = \sqrt{\frac{\tau_w}{\rho}} \tag{3.210}$$

where: τ_w – stress; and ρ - solution density, and expressed as two adimensional terms: $u^* = \frac{u}{u^*}$ and $y^+ = \frac{yu^*}{\nu}$ where: u – local velocity, varying with y – the distance from the wall; ν –flow rate component in the sense of the x axis. The results obtained are illustrated in Figure 3.383. The correlation of the obtained results with the pipes diameter is depicted in Figure 3.384. It can be seen that irrespective of pipe diameters, drag reduction always starts from the pipe wall. In bigger pipes drag reduction occurs at higher values of Reynolds number.

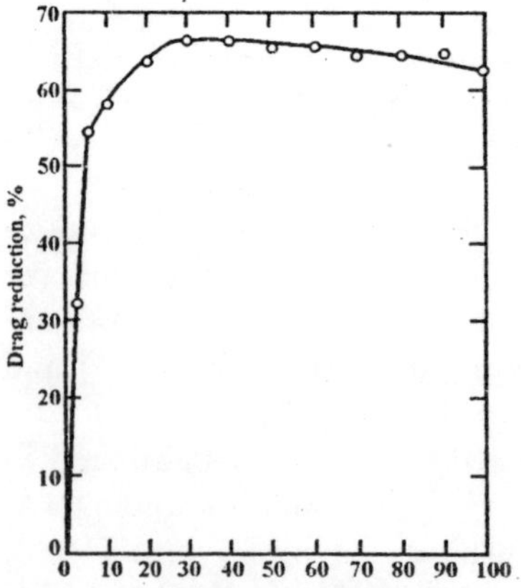

Fig.3.382. Effect of increased poly(ethylene oxide) viscosity on drag reduction in a 1.09-mm-dia pipe, [1080].

Table 3.124. Commercial drag-reducing polymers [1080]

Type	Manufacturer
Poly(ethylene oxide)	
• Polyox	Union Carbide Corp.
Polyacrylamides	
• Separan	Dow Chemical Co.
• TRO	Calgon Corp.
• Magnafloc	Allied Colloids
• Superwater	Berkeley Chemical
Xanthans	
• Kelzan	Kelko Co.
• Rhodopol	Rhone-Poulenc
Guar	
• Jaguar	Stein, Hall & Co., Inc.
Oil-soluble polymers	
• CDR	Conoco, Inc.
• FLO	ChemLink Petroleum, Inc.

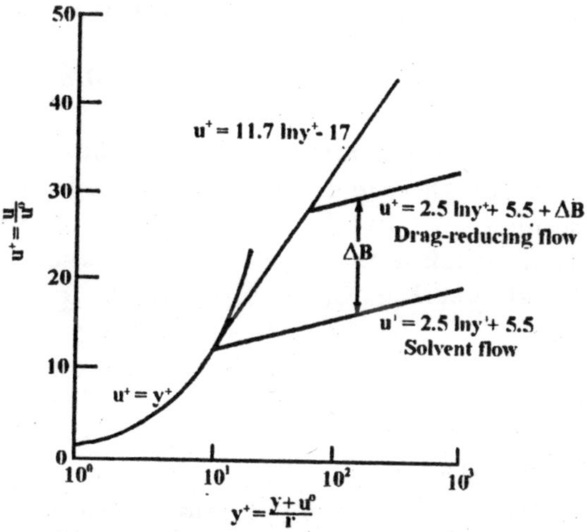

Figure 3.383. Velocity profile relationships near the wall of a pipe or in the boundary layer near the wall in an external flow, [1080].

Some commercial products with drag reduction properties are presented in Table 3.124.

The most important application fields of drag reduction effect are: fire-fighting, storm-sewer, flow augmentation, and pipes design for the transport of different fluids, with the performances and limits described by J.W. Hoyt in ref. [1080].

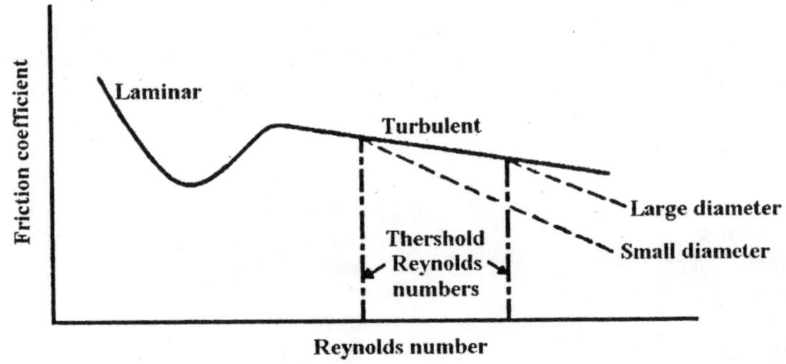

Figure 3.384. Pipe diameter effect. Friction reduction begins at the same pipe-wall shear stress for all pipe diameters; small pipes thus show the effect at a lower Reynolds number than large pipes, [1080].

3.9.6. Mechano-chemical destruction under ultrasound action

3.9.6.1. The mechanism of physical and chemical action of ultrasound waves

Mechano-chemical degradation of dissolved polymers should be released not only by the simple mechanical solicitations of low frequency, as those previously analyses, but also by the more complexes actions, such as under ultrasound waves action. In this case the chemical reactions are determined by many reasons, as follows:

Cavitation phenomenon

The propagation of ultrasound waves through liquid media occurs by its successive compressions and decompressions. In the case of high intensity waves, in the decompression phase (especially in the vicinity of some suspended particles, air bubbles, or vapours) the microscopic fractures of the liquid occur, generating voids or cavities, toward which the dissolved gaseous molecules migrates or the liquid vapours resulted by liquid evaporation. In the next step, the compression, the formed cavity diminishes its volume, locally the pressure increases drastically, reaching up to several thousandth of atmospheres. Cavitation phenomenon is also accompanied by high local increase of temperature, electric discharges, or acoustic waves release, whose intensity permits the evaluation of hydraulic shock waves intensity that are initiated in the moment of broken of cavity bubbles.

The initiation and development of cavitation bubbles into a liquid medium is influenced by several factors, such as: frequency of ultrasound vibration [1088–1092], ultrasound intensity [1093, 1094], duration [1095–1098], external pressure exerted on the liquid medium [1099, 1010], the nature of dissolved gasses into the cavitation bubbles [1088, 1101–1104], solution viscosity, geometry of the working cell [1088].

Hydraulic shock waves

Hydraulic shock waves represent intensive impulses of pressure that appear in the moment of cavitation bubbles breaking, in the compression phase of ultrasonic wave. The intensity of hydraulic shock waves increases with the decrease of ultrasonic waves intensity. The increase of medium temperature diminishes the intensity of hydraulic shock waves [1088].

Ultrasonic waves determine the splitting the chemical bonds of the molecular structures that are intercepted during their propagation. Consequently, electrons, ions, and radicals appear at the surface of the cavitation bubble, which are able to react with the substances localised at the cavity-liquid interface. The nature of released chemical reactions depends by the nature of the gaseous atmosphere that saturates the ultrasonic cell and appear as result of the mechanical forces effects, being "genetically" related to shock wave appearance at the moment of breaking of cavitation bubbles [1105].

Resonance pulsating bubbles

The gaseous bubbles suspended in liquid continuous phase suffer periodical compressions and dilatations, pulsations, characterised by specific frequencies [1106]. The bubbles amplitude and frequency become maximal whenever their own frequency coincides with the ultrasound frequency, therefore when together give rise to system in resonance. The resonance oscillations can generate local pressures that overtake up to 150,000 times the hydrostatic pressure.

It was accepted that the gaseous bubbles keep their spherical shape up to radius value of about 0.003 mm, but under the ultrasound action may suffer important deformations. The number of bubbles in resonance is governed by the amplitude and frequency of ultrasound waves and it is decisive in the release of chemical reactions [1107, 1108]. In their absence the chemical reactions do not occur [1109].

The process of ultrasound waves propagation is an adiabatic one, since the pressure and density modification in the oscillating layers occurs so quickly that the caloric transfer from the compressed zone of gas or liquid toward the surrounding medium and conversely does not occur. In function of the dissolved gas nature in the resonance pulsating bubbles, high temperatures can arise whose value, under the conditions of adiabatic compression and of relative decrease of the gas volume, will be determined by the ratio $\gamma = c_p/c_v$. Thus, temperatures of 1000 K and even more have been calculated [1088, 1109, 1110]. In the minuscule cavity, filled by gas, such high temperatures are practically able to generate all types of active centres. On their turn, these ones can initiate almost all-chemical reactions that occur in ultrasonic field. Subsequently, the reaction products that occurred in gaseous phase diffuse into liquid medium where react with the dissolved substances, releasing a series of secondary transformations.

Electrical phenomena in cavitation bubbles
Chemical reactions activated by ultrasounds are conditioned not only by the mechanical forces developed during the dilatation phase but also by the electrochemical or photochemical phenomena that initiate into cavitation bubble, starting with its first stage of formation [1088]. In this stage, suitable conditions for the release of electrical discharges are created, i.e. low pressures and the presence of gases and vaporised substances that by ionisation pass in activated states. Subsequent development of cavitation bubble is accompanied by the increase of internal pressure, followed by decompression, when in the continuous liquid medium the already formed radicals and ions will diffuse.

3.9.6.2. Classification of the reactions occurring in ultrasonic field
It was established that into an ultrasonic field chemical reactions occur in two stages, each of them being related to the development stage of cavitation bubble.

The first stage consists of the development of photo- and electrochemical phenomena inside the bubble, the pressure in cavity being relatively low. As it was already described, these conditions are suitable for release of electrical discharges, accompanied by the formation of ionised or activated particles. The process of active species generation is finished by electronic penetration of cavity.

The second step coincides to the cavity dilatation, accompanied by the increase of pressure inside of it and finally by its crack. In this moment, the previously formed active particles, representing ions or other active species "are thrown out" in the liquid medium. Concomitantly with the broken of cavitation bubble, the hydraulic shock wave is released, these one being essential for the activation of the mechano-chemical reactions.

In accordance to this succession of events, the following types of chemical reactions occur in the medium:

1. Reactions, in gaseous phase, inside of cavitation bubble;
2. Reactions at the interface gaseous bubble-liquid, initiated by the radicals and active species that appeared inside of cavitation bubble, at the level of monomolecular layer from its surface;
3. Chemical reactions initiated by the active species that appeared in gaseous and that were already entered in the liquid phase in the moment of the broken of cavitation bubble; and
4. Reactions in liquid medium, released under the action of hydraulic shock waves, which arise in the moment of the broken of cavitation bubble.

The chemical efficiency of the ultrasonic field is usually evaluated by the chemo-acoustic coefficient, $\eta_{ch\text{-}a}$, that represent the product between the utilising degree of acoustic energy, x, and the coefficient of cavity chemical activity, χ, [1110]:

$$\eta_{ch-a} = x \cdot \chi = (E_x . E) \cdot (E_{ch-a}) / E_k = E_{ch-a} / E \qquad (3.211)$$

where: E – amount of acoustic energy absorbed by the liquid unitary volume; E_k - amount of chemical energy consumed for the increase of cavitation bubble; E_{ch-a} – amount of acoustic energy consumed for the generation of free radicals, called chemo-acoustic energy.

The chemo-acoustic efficiency of ultrasound waves depends by their frequency. In the majority of chemical reaction, its optimum values are located in the frequency range from 200 to 400 KHz.

Whenever in system are also substances with high tension of saturated vapours, these ones penetrating the cavitation bubble cause its more rapid dilatation than the medium vapours, frequently water, impeding, in this way, the electronic penetration of bubble. In addition, the modification of system temperature concomitantly influences the cavitation efficiency [1111] and the process kinetic.

Due to the high number of factors affecting the chemo-acoustic reactions, it is advisable to be carried out in the limits of the following values of the fundamental parameters:

- frequency: 400 KHz ÷ 1 MHz;
- specific power: 2÷5 V/cm^2;
- temperature: close to room temperature;
- the absence of the substances with high values of vapours tension;
- small volumes of liquid, for which the ultrasound field may be considered uniform; and
- low periods of ultrasound waves action, when both the variation of the gas quantity dissolved in liquid and the influence of new formed products can be neglected.

In these conditions, the proportionality between the yield of chemo-acoustic reaction and the total power consumed during ultrasonic treatment is respected.

The degree of utilising acoustic energy was defined, in order to evaluate the energetic factor in the release of chemical reactions. As equation (3.212) it reflects the intercorrelated action of some important factors [1112]:

$$q = \frac{I \cdot t}{c \cdot V} \tag{3.212}$$

where: q – acoustic energy utilisation degree; I – ultrasound intensity; t –period of ultrasound action; V – sample volume; and c – solution concentration.

The maximal efficiency of the chemo-acoustic reaction is assured by the maximal absorption of acoustic energy, the fact demonstrated by the energetic and kinetic study of the chemo-acoustic degradation of polystyrene solutions in benzene, varying the conditions of ultrasounds absorption by changing the geometry of the reaction vessel, Figure 3.385. One can see that for the identical values of factor q, the acoustic destruction efficiency depends by the concentration of acoustic energy, which, in turn, is assured by the dimensions and shape of the reaction vessel. Whenever the ultrasounds energy is generated outside of the reaction vessel, the destruction effects are minimal, curve 1 in Figure 3.385; contrary, when the acoustic energy is concentrated in a 'well', therefore, directly taken out in the reaction medium, the mechano-degradation is stimulated, curve 2 in Figure 385. Furthermore, when the generator well is placed just in the reaction medium, the absorbed acoustical energy is about 90% and the decrease of the polymerisation degree is higher, curve 3 in Figure 3.385. The maximal effect is achieved working in a parabolic vessel, which realises the focal concentration of ultrasound energy, curve 4 in Figure 3.385.

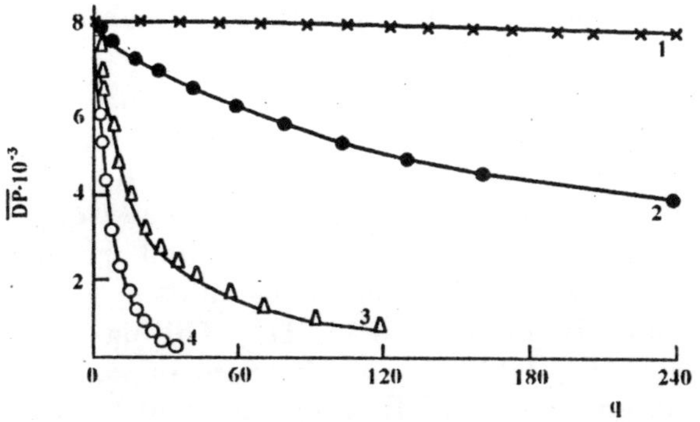

Figure 3.385. Dependence of polymerisation degree on the value of q for different conditions of acoustic energy absorption: 1) outside of reaction vessel; 2) reaction vessel placed in the generator well; 3) generator well placed inside of reaction vessel; 4) local absorption of acoustic energy inside of parabolic vessel (benzenic solutions of polystyrene) [1112].

The concentration influence, reflected by equation (3.212) is manifested, in fact, by the modification of medium viscosity, which determines the efficiency of chemo-acoustic transformation during the appearance and broken of cavitation bubble, respectively.

By maintaining constant the vessel section and modifying the height of liquid column, a direct dependence of the chemo-acoustic degradation and degradation parameters, as effect of the different distribution of the elastic energy in the reaction medium [1112].

Evidently, the rate of possible reactions in ultrasonic field depends by the stressed polymers structure.

By analogy with radiation chemistry, in sono-chemistry we also define the sono-chemical yield, $F(Y)$ [1094], which represents the number of formed molecules or radicals, or which are consumed for each 100 eV of absorbed energy

$$F(Y)=\frac{c\cdot N\cdot 100}{M\cdot E\cdot 6.25\cdot 10^{15}\eta_{ch-a}} \qquad (3.213)$$

where: c – concentration of obtained product, g/cm^3; E – acoustic energy absorbed by the volume unit, V·s·cm^{-3}; $\eta_{ch\text{-}a}$ – chemo-acoustic coefficient ($\eta_{ch\text{-}a} \cong 0.01$); N – Avogadro's number; and $6.25\cdot 10^{18}$ – coefficient of energy conversion from V·s in eV. The use of this coefficient has the advantage that allows comparing the results of different types of reaction and therefore of the mechanisms characterising them.

3.9.6.3. *Chemo-acoustical degradation of polymers in solution*

The pioneering researches in the field of chemo-acoustic degradation of polymers belong to E. Flasdorf and L.A. Chambers [1113], A.Z. Szent-Gyorgyi [1114], and A. Szalay [1115], respectively. Supposing to ultrasounds action, at different frequencies, a series of natural polymers solutions, the authors found the decrease of viscosity, which was ascribed to the chains splitting.

On the other side, H. Freundlich and D.V. Gillings [1116] ultrasonically treating colloidal solutions observed the same tendency. The authors related this behaviour to the modification of the polymer structure under the action of ultrasound.

An ample and systematic study was carried out by G. Schmid and O. Rommel, which followed the ultrasounds action on the diluted solutions of many polymers, such as: poly(vinyl acetate), poly(methyl methacrylate), cellulose nitrate in organic solvents [1117–1122]. Analysing the behaviour of fractionated and unfractionated polystyrene solutions under ultrasonic field, in all cases they found the decrease of molecular weight in time and proved that, finally, all samples tend to the same limit value of the molecular weight, Figure 3.386, [1117]. The authors reported the same results for a series of polyacrylates.

Other researches that contributed to the clarification of the mechano-chemical nature of the polymers ultrasonic degradation, performed both on natural and synthetic polymers, belong to I.I. Zhukov and M.A. Klenoch [1123], and H. Sobue and co-workers [1124, 1125].

H. Mark considered that in the ultrasonic field ân important lagging exists between the macromolecules and solvent molecules dis-

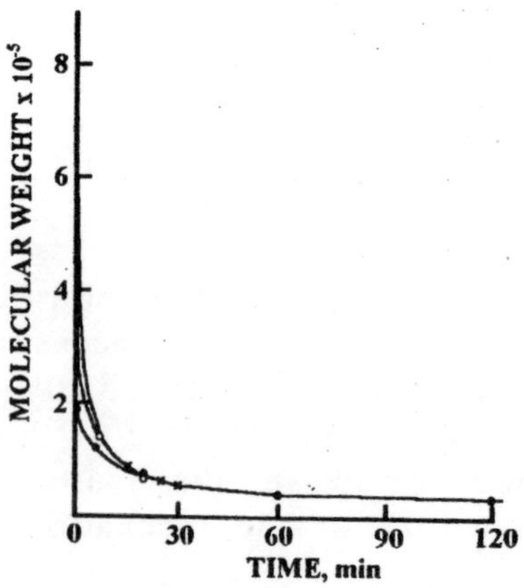

Figure 3.386. Ultrasonic degradation of fractionated polystyrene samples in toluene solution as a function of time: M^{in}: (o) 850 000; (×) 350,000; (•) 192,000. Conditions: frequency, 284 kHz; intensity, 5.8 W/cm² [1117].

placement. The shear forces developed by this reason are enough strong to well overtake the energy of chemical bonds, therefore, being able to cause the splitting of any macromolecule [1126]. The experimental results accumulated in the level of 45–50 years evidenced a series of common characteristics of sono-chemical degradation with other types of mechano-chemical degradations, which occur under the action of other types mechanical solicitations. It was proved that sono-chemical degradation is released both in compression or dilation stage of cavitation bubble, on the trajectory of ultrasound propagation, its wavelength overtaking by far the macromolecule length. The process occurs with high speed in the first moments of solicitation, after that degradation rate decays and is stopped for a certain limit value of the molecular weight (see Figure 3.386). The existence of the destruction limit in the case of sono-chemical degradation was clearly evidenced in many publications [1127–1130].

3.9.6.3.1. Kinetic aspects

In the field of polymer sono-chemical degradation, the kinetic evaluations are based on the phenomena initiated by breaking of the cavitation bubble. It is worth mentioning the contributions of H.H. Jellinek and G. White [148–151], M.A.K. Mostafa [1131–1134], and D.W. Ovenall and co-workers [1136, 1137].

Taking the number of split bonds as a criterion of mechanodegradation, the destruction rate is given by the following relationship:

$$\frac{dx}{dt} = k \cdot F \cdot n_1 \cdot n_c \cdot v \tag{3.214}$$

where: x – the number of split bonds at the moment t; k – proportionality constant; F – factor depending by the mechanical force generated in the moment of cavitation bubble opening; n_1 – the number of chemical bonds at a given moment, t; n_c – the number of gaseous bubbles in system; and ν is the frequency of ultrasounds waves.

For the initial stage of the process, the equation (3.214) may be written as:

$$\begin{aligned} -\frac{dn_1}{dt} &= -dn_n(P_n - 1)/dt = k_1 \cdot n_c \cdot n_n \cdot (P_n - 1) \\ -\frac{dn_n}{dt} &= k_n \cdot n_c \cdot n_n \cdot \upsilon \cdot (P_n - 1) \end{aligned} \tag{3.215}$$

where: n_n – the initial number of molecules with the length equal to P_n; k_1 – rate constant; $k_1/P_{n-1} = k_n$ – constant determined by the cavitation parameters (intensity, frequency, initial radius); both k_n and k_1 are related to chain length.

In ultrasounds field, the most important chemical reactions occur in the proximity of the cavitation bubble, were its opening takes place and, therefore, the release of maximal speed and pressure gradients. Admitting that a macromolecular chain in this zone is rigid fixed at one of its extremities and in free movement at the other one and is tangentially disposed to the surface of spherical cavity in dilatation, when its radius is maximal, and also considering the liquid medium is incompressible, according to Stock's law, the friction force that arises is equal to:

$$f = 6\pi \cdot r \cdot \eta \cdot P_n \cdot \mu \tag{3.216}$$

where: η - solvent viscosity; r – half of lenght a structural unit, which are considered of spherical shape; P_n – chain length and μ - the speed of sound at a given moment. Assuming that before dilatation, the both chain ends tangentially remain at the cavity limit, the force acting on them is equal to:

$$f = p \cdot P_n \cdot l \cdot d \tag{3.217}$$

where: d – macromolecular chain diameter; l - structural unit length; $P_n \cdot l \cdot d$ – chain cross section; p – liquid pressure, at a given moment. Considering that the force that acts on the macromolecular chain is sinusoidal [151], it can be written:

$$f = f_0(\sin \omega t - \phi) \tag{3.218}$$

where: f_0 – force amplitude; ω - angular frequency; ϕ - phase angle.

The reaction rate constant may be expressed as:

$$k_n = \frac{2\beta}{\pi P_n}\left[\sqrt{f_0^2 - r^2} - r \cdot \cos^{-1}\frac{\upsilon}{f_0}\right] \tag{3.219}$$

where: β - proportionality constant; r – the force required for chemical bond splitting.

Considering now a cavitation bubble in dilatation into an incompressible fluid and neglecting the interfacial tension and its viscosity, the value μ of cavitation breaking is given by the following expression:

$$\left[\frac{dR}{dt}\right]^2 = u^2 = \frac{2}{3\rho}\left[P_A(Z-1) - \frac{Q}{1-\gamma}\left(Z - Z^{\gamma}\right)\right] \tag{3.220}$$

where: ρ - medium density; P_A – hydrostatic pressure; $\gamma Z = \left(\frac{R_m}{R}\right)^3$ is the compression ratio in the moment of cavitation bubble breaking, characterised by the maximal radius, R_m, and radius, R, at a given moment, t; Q – gas pressure inside of cavity with the radius R_m; and γ' – ratio of the gas specific heats. Assuming now the following equality:

$$\mu^2 / u = \frac{1}{Z^{u/3}} \tag{3.221}$$

the relation between μ and u, as a function of Z and duration will be: $udt = -dR$, therefore $t / R_m = -\int^{R/R_m} \frac{1}{u} d\left(\frac{R}{R_m}\right) = -\int^{Z^{-1/3}} \frac{1}{4} dZ^{1/3}$, the value t/R_m as a function of u and μ is obtained from numerical solutions, calculated for a series of concrete cases; u is calculated as a function of Z, and the average velocities $\mu_{1,2}$, for low values of Z, are obtained using the following relation:

$$\frac{\Delta t}{R_m} = \frac{1}{\bar{\mu}_{1,2}}\left(\frac{1}{Z_1^{1/3}} - \frac{1}{Z_2^{1/3}}\right) \tag{3.222}$$

From the equations (3.220) and (3.221) the maximal value μ_0 of the velocity μ may be calculated:

$$\mu_0^2 = \frac{2}{3\rho}\left\{P_A Z_0^{-4/3}(Z_0 - 1) - \frac{Q}{1-\gamma'} Z_0^{-4/3}\left(Z_0 - Z_0^{\gamma'}\right)\right\} \tag{3.223}$$

for $Z_0 = 4$; $P_A = 10^6$ dyn/cm^2, $\rho = 1$ g/cm^3, the value of μ_0^2 is equal to $5.65 \cdot 10^2$ cm/s.

In the limits $t/R_m = 0$ and t/R_m = maximum, the expression for calculating μ becomes:

$$\mu = \mu_0 \sin 3.3 \cdot 10^2 \cdot 2\pi t / R_m \tag{3.224}$$

Based on the equations (3.216) and (3.224), the value of force f, exerted on the macromolecular chain will be:

$$f = 6\pi \cdot \eta \cdot r \cdot P_n \cdot \mu_0 \cdot \sin \omega(t / R_m) \tag{3.225}$$

Considering that the number of split chemical bonds, in the time interval dt, is proportional to $f - \gamma$ (f – representing the friction force that acts on the chain at a certain moment, t; γ – chemical bond resistance, under the assumption that all chemical bonds are identical), for one action cycle of ultrasound wave it can be written:

$$n_n(-1) = \beta \cdot n_n \cdot n_c \int_{t_1}^{t_2} (f - \gamma) dt \qquad (3.226)$$

where β is the proportionality constant.

If the angular frequency of ultrasound takes the expression: $\omega_1 = 2\pi/T$ (T is the wave period), the chain splitting rate is given by the following relation:

$$dn_n / dt = -\beta n_c n_n \omega_1 / 2\pi \int_{t_1}^{t_2} (f - \gamma) dt \qquad (3.227)$$

from equation (3.219) and $\upsilon = \omega_1 / 2\pi$, the following expression is obtained for k_n:

$$k_n = (\beta / P_n - 1) \int_{t_1}^{t_2} (f - \gamma) dt \qquad (3.228)$$

where: t_1 – the moment when $f = \gamma$, and t_2 – the time defining the rapid rate decrease; t_2/R_m = constant and therefore:

$$k_n = \frac{\beta}{(P_n - 1)} \int_{t_1}^{const \cdot R_m} \left\{ f_0 \cdot \sin\left[\omega_2 (t / R_m) - \gamma\right] \right\} dt \qquad (3.229)$$

and:

$$k_n = \frac{\beta R_m}{(P_n - 1)\omega_2} \begin{bmatrix} \left(f_0^2 - r^2\right)^{1/2} + \left(f_0^2 - f_2^2\right)^{1/2} - \\ -\gamma\left(\cos^{-1}\frac{f_2}{f_0} + \cos^{-1}\frac{\gamma}{f_0}\right) \end{bmatrix} \qquad (3.230)$$

This equation is valuable only for $f_2 > \gamma$.

Transcribing equation (3.230) in function of the initial and final terms of the chain, P_1 and P_n, respectively, it results: $f_0 = 6\pi\eta\mu_0 r P_n$, $f_2 = 6\pi\eta\mu_2 r P_n$, and $\gamma = 6\pi\eta\mu_2 e P_n$, and finally:

$$k_n = \frac{\beta R_m \gamma}{(P_n - 1)\omega} \left[\left(\frac{P_n^2}{P_1^2} - 1\right)^{1/2} + \frac{P_n}{P_1}\left(1 - \frac{\mu_2^2}{\mu_0^2}\right)^{1/2} - \cos^{-1}\frac{\mu_2}{\mu_0} + \cos^{-1}\frac{P_1}{P_2} \right] \qquad (3.231)$$

the ratio $\dfrac{\mu_2}{\mu_0} = \dfrac{\mu_0 \sin \omega_2 (t_2 / R_m)}{\mu_0} = \text{const}$ does not depend neither on the chain length nor on the ultrasound wave frequency and intensity or hydrostatic pressure.

The variation of destruction rate constant with chain length was calculated for constant intensity and frequency of ultrasound waves, and P_1 = 330; μ_2/μ_0 = 0.86; and $\frac{\beta R_m \gamma}{\omega_2}=1$, therefore:

$$k_n P_n = k(P_n - P_1) \tag{3.232}$$

Replacing this expression in equation (215) and taking into account that n_c, n_n, and γ are constant, we obtain:

$$-\frac{dn}{dt} = k'(P_n - P_1) \tag{3.233}$$

Equation (224) can be used for the calculation of k_n as a function of ultrasound waves intensity, I. The relation between I and R_m is the following: $R_m = k \cdot I^{1/2}$. For a benzenic solution of polystyrene, when P_n = 1620; η = 0.0062 Poise; $r = 3{\cdot}10^3$ cm; μ_2/μ_1 = 1 g/cm³; $Q = 1{\cdot}10^5$ dyne/cm²; $P_A = 1{\cdot}10^6$ dyne/cm²; $R_m = 5{\cdot}10^{-3}$ cm; I_1 = 2.12 (relative value); $\gamma = 5.64{\cdot}10^4$ dyne (-C-C- bond); and $\omega_2 = 3.3\times10^2\times2$ cm, the obtained data are collected in Table 3.125 [151].

For establishing the relation between the destruction rate and the ultrasound frequency it was accepted that the minimal radius of the cavitation bubble is inversely proportional to ν, $R_m \frac{k}{\upsilon}$, and k – the constant of proportionality.

The following relation expresses the process rate:

$$dn_n / dt = k_n \cdot P_n \cdot n_c \cdot n_n = k_n' \tag{3.234}$$

Table 3.125. The correlation between destruction rate constant and relative intensity [51]

Relative intensity	R_m (cm·10^3)	Q (dyne/cm²·10^{-5})	μ_0 (cm/s·10^{-2})	P_1	$I^{1/2} \cdot Y = K'_n$
0.6	2.7	6.7	1.3	1 110	0.34
0.7	2.9	5.4	1.5	564	2.0
1.0	3.4	3.1	3.8	421	3.1
1.42	4.1	1.9	4.4	366	5.5
2.12	5.0	1.0	5.0	322	8.0
4.40	7.2	0.34	5.4	300	12.8

Table 3.126. The correlation between hydrostatic pressure and the radius of cavity bubble

P_A (dyne/cm$^3 \cdot 10^{-6}$)	R_m (cm$\cdot 10^3$)	Q (dyne/cm$^2 \cdot 10^{-5}$)	μ_0 (cm/s$\cdot 10^{-2}$)	P_1	Y	$Y \cdot R_m \cdot 10^{-4}$
0.5	5.8	0.64	1.1	1 520	0.015	0.01
1.0	5.0	1.0	5.0	320	5.5	2.7
1.5	4.2	1.7	7.3	224	8.8	3.7
2.0	3.3	3.4	6.4	254	7.5	2.5
2.5	2.5	8.0	5.8	280	6.7	1.7

Finally, the relationship between R_m and hydrostatic pressure is:

$$P_m = -k(P_a - P_0) \qquad (3.235)$$

where P_A is hydrostatic pressure; and P_0 – pressure of ultrasound wave amplitude.

The dependence of hydrostatic pressure on the cavitation bubble radius is illustrated in Table 3.126.

3.9.6.3.2. Aspects of the sono-chemical degradation mechanism

After the establish of the ultrasound waves physical and chemical components, the mechanisms of sono-chemical destruction have dominantly been interpreted on this basis. Thus, in the cavitation bubble collapse moment, N.Sata and H. Okuyama defined four types of possible reactions, namely: 1) collision of the polymer - polymer molecules, occurring by a bimolecular reaction; 2) collision of the polymer–solvent molecules, which proceeds by a pseudo-monomolecular reaction; 3) intramolecular collision that occurs by a monomolecular reaction; and 4) tearing resulting from the entanglement points that gives a monomolecular reaction. The authors accepted that the active centres of this type of reaction are free radicals and that about 30% from the efficiency of this process is assured by the subsequent reactions of the formed macroradicals with the polymer chains [1130, 1139].

T. Hirose ascribed the main role to the mechanical forces that arise by successive contractions and dilatations of cavitation bubbles, which suppose to a shear stress the polymer–solvent molecules, process finished by macromolecules splitting.

In the case of polystyrene solutions ultrasonation, G. Gooberman accorded more importance to the rapid variations of pressure that accompany the shock wave, initiated in the moment of cavitation bubble crashing. The shock wave itself may be regarded as a suddenly increase of pressure, followed by its rapid exponentially decrease. In the period of pressure increase, the number of solvent molecules founded inside of macromolecular coil is higher than at the atmospheric pressure. Consequently, the chains conformation should not essentially change. The value of sono-degradation rate constant of the polystyrene solutions in benzene, as a function of the molecular weight, suggested that all cavities are similar ones. According to Gooberman's theory, the maximal pressure developed during the broken of cavity bubble is of about 1600 atm. [216–218]. The study of ultrasonically treated polymer polydispersity is based on the assumption that firstly the chemical bonds are split. The bonds are located in the plane crossing through the molecule centre and normal to the propagation direction of shock wave, generated in the moment of cavity bubble collapsing.

All the studies dealing with the mechanism of sono-degradation of the most different classes of polymers, unanimous accepted the radical nature of the active centres appeared in these conditions. This supposition was especially demonstrated using radical acceptors, DPPH being widely used in this scope [222–227, 1142–1144].

3.9.6.3.3 Factors of influence

It was proved that the factors affecting the most important characteristics of the sono-degradation process (destruction rate) and of ultrasonically treated polymer (chains' length, molecular weight distribution) are just those that govern both the cavitation phenomena and the generation of shock wave, namely: duration, frequency, nature and quality of dissolved gases in liquid phase, ultrasound propagation rate, intensity, and power. Adjacently, the solvent nature and quantity, solution concentration, and not in the last instance the polymer chemical structure are important factors of this process.

A. Frequency

Ultrasound waves frequency is an important factor that controls the sono-degradation rate. Form this point of view an optimal value was found of about 1 MHz, value at which the cavitation and degradation rate constant reach their maximal values, Table 3.127, and the chains limit length becomes minimal, respectively, Figure 3.387. It can be observed that to a frequency of 1 MHz the value of destruction rate is maximal, 13.0, comparatively with only 0.65 corresponding to a frequency of 20MHz [1135].

Table 3.127. Rate constant of degradation polystyrene samples degradation with different initial average chain lengths at different frequencies

Initial chain length	*k* at different frequencies of ultrasonic waves				
$\overline{DP}_{in}$	0.75	1.00	1.25	1.50	2.00
3744	7.97	13.0	7.20	3.36	0.65
3260	6.77	12.2	8.49	3.92	0.95
2460	7.59	9.82	8.18	3.94	1.38
1505	4.49	9.40	5.22	1.93	0.72

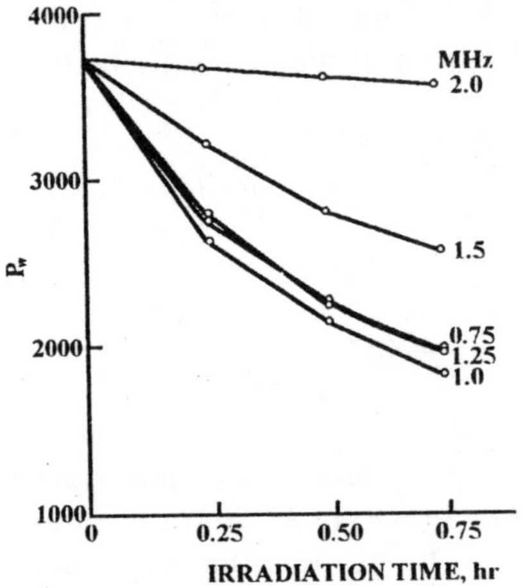

Figure 3.387. Degradation of 1% polystyrene in benzene at different ultrasonic frequencies [1135]. Conditions: P_w = 3744; intensity =12.5 W/cm².

B. The presence of dissolved gases bubbles in liquid medium

The first observations concerning the role of the gas bubbles existent into a liquid supposed to the ultrasounds action belong to Boyle and Lemann (1923) and V.C. Sorensen (1936). The first ones underlined that the gaseous bubbles can penetrate into the spaces formed by cavitation and discovered the tendency of sound to remove the gases dissolved in the liquid medium. The cavitation phenomenon depends on the external pressure that can increase up to a certain limit from which gas removal starts [1145]. According to V.C. Sorensen the removing of 1 cm³ of gas from saturated water requires 51.2 kW at 197 kHz, 72.6 and 87.4 kW at 380 kW and 530 kHz, respectively. Therefore, the gas removing essentially depends by the ultrasound intensity [1146].

It was A. Weissler who recognised the cavitation role and experimentally proved its effect on the ultrasonic degradation of polymers. Thus, for a 1% toluenic solution of polystyrene, he found a decrease of the solution viscosity under the conditions of ultrasonic treatment without degassing, Figure 3.388 a and b.

Working with hydroxyethyl cellulose the same author showed that gas removal starts concomitantly with cavitation, the fact proved by the luminescence effect, Figure 3.389 [1147, 1148].

H.W. Melville and A.J.R. Murray succeeded to release the sonochemical degradation even in the cavitation absence. In order to avoid this phenomenon they applied increasing pressures, using the nitrogen as pressurising agent, the process occurring in the oxygen absence. Even if the degradation effects were recorded these ones were not so intense as in the cavitation presence. Another way to avoid the cavitation is working under vacuum. In this case, the degradation effects are also small as compared to those stimulated by cavitation effect [1150].

The rate of ultrasonic degradation directly depends by the nature of the gas dissolved into the continuous liquid medium of ultrasounds propagation [1151, 1152]. Table 3.128 shows the influ-

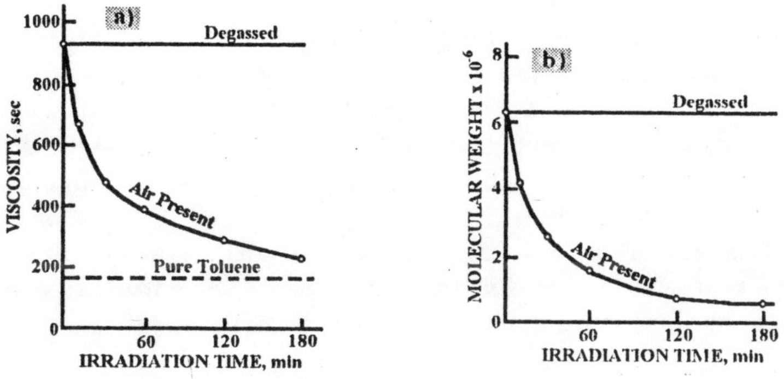

Figure 3.388. Influence of gaseous bubbles presence on the decay of viscosity (a) and of average molecular weight (b) of polystyrene ultrasonically irradiated in 0.5% solution in toluene [1149].

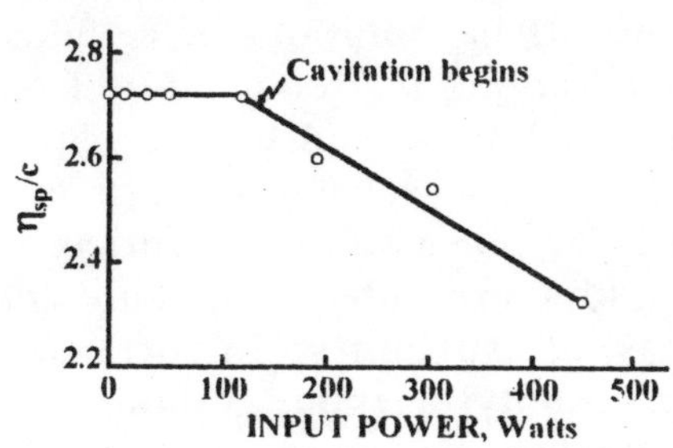

Figure 3.389. Effect of input power on hydroxyethyl cellulose degradation. Conditions: 0.8% solution, frequency 400 kHz, [1149].

Table 3.128. Intrinsic viscosity obtained on irradiation of 1% polystyrene in benzene containing different gases [1152]

Dissolved gas	Time (min) 0	10	20	40	60	120	240
			1.00				
Air	1.81	1.22	0.95	0.81	0.68	0.60	0.47
			0.89				
N_2	1.82	-	1.63	-	0.80	-	-
O_2	1.79	-	1.04	0.82	0.80	-	-
H_2	1.81	-	1.09	-	0.68	-	-
A	1.81	1.30	1.21	0.95	0.77	0.71	0.56
			1.18				
NH_3	1.81	1.67	1.58	1.28	1.06	-	-
					1.55		-
CO_2	1.80	1.66	1.65	1.50	1.28	-	-
					1.24		
SO_2	1.90	-	1.72	-	1.86	-	-

ence of the nature of dissolved gas on the ultrasonic degradation of 1% polystyrene solution in benzene. One can note that degradation rate decreases in accordance to the increase of gas solubility in liquid. For the same compression degree, gases with different chemical structure of molecules develop different temperatures and pressures, due to their different specific heats. Based on these conclusions and taking into account that adiabatic cavitation breaking occurs, it was concluded that the sono-degradation is a process of mechanical origin.

C. Temperature

B.B. Thomas and W.J. Alexander demonstrated the influence of temperature, in correlation with the occurrence of cavitation phenomenon [1154]. They found that the cavitation phenomenon and therefore sono-degradation do not occur below and above certain limits of temperature. Using solutions of cellulose nitrate in ethyl or butyl acetate and varying the temperature from 0 to 85 °C, the results given in Figure3.390 a and b were obtained. Thus, in the case of using ethyl acetate as solvent, the extreme limits are –1 and 64 °C, respectively. The process faster occurred in the rage from 21 to 51 °C. The results presented in Figure 390b confirm this conclusion, stressing solvent's importance. At the same concentration, but in butyl acetate, cellulose nitrate degradation presents other limit values, namely +1 and 12 ÷ 14 °C, respectively [1154].

N.H. Langton and P.Vaughan accept that the polymer macromol-

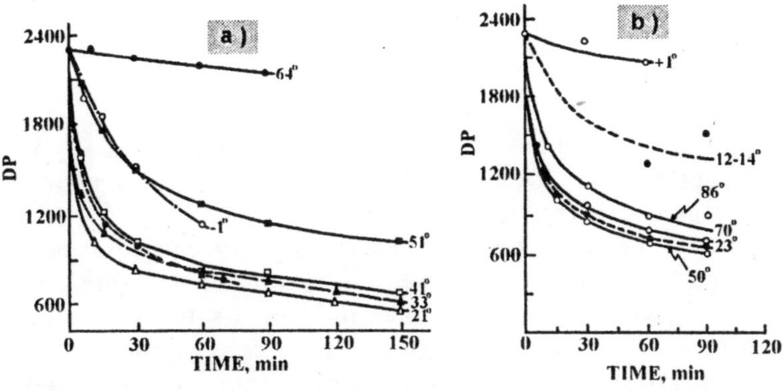

Figure 3.390. Ultrasonic degradation of cellulose nitrate [1150]: (a) in ethyl acetate solution; (b) in n-butyl acetate solution. (Frequency, 400 KHz; power, 50 W).

ecules act as cavitation nucleation centres, whenever the solution is ultrasonically irradiated. The dimensions of the pârticles able to nucleate the cavitation were found to be in good agreement to those of macromolecules. In the mechanism explanation they accept that in the moment of violent collapse of thus nucleated cavity, the local pressure and shock waves intensity suddenly increase and at high intensities the cavities arė filled with the solvent vapours. Under the action of ultrasound wave cyclical movement compression-decompression occurs, favouring the cavitation. The founded limits for the polymerisation degree, $\overline{DP_{\text{lim}}}$, also confirm the importance of nucleating centres dimension. Below this value the degradation does not occurs, since the nucleus are to small for the given pressure [1155, 1156].

The manifestation of luminescence phenomenon, concomitantly with sono-degradation, with the increase of temperature was evidenced by A.V. Borgstead [1157]. It was established that the sono-degradation increases with increasing the temperature, while the sono-luminescence decreases. The polymer degradation does not appear as a secondary effect of cavitation but is caused by the high intensity shock wave that was generated by the implosion of the cavitation bubble [469]. On the other side, it is well recognised that in the mechano-chemical processes, the luminescence is manifested in the moment of free macroradicals recombination. This think suggested that sono-degradation occurs by successive acts macromolecules scission initiation, which function as nucleus inside of cavity, and of macroradicals recombination, during the elimination of active fragments of destruction into the liquid medium. A fraction of these fragments by recombination reaction release the luminescerrce phenomenon.

D. Pressure

G. Schmidt and O. Rommel [1117], S. Ono [1158], and H.H. Brett and H.H.G. Jellinek [1159] reported a series of results concerning the effect of pressure on sono-degradation.

A series of experiments performed on polystyrene solutions in benzene offered valuable results. Based on the results obtained in two variants of work, namely: 1) the study of a saturated solution with nitrogen over which a constant pressure was maintained during the whole ultrasonic irradiation period, Figure 3.391 a; and 2) firstly, the solution was saturated in nitrogen and subsequently the pressure was exerted over it directly using a mercury column, Figure 3.391 b. In the both cases the destruction rate increases, reaching a maximal value, and after that decreases, Figure 3.391 c. It seem that on the increasing part of the curves from Figure 3.391 c; the pressure favours the development of cavitation bubbles and, consequently of mechano-degradation, while in the range of high pressures the cavitation is impeded to occur and the mechano-degradation rate decays [1099, 1100].

E. Ultrasound intensity

The accumulated results related to the elucidation ultrasound intensity effect on the polymer degradation are enough contradictory [1103, 1147, 1150, 1160–1163]. A more veridical explanation is obtained by its correlation with temperature variation. T. Hirose established that above a certain value of intensity the cavitation bubbles do not completely collapse. Most likely, between the cavities growth and ultrasounds intensity do not exist a linear dependence. In the range of low temperatures, the solutions viscosity increases, reaching enough high values to prevent the cavitation. At higher temperatures the degradation rate increases, existing an optimal temperature for which the cavitation intensity reaches its maximal value [1141].

B.B. Thomas and W.J. Alexander analysing the relation intensity-cavitation-temperature founded that the degradation is most sensible is more sensitive to the power changes, especially at boundary conditions. The increase of ultrasound power expands the range of temperature within which the polymers degradation occur [1154].

3.9.7. Mechano-chemical reaction during the cryolitic cycle freezing – thawing

During polymer solicitation as solid solutions, in water or in organic solvents, by cryolitic cycle frozen-unfrozen, mechanical forces high enough to causes the splitting of macromolecules chemical bonds are developed [1164–1170].

In aqueous solutions the mechanical force is the result of concomitant act the forces developed to the transformation of

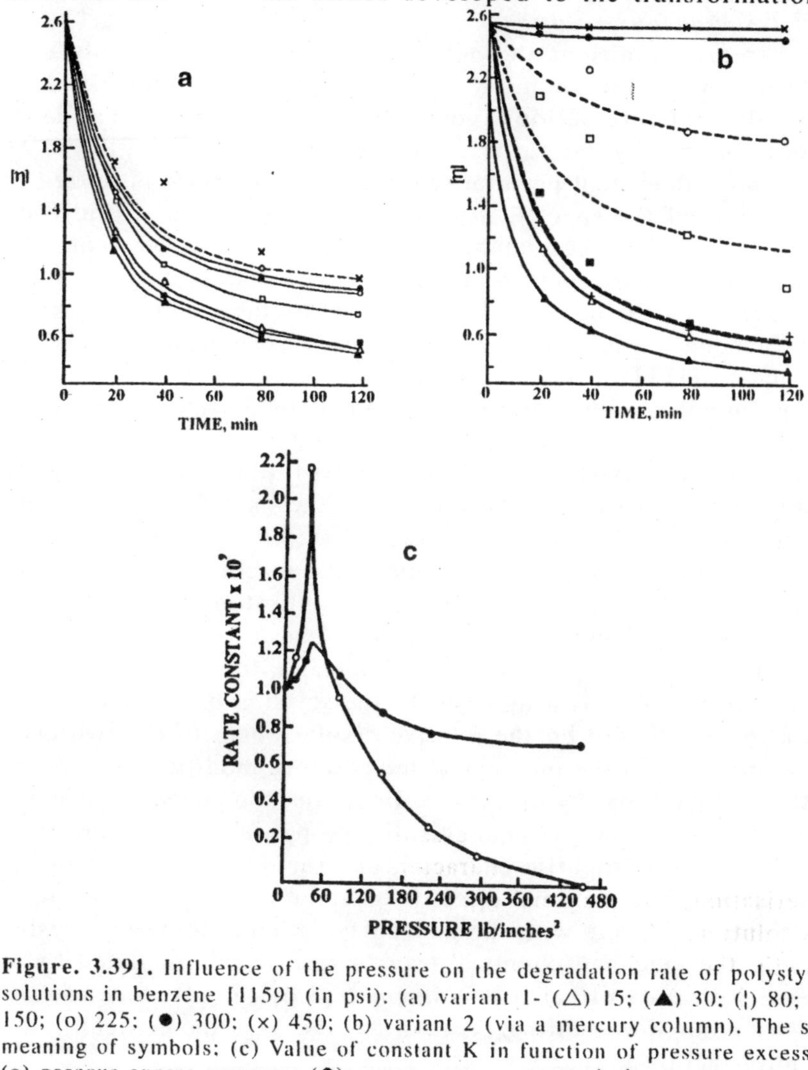

Figure. 3.391. Influence of the pressure on the degradation rate of polystyrene solutions in benzene [1159] (in psi): (a) variant 1- (△) 15; (▲) 30; (¦) 80; (□) 150; (o) 225; (●) 300; (×) 450; (b) variant 2 (via a mercury column). The same meaning of symbols; (c) Value of constant K in function of pressure excess for (o) gaseous excess pressure; (●) excess pressure exerted via a mercury column. Condition: frequency 500 KHz; intensity 10 W/cm².

water in ice, accompanied by a considerable increase of specific volume, and of the forces generated by the pressure exerted on the macromolecular chains, due to the frontal development of ice crystals.

3.9.7.1. Behaviour of the linear polymers as solutions during the cryolitic cycle

3.9.7.1.1. Aqueous solutions

Under freeze conditions the polymer macromolecules dissolved in water diminish their mobility, become more rigid, intensify their intramolecular interactions, generate some supramolecular interactions, and the hydration degree decreases.

The most important peculiarity of aqueous solutions is related to the increase of the specific volume by freezing. As it was already mentioned, during the change of aggregation state a mechanical force develops the superposed on the force generated by the pressure exerted on the macromolecular chains by the ice crystal growth is high enough to produce the scission of the chemical bonds of the main chain [1171–1173].

The forces developed during freezing stage are transmitted to the polymer chains, in whose bulk the ice crystals growth, and are located on the relative short chain segments, initially determining the modification of valence angles and deformation of chemical bonds, and finally their splitting. In the second step of the cryolitic cycle, the pressure of unfrozen liquid solicits the macromolecular chains, which due to the liquid exclusion by crystallisation are organised in vitreous supramolecular formations. Thus, series of chemical bonds scissions are possible to occur on this stage, too. If a chain is located between two growing crystals, the scission of chemical bonds can also be released by the relative displacement of the two crystalline regions, in the moment of temperature modification.

It was found that by cryolitic solicitation, the aqueous basic cellulose solutions change some essential properties of polymers, such as solubility, hydrophilic character, and the average degree of polymerisation. The maximal effect is obtained for in 8% NaOH aqueous solutions. Along with the molecules solicitation by ice crystals growth, the mechano-chemical transformation is stimulated by an intense hydration effect that is intensified by the presence of ions, which favour the swelling process and increase the content of the soluble fraction [1165].

Diluted solutions of water-soluble cellulose derivatives, such as carboxymethyl cellulose and methyl cellulose cryolitically stressed at –80 °C (frozen stage) and 30 (unfrozen stage) evidenced a similar behaviour [1174, 1175]. The decrease of polymerisation degree depends by the number of cryolitic cycles, n, and the polymer concentration. The degradation curve, plotted in co-ordinates η_{sp}/η_{sp0} vs. n (η_{sp}/η_{sp0}- destruction degree, calculated as the ratio between the specific viscosity at a given moment and the initial specific viscosity), presents all particularities of the mechano-chemical processes. As expected, the destruction degree faster decays in the first cycles, after which the curve tent slowly to a limit value.

The profound transformations were recorded by treating in cryolitic cycle the starch in the range of temperatures –20÷–70 °C [1166, 1167]. Thus, the aqueous solutions containing 2÷5 % potatoes starch increase by ten times their content of soluble fractions by cryolitic treatment concomitantly with the separation of a fibrous precipitate, insoluble in water. The soluble product differs by the initial sample, its indexes of copper and iodine being of 5 to 10 times higher and smaller values of η_{sp}/c. In turn, the insoluble fraction is characterised iodine index of 2 to 2.5 lower, higher values of η_{sp}/c, and higher asymmetry degree of macromolecules, which are organised in fascicles with a fibrous aspect.

3.9.7.1.2. Cryolitic behaviour of polymer solutions in organic solvents

F. Patat and W. Högner studied the cryolitic behaviour of organic solutions of polystyrene, polyisobutene, poly(vinyl pirrolidone), poly(vinyl acetate), polyamide 6, poly(ethylene terephthalate), evidencing the polymers destruction in these conditions. They shown that the most important factors affecting the destruction are the following: freezing rate, solution concentration, nature of solvent, and oxygen presence [1170].

It was established that the cryolitic destruction occurs only at high freezing rates, due to both of the reduced motility of polymer chains, in the highly viscous medium, and of reduced possibility to avoid the growing crystals. The effects founded at high freezing rates have been interpreted in the light of the visco-elastic behaviour of polymers.

By freezing, in the initial stage a partial crystallisation of solvent occurs, the rest of solution becoming more concentrated. The relaxation times of this system are so small that the macromolecules in the solution behave as a plasticized system. Subsequent development of the crystallisation process intensifies the increase of solu-

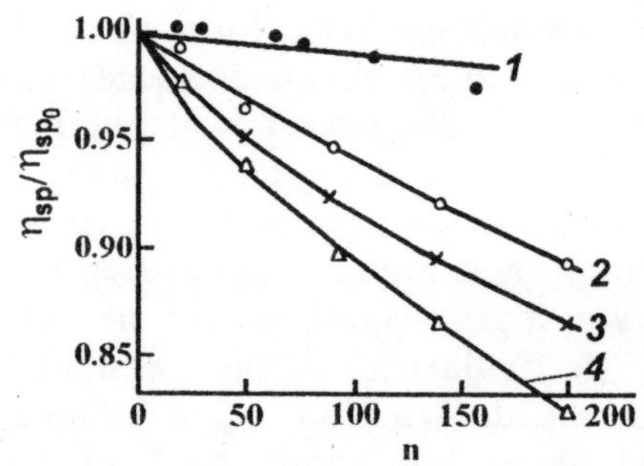

Figure 3.392. Influence of freezing rate on the cryolitic destruction of polyisobutylene [1170]: 1) in the oxygen presence, low freezing rate; 2) idem, a higher rate; 3) in the absence of oxygen, high freezing rate; 4) in the oxygen presence, high freezing rate.

tion viscosity and the plasticizing power decreases. Due to the frontal growth of crystals, the macromolecules suffer solicitations that are enough high in order to split the chemical bonds and to initiate the mechano-chemical destruction.

The essential role of the freezing rate in the achievement of a maximal effect of cryolitic destruction was proved on the polyisobutene solution both in the oxygen presence or absence, Figure 3.392. It is clear that the highest destruction corresponds to the highest freezing rate. At low freezing rate, even in the oxygen presence, the mechano-degradation is insignificant, Figure 3.392, curve 1; associating the oxygen presence to higher freezing rate the mechano-degradation intensifies, curve 2 in Figure 3.392. As the shape of curve 3 indicates, increasing the freezing rate even in the

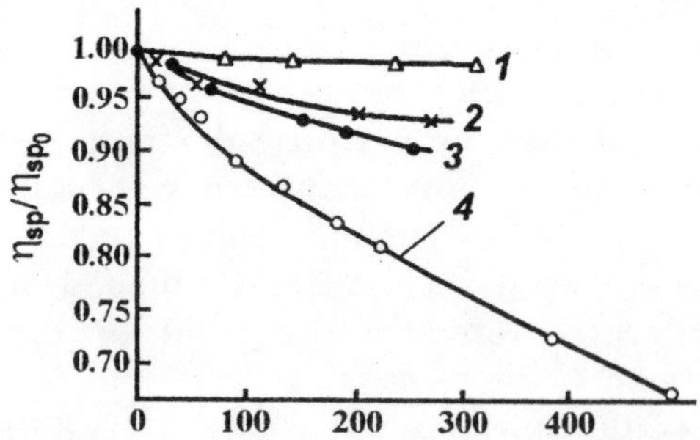

Figure 3.393. Influence of concentration and of initial molecular weight on the cryolitic destruction of polystyrene [1170]: 1) $\bar{M}$, C = 0.1 g/ml; 2) PS-36: $\bar{M}$ = 7×10^4; C = 0.05 g/ml; 3) PS-VI: $\bar{M}$ = 3×10; C = 0.020 g/ml; 4) PS-98: $\bar{M}$ = 6×10; C = 0.010 g/ml.

oxygen absence the mechano-degradation is higher; the maximal effect is obtained by combining the highest freezing rates with the oxygen presence (curve 4 in Figure 3.392).

The effect of polymer concentration and of its initial molecular weight is illustrated in Figure 3.393. Polystyrene samples with increasing molecular weight and decreasing concentrations have been used for the curves 1 ÷ 4. The both factors stimulate the mechano-degradation process, the most pronounced effect being obtained for the sample with the highest molecular weight, PSt-98, $\overline{M_w} = 6 \cdot 10^5$ and at lowest concentration c = 0.01g/ml, curve 4 in Figure 3.393, comparatively to PSt-20, characterised by $\overline{M_w} = 1.35 \cdot 10^4$ and c = 0.1 g/ml, curve 1 in Figure 3.393.

The radicalic mechanism of the destruction proceeding by frozen-unfrozen cycles was proved on the benzyl cellulose example in dioxane, whose macroradicals initiated the grafting of acrylonitrile [1175].

3.9.7.2. Behaviour of polymer networks during the cryolitic cycle

Studying the decrease of melting point of polymer gels, W. Kuhn and co-workers described their behaviour during the cryolisis process, postulating that these ones suffer a destruction, which is mainly caused by thermodynamic actions. Let us consider a polymer gel based on poly(vinyl alcohol) and poly(acrylic acid), in the shape of a fine spatial network, the aqueous phase being distributed through macromolecules, Figure 3.394. A crystalline intact surface will be impeded in its growth by the macromolecular network of the length a. The ice cubes with the surfaces equal to a^2 will grow through the points A an A′, existing the possibility of unification into a single crystal that subsequently grows until it

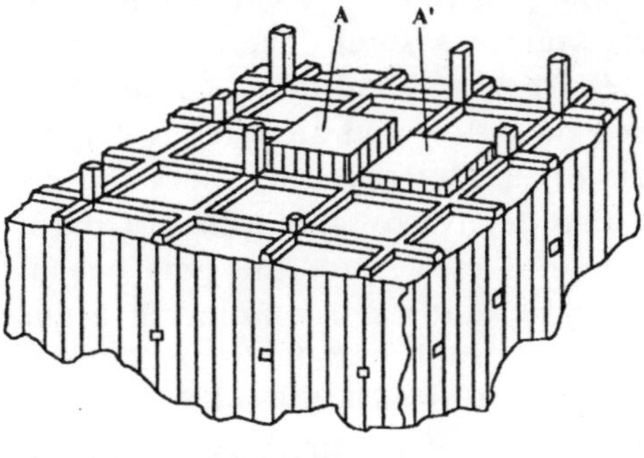

Figure 3.394. Schematic representation of the freezing process into a polymer gel [1164].

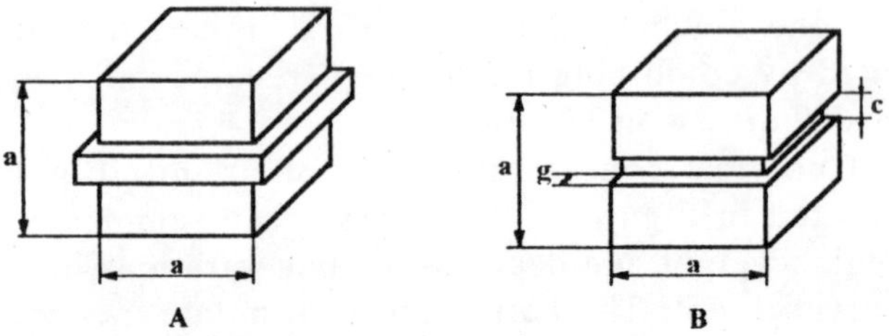

Figure 3.395. Growing of a cubic crystal within a polymer network [1164]: (a) crystal tensioning the macromolecular chain; (b) subsequent growing of the crystal containing a macromolecular chain.

meets once again the macromolecular chains where two situations are expected, namely: 1) the limitation of crystal growth; and 2) crystal development, having as result the tensioning of network. The two situations are illustrated in Figure 3.395. In a certain stage of growth, a canal marks the position of the macromolecular chain, its depth and breadth being noted *g* and *c*, respectively. The force that should act on the chain can be calculated with the following relation [1164]:

$$\Omega = a \cdot \sigma_{cr,liq} \tag{3.236}$$

where: Ω - the stressing force exerted by the crystal on the chain, expressed in dyne, a – chain length; and $\sigma_{cr,\ liq}$ – crystal–chain superficial energy.

It is observed that Ω is independent of the chain thickness, c, and it is proportional with the length of tensioned chain. For the poly(vinyl alcohol) or poly(acrylic acid) in water, where $\sigma_{cr,\ liq}$ = 10 erg·cm^{-2} and a = $3 \cdot 10^{-6}$ cm the value of Ω = $3 \cdot 10^{-5}$ dyne is obtained. A force of $5 \cdot 10^{10}$ dyne/cm^2 corresponds to a chain segment with the cross section equal to $6 \cdot 10^{-6}$. This value exceeds the tensile strength of the best stainless steel. Therefore, the scission of the network chains under freezing conditions can not be avoided. In the case of the gel based on poly(vinyl alcohol) it was found that about 15÷20 % of the network is destroyed by mechano-degradation. The elasticity modulus also decays by about 20 % from its initial value [1172]. The experimental curves for the gels based on poly(vinyl alcohol) and poly(acrylic acid) are depicted in Figure 3.396.

G. Konig studying the cryolitic behaviour of the polystyrene and of crosslinked copolymer styrene – divinylbenzene swollen in naphthalene proposed a series of comparative models. Figure 3.397 shows a gel that is a swollen unit in saturation, for which is ac-

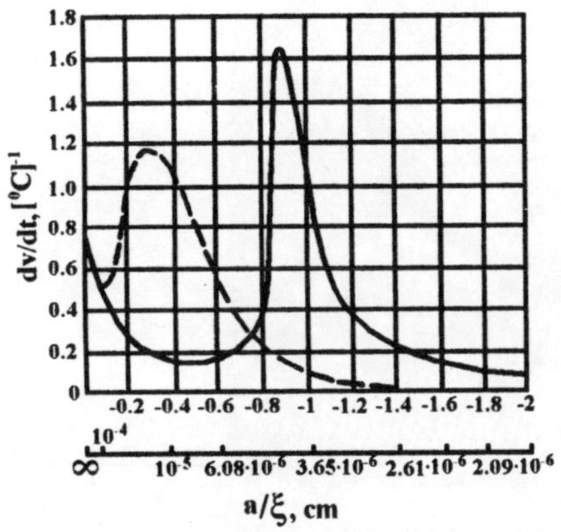

Figure 3.396. Destruction by freezing of the dissociated gels based on poly(vinyl alcohol) and poly(acrylic acid) [1177]. dv/dt is the relative amount of water required for crystallisation per centigrade. ( ) frozen curve; (———) frozen curve for the same sample, firstly defrost and filled in the same conditions and refrozen.

cepted a statistic distribution of the internodal loops. The microcrystals growth inside of gel generates tensions in network and depending on the swelling degree macromolecular strips are separated between the individual crystals, Figure 3.398. The manner in which a growing crystal stresses a macromolecular chain or a loop from a network is illustrated in Figures 3.399 and 3.400.

The use of electronic microscopy allowed the establishing the crystals average dimensions, which cause the network fracture due to the complete stretching of the surrounding chains.

3.9.7.3. The mechanism of cryolitic destruction

Under the action of the mechanical forces developed during cryolitic treatment, the covalent or ionic chemical bonds suffer the mechanocracking process, generating radicals, ions or combined particles

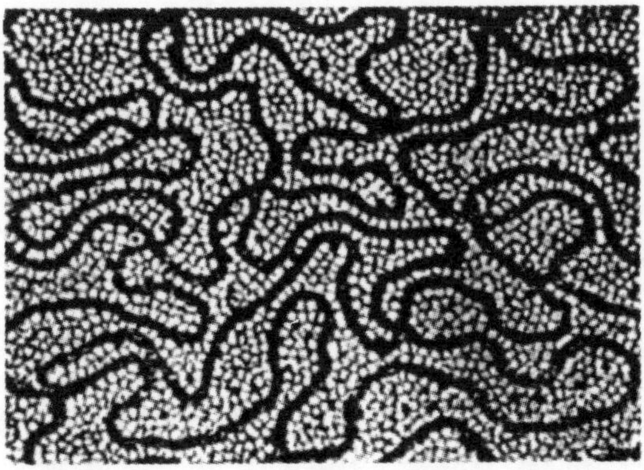

Figure 3.397. Two-dimensional model of a polymer gel swollen to saturation [1177].

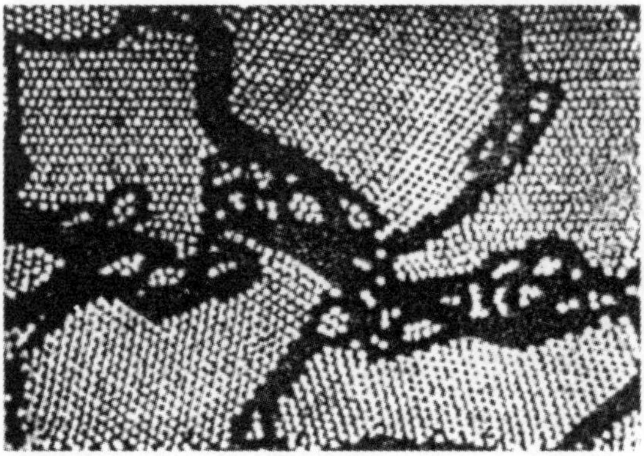

Figure 3.398. Microscrystals within gel [1177].

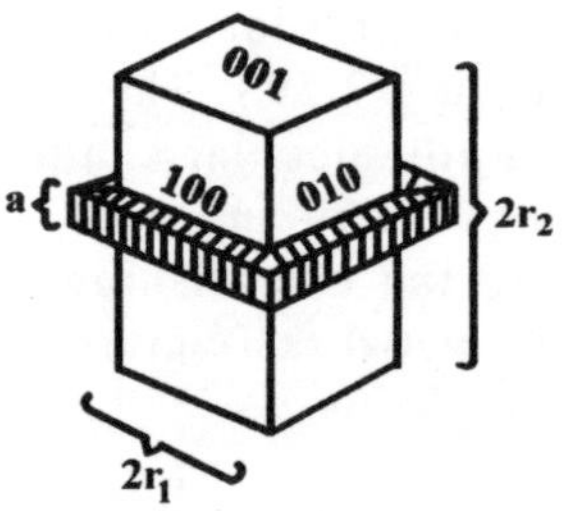

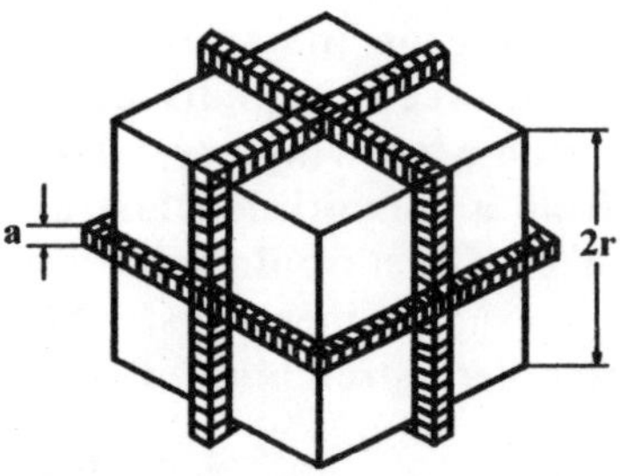

Figure 3.399. Microcrystal locked by a macromolecular chain [1117].
Figure 3.400. Microcrystal surrounded by three macromolecular chains [1117].

ion-radicals [1178]. In the throwing stage, the mechano-chemical hydrolysis can be activated, too.

Subsequently, the generated macroradicals react either with the chains that were not affected by destruction or with other components of the surrounding medium. In the case of polysaccharides, the glucosidic bonds are split with the highest probability and the amide bonds, -HN-CO-, in the case of polypeptides or proteins. If the cryolitic degradation occurs in the presence of oxygen, this one stimulates the formation of peroxy and hydroperoxy compounds.

The reaction mechanism includes the following stages:

a) Mechanochemical cracking of the polymer chain:

$$R_m{-}R_n \longrightarrow R_m^{\bullet} + R_n^{\bullet} \xrightarrow{+O_2} R_m{-}OO^{\bullet} + R_n{-}OO^{\bullet}$$

b) Chain transfer and formation of hydroperoxides:

$$R_m{-}OO^{\bullet} + R_m{-}R_n \longrightarrow R_m{-}OOH + R_m{-}\dot{C}H{-}R_{n-1}$$

c) Hydroperoxides decomposition and the initiation of mechano-degradation processes or reticulation:

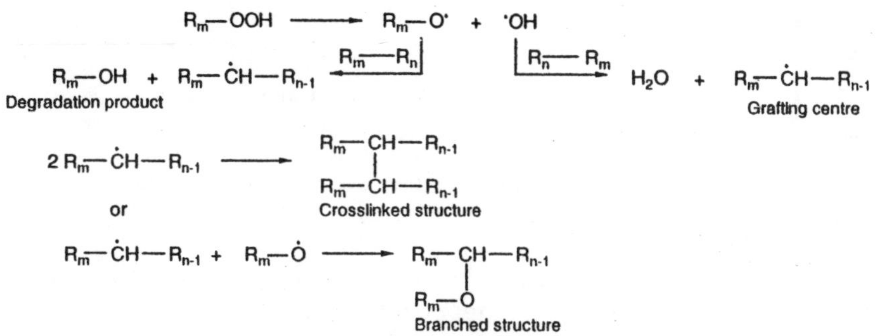

d) Interruption of chained mechanism by: recombination

$$R_m{-}OO^{\bullet} + R_m^{\bullet} \longrightarrow R_m{-}OO{-}R_m$$

$$R_m{-}OO^{\bullet} + R_n^{\bullet} \longrightarrow R_m{-}OO{-}R_n$$

and disproportion

$$R_m^{\bullet} + R_n^{\bullet} \longrightarrow R_mH + R_n$$

Most likely the cryolitic cycle includes the first stage (a). The reactions (b) and (d) are favoured under the conditions of maintaining frozen the solution or for a short heating period.

This scheme of reactions is sustained by the experimental data, such as those obtained by A.A. Berlin for starch cryolisis [1178]. Thus, by cryolisis the content of reducing substances from starch, expressed by copper index, decreases, to increase strongly in the subsequent heating stage, Figure 3.401. In this stage, the macroradicals recombination is the dominant process and the mechano-degradation process does not yet play an important role. By continuing heating, mechano-destruction is stimulated, accompanied by the increase of the amount of reducing sugars.

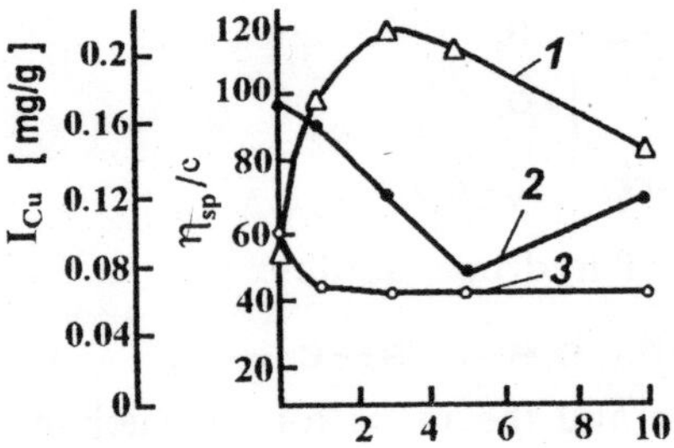

Figure 3.401. Dependence of the heating period on: viscosity (η_{sp}/c) of cryolised starch solution; 2) reduction capacity – copper index of cryolised starch; 3) η_{sp}/c of starch soluble fractions obtained after cryolisis [1178].

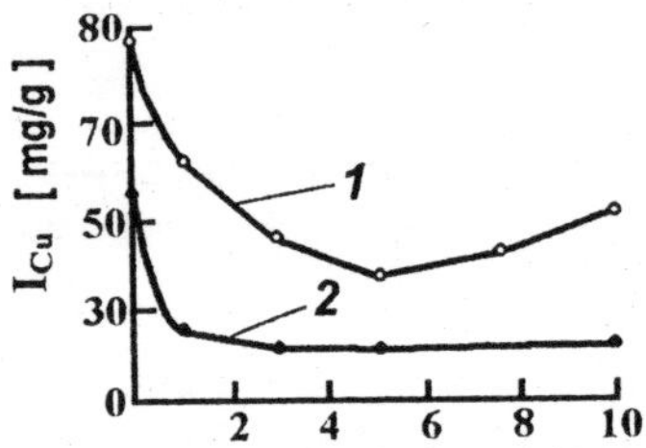

Figure 3.402. Dependence of the copper index of the water-soluble starch fraction: 1) starch cryolised in the presence of air; 2) starch cryolised in argon atmosphere [1178].

As Figure 3.402 shows, if the frozen starch solutions are heated in an inert atmosphere, the copper index drastically decays and the reducing capacity does not increase again. In the absence of peroxy radicals, the recombination reaction prevails, the chained reactions, which contribute to the starch degradation, being suppressed. In the frozen stage, the presence of the radicalic centres, in concentrations of about 10^{17} paramagnetic units/g polymer was proved using ESR spectrometry. By heating at 20 °C, the ESR signal disappears, in this stage the recorded degradation being the result of mechanochemically activated hydrolysis.

Important contributions to the clarification of mechano-chemical mechanism of polymers degradation at low temperatures belong to P.Iu. Butjagin and co-workers [1179, 1180], S.E. Bressler and co-workers [980], and S.N. Jurkov's team [1189].

The radicalic stage of the cryolitic mechanism is also sustained by the grating and block copolymerisation reactions released, for instance those carried out on the cryolised starch solution in the presence of different polymers (polystyrene) or monomers (styrene, acrylonitrile) [1169, 1182].

By increasing the number of cryolitic cycles, the polymerisation degree, decreases until a limit value, as it happens for any mechano-chemical process, Figure 3.403 [1174, 1184], and which can be expressed as:

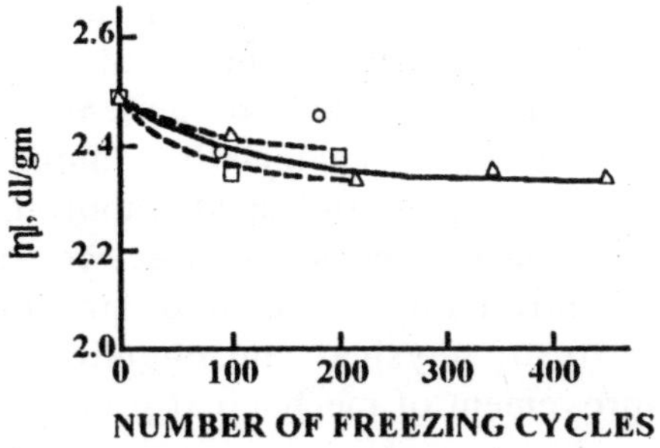

Figure 3.403. Influence of number of cryolitic cycles on the degradation by cryolisis of the aqueous solution of poly(vinyl pyrrolidone) [1183]: (○) 1; (△) 0.5; (□) 0.25. Concentration in %vol.

$$P_{\lim}\left(P_{in}^{-1} - P_t^{-1}\right) + \ln\left(1 - P_{\lim}/P_{in}\right) - \ln\left(1 - P_{in}/P_t\right) = kt \qquad (3.237)$$

where: P_{in} – the initial length of the macromolecular chain; P_t – the chain length in the moment t; $P_{\lim}$ – the chain length at the end of mechano-degradation.

H.H.G. Jellinek modelled the cryolitic degradation by assuming: 1) an impact mechanism (polymer molecules fixed at either end and lying at a right angle to the direction of advancing freezing fronts; or 2) a friction mechanism (molecules fixed at one end and lying in the direction of the streaming liquid). A rate equation has been formulated for both cases:

$$d\left(\left[\eta_{\lim}\right] - \left[n_1\right]\right)/dt = k\left(\bar{f} - f_b\right)n_{in} \qquad (3.238)$$

where: k – destruction rate constant; $\bar{f}$ - the average force acting on the long-chain molecules; f_b –the necessary force to split one bond from the main chain; $[n_n]$ and $[n_i]$ – the number of bonds at time t, and initial on the main chain.

If $\bar{f} \geq f_b$ the chain splitting will occur. The calculation of minimal value of rate required for the main chain splitting evidenced that the second mechanism is more probable.

3.10. MECHANO-CHEMICAL REACTIONS OF COMPOSITE MATERIALS WITH POLYMER MATRIX

Composite materials, initially called by H. Mark as structural-engineering materials, represent multiphase materials, containing components which together are able to induce synergic effects, which cause the enhancement of material ultimate properties, as compared to those of any individual component.

At the interface between the material's components, under thermal and/or mechanical actions, the most important physical interactions and chemical reaction occur. A direct consequence of these actions is the substantial improvement of properties of the new material. The totality of physical and chemical interactions that determines the improvement of mechanical properties are defined as reinforcement.

The essential components of composite materials are the following:

1) The matrix (a thermoplastic polymer of a thermoreactive oligomer), light, ductile, easily deforming, which constitutes the continuous phase that includes all the other components;

2) Reinforcing agent, rigid but with high mechanical characteristics;

3) Filling material, cheap, which can be included in high quantities in matrix, up to 50%, thus decreasing the cost of ultimate material;

4) Additives, representing components that play a strictly determined role, such as: catalysts, pigments, colorants, etc. usually the additives are present in small quantities into a composite material.

The polymeric composite materials are frequently produced as: 1) particle containing, whenever the matrix contains included powders or flakes; 2) fibrous, when containing long or short fibres, with standard or high performance properties; 3) laminates, especially representing layers of many materials; and 4) hybrid materials, i.e. combinations of the first three types.

The interface assures the structural integrity of the composite material and it represents an important criterion in the design of this type of materials. The correct selection of the components and their adequate processing are prerequisites in the achievement of strong bonds at interface, which, in the last instance, assure the obtaining of a material with structural integrity and good properties.

3.10.1. Elastomer reinforcement

The understanding of the reinforcing mechanism of the ingredients constitutes the base instrument for the improvement of the physico-mechanical properties ot polymers. Due to their different physical states, the reinforcing mechanism is different for thermoplastic material and for elastomers.

The activation of the reinforcing mechanism is conditioned by the whole factors influencing the material properties: chemical structures both of the polymer matrix and reinforcing reagents, their physical state and shape (highly dispersed powders, fibrous materiāls, layer rowings), polymer adhesion to the solid surfaces of other components of the charge, and the conditions of subsequent processing. In essence, the reinforcing ability is related to the adhesive properties of polymers that are concretised by the formation of the physical, adsorption, and chemical bonds between the surface of reinforcing reagent and the polymer matrix. The matrix interaction with the ingredient active surface determines the deformation and fracture character of the reinforced material, conferring to the final product a set of new and valuable properties.

Rubber reinforcement is a process of mechano-chemical nature, consisting of elastomer structuration from the crude mixture under the action of the mechanical forces. It depends both by the polymer destruction degree and mechanical regime applied during the processing. The absence of radical acceptors in the reaction medium, especially of the oxygen, does not modify the nature of the transformations produced on the properties of mechanically processed crude mixtures.

The determination of the structural transformations, which determine the modification of tensile strength and elasticity as well as the utilisation of some physico-chemical methods, allows the quantitative evaluation of the mechano-chemical phenomenon that occur during processing period, evidencing their contribution to the modification of rubber properties at the end of processing. The structural modifications that occur during processing are stable in time, being maintained under the temperature and cross-linking agents action during elastomer vulcanisation.

Depending on the processing temperature and deformation rate, the mechano-degradation mechanism should be different, which imparts different and specific properties to the fracture surface. The modification of the fracture surface characteristics during the mechanical processing was followed on the elastomer mixtures with different ingredient (black carbon, chalk), firstly in the vitreous range of temperature and subsequently the investigations have been extended in a large range of temperature [462–465].

In all cases, the presence of the free macroradicals at the new-formed surfaces by fracture was proved. It was established that these radicals are able to react with the active compound, for in-

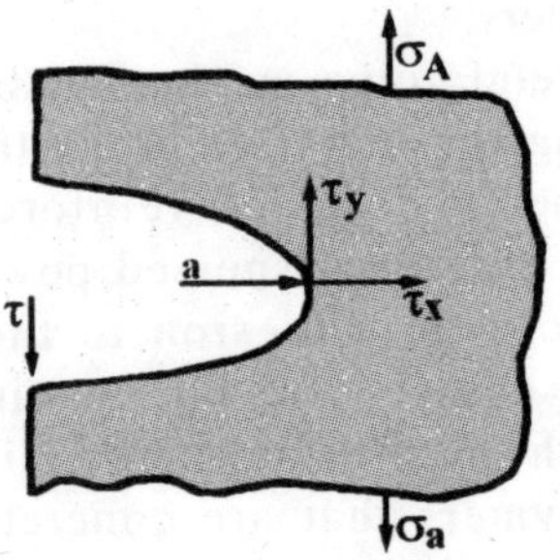

Figure 3.404. Stress concentration in the tip of an elliptical crack within an izotropic material [1184].

stance black carbon that play role of a radical acceptor. A significant increase of the mechanical properties can be seen at the end of the process.

The developed effort during mechanical solicitation is located at the top of the cracks, formed by fracture, or on the other faults from the polymer material. Thus, for a crack of elliptical shape, the effort concentration is illustrated in Figure 3.404.

As the figure shows, the maximal values of the tensile component at the crack tip, can by expressed as [1184]:

$$\sigma_{y_{max}} = \sigma_A \left[1 + 2\left(\frac{a}{r}\right)^{1/2} \right] \tag{3.239}$$

$$\sigma_{x_{max}} = 1/5 \cdot y_{max} \tag{3.240}$$

In elastomers, the crack depth, a, has values of about $3 \cdot 10^{-3}$ cm, and the maximal value or the crack radius, r, is 10^{-6} cm. For these values, the tension at the crack tip increases more than 10 times. For larger cracks or other types of faults, even greater increases of tension were recorded. However, if the crack presents a flattened top, σ_{max} decreases and the materials manifest an enhanced crack growth resistance. The faults are statistically distributed in the material volume, having specific shapes and dimensions, in accordance with the chemical structure of the elastomer and composition of the ingredients mixture [1185].

The structuration by reinforcement is manifested to the strictly determined values of temperature and deformation rate. Low resistance rubbers show thick ripped surfaces, while striated surfaces characterise the high resistance ones. The striation effect arises as result of increase of diameter in the crack top and it represents the most important factor that determines the increase of tensile strength in the presence of reinforcing agents.

The investigation by electronic fractography of the mechanism of striated surfaces formation, under the ripped/shear conditions, revealed that in the rubber mixtures with reinforcing reagents, the fracture propagates after a sinuous trajectory, crossing from a rubber-reinforcing interface to other one. The surface of the active ingredient particles, in fact the parts involved in the fracture process represents the "weak points" from which this one starts [1186]. Starting from this conclusion, the structuration action induced by the active ingredient and therefore the reinforcing effect were interpreted. The numerous internal faults, which determine energy dissipation, due to the increase of rubber volume that inevitably suffer strong stretching, induces the fracture process. The volume involved in deformation depends on the strength of rubber adhesion to the filling agent surface. An additional cause that determines the reinforcing effect is related to the high value of the tensions relaxation rate, in the range of high deformations from the growing crack proximity. On its turn, the degree of tensions relaxation is governed by the rubber mechanical hysteresis and especially by the difference between the character of the strength–strain curve in the materials stretched and compressed phases, respectively. It was established that the reinforced rubbers presenting a maximal mechanical hysteresis assure a maximal relaxation of tensions.

Using uncharged but vulcanised natural rubbers, the influence of hysteresis that appears as result of crystallisation caused by the material stretching, just in front of the crack top, was proved. The crystallised zone appears as the most important and closest inhomogeneity on which the mechanical energy concentration is focused, after which the crack enters into a zone of lower tension and its growth by impulses follows. In this stage, the crack diameter strongly increases, Figure 3.405.

The sharpening of the crack tip and its growing rate are interconnected with each other and depend on the elastomer hysteresis behaviour. It was found that the hysteresis of charged rubber is higher than that corresponding to the elastomer matrix. The main reason of the increase of hysteresis in the released energy during the splitting of physical bonds that keep in shape the particles of active ingredients. In addition, the chains mobility at the particles surface is drastically diminished, due to hydrodynamic reasons. The use of black carbon determines the shifting and broadening of the curve that describes the dependence with time of energy and hysteresis,

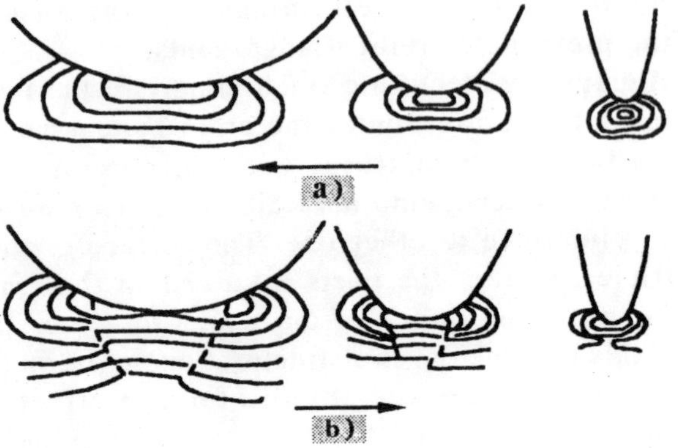

Figure 3.405. Influence of mechanical hysteresis on the profile of crack tip and of fracture surface for crystallisable rubbers [1184]: (a) stretching; (b) compression.

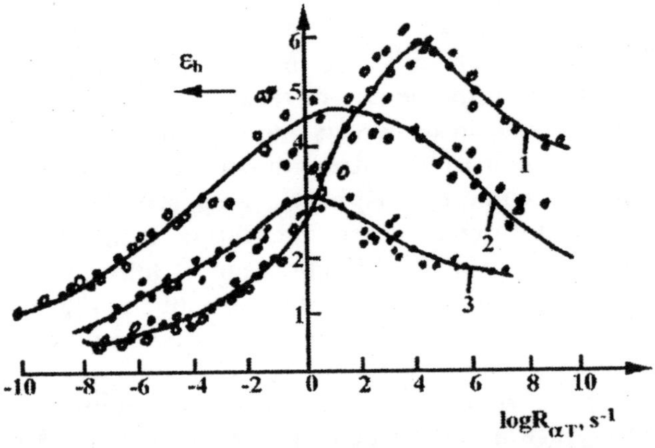

Figure 3.406. Dependence of strain, ε_b, by the deformation rate $R_{\alpha T}$ at 294 K: 1) poly(butadiene-co-styrene) vulcanised rubber without filler; 2) idem, containing 30 %vol. black carbon; 3) idem, containing 60 %vol. black carbon.

whenever the fracture occurs as effect of the increase of relaxation times, Figure 3.406.

At low degrees of deformation, the increase rubber rigidity, due to the use of reinforcing agent, can be calculated using the classical equations that describe the increase of viscosity of a liquid within which solid particles have been introduced. Young modulus, expressed as stress to strain ratio, increases in the presence of fillers with a coefficient X, so called "deformation increase" [1187, 1188].

Under the action of stretching forces, the vulcanised rubbers, with or without fillers, suffer a softening effect that was interpreted as result of modification of network configuration, due to the displacement, during deformation process, and incomplete rotation of the junction points, in the recovering step, to the initial position. Consequently, the subsequent deformations will take place more and more easy with respect to the previous ones [1184]. Both by increasing the temperature or swelling of the non-reinforced vulcanised rubbers, after deformation the initial rigidity is almost totally recovered, while in the case of reinforced rubbers this effect is only partially obtained and much more slowly. Under the deformation conditions, the filler particles or their agglomerates are displaced through the rubber network, thus limiting the material ability to regain the initial configuration of equilibrium. Consequently, the elastomer hysteresis and capacity of energy absorption increase, which are responsive for the differences observed in the shape of stress–strain curves [1187–1190].

The study of the fracture process of the elastomer-based systems containing fillers evidenced that the weakest zone is the interface elastomer-particle [1191]. The fracture occurs by the destruction of the existing bonds in the contact area, especially of the adhesive bonds. In means that the material structuration by reinforcement is intensified by the increase of adhesive forces. Therefore, using black carbon as reinforcing agent, the maximal effect is achieved whenever the elastomer is in highly elastic state [1192]. Consequently, it results that the reinforcement sensibly depends on temperature [464].

A reinforced elastomer with black carbon, cooled below the flowing temperature, suffers an instabilisation, due to the adhesive bonds weakness at the elastomer-particle interface, the released contraction energy being mainly concentrates in the contact zone of the two components. Due to the difference between the values of dilatation coefficients that characterise the two materials, additional tensions are generated, which supplementary will stress the interface. With the decrease of the difference between the dilatation coefficients decreases, the destruction of the contact zone will occur at lower temperatures. The decay of polymer adhesion determines the system weakness [1192]. It results that in case the elastomers reinforced with black carbon, at exploitation temperatures lower than the elastomer flowing temperature, the black carbon is not a suitable reinforcing reagent.

The systems based on poly(butadiene-co-styrene) elastomer and

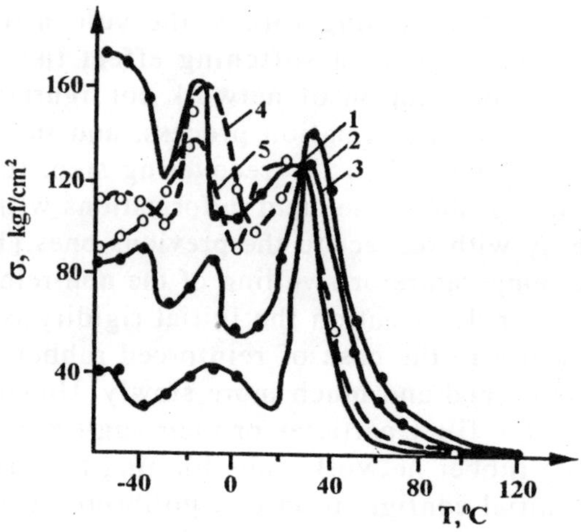

Figure 3.407. Dependence of tear tensile strength on temperature: 1) poly(butadiene-co-styrene) rubber; 2) idem with black carbon; 3) idem with chalk; 4) idem mixture with polyamide; 5) idem mixture with cellophane [464].

polymeric fillers, such as polyamides or cellophane, shown in the range of temperature from –60 to 110 °C an exponentially dependence of the ripped tension with temperature, bearing the characteristics of highly elastic domain, Figure 3.407. In the vitreous range of temperature, the curves show the same shape and differ each to another only by the absolute values of σ. The decrease of temperature belows the elastomer brittle temperature practically does not determine significant changes of the material resistance. Comparatively, it was established that in the rubbery domain the classical ingredients, such as black carbon and even chalk, assure the maximal stability. In the vitreous domain, the macromolecular ingredients are most active ones.

Under cooling conditions, admitting the adhesive nature of the reinforcing phenomenon and taking into account the importance of differences between the values of dilatation coefficients, it results that the tensions concentration at the interface is lower for the mixtures containing macromolecular reinforcing agents than that corresponding to the traditional materials, Figure 3.408.

In accordance to these results, the black carbon, the most efficient reinforcing reagent in the elastomer rubbery domain, determines the most pronounced weakness effect in the vitreous range of temperature, while the polyamides, at 10 °C, increase the total work to fracture with 80%. Electron fractography confirmed that in the

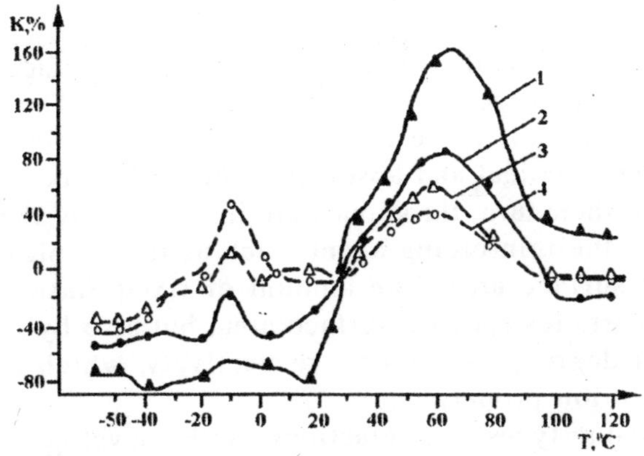

Figure 3.408. Dependence of reinforcing effect, under tearing conditions, by temperature: 1) poly(butadiene-co-styrene) rubber with black carbon; 2) idem with chalk; 3) idem with cellophane; 4) idem mixture with polyamide [464].

range of low temperatures, in the case of inorganic fillers, the tension concentration occurs around of reinforcing reagents particles, which is not case for the macromolecular fillers.

Concluding, the currently used filling materials at industrial scale are not adequate for the obtaining of rubbers that follow to be used at temperatures below flowing temperature of the matrix.

In the case of elastomers according to E.M Dannenberg, the reinforcement represent the phenomenon of the increase of rigidity, modulus, fracture energy, resistance to fatigue, cracking, and abrasion [1193]. From the practical point of view, it is defined as the process that determines the increase of the exploitation period of polymers, due to the increase of their physico-mechanical properties.

The dimension and shape of ingredient particles, their tendency of agglomeration, deformation of structures, dispersion capacity, and chemical nature of the external surface are characteristics that decisively influence the ultimate properties of reinforced rubber.

G. Kraus investigated in detail the types of interaction elastomer-reinforcing agent. In the first place, he found that the type of interactions rubber-black carbon depends on the characteristics of the black carbon surface. He admitted that the macromolecular fragments formed by scission of the chains under the action of mechanical energy, during the processing period, can be grafted on the black carbon surface. These ones can react with the continuously appearing macroradicals at the newly formed surfaces, either by their free electrons or by different functional groups, which appears by radicals stabilisation.

It was established that at the black carbon surface about 10% of the existent hydrogen is highly reactive [1195]. The strong binding of the elastomer to the black carbon is determined by the intensity of the chemical and absorbing interactions developed at the interface. On their turn, these ones are influenced by the some characteristics of the reinforcing agent, such as its chemical nature and the specific surface area. The amount of fixed elastomer depends just by the particles specific surface area, but also by the elastomer unsaturation degree, its polarity, respectively, which is responsive by the absorption properties.

The following types of interactions can be developed between the elastomer and reinforcing agent: 1) elastomer adsorption on the surface of black carbon; 2) the reaction between the ingredient and the macroradicals formed during processing; 3) retarded reaction of the secondary radicals, occurring after processing; 4) reaction with the macroradicals that appear as result of polymer thermal degradation; and 5) the reaction with functional groups formed after stabilisation of the free macroradicals.

Accepting the formation of strong bonds of elastomer-ingredient particle type, L. Mullins proposed the modification of the kinetic theory of elasticity for adequately describing the behaviour of filled elastomers [1196].

In accordance with kinetic theory, the uncharged elastomer network is deformed under the action of uniaxial stress, σ, the elongation ε being correlated with the applied stress by the following relation:

$$\sigma = \upsilon kT\left(\varepsilon - \frac{1}{\varepsilon^2}\right) \qquad (3.241)$$

where: ν is the crosslinking density; k – Boltzmann constant; T – temperature in Kelvin degrees; ε - elongation, %.

In order to adapt this equation to the system representing the charged elastomer, the following simplified hypotheses, valuable for the majority of elastomers, were formulated: 1) the volume of material is not changed by deformation; 2) the initial distribution of the chains length is of Gaussian type; 3) chain segments of the network are characterised by the finite value of deformation, therefore of elongation; 4) during elongation, the internal energy of the network does not vary significantly; and 5) the adhesive forces elastomer-filler are enough strong so that during stretching the bubbles do not appear in the processed samples, otherwise the first hypothesis is not valuable. The volume measurements performed on the natural rubber and poly(butadiene-co-styrene) copolymer proved that the difference in the volume variation of the unfilled and filled rubbers under stretching conditions can be considered equal to zero.

A.M. Bueche was one of the first scientists that attempted to translate the theory of high elasticity to the charged rubbers. He assumed the existence of the crosslinking centres at the black carbon surface, which contributes to the increase of network crosslinking degree. In this case, the equation (3.241) becomes:

$$\sigma=\left(\nu+\nu_f\right)\cdot kT\cdot\left(\varepsilon-\frac{1}{\varepsilon^2}\right) \tag{3.242}$$

where: ν is the number of transversal chains in the unfilled elastomer; ν_f – the number of additional bridges, introduced due to the presence of active ingredients.

The value of ν_f is correlated both to the ingredient surface of that is in direct contact with the elastomer and the average surface divided to the number of reticulation centres. On the other side, ν_f is proportional to the volume fraction of the active ingredient, ν_2, and similarly ν_f is proportional to the volume fraction of the elastomer, ν_1. Consequently, equation (3.242) becomes:

$$\sigma=\left(\nu\nu_1+\beta\cdot\nu_2\right)\cdot kT\cdot\left(\varepsilon-\frac{1}{\varepsilon^2}\right) \tag{3.243}$$

where β is the proportionality constant.

Considering that S is the available charge surface of the rubber unitary volume and s is the average surface per number of crosslinking centres, then:

$$\nu_f=\left(S/s\right)\nu_2 \qquad \text{or} \qquad \nu_f=\beta\cdot\nu_2 \tag{3.244}$$

The graphical representation of $\sigma/\nu_1KT(\varepsilon - 1/\varepsilon^2)$ *vs.* ν_1/n_2 leads to a straight line in the case of silicone elastomers reinforced with silica. The experimental results, reported by A.M. Bueche, are really located on a straight line, in the limits of performed measurements, *i.e.* $0 \leq \nu_2 \leq 0.25$ and $1 \leq \varepsilon \leq 1.5$, respectively, Figure 3.409. Using the graphically calculated values of b, the author found the value s (the surface corresponding to a crosslinking centre) varying from 2 to 40 Å, as a function of the type of used ingredient.

G. Krauss graphically represented the Mooney–Rivlin equation as $\sigma/(\varepsilon - 1/\varepsilon^2)$ vs. σ^{-1}, for the ethylene-co-propylene rubber, uncharged or charged with 15% black carbon respectively, Figure 3.410.

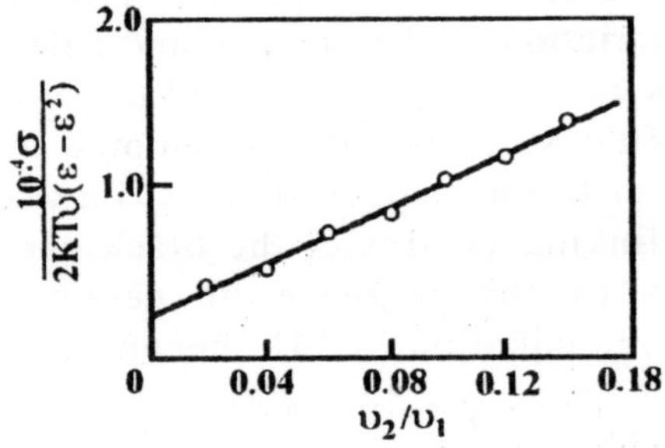

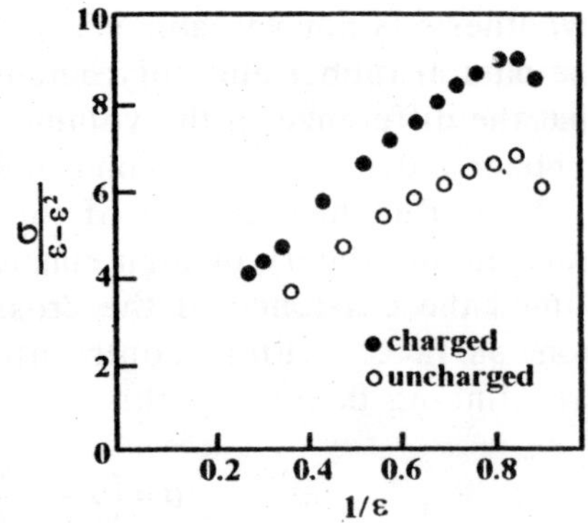

Figure 3.409. Verification of equation (241) in the case of silicon rubbers reinforced with silica (σ is given in dyne/cm²) [1197].

Figure 3.410. Graphical representation of Mooney–Rivlin equation (244) for the unreinforced and reinforced with 30% black carbon ethylene-propylene rubber at 30°C (σ in dyne/cm²) [1198].

In its general form, the Mooney–Rivlin equation is expressed as:

$$\sigma=\left(2C_1-2C_2/\varepsilon\right)\left(\varepsilon-1/\varepsilon^2\right) \qquad (3.245)$$

or

$$\sigma\Big/\left(\varepsilon-\frac{1}{\varepsilon^2}\right)=\left(2C_1+2C_2/\varepsilon\right) \qquad (3.246)$$

Graphical representation of equation (3.246) gives a straight line with the slope equal to $2C_2$ and the intersection point with the ordinate equal to $2C_1$. From equations (3.231) and (3.246) results that $2C_1 = \nu kT$ and $2C_2 = 0$, the condition that is not fulfilled in Figure 3.410. If the same samples are swollen in benzene, until $\nu_r = 0.3$ (ν_r is the volume fraction of rubber in the swollen network), as is seen in Figure 3.411, the value of C_2 tends to zero. These results confirms the theory of A. Cifferi and J.P. Flory, which postulated that C_2 is equal to zero in the equilibrium conditions. The same researchers have foreseen that "viscous" displacement of the

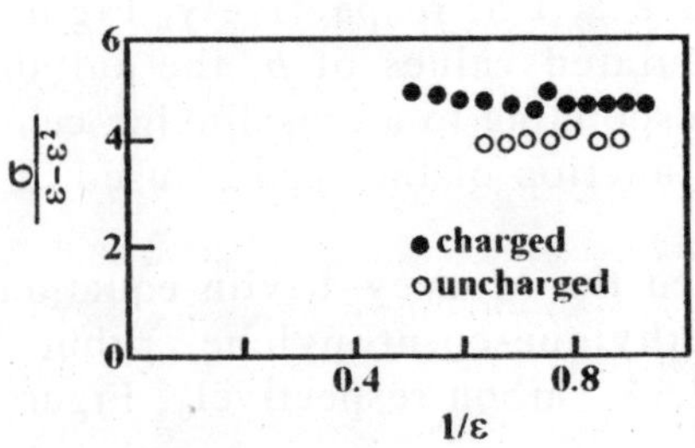

Figure 3.411. Graphical representation of the Mooney–Rivlin equation for the ethylene-propylene rubber swollen in benzene [1198].

network junctions and knots as well as of the charge's particles will give rise to the differences from the kinetic theory of rubbers elasticity. In the case of swollen elastomer all these movements occur almost instantaneously, due to the solvent presence, then the equilibrium conditions prevail. This interpretation is also sustained by the fact that whenever the molecular weight of a chain between two knots is high and the temperature is well above the vitreous temperature, the value of C_2 is small, too [1198].

The reinforcement induces important changes of the elastomer network, due to the formation of strong and weak bonds between the chain segments and reticulation centres located on the surface of active charge particles [1199]. The Mullins effect or the tension relaxation in networks is ascribed to the splitting of weak bonds.

The mechanical resistance of the charged rubbers was ascribed to the strong bonds that are not split until to the fracture occurrence. The notion of "strain toughening" was also introduced, which characterises the strongly deformed elastomers, by the collective contribution of the following factors: 1) the compacting of short chain segments between the active particles; 2) the orientation of the same segments, by stretching, between charge's particles; and 3) the rearrangement of macromolecular chains, in conditions of their sliding, which determines the an increase of orientation and a more uniform distribution of the stress on the solicited chains.

It is believed that the "strong" bonds are those responsible for elastomer reinforcement, the "weak" ones being split at low values of elongation and therefore their effect on the final result of reinforcement can be neglected.

For the explanation of reinforcement phenomenon, F. Bueche proposed a mechanism based on the splitting of the network chains between two adjacent particles that occur during network deformation. As is seen in Figure 3.412, due to the statistic character of the elastomer network formation, between two neighbour particles of ingredient, three types of bonds should appear [1200].

Whenever, during the rubber deformation, the charge particles are forced to move away each to others, the chain A will be instantaneous split, being just from the beginning strong stretched. On the other side, the chain B will be split at lower deformation and the chain C just before the elastomer to reach a very close deformation to that of fracture. In results that whenever during the stretching the ingredient particles are moved away each to another, the chain segment between them will be mechanically solicited and firstly split those with stretched conformations, even at low deformations of

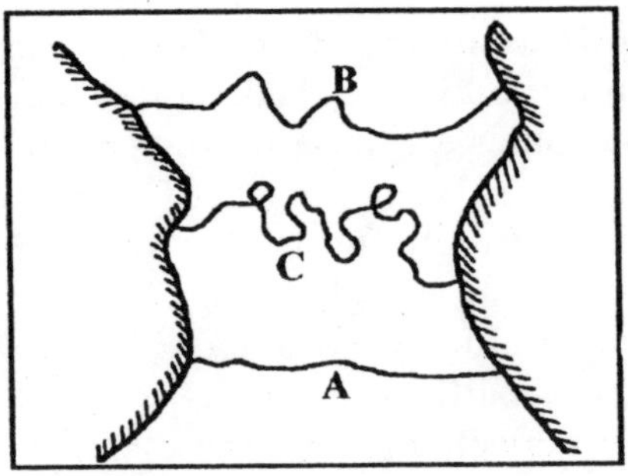

Figure 3.412. Schematic representation of the bonds between two neighbour reinforcing particles [1200].

the sample. The broken chains will not contribute to the sample mechanical resistance in the subsequent steps of solicitation and a relaxation effect will occur. The initial statistic distribution of the chains segments disappears during the first solicitation and the tensions relaxation is accomplished by splitting of some chains and the recovering to a randomly distribution of the other ones. During the splitting of the chains between the filler particles, an additional reason of stress relaxation is taken in consideration, namely the displacement of the charge particles in the elastomer matrix. The particles displacement is favoured by high temperatures and deformations occurring at low rates. However, the limited movements of particles are enough high in order to cause the relaxation of solicitation stress of the strongly stretched bonds. It results that the charge particles can increase elastomer's mechanical resistance, by a more uniform stress distribution on the network chains. This mechanism of stress distribution becomes very important, if it is known that in the case of an uncharged elastomer the chains splitting does not realise the stress transfer between the neighbour chains, but it constitutes the main reason that released of fracture process.

The mechanism proposed by F. Bueche comprises the following fundamental steps: 1) the splitting of the chain fragments located between two particles of filler; 2) recovering to a statistical distribution of tensions, after; 3) tensions relaxation; and 4) stress redistribution on the network chains.

According to this mechanism of the reinforcing action of the active ingredients, it results that the high value of the reinforced rubbers modulus is, at list partially, the result of the strong bonds between the filler particles and elastomer. The measurements of modulus, performed at high deformations, for different temperatures, proved that it suddenly diminishes at temperature values higher than that temperature for which the splitting of the elastomer-ingredient particle occurs. In the case of the poly(butadiene-co-styrene) rubber-black carbon systems and the same rubber-silica, the measurements shown that below 90 °C and for high deformations, only small variations of modulus occur. This behaviour proves that the elastomer-particles bonds are stable until this temperature.

Under the conditions of mechanical solicitation, firstly are split the most tensioned bond between particles that behave as authentic centres of crosslinking, which should be assimilated to the knots of chemical bonds; subsequently these ones do not contribute to the material resistance,when exposed to the subsequent solicitation. In this case, the chains distribution between the particles is not statistic one, since the weakened bonds have been split. At high temperatures, the scissions of and remakes of the chemical bond occur in the elastomer network. Consequently, after a certain period the chains distribution will regain its statistic character and the split chains during the first stressing period are replaced with other ones identical. Therefore, a recovering of the modulus to the initial values is expected. However, at small deformations and especially in the case of a poor dispersion of the particles, a decay of modulus is recorded, which was not recuperated during the elastomer thermal treatment.

Based on these observations, it can be concluded that in the case of normal charges, well dispersed, the reinforcement, evaluated by the increase of rubber modulus, is the result of appearance of strong elastomer–charge bonds. The high modulus, found for moderate deformations is the result of the existence of a relatively small number of chains with stretched conformation between the particles of the reinforcing reagent. These chains will be split at the attainment of their maximal stretching, when the particles from which the chain are fixed are moved away. As a consequence, the phenomenon of tensions relaxation "Mullins effect" arises. This effect will be more pronounced with the increase of number and mechanical resistance of the elastomer-particle bonds. A similar effect occurs in the case of poorly dispersed fillers, due to the destruction of the particle agglomerates. However, it is only manifested at relatively small deformations, contributing in a small measure to the rubber's mechanical properties in the case of moderate and high deformations.

In the explanation of the reinforcement mechanism, E.M. Dannenberg assumed that the segments of elastomer chains, bounded to the surface of charge particles by adsorption, can suffer, by rubber tensioning, a movement of molecular sliding, in this way accommodating to the imposed tension and thus preventing their scission.

Taking into consideration the model illustrated in Figure 3.412, during the effort application the shortest chain will be the most stretched one. Continuing the mechanical solicitation, the chain will be detached from the particle surface, two situations being expectable, namely: 1) chain splitting; or 2) its sliding. The effort period of action and its magnitude governs which from the two above-mentioned mechanisms will acts. When the chains sliding occurs, the chain bonds are not prematurely split on the incipient stages of the deformation. In the final stage, all the three segments are stretched at maximum, favouring a uniform distribution of the effort, therefore a significant improvement of the elastomer resistance and of modulus, too.

At the end of tensioning, the materials suffers a contractile process. Consequently, due to the sliding all the chain segments between the charge particles have the same length and to a new solicitation the elastomer will present a lower modulus than before of the first solicitation, since the required energy for chains sliding is not the same as in the initial state. According to E.M. Dannenberg, just this mechanism explains the Mullins effect. With other words, the sliding process appears as a consequence of energy dissipation, which otherwise should be used for the chemical bonds scission, fixed on the particles surface. For instance, it was calculated that the abrasion resistance of a charged rubber is well diminished when the absorption centres of high energy are annihilated, as it is happened using the graphited black carbon, when the abrasion resistance decreases to one fourth from the characteristic value of uncharged elastomer.

In essence, the reinforcement realised by chain sliding is based on the stress redistribution at the level of chain segments and their alignment, which contribute to the increase of vulcanised rubber resistance. The rearrangement of the macromolecular chains, under the action of applied effort, prevents their premature scission on the incipient stages of the deformation process.

The correct interpretation of the reinforcing action of the used charges must also take into account their effect on the mobility of chain segments belonging to polymer network. The elastomer layer immobilised on the surface of the black carbon particles is characterised by a higher rigidity and vitreous temperature than that of non-reinforced elastomer. The existence of a such quasi-vitreous

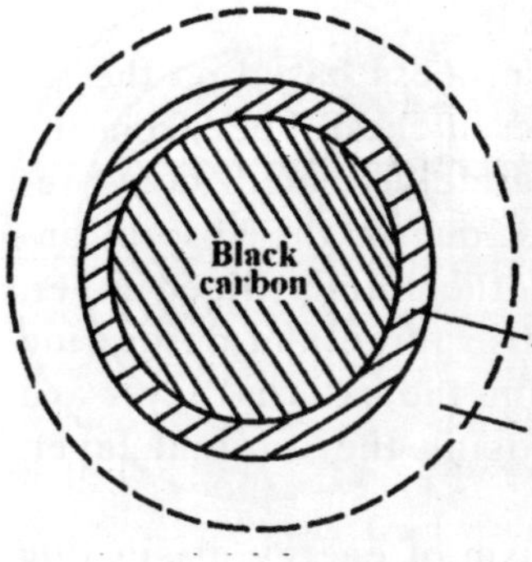

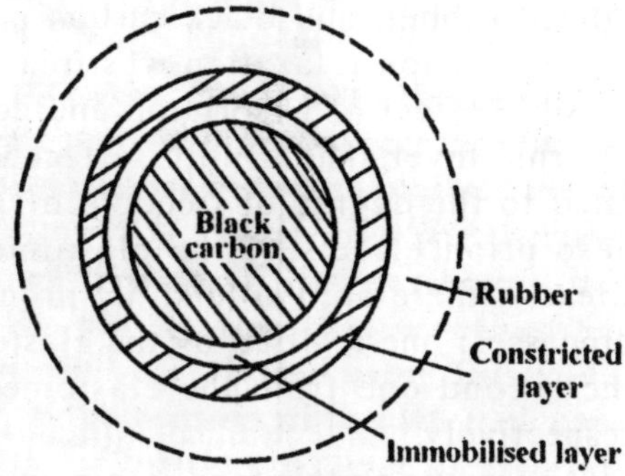

Figure 3.413. Single layer model [1201].
Figure 3.414. Double layers model [1201].

surface was proved using dynamo-mechanical spectroscopy and NMR techniques [1201]. In this way, the concept of "immobilised layer" was confirmed, being demonstrated that the black carbon induces, in the case of usual rubbers, an important limitation of the network chain segments displacement into a superficial layer of 5 Å thickness. The analysis of the genesis of the elastomer layer immobilised on the surface of black carbon particles is based on two models - with a single layer, Figure 3.413, and with double layer, Figure 3.414, respectively.

According to the first model, the system elastomer-black carbon consists in two components: 1) the rubber whose properties are slightly affected by the black carbon presence; and 2) black-carbon particles, surrounded by a thin layer of immobilised rubber, of uniform thickness. In the second model, the black carbon particle in surrounded by two layers of elastomer, namely - an immobilised layer and the second one with a relatively "constricted" mobility of the chains segments, as compared to those of uncharged elastomer. Once again, using DMA and NMR techniques, the thickness of the two layers was established as being of 5 Å and 35 Å, for the immobilised layer and constricted one, respectively.

The existence of the two layers with different mobility around of black carbon particles has important practical consequences. The mobility of the internal layer does not suffer important modifications until to 200 °C, while the constricted ones is continuously modified, behaving a limit value at 100 °C. This means that in the majority of cases, at room's temperature and even at higher temperatures, the reinforced elastomers represent systems containing four phases: pure rubber, rubber with constricted molecules, immobilized rubber, and black carbon particles.

The tension relaxation was treated as a process based on the reversible effect of absorption and desorption of the rubber from the external layer. The Mullins effect was separately treated, being related to the degree of freedom of particles, due to the absorption-desorption effects of the elastomer from the immobilised layer. Clearly, there are two distinct processes absorption and desorption processes, one realised by the elastomer from the external layer and the second one from the elastomer composing the internal layer, respectively.

According to this model, a new mechanism of energy dissipation in the rubber charged with black carbon was proposed. It is based on the manifestation of the hysteresis phenomenon by a specific mechanism of rubber absorption-desorption, in the two layers. Thus, a vulcanised rubber has the following structure: a matrix, representing the pure elastomer, which normally vulcanises; 2) inside of this rubber there are black carbon particles surrounded by the two concentric layers: a first non-vulcanised, but immobilised layer of about 5 Å and the second, external one, with a constricted motility of chains, with a lower crosslinking degree as compared to the matrix of about 30–40 Å thickness. The diffusion of the vulcanisation agents is limited by the adsorbed rubber layer that acts as a barrier, impeding their migration to the surface of black carbon.

The charged vulcanised elastomers show a higher hysteresis that the non-charged vulcanised rubbers. The introduction of black carbon particles in rubber determines the increase both of hydrodynamic and viscoelastic hystereses. The two contributions are additive. Under the conditions of the deformation rate and temperature variation, the first one remains constant, but the second one decays with the decrease of temperature. The viscoelastic contribution was ascribed to the macromolecular chains, immobilised on the black carbon surface, which modify the viscoelastic response of the elastomer network.

It was also established that the black carbon particles included in the elastomer matrix show a different behavior as a function of deformation [1202]. It is assumed that the tensions relaxation in reinforced rubbers, which represents a energy dissipation at high deformations, is not associated with any change of rubber-particles bonds, and it should by explained only by the modifications occurring in the highly elastic phase. Generally, the vulcanised or non-vulcanised elastomers present a common feature, namely to reach

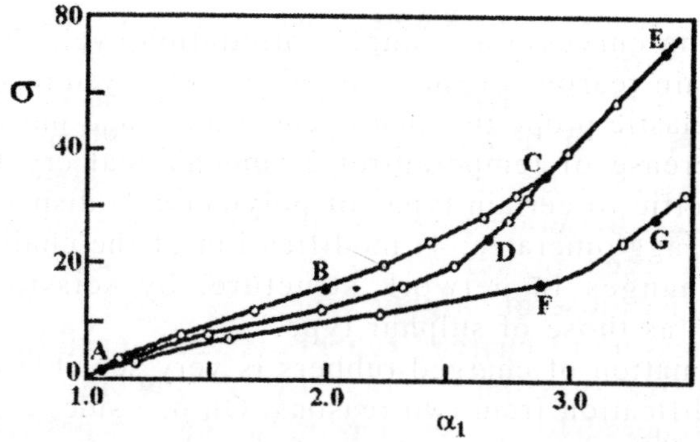

Figure 3.415. Curves of stress relaxation, at 30 °C of the vulcanised poly(butadiene-co-styrene) rubber reinforced with 30% black carbon [1202].

the same deformation, both of them require a solicitation effort higher in the first loading–unloading cycle, as compared to the next ones. The intimate interpretation of the all these phenomena can be realised by the analysis of the stress–strain curves. As Figure 3.415 shows, after the first elongation, a poly(butadiene-co-styrene) elastomer, vulcanised and charged with 30% black carbon, follows the curve ABC. At the point C the stress was removed. In the second cycle of deformation, the increase of deformation covers the route ADCE and in the point E the stress is removed. After the relaxation period the sample is stretched for the third time. In this case, the deformation evolves on the curve AFG. One can conclude that by mechanical solicitation, in the range of high deformations, the charged elastomers suffer a considerable "softening" effect.

In the chase of charged elastomers, characterised by reduced vulcanisation degrees, the tensions relaxation is based on a mechanism for which the following three steps were assumed: 1) rearrangement of the ordered regions that is associated to the network knots and junction points displacement within the elastomer matrix; 2) modification of the black carbon agglomerates structure, accompanied by deformations and the formation of new surfaces; and 3) splitting of the weak crosslinkings, such as: hydrogen bonds; ionic or polysulphidic bridges. Whenever an elastomer is deformed under the applied effort, a part of the required mechanical work is accumulated in material and the other part is dissipated as heat. This energy dissipation or hysteresis is the result of some different processes, which depend on the type of material and the processing regime. A correct evaluation of the amount of dissipated energy can be realised by measuring of the surface area between the elongation-recovering curves, for a single solicitation cycle. The following are he main reasons of the charged rubbers hysteresis: 1) polymer's visco-elastic properties that become more and more important with the decrease of temperature; 2) mechanical crystallisation, which is specific to certain types of polymers; 3) dispersion of the black carbon agglomerates; 4) modification of the chains configuration; 5) changes in network structure, by scission of weak bridges, such as those of sulphur type.

The deformation of charged rubbers is very sensible to the temperature modification from two reasons. On one side, the elastomer chains that adhered to the surface of the charge particles are detached with the increase of temperature and, on the other side, the existence of a "cage", made by elastomer macromolecular chains, immobilised around a polymer particle, which diminishes its thickness concomitantly with the increase of temperature.

At the present time, all the theories concerning elastomers reinforcement admit that this process is rigorously dependent by a series of variables, such as: the structure of elastomer network, dimension and shape of the black carbon that governs the characteristics of its specific surface, the characteristics of the elastomer-reinforcing agent interface, the nature of polymer-filler interactions at their interface. In was found that even the Van der Waals fórces are enough strong to determine reinforcing effects, but the strongest effect arise to the formation of chemical bonds between elastomer and particles. The black carbon surface contains sufficient functional groups able to react with polymer macromolecules, both during its processing and vulcanisation, respectively. During rubber processing by two-roll mixing, the develóped shear forces determine the destruction of black carbon agglomerates, accompanied by the apparition of radical centres. These ones by combination with the elastomer macroradicals, formed by mechano-cracking, lead finally to the grafting on the black carbon surface as well as to crosslinking effects, the last ones responding by the formation of insoluble "bounded rubber"

While some types of carbon (graphite, diamond) are inert from chemical point of view, the black carbon is characterised by a pronounced reactivity. This one is not only caused by the high specific surface area of the black carbon, but especially by the presence on the particles' surface of numerous oxygen containing functional groups of hydroxyl, carbonyl, carboxyl, hydroperoxy, phenol, and quinone type. Finally, the presence of some active hydrogen atoms also contributes to this effect [1203, 1204]. It is known that by graphitising, the black carbon looses almost completely its reinforcement ability.

The reactivity of the systems supposed to mechanical processing was proved not only with respect to the reinforcing agent, black carbon, but also concerning the elastomer itself reactivity. In the absence of charges, the majority of elastomers are thermodynamically non-miscible, therefore incompatible each another. In fact, thermodynamic compatibility of elastomers couples is a exception, but not a rule. However, the couples of incompatible elastomers, in function of their morphology, can reach maximal values of some properties as compared to those characterising the both pure elastomers. Therefore, in the practice of the obtaining and exploitation of macromolecular mixtures can be used the notion of "technological compatibility". Thus, the total incompatible thermodynamic polymers can manifest, in certain conditions of mechanical processing, technological compatibility, if it is especially associated with the use of an adequate reticulation system, able to assure the co-vulcanisation of the two components.

The studies concerning the reciprocal polymers' diffusion at the phases interface shown that the local diffusion strongly diminishes the interfacial tension, determining the intimate mixing of the chain segments of the two phases in contact and leading to the apparition of a transition phase. This one becomes the optimal place of occurrence of the phenomena based on the mixed combination of the macroradicals generated by mechano-cracking, under the processing conditions. The presence of this zone to the interface of the two elastomers demonstrates why the thermodynamically non-miscible polymers can sometime lead to the end products with high performances during exploitation. Whenever the physico-mechanical properties satisfy the processing complex that assures end product formation and guarantees the long time exploitation, it means the technological compatibility of the elastomers pair involved was achieved. Technological compatibility is conditioned by the correct selection of the preparation parameters of the binary mixtures as well as of the processing ones. It presents a practical importance higher then the thermodynamic compatibility [1205].

From the ensemble of presented results, it results that the elastomers reinforcement and fracture are independent phenomenon that occur between the matrix and the reinforcing reagent, under specific mechanical and thermal conditions. For a given material and into a given processing regime, the reinforcement end result is the obtaining of good mechanical properties. In fact, under the conditions of mechanical processing, the elastomer and ingredients form a new complex system of reaction, within which the components are involved in the "re-synthesis" of a new material - the composite material – whose performance is obtained for optimal composition and thermo-mechanical parameters.

The elastomers reinforcement occurs anytime when the macromolecules flow at phases interface becomes the dominant effect as compared to the chains splitting. Contrary, when the last effect prevails, the result of mechanical solicitation is the material destruction, which thus looses its exploitation properties, just the reason of its fabrication.

As a function of the technological conditions, the mechanical properties of material are differently modified. The large scale process of rubber manufacturing implies, on each of its stages, the specific mechano-chemical reactions, which, in any moment of processing, bring the macromolecules, into a physical state corresponding to the concrete mechanical and thermal regime.

It is well known that in the tyres' industry, the mixtures prepared for processing by calandering or extrusion and cooled down immediately after obtainement suffers an additional process of heating, after which their technological properties increase. By replacing this operation of supplementary heating with a mechanical processing by two-roll mixing, a new increase of the adhesive and cohesive properties, and of reaction ability is obtained as well as an improvement of the properties of final products. All these effects are the result of the mechano-chemical reactions occurrence under the conditions of supplementary processing by two-roll mixing [1206].

It was found that the elastomer properties are differently modi-

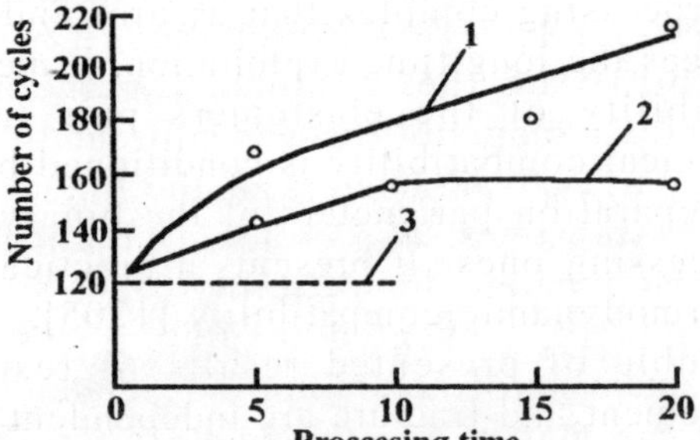

Figure 3.416. Change of mechanical properties in function of two-roll mastication period [1206]: 1) 30 min, cooled mixture after obtaining; 2) processed after slow cooling and keeping for 24 h; 3) hot processing.

fied whenever the supplementary processing by two-roll mixing is made at cold or at temperature, respectively. During cold mastication, the resistance appreciably increases, i.e. 35% for 5 min of processing and 50% for a duration of 15 min. As Figure 3.416 shows at high temperature these changes do not appear.

Once again, the use of inhibitors proved the splitting of the macromolecular chains and the formation of free radicals, during this processing step. In their presence, the plasto-elastic properties change greatly due the reaction of the formed macroradicals with the introduced additives, which, in fact, favour the decay of the elastomer molecu'ar weight.

3.10.2. Mechano-chemical reactions during vulcanisation

It was found that the mechano-chemical processes occur during vulcanisation, too. In this case, concomitantly with crosslinking reactions, some destructive processes are released both to the level of internodal segments of the network and of the transversal bridges [1207]. A such result, was recorded to the vulcanisation with sulphur, in the diphenylquanidine and activators presence, when the crosslinking degree decreases, mainly due to the transversal bridges splitting [1208].

The evolution of the destruction process that accompanies the rubber crosslinking by vulcanisation can be followed by kinetic calculations, starting from the number of reticulated structural units on each elastomer macromolecule, which represents the fundamental characteristic of this process. Denoting this parameter by r_τ it can be written [1209]:

$$r_\tau = \frac{M_{n\tau}}{M_{chem\tau}} \tag{3.247}$$

where $M_{n\tau}$ is the chain's molecular weight at a given moment, t; M_{chemt} - is the molecular weight of the chain between two transversal chemical bonds; t - duration of vulcanisation process.

The value of r_τ can be deduced from the equation:

$$\frac{1}{r_\tau} = S_z + \sqrt{S_\tau} \tag{3.248}$$

where S_τ is the the content of the soluble elastomer at moment t.

$M_{chem\tau}$ is calculated using the following expressions:

$$\frac{1}{2M_{fiz\tau}} = 0.32 \cdot 10^{-4} + \frac{1.2}{2M_{chem\tau}} \tag{3.249}$$

and

$$\frac{1}{M_{fiz\tau}} = \frac{1}{M_{c\tau}} + \frac{5.44}{M_{n\tau}} + 0.64 \cdot 10^{-4} \tag{3.250}$$

where: $M_{fiz\tau}$ - molecular weight of the segment comprised between two transversal bridges, corrected for the whole length of chain; $M_{c\tau}$ - molecular weight of the internodal segments, deduced from the swelling data and using the Flory–Rehner equation.

From the equations (3.247) – (3.250) we obtain:

$$\frac{1}{M_{c\tau}} = \frac{1}{M_{chem\tau}}\left(1.20 - \frac{5.44}{r_\tau}\right) \tag{3.251}$$

This equation permits the calculation of $M_{chem\tau}$ and $M_{c\tau}$, based on the swelling data and the quantity of soluble fraction in any moment of vulcanisation.

Under the ideal conditions, when the vulcanisation is not accompanied by mechano-cracking, the crosslinking progresses by the value $\frac{1}{2M^*_{chem}}$. The secondary reactions can be correlated to the decay of the number of *cis*-double bonds, which for the initial phase of vulcanisation is expressed as:

$$\frac{1}{2M_{chem\tau}} = \frac{1}{2M^*_{chem\tau}} \tag{3.252}$$

For a given moment of vulcanisation the following relation is valid:

$$\frac{S_\tau}{S_z} = \frac{1/2M_{chem\tau}}{1/2M^*_{chem\tau}} \tag{3.253}$$

therefore $M^*_{chem\tau}$ can be determined in any moment of vulcanisation.

Under the assumption that in no secondary reactions occur in the system, knowing the initial values of M_{n0} and $M^*_{chem\tau}$, and using the equations (3.247) – (3.250), the value of 1/2 M^*_c can be calculated. In these conditions, the kinetic curve of reticulation can be plotted.

Analysing now the real case, one can conclude that it is disturbed by the mechano-cracking of macromolecular chains and of sulphur crossbridges. Two elementary steps, components of mechano-degradation, have been considered here, namely: 1) the

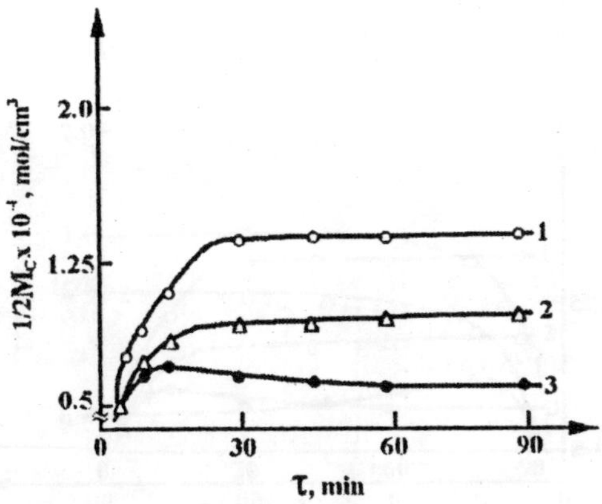

Figure 3.417. Crosslinking kinetic curves of polyisoprene rubber with molecular weight M_w=123,000 [1209]: 1) ½ M_c^*; 2) ½ M_c^{**}; 3) ½ M_c.

chains splitting; and 2) the transversal bridges splitting. In the first case, the value of M_{n0} changes and that corresponding to $M^*_{chem\tau}$ remains constant. At any moment of vulcanisation, the system will be characterised by $M^*_{chem\tau}$ and different values $M_{n\tau} = M_{chem\tau} \times r_\tau$. Using the same system of equations as in the ideal case, the value 1/2 M^*_c is calculated and the kinetic curve of reticulation, when this process is accompanied by chain mechanocracking can be constructed, too, curve 2 in Figure 3.417. In a similar manner, the kinetic curves describing the reticulation with sulphur were obtained, Figures 3.418 and 3.419. The founded differences between the corresponding values from curves 1 and 3 are related to the mechano-degradation that occurs during vulcanisation and those between curves 1 and 2 are related to the destruction of the transversal bridges, respectively. It is clear that the destruction of the macromolecular chains prevails, those belonging to the splitting transversal bridges being insignificant one. However, this one becomes important for the mixtures containing high amounts of sulphur.

3.10.3. Mechano-chemical reactions during elastomer regeneration

Subsequently, some considerations concerning the mechano-chemical reactions that accompany the process of regeneration of vulcanised rubbers are reviewed.

The vulcanised, rubbers supposed to regeneration, by two-roll mixing, at temperatures between 50 and 60 °C, by the combination of mechanical and chemico-colloidal actions, regain the final prop-

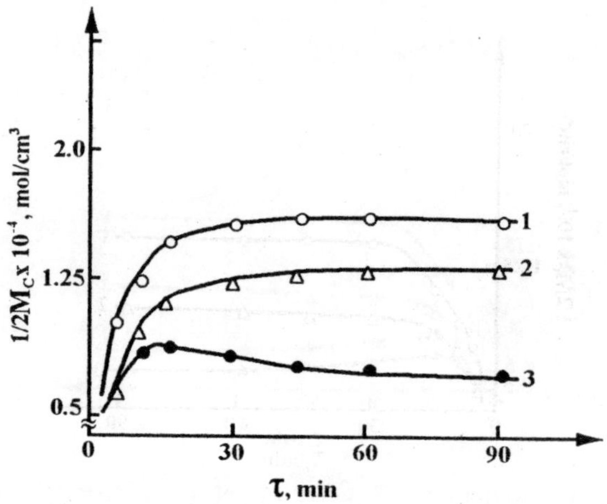

Figure 3.418. Crosslinking kinetic curves of polyisoprene rubber with molecular weight M_w=126,000 [1209]: 1) ½ M_c^*; 2) ½ M_c^{**}; 3) ½ M_c.

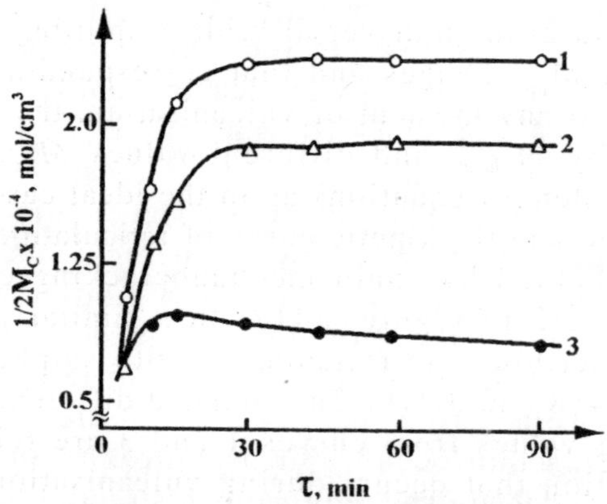

Figure 3.419. Crosslinking kinetic curves of polyisoprene rubber with initial molecular weight M_{n0}=140,000 [1209]: 1) ½ M_c^*; 2) ½ M_c^{**}; 3) ½ M_c.

erties almost similar to the initial rubber. From this reason, the finding of new methods of improving the quality of the regenerated elastomer obtained by mechano-degradation as well as of optimal utilisation of the energy consumed for this purpose present a peculiar practical interest. These requirements suppose the knowing of influence of the initial elastomer nature, type of transversal crossbridges, nature of filler, plasticizer, radicals acceptors on the mechano-degradation mechanism as well as the determination of the changes that occurred in the network structure by mechanical solicitation [1210].

Comparing the behaviour of the vulcanised natural rubber to those of other synthetic rubbers, reticulated with peroxides or with sulphur, the following conclusions were imposed: 1) the vulcanised rubbers containing peroxy bridges are more intensely mechano-chemically destructed that those containing sulphur bridges; 2) the number of crossbridges represents an important structural factor for tuning of the mechano-chemical reactions. Thus, under the conditions of regeneration, the vulcanised rubbers containing approximately $2 \cdot 10^{19}$ crossbridges/cm^3 mainly suffer the structuring effects, destruction remaining a secondary effect. For a lower number of crossbridges, for instance, $1 \cdot 10^{19}$ crossbridges/cm^3, the effects are inverted. However, in the case of filler containing vulcanised rubbers, the destruction prevails anytime; 3) The plasticizer presence determines the same effect to the sulphur vulcanised rubbers and it does not affect the process in the mechano-cracking kinetic of those containing peroxy crossbridges [1211].

The increase of the number of crosslinked molecules by mechano-degradation of vulcanised rubbers both with sulphur and peroxides proves that in the two cases the elastomer chains are those suffering the splitting process. The determination of the rate constants of splitting of the reticulation points (k_1) and of the macromolecular chains (k_2) was carried out on a peroxy-type vulcanised rubber. In this study, the secondary structuring effect was neglected and the condition as the energetic consumption of 1kW/hr to be enough for a long period, τ. It was found that under the investigated conditions, the network knots are also supposed to an intense process of splitting, the value obtained for k_1 being higher more than a magnitude order than those of k_2, i.e. $k_1 = \frac{8}{\tau} \cdot 10^{-2}\ \text{min}^{-1}$ and $k_2 = \frac{1.35}{\tau} \cdot 10^{-3}\ \text{min}^{-1}$, respectively.

In the case of sulphur-based vulcanised rubber, even if the mechano-cracking of the sulphur bridges was observed, it occurs much more slowly than the thermal one.

By vibratory milling of sulphur containing vulcanised, at low temperatures, 80 °C, and under vacuum, it was proved that the mechano-cracking process is accompanied by the migration of the carbon free electron to the sulphur atom [1212].

The studies performed on the vulcanised or non-vulcanised natural and synthetic elastomers, of polyisoprene and polybutadiene type, allowed the elucidation of the molecular-structural transformations, recorded in different time–temperature conditions. On their turn, these transformations were correlated with the resistance, fatigue, and relaxation specific properties of the investigated elastomer. The two important effects simulated during elastomers processing, the structuring and mechano-degradation, were analysed by IR spectroscopy [1213]. The intensive mechanical processing of the elastomers determines important changes of the IR spectrum, especially in the range from 1100 to 700 cm^{-1} for polybutadiene rubber. The changes are less important in the case of natural rubber. The structural modifications of the polybutadiene rubber insoluble fraction are reflected by the increase of the absorption bands at 965 and 1015 cm^{-1} and the decay of that located at 720 cm^{-1}, expressing the cis-trans isomerization process that affects the chain microstructure (structural regularity). As expected, these transformations occurring at molecular level are transmitted to the supramolecular level, being evidenced in the images obtained by electronic microscopy, but more evident at the macroscopic level by the modification some properties, such as the tensile strength, fatigue, and relaxation.

The conclusions concerning elastomers mechano-chemistry and fracture are sustained by concordant results obtained by G. Ivan and co-workers in Romania [1214–1230].

3.10.4. Reinforcement of the thermoplastic polymers and thermoreactive oligomers

The reinforcement of thermoplastic or thermoreactive polymers, by using different ingredients, represents the fundamental principle on which the design and obtaining of the composite materials with superior properties are based. Their manufacturing implies the use of processes and devices that are typical for the mechanical processing of polymers, in order obtain the end products.

The main characteristic of a composite system is its heterogeneity, expressed by the existence of the interface between components, which concentrates the most important transformations that are responsive for the obtaining of a material with superior properties. The adhesion, mechano-chemical and thermo-chemical reactions, etc., initiated in the interfacial (of transition) layers between layers, which occur by specific mechanisms, will determine or not the increase of physico-mechanical properties.

3.10.4.1. The role of the adhesive contact

From thermodynamic considerations, the formation of the adhesive contact between polymers is related to the appearance of surface excess. Any thermodynamic parameter Y, arbitrary and extensive, and y an intensive parameter, characteristic to the material volume, reported to a single molecule, determine the excess of energy at surface, which can be expressed as:

$$\Delta^S = Y - Ny \tag{3.254}$$

where: N is the number of constituent molecules of the phase.

In accordance to Gugenheim approximation the surface layer represents a zone with a certain thickness, within which the interfacial forces are not in equilibrium. Such regions appear when at least two phases interact in the following combinations: liquid-gas; solid-gas; solid-liquid; and liquid-liquid. Based on statistical physics, for the interfacial layers was defined a state function (3.224) and a probability of distribution (3.225), into a wider canonized ensamble:

$$Z = \sum_N \frac{(j^N)}{N/h^{3N}} \cdot \left[\int \exp\left(-\frac{i}{2MkT}\right) \cdot di_x \cdot di_y \cdot di_z \right]^N \cdot \left[\int \exp\left(-\frac{E-\mu N}{kT}\right) dV_1 .....dV_N \right] \tag{3.255}$$

where $i(i_x, i_y, i_z)$ – the impulse of a microparticle and its projections on the co-ordinates axes; M – particle's mass; E – potential energy of the direct interactions between system particles; $dV_1 ......dV_N$ the microvolumes occupied by the particle; j – the increment of the rotation and vibration movements; μ - microparticle chemical potential; N – the number of particles; T – temperature in K; k – Boltzmann constant; and h – Planck constant.

Respectively:

$$W = Z^{-1} \cdot \exp\left[-(kT)^{-1}(H_N - \mu N)\right] \tag{3.256}$$

where $H = T + E + \varphi$; $T = \sum_1^N \frac{i^2}{2M}$; $E = (r_1 .....r_n)$; $\varphi = \sum \varphi'(r_n)$; $r_1 ...r_n$ – the vectors of microparticles radius; and $\varphi'(r_N)$ – the potential of the external field forces that affect the system.

Using quasi-thermodynamic treatment, according to which to the evaluation of the material density in the interfacial zone, the macromolecule orientation and length, alongside of co-ordinates axes, is neglected, according to the Bakker equation [1231] it can be written:

$$\Delta\Psi - \frac{\Psi}{k_1^{\ 2}} = 4\pi k_2 \rho(x, y, z) \tag{3.257}$$

where: ρ - density; ψ - the specific potential of interaction between particles; k_1 and k_2 – constants.

Therefore, the interfacial layer thickness, between the pure liquid ant its vapours, could be determined only be the surface energy of a condensed phase and the corresponding energies:

$$d = 8\pi \frac{\sigma}{RT(\rho_1 - \rho_2)} \ln \frac{\rho_1}{\rho_2} \tag{3.258}$$

The evaluation of the solid energy of the surface is a more complex process. The separation of a molecule from a fluid surface requires, on the one side, the splitting of its bonds with the neighbour molecules and on the other side, the splitting of the bonds with the inferior layer. Consequently:

$$\sigma = F + A\left(\frac{dF}{dA}\right)_T \tag{3.259}$$

where: A – surface area; and F – surface free energy.

Since, for liquids $A(dF/dA) = 0$, $\sigma = \sigma_0$, for solid substances the values F and A are of the same order. Thus, the concept of tension is devoid of any physical sense and, consequently, is replaced with the surface energy.

$$\sigma_{sa} = \sigma_0 - \sigma_s \tag{3.260}$$

where: σ_0 – the required energy for dislocation of an atom from surface; σ_s – the required energy for atom movement on the position from which it will be subsequently dislocated. On the surface of a condensed phase, a certain number of molecules in excess exists, whose specific value is:

$$\Gamma = \Delta^S N / A \tag{3.261}$$

and describes "the mass accumulation at surface", therefore, the performing of the absorption process. With other words, excepting the vacuum, the surface is continuously enriched with molecules from surrounding medium.

For a bicomponent system and the three usual parameters, namely – temperature, concentration at surface C^σ, and chemical potential μ, in fact the activity, expressed by exp (μ/kT) – a fourth parameter must be added, which describes the difference between the surface energy of the pure absorbent and which arises from the bicomponent system:

$$\Pi = \int_{p=0}^{P} \frac{\Gamma}{kT} d\ln p \tag{3.262}$$

By definition:

$$\Pi = \sigma_s - \sigma_0^s \tag{3.263}$$

for the measurements performed in equilibrium conditions, T.L. Hill obtained [1232]:

$$\Pi = \frac{RT}{AV_m} \int_{p=0}^{p} V_p dp \tag{3.264}$$

where: V_m and V_p – the molar volumes of the adsorbent, at surface and in normal conditions, respectively; R – universal gas constant.

For the whole phase volume, the thermodynamic potential, expressed by the Gibbs equation is:

$$G^V = E^V - TS^V + pV = F^V + pV = \sum_{\rm I} \mu_i N_i \tag{3.265}$$

and the expression for the Helmholtz free energy is:

$$F^V = E^V - TS^V = G^V - pV = \sum_{\rm I} \mu_i N_i \tag{3.266}$$

or the Gerring function:

$$F^V = E^V - TS^V = G^V - pV = \sum_{\rm I} \mu_i N_i \tag{3.267}$$

For the systems including the interface, these functions become:

$$G = E - TS + pV - \sigma A \tag{3.268}$$

$$F = E - TS \tag{3.269}$$

$$\Omega = F - G = \sigma A - pV \tag{3.270}$$

The surface generation implies a certain mechanical work, which for an isothermal and isobar process is equal to σA. Taking into consideration the solid–liquid interaction and neglecting that the liquid is shape-dependent, from equation (3.266) we obtain:

$$U = TS - pV + \sum \mu_i N_i \tag{3.271}$$

or

$$U = T(S_{sl} - S_s - S_l) + \mu(N_{sl} - N_s - N_l) + \sigma\Delta A \tag{3.272}$$

dividing the last equation by A, we obtain:

$$U_{sl} - TS_{sl} + \sigma + \mu\sigma \tag{3.273}$$

where: U_{sl} – the maximal, total, excess of energy; S_{sl} – the corresponding excess of entropy.

The equations (3.271)–(3.273) represent the relations between the surface energy and absorption in the model of energetic spectrum of adhesive joining. The total energy of surface U^s of a system composed by the pair adhesive–substrate, at a given moment t can be calculated with the following relation:

$$U^s = TS^s + \sigma_s + \mu T \tag{3.274}$$

At $t = 0$, when the absorption on the substrate surface does not occurred yet, the expression of the total energy of surface is equal to:

$$U_s^s = T_s^s + \sigma_s \tag{3.275}$$

according to the energy conservation law, subsequently, it can be written:

$$U^s \pm c\Delta T = U_l^s + U_s^s \tag{3.276}$$

On the other side, the excess of enthalpy, related to the fact that the molecules are arranged in the adhesive interaction, will be influenced by the ratio of average frequencies of the thermal vibration:

$$\Delta S = \Delta N_{sl} \ln(\nu/\bar{\nu}_s) \tag{3.277}$$

where: N_{sl} – the number of molecules from the interface region, additive considered, which can be expressed as:

$$N_{sl} = -k_r\Gamma \tag{3.278}$$

The last relation allows the deduction of the equations that describe the isothermal or non-isothermal formation of adhesive joining:

$$\Gamma_{T=0}^{\max} = \frac{\sigma_s + \sigma_l}{-k_\Gamma T \ln\left(\bar{\nu}/\nu_s\right) + \mu} \qquad (3.279)$$

$$\Gamma_T^{\max} = \frac{\nu_s + \sigma_l}{\mu + k_\Gamma T \ln\left(\bar{\nu}/\nu_s\right)} + \Delta T \frac{c + S_l^s + S_s^s}{\mu - k_\Gamma T \ln\left(\bar{\nu}/\nu_s\right)} \qquad (3.280)$$

Equation (3.279) shows that whenever the chemical interactions do not occur at the interface, due to the reduced mobility of the adhesive molecules in contact with the substrate, the first term of the denominator faster increases than the numerator. According to the equation (3.280), a non-linear increase of adsorption occurs with the increase of temperature.

The energetic spectrum of adhesive joining U_{sl} is defined by equation (3.273). One can observe that the total excess of energy of the interfacial region overtakes the energy values of every phase in contact, therefore, it represents the parameter, which from the thermodynamic viewpoint, governs the result of adhesive interaction.

The total surface energy of an adhesive, U_s, is composed by the interactions energy inside the phase volume, $U_1\mu$, and the energy, $U_{s,l}$, related to the field asymmetry at the surface. The first term is determined by the nature of interacting molecules, representing the measure of the mechanical work required for molecules separation. The total energy of interaction, resulted from the solid–liquid intermolecular contact, U_{sl}, includes the energy of attraction interactions from the volume of every phases and the interaction energy governed by the symmetry gradient of the forces field from the interfacial region. The first component is measured by the mechanical work required for the detachment of the adhesive molecules from its own volume or from the adhesive-substrate limit. The second component is quantified by mechanical work, which could be performed to the molecules separation from the two phases, under the assumption that their interaction is related only to the asymmetry of force field at the surface.

The interfacial layer that arises from the adhesive interaction is characterised by its own field force and the energy U_{sl}, of whose value can overtake the cohesion energies, specific to every phases, or can take an intermediate average value.

If two separated parts of the same phase are brought in molecular contact, an interaction of $E_{1\,coh} = E_{2\,coh}$ type results; therefore,

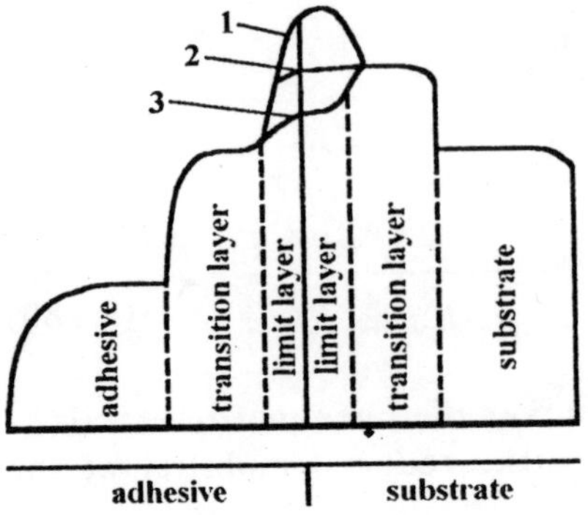

Figure 3.420. Energy spectrum of an adhesive joining [1233].

$U_{sl} > E_{1\ coh}$ $(E_{2\ coh})$ and the total energy overtakes the cohesive interaction of every phase, $U_{s,l}$, reaching its maximal value at the geometrical interface between the two phases. With other words, the splitting of the adhesive bond inevitably implies the scission of some cohesive bonds (chemical bonds from one or other phase, respectively). This parameter gives the possibility of examination of the energetic spectrum of an adhesive joining, which consists from two different layers and an adhesive one between them. Figure 3.420 illustrates the typical energetic spectrum of an adhesive joining [1231].

Another phenomenon, also of thermodynamic origin, is the wetting that occurs during the adhesive interaction. The wetting is defined as the spontaneous movement (or forced movement whenever the process is deliberately caused) of the system energy, which accompanies the components interactions. It has as quantitative parameter, the contact angle between the drop surface and that of solid one, respectively. The contact angle is noted by q for the systems in equilibrium and by f for those far from equilibrium [1233].

In the case of adhesion, the wetting by contact implies three phases, each of them is considered homogeneous and which, as is

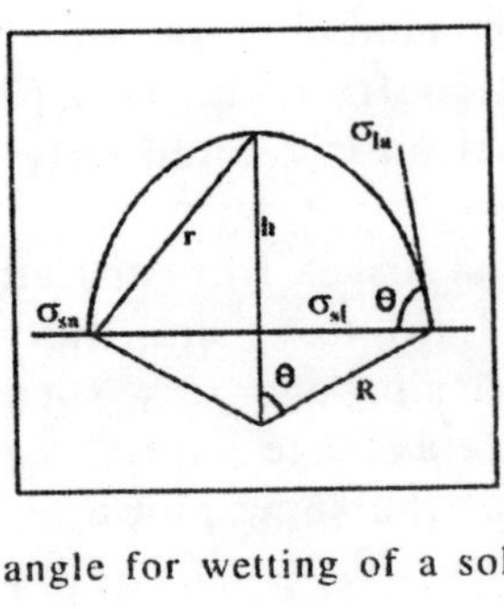

Figure 3.421. Contact angle for wetting of a solid surface by a liquid [1233].

depicted in Figure 3.421, at limit are intersected. One of the phases, the liquid one, has a lens-like shape, which takes such dimensions than the gravitational force does not significantly affect the interfaces shape (surfaces of discontinuity).

When the temperature and the chemical potentials are constant and equals, and the phases volumes are not changed at reversible variation of the position of discontinuity surfaces, the Gibbs relation becomes:

$$\sum_i \sigma_i dA_l + l \cdot \sum \sigma_{i+1} \cdot dT_{i+1} = 0 \tag{3.281}$$

where: l – the wetting perimeter; σ_{i+1} and dT_{i+1} – increments of the corresponding parameters of dislocation of any element of line l, in normal direction to this one and in a tangential plane to the ith surface. If the discontinuity surfaces form a continuous plane, the equilibrium condition is given by the expression firstly proposed by Young:

$$l \sum_i \sigma_{i+l} dT_{i+l} = \sigma_{AB} - \sigma_{BC} - \sigma_{AC} \cos\theta \tag{3.282}$$

The equilibrium condition (3.282) can be written as:

$$\frac{dF}{dA_{BC}} = 0 \tag{3.283}$$

Under the conditions of surface energy changes, at the interfaces between the three phases we have:

$$\frac{dA_{la}}{dA_{sl}} = \frac{\sigma_{sa} - \sigma_{sl}}{\sigma_{la}} \tag{3.284}$$

In geometrical terms, according to Figure 3.421, it results:

$$\frac{R-r}{R} = \cos\theta \tag{3.285}$$

therefore

$$\frac{dA_{la}}{dA_{sl}} = \cos\theta \tag{3.286}$$

and the equilibrium condition takes the form of the Young equation:

$$\frac{\sigma_{sa} - \sigma_{sl}}{\sigma_{la}} = \cos\theta \tag{3.287}$$

The parameters from equation (3.286) have a perfect physical correspondence. In agreement to equation (3.269) represent the partial derivatives of the Helmholtz free energy:

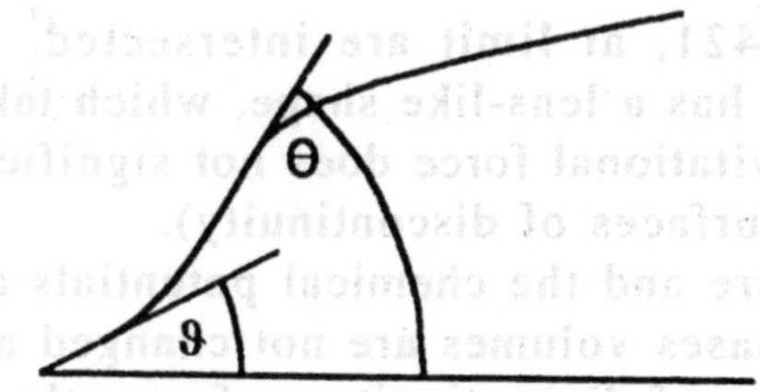

Figure 3.422. Micro- and macroangles of contact [1233].

$$\sigma_{sa} = (\partial F/dA_{sa})_{T,\mu_i} \tag{3.288}$$

$$\sigma_{la} = (\partial F/dA_{la})_{T,\mu_i} \tag{3.289}$$

$$\sigma_{sl} = (\partial F/dA_{sl})_{T,\mu_i} \tag{3.290}$$

The equation (3.286) describes the state of hydrostatic equilibrium at the interphase of a three-phase system. It is not valuable in the case of spreading. A drop of polymeric adhesive on a solid surface is only rarely of spherical shape. Usually, it asymptotically tends toward the solid surface. Consequently the contact angles, i.e. macroscopic angle, θ, and microscopic one, ν, must be distinct each to other, Figure 3.422. The first angle can be calculated from Young's equation. For the calculation of the microscopic angle, the capillary effects are considered, which are related to the curvature radius, r; these effects can not be neglected when the particles' dimensions are very small. The following relation was established between the two contact angles:

$$\cos\theta = \cos\nu \frac{\sigma_{la}}{\sigma_{la} - r(d\sigma_{la}/dr)} \tag{3.291}$$

The necessity of taking into consideration the two angles complicates the thermodynamic treatment of contact wetting. It must be considered that in the adhesive's concrete case, the liquid absorption on the solid surface determines the decrease of σ_{sa}. Introducing π, the following expression is obtained:

$$\cos\theta = \frac{\sigma_{sa}^0 - \sigma_{sl}}{\sigma_{la}} = \frac{\sigma_{sa} - \sigma_{sl} - \pi}{\sigma_{la}} \tag{3.292}$$

In the case of hydrophobic surfaces, like that of polyethylene or poly(tetrafluoroethylene), this effect is apparently negligible, when π = 0. For highly energetic surfaces, the degree of manifestation of this effect is difficult to determine, due to the difficulties of measuring σ_{sa} [1235].

The real surfaces of the solid bodies are characterised by a developed topography, which imposes the considering of this factor. In this case, the wetting rate must depend on the reciprocal position of the liquid spreading front and the faults profile (irregularities), which inevitably exist at the surface of solid body. The distance between the irregularity cavities, randomly disposed, diminishes the wetting rate, due to the mechanical barriers encountered by the moving liquid. Two distinct steps can be distinguished, namely a short step (hundredth of second) that corresponds to the formation of pedunculare drop and the second step (several tenth of second), corresponding the liquid flowing outside of drop, alongside of rugged surface that intersects the perimeter of the drop base [1236].

The general condition for a liquid in equilibrium on a rugged surface is described by Wentzel's equation:

$$\cos\theta_r = K_r \cos\theta \tag{3.293}$$

where: K_r is the ruggedness coefficient.

The substrate surface roughness has tendency to diminish the values of θ when this one is lower than 90° and to increase its values for θ > 90°.

Beside of mechanical heterogeneity, the chemical heterogeneity of the solid surfaces must be considered. For the existence of two microscopic surfaces, the surface energies at the limit solid-liquid and solid-gas are given by the following expressions:

$$\sigma_{sl} = A' \cdot \sigma'_{sl} + A''_l \cdot \sigma''_{sl} \tag{3.294}$$

$$\sigma_{sa} = A' \cdot \sigma'_{sa} + A'' \cdot \sigma''_{sa} \tag{3.295}$$

where: A′ is the function of the total surface occupied by the microscopic surfaces of the same type. By substituting the equations (3.294) and (3.295) in equation (3.287) we obtain:

$$\cos\theta = A'\left(\sigma'_{sa} - \sigma'_{sl}\right) + A''\left(\sigma''_{sa} - \sigma''_{sl}\right)/\sigma_{la} \tag{3.296}$$

considering now that:

$$\cos\theta' = \left(\sigma'_{sa} - \sigma'_{sl}\right)/\sigma_{la} \tag{3.297}$$

$$\cos\theta'' = \left(\sigma''_{sa} - \sigma''_{sl}\right)/\sigma_{la} \tag{3.298}$$

where σ′ and σ″ are the contact angles at equilibrium that corresponds to the homogeneous surfaces of any type.

Subsequently, it can be written:

$$\cos\theta = A'\cos\theta' + A''\cos\theta'' \qquad (3.299)$$

or in the generalised form:

$$\cos\theta = \sum_{i+l}^{N} A^{i}\cos\theta^{i} \qquad (3.300)$$

According to Good [1237]:

$$\sum_{i+l}^{N} K_r = 1 \qquad (3.301)$$

In the case of real bodies, the other factor affecting the value of the contact angle is that of substrate deformation. For instance, an elongation of 30% of polyesteramide diminishes the value of θ with 5.5°, when the wetting liquid is water, Figure 3.423.

In order to consider this influence, the Young equation was modified, by introducing of the notion of reversible mechanical work of adhesion:

$$W_{Ad} = \sigma_{la}\left(1+\cos\theta\right) \qquad (3.302)$$

The major consequence of the wetting in the system adhesive-substrate is the adhesive spreading on the substrate, which allows the formation of an intimate interfacial contact.

The spreading can be regarded as the effect of flowing of a condensed phase over the other one, having as a decay of the free energy of the system surface if it does not suffer an external influence. In accordance to the liquid advancement the interface increases or diminishes in the case of its withdraws.

The spreading coefficient is given by the difference:

$$\chi = \sigma_{sa} - \left(\sigma_{sl} + \sigma_{la}\right) \qquad (3.303)$$

this one can also be expressed as the partial derivative if the free energy, according to Helmholtz expression:

$$\chi = -\left(\partial F/\partial A\right)_{T,\mu_i} \qquad (3.304)$$

or

$$\chi = W_{Ad} - W_{coh} \qquad (3.305)$$

χ is a parameter with physical meaning. Thus, whenever $\chi > 0$ the spreading is complete, by the formation of a liquid film on the solid surface; when $\chi < 0$, only a limited spreading is manifested, for a certain value of the contact angle. The spreading coefficient allows

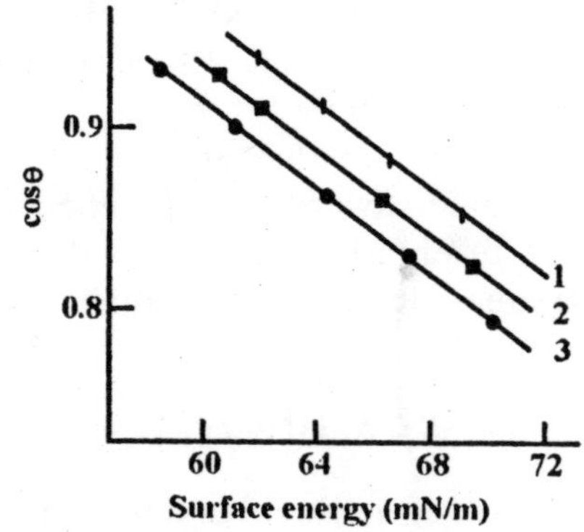

Fig. 3.423. Contact angles of polyesteramide as a function of the surface energy of wetting by a liquid [1238]: 1) water, 2,3) more or less polar liquids that wet the surface σ_{la} = 61–62 mN/m.

the thermodynamic treatment of the tendency of formation of adhesive joining and, in this sense, it is useful for some practical reasons. Expressing the adhesion work by the relation:

$$W_{Ad} = \sigma_{la} + \sigma_{sl} = 2(\sigma_{sa}\sigma_{la})^{1/2} \tag{3.306}$$

and considering for liquids that $W_{coh} = 2\sigma_{la}$, substituting in equation (3.305) the last two equations, we obtain:

$$\chi = 2\left[(\sigma_{sa}\sigma_{la})^{1/2} - \sigma_{la}\right] \tag{3.307}$$

The last equation allows the calculation of χ values for different materials.

In accordance to the accepted terminology in the molecular physic field, the region between the volume of a condensed phase and its geometric surface is defined as "transition layer", while its adjacent portion with the surface is defined as "limit layer", this one being of adsorbent nature. Between the specific properties of the phase volume and those of the limit and transition layers significant differences exist. For instance, abnormalities of viscosity have been detected at distances up to 500 μm of the solid surface.

In was proved that a macromolecule from the limit layer is retained with a lower force than other one from the transition layer and, especially, than a macromolecule belonging to the phase volume. Thus, from molecular-kinetic viewpoint the adhesive should be inert with respect to substrate and the adhesive interaction should only be limited to the equilibrium between the surface energies of the phases in contact. In order to avoid this situation, the adhesives are used only as solutions or melts. When the limit layers of the adhesive interact with the substrate, an orientation of the macromolecules occurs. The energy gain that accompanies molecule transfer from the phase volume towards the geometrical surface, characterised by enhanced orientation effects, is determined by the ratio of the average frequencies of the thermal vibrations [1233]:

$$\Delta S = N_s \ln\left(\overline{\nu}_V^* / \overline{\nu}_S\right) \tag{3.308}$$

where: N_s – the number of molecules from limit and transition layers; the inferior indexes represent the phase volume, V, and its surface, S, respectively.

The effect of the solid surface on the transition layers in the polymers case is manifested both spatially and energetically. From the geometric point of view, the substrate imposes restrictions to the material in contact with it, which are manifested at three structural levels, namely: chain segment level, macromolecular, and supramolecular level, respectively. From energetic viewpoint, the substrate role consists in the modification of the interphasic interaction character. Consequently, the motility of the structural elements into the transition layers is well impeded, by comparison with the material volume, due to the decrease of the number of possible conformations of a macromolecule in contact with a solid surface [1239,1240].

The limit layer of the polymer, founded in intimate contact with the solid body surface accomplishes an interaction of adhesive origin, being constituted from several monolayers. It is important that not the individual macromolecules are adsorbed on surface, but – due to their associative character – the supramolecular formations materialise this effect. In the case of multicomponent systems, the adsorption is selective. For the systems without solvents, the conform-ational spectrum of its macromolecules, within the transition layer is fundamentally modified. In this case, the profile of unitary distribution density of the chains within the adsorbed layer, below the point θ, is characterised by three domains: the first one, near the interface, a rapid decay of density, according to the relation $\rho = (1 + r_x / d)^{-1}$ (r_x is the distance from the interface); the second region is characterised by a constant value; and the third region is characterised by an exponential decay of density, until the value characterising the phase volume [1241].

Another important aspect is that concerning the sliding modulus of the associated macromolecules on the substrate surface. When the behaviour of the chains placed between two, adsorbent or not, solid surfaces is analysed, the following types of macromolecules fixing are expected: 1) with the both ends at the same substrate, in the case of "loop model"; 2) each of the macromolecule ends is fixes at different layers, the "bridge model"; and 3) only one end is fixed at substrate, in the case of "floating model".

The physical properties of polymers, correlated to the effect of the above-mentioned factors, results as functions of distance and surface. The transition layer and, especially, the one located at the interface, are characterised by an energy excess, therefore, by a thermodynamic predisposition for the adhesive properties. These ones are characterised by the change of chain segments flexibility, which corresponding modify the relaxation times [1243]. The mobility of kinetic elements (chain segments, macromolecules, supramolecular formations), is the determinant parameter of the properties of transition layer.

At the same time, the surface energy of the polymer limit layers is another very important factor influencing the adhesive properties. It was established that the surface energy of the limit layers of the condensed phase is higher than those of internal layers, which induces a negative effect on the adhesive properties. Thus, in the first place, as a consequence of the intensified sorption phenomenon of the polymer, with respect to the medium and its own volume. For instance, the sorption of impurities influences the wetting capacity, on the one side, due to the changes of the energetic spectrum and the real decrease of the interface, on the other side. The incomplete wetting is one of the important technological factors that impedes the attainment of the adhesion maximal resistance [1244].

The adsorptive interaction can generate structural inhomogeneity at molecular and supramolecular levels [1245], which are decisive for the concentration in these regions of the stress and the release of the deformation and fracture phenomena [1246].

Under the conditions of mechanical solicitation, the tensions are mainly concentrated on the structural faults, therefore in the transition and surface layers. With other words, the "most dangerous" zone of a solid body is located just at its surface, where the mechano-cracking and fracture processes can be activated with higher rate as in body's volume.

In conclusion, the interaction at the adhesive-substrate interface, from the thermodynamic viewpoint is determined by the energy of the phases in contact and from the molecular-kinetic viewpoint it is governed by the mobility of the kinetic elements.

During polymer absorption, the interaction energy continuously varies, showing the characteristics of a second order transition. The transfer of a macromolecule from a liquid volume into a limited space of a microfault is a discrete process from the energetic viewpoint and it is released at a certain threshold of the contact energy between the polymer's kinetic element and substrate. This transfer represents a first order phase transition, its enthalpic effect being proportional to the microfault dimension [1248].

In the mechanics of composite materials, the interpretation of adhesion characteristics is based on the Griffith's theory concerning to the fracture process. The Griffith fracture energy, U_G, which determines the critical tension that starts crack propagation, with a length equal to l_G is given by the relation:

$$(2EU_G/\pi l_G)^{1/2} \tag{3.309}$$

and can be expressed in terms of surface energy.

In terms of the kinetic theory of the resistance and fracture of the adhesive joint, the relation between the stress, γ, and durability, γ, is given:

$$\tau = \tau_0 \exp(U_0 - \lambda \cdot \gamma/RT) \tag{3.310}$$

where: U_0 – the activation energy of the process; λ – the structural coefficient.

In the case of heterogeneous systems, the internal tensions, γ_0, induced during the material processing, must be considered:

$$\tau = \tau_0 \exp\left[U_0 - \lambda(\gamma \pm \gamma_0)/KT\right] \tag{3.311}$$

The parameters from equations (3.310) and (3.311) are determined from the thermomechanical curves of system durability.

Taking into account that the variation in time of the factor γ_0 is also described by:

$$d\gamma_0 = E_a d\xi \tag{3.312}$$

where: E_a – the contraction modulus of the adhesive; and ξ is the relative contraction of the adhesive, due to reticulation. Since E_a and ξ are correlated to the kinetic parameters that describe the increase of adhesive joining, it results that the internal tensions mast depend by the apparent energy of crosslinking activation U_a [1248].

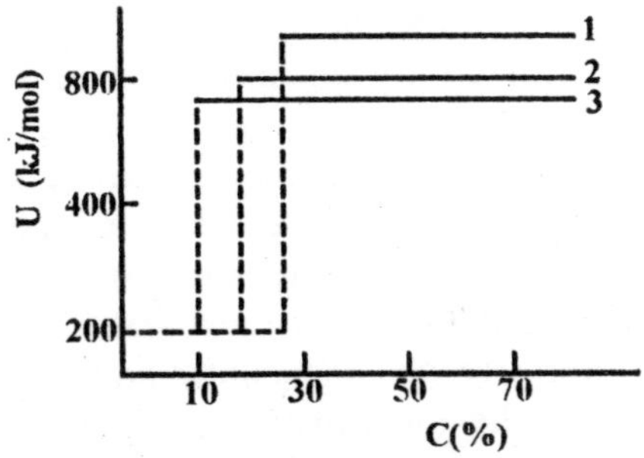

Figure 3.424. Activation energy of fracture for the three types of composite materials [1248].

The fracture activation energy is mainly determined by the substrate nature. Taking as typical example a composite material, reinforced with fibres, for which it is accepted that the fracture energies coincide with those of matrix fracture, the curves presented in Figure 3.424 are typical ones.

The analysis of the discrete values of fracture activation energy of the composite system, with polymer matrix and glass fibres as filler, evidenced that these ones are multiples of the activation energy of the mechano-chemical processes of substrate fracture:

$$U_0/U_S = N_i \tag{3.313}$$

where N_i is the number of split bonds into a composite supposed to stress action concomitantly with thermal fluctuation. A certain variation of the activation energy can be ascribed to the presence on the material surface of some impurities or the water adsorbed on the glass fibre surface. Consequently, the fracture activation energy of adhesive joining should constitute a criterion for the evaluation of its efficiency.

The formation and the thickness of the limit and transition layers are essential influenced by the polymer structural characteristics. In the first place, the chains mobility, able to be changed during the interaction period, is an very important factor. Based on this characteristic, the interpenetrations and mutual coilings between phases can occur, giving rise to the interfacial bonds that influence the diffusion in the limit region of the chains or chain segments. It was found that the polarity is directly related to this behaviour. In the case of polymers with similar or close polarities the interfacial interaction favours a reduced diffusion, the effect remaining at the surface. In the case of the adhesive interaction between the polymers with different polarities, a mechanism of "local diffusion" was proposed, based on the microheterogeneity of polar materials. In essence, it is assumed that only certain structural units can cross the interface. Specific for polymers, the chain segments through which the interdiffusion is assured must overtake strong intra- and intermolecular interactions, structural barriers, consisting in network's knots, loops, and chains connective textures. From all these reasons, the chain segments are characterised by a reduced mobility potential whose diminishing also contribute to both the internal tensions, manifested at the molecular level, and the co-operative character of diffusion. Whenever the two polymers are miscible, this factor can affect the formation of some supramolecular formations, in the proximity of the limit layers. Usually the region with modified structure is limited to a thickness ranging from 10 to 100 nm. It is considered that the thickness of these zones is of the same order of magnitude as the chain segments length [1249–1251].

It was demonstrated that beside of thermodynamic and molecular-kinetic interpretations, the rheologic ones, which introduce the influence of external field of forces, are very important in the case of adhesive joining formation. Due to the macromolecular character, the majority of interdependencies have a non-linear character. Their analysis, from rheologic viewpoint, represents the connection bridgė between the theory that describes a formal behaviour and practice, which tries to explain the reality of the observed phenomena.

3.10.4.2. The role of filling agents

The obtaining of the composite materials with advances properties has been related in a small measure to the structural modifications of the matrix and reinforcing agents. A real progress was achieved whenever the coupling reagents were used in order to enhance the resistance of the couple matrix-reinforcing agent [1252].

The structural variations existent to the matrix-reinforcing agent interface have been investigated from qualitative and quantitative viewpoints; it was found that these variations depend by a series of factors, such as: pH, filler concentration, organic-functionality, solvent nature, method of application, treatment temperature, etc.

The first composite system used in this scope was that constituted by a polymer matrix, reinforced with glass fibres, for which the coupling reagents of silane-type were used. The silicone monomers have been used at large scale for chemically bonding the polymers, especially to the inorganic materials, glass fibres, mineral fillers, metals, metallic oxides, having as effect the improvement of electrical and mechanical properties for a long exploitation period.

The silane-based coupling agents are compatible with the majority of organic polymers, from the thermo-reactive resins to elastomers and thermoplastic ones. The filling agent can be of fibrous or particulate nature and the silane may be applied on the substrate as pre-treatment or, in many systems, directly to the resin (the method of the integral mixture). In the last case, the coupling reagent migrates to the substrate during the mixing and mechanical processing.

A simplified representation of the coupling agent interaction, using organo-functional silanes, assumes the formation of a silane monolayer, which realises a covalent bridge between the polymer matrix and filler. In fact, the real situation is much more complex; a multilayer model adequately describes the covering with hydrolysed silanes. The hydrolysed silanes by condensation reactions give rise to soluble and fusible silanole oligomers, with enhanced crosslinking ability. The contact with the polymer matrix is realised in that step within which they still keep certain solubility. Consequently the bounding to the polymer matrix can be realised as follows: 1) when the silanol oligomer layer is compatible with the matrix, during the thermal treatment, by the activation of the copolymerisation process; 2) if the oligomer layer is only partially compatible with the liquid matrix, an interpenetrated network can be generated; 3) a diffusion process can occur between the silane oligomer and the thermoplastic matrix, followed by crosslinking; this process is just stimulated by the conditions of the composite material formation, the net result being an interpenetrated pseudo-network; and 4) siloxane segments by diffusion enter in the liquid matrix, without any crosslinking reaction to accompany this process.

The performances of the coupling reagents in the reinforced composites practically depend in equal measure by the physical properties that are related to the method used for the coupling agent application as well as by the chemical properties of the coupling agent. Some of the most common types of coupling agents are presented in Table 3.129.

Using a coupling agent of methacrylate type, with the chemical structure $H_2C{=}C(CH_3)COOCH_2CH_2CH_2Si(OCH_3)_3$, was applied to the glass fibre of type E from an aqueous solution. It was found that its performance strongly depends on the conditions of fibre treatment. As Figure 3.425 shows, the best results are obtained when it was dried for 7 min at 110 °C. A more pronounced dry-

Table 3.129. The most common types of coupling agents

The organo-functional group	Chemical structure
A. Vinyl	$CH_2=CHSi(OCH_3)_3$
B. Chloropropyl	$ClCH_2CH_2CH_2Si(OCH_3)_3$
C. Epoxy	$CH_2(O)CHCH_2OCH_2CH_2Si(OCH_3)_3$
D. Methacrylate	$CH_2=C(CH_3)COOCH_2CH_2CH_2Si(OCH_3)_3$
E. Primary amine	$H_2NCH_2CH_2CH_2Si(OCH_3)_3$
F. Diamine	$H_2NCH_2CH_2NHCH_2CH_2CH_2Si(OCH_3)_3$
G. Mercapto	$HSCH_2CH_2CH_2Si(OCH_3)_3$
H. Cationic styryl	$CH_2=CHC_6H_4CH_2NHCH_2CH_2NH(CH_2)_3Si(OCH_3)_3 \cdot HCl$
I. Chromium complex	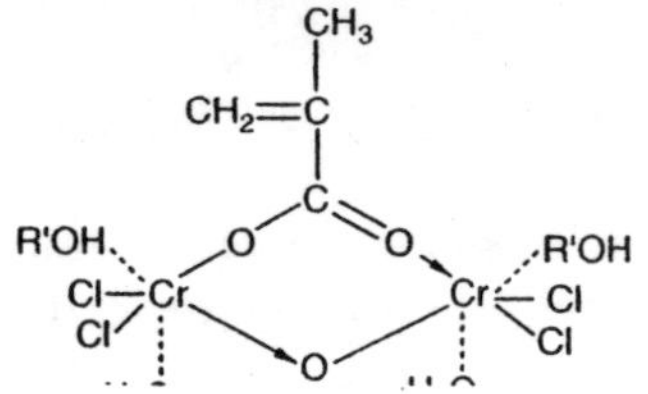

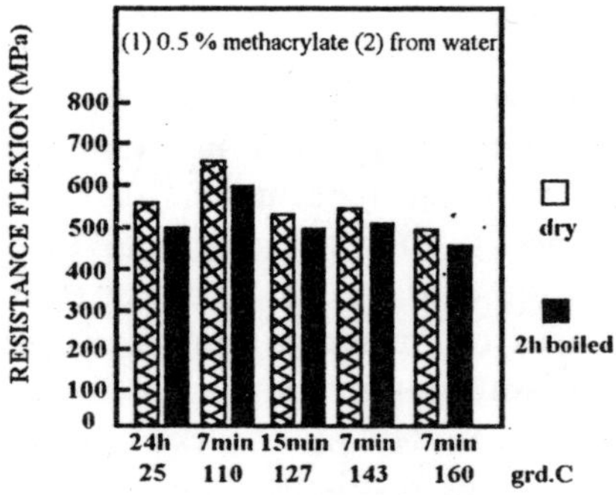

Figure 3.425. Performances of a composite material in function of the drying time of the glass fibre treated with siloxanes [1252].

ing can cause the loss of some methacrylic groups by oxidation or polymerisation reactions. A prehydrolysed amino-functional silane, i.e. $H_2NCH_2CH_2NHCH_2CH_2CH_2Si(OCH_3)_3$, is suitable for reinforcing PVC plastisoles to the glass or metals, respectively.

The composite materials performance is also correlated with the matrix morphology to the interface.

3.10.4.3. The bonds at matrix-filler interface and mechanical models of load transfer

Within a composite material, the reinforcing agent-matrix interface represents the common surface of the two elements along with the

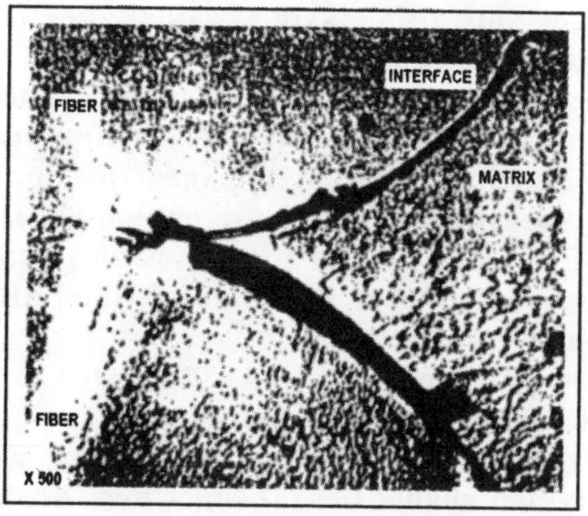

Figure 3.426. Crack initiation at the contact between filaments (Specimen loaded 16 hr at 80% compressive force) [1245].

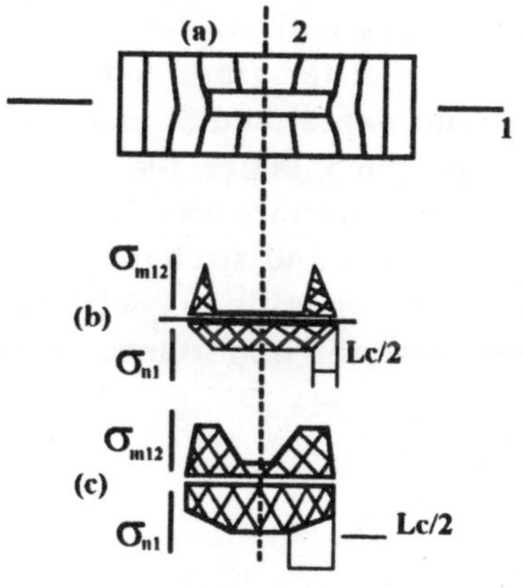

Figure 3.427. Model of load transfer from the matrix to fibre [1245]: (a) deformation model; (b) distribution of shear and stretching tensions for the elastic transfer; (c) idem, for unelastic transfer.

region of its proximity. The interface possesses its own physical and mechanical properties which differ from those of the matrix and the filler, respectively. For an interface of the fibre–matrix type, the most important characteristics are depicted in Figure 3.426. One can be seen that the fibre is included in matrix. The fibre surface and adjacent to the matrix defines the interface.

For this model, the load (charge) transfer is realised in accordance to Figure 3.427. Three interesting conclusions are offered by the analysis of this Figure, namely: 1) the shear tension at surface rapidly increases until to a maximal value, after which quickly diminishes toward the fibre extremity; 2) the axial tension strongly to an average value, that attained by fibre in composite material and in the case of an elastic behaviour, which remains constant; an 3) for an non-elastic behaviour, the shear tension at interface increases until to a value the will cause on non-elastic behaviour, i.e. plastic one, of the matrix. The shear tension is maintained at this value for a certain distance, alongside of fibre, after which it will quickly decrease.

Another model of load transfer is illustrated in Figure 3.428. It allows the prediction of the charge transfer through the interface. One can note that the shear tension evolves into a direction at one of its extremity and changes at the other one. At the fibre ends the tension reaches the maximal value. Alongside of fibre it decreases rapidly to zero there where the normal tension from composite has an average value. In the proximity of the other end of the fibre, the shear tension at the interface firstly increases gradually to attain finally the maximal value at the fibre end.

A vast theoretical and experimental material was devoted to the reinforcement of the polymer materials with short fibres. Thus, a new class of engineering materials was obtained, which are characterised by improved stiffness and strength, heat distorsion temperature, comparatively to the analogous without fillers, and which gained a peculiar commercial importance in the last 2–3 decades [1253–1256].

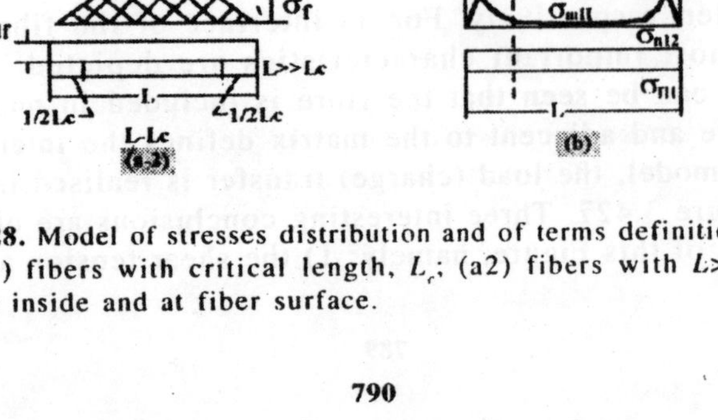

Figure 3.428. Model of stresses distribution and of terms definition at interface [1245]: (a1) fibers with critical length, L_c; (a2) fibers with $L>L_c$; (b) stress distribution inside and at fiber surface.

The use of short fibres, $L \approx 200$ μm, 30 weight %) as reinforcing agents of the thermoplastic matrix determines the improvement of almost all properties [1256–1261], such as:

- increase in modulus (by a factor of 3 to 6);
- higher hardness (by a factor of 1.1 to 2);
- improved tensile strength (by a factor of 1.1 to 3);
- higher thermal stability (by a factor of 1.1 to 3.1);
- lower thermal expansion coefficient (by a factor of 0.2 to 0.5)

The relation between the characteristics of the composite material microstructure, the mechanical, thermal, and environmental factors, which determine the performance of the mechanical properties by reinforcement, is schematically illustrated in Figure 3.429 [1253].

The effects of the microstructural parameters of the composite materials with the thermoplastic matrix, reinforced with short glass fibres, should be ascribed to the three fundamental elements of the material: 1) polymer matrix; 2) glass fibre; and 3) their interface.

The introduction of the short fibres within the thermoplastic matrix stimulates the action of a series of factors, such as: the flow mould conditions during processing procedure of the final parts (for instance, plaque), the fibres geometry, type, and volume

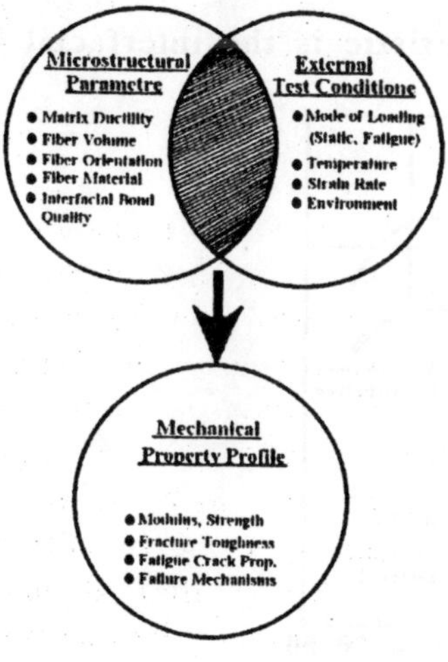

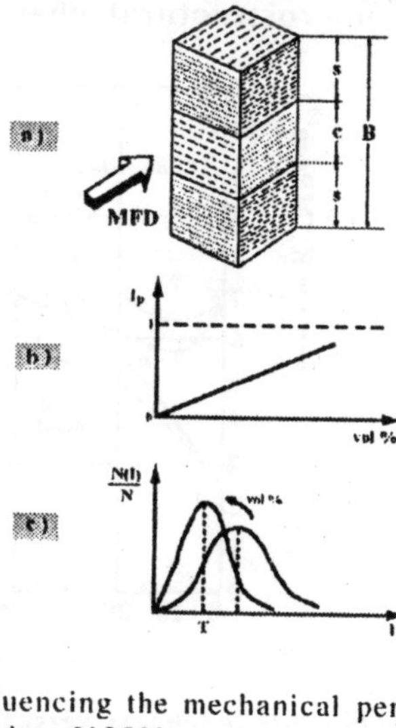

Figure 3.429. Schematic of factors influencing the mechanical performance of short fiber/ thermoplastic matix composites [1253].

Figure 3.430. Fiber orientation pattern across the plaque thickness B. Fiber orientation factor f_p as a function of volume fraation V_f, and change in fiber length distribution with V_f (schematic) [1253].

fraction. Figure 3.430 a illustrates the typical fibre orientation distribution across a plaque thickness (B). A natural triple laminate is formed by fibres preferably orientated parallel to the mould filling direction (MFD) in the two surface layers (S), located near the mould walls (shear flow), and a central layer (C) within which a transverse fibres orientation (as result of the extension flow in the middle of the melt) exists. The relative thickness of the central layer reported to the total plaque, i.e. *C*/*B* ratio, increases with total plaque thickness, and is reduced with higher melt velocity and temperature, and with lower mould temperature. Within each of three major layers, a fibre orientation factor, f_p, can be defined:

$$f_p = 2\langle\cos^2\theta\rangle - 1 \qquad (3.314)$$

where: θ – the angle between individual fibres and a reference direction (for example MFD). In principle, the absolute value of f_p varies between zero (random distribution) and one (perfect alignment) and is strongly dependent on the fibre volume fraction V_f, Figure 3.430 b). Another factor which may influence f_p are the fibre aspect ratio (*l*/*d*–length/diameter) and certain processing parameters. With respect to the fibre length, *l*, it was found that the average length, $\bar{l}$, usually decreases with the increase of fibre volume fraction and, sometimes, the fibre length distribution also becomes narrower, Figure 3.430 c.

Another microstructural characteristic is the interfacial bond

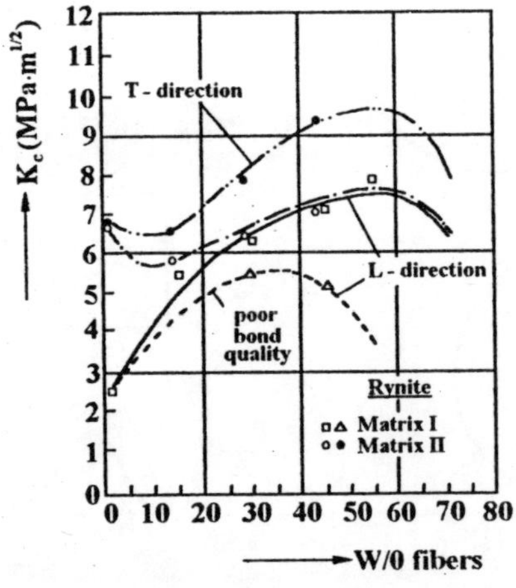

Figure 3.431. Fracture toughness versus fibre weight fraction of various GF-PET composites cracked in L- and T-direction, respectively. Matrix I contains about 6 vol %, while matrix II contains about 20 vol % impact modifier of ionomer type (Rynite, DuPont Inc) [1254].

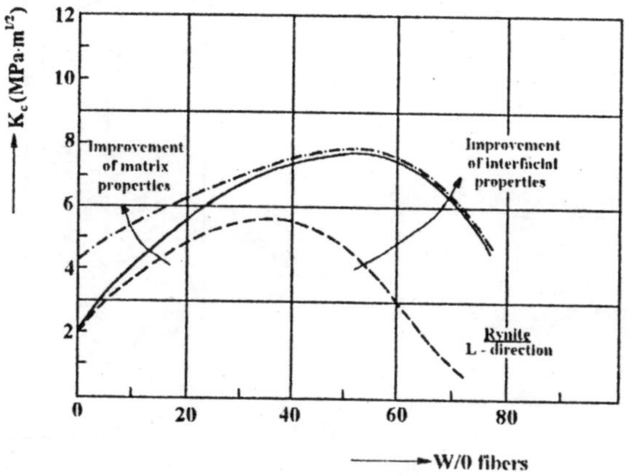

Figure 3.432. Qualitative representation of the effect of matrix and interfacial properties on the course of fracture toughness as a function of fibre weight fraction for GF-PET (Rynite, Dupont Inc) [1254].

quality. As criterion of its evaluation was used the factor K_c – the fracture toughness (the resistance to the material fracture by crack propagation). For a GF-PET composite material, poly(ethylene terephthalate) reinforced with glass fibres, it was found that a good interfacial bond quality can be achieved by using an optimal coupling agent yields, which gives much higher K_c values than those of poor fibre matrix adhesion, Figures 3.431 and 3.432. The effect is much more pronounced with the increase of fibres fraction in the composite. Especially for the L-cracked specimens, the fracture is dominated by the failure along the fibre/matrix interfaces. Under these conditions the interfacial adhesion becomes the critical factor which determines whether an optimum in the fracture toughness of the composite can be reached.

The mechanical properties of the polymer-based composites are dependent on the short fibres' orientation, and, on its turn, this one is influenced by the characteristics of melt flow, injection conditions, and mould geometry [1254, 1256].

The fibres' orientation in polymer composite determines the anisotropy of the mechanical properties. Those composites with perfectly aligned fibres show values of Young modulus much higher in their direction than in transversal one. Figure 3.433 shows a schematic representation of the filling of mold cavity during molding and Fig. 3.434 a schematic of the short-fibre-alignment in the three principal layers (c and s) in an injection molded rectangular specimen. It is noticed that in molded composites, however, the fibers

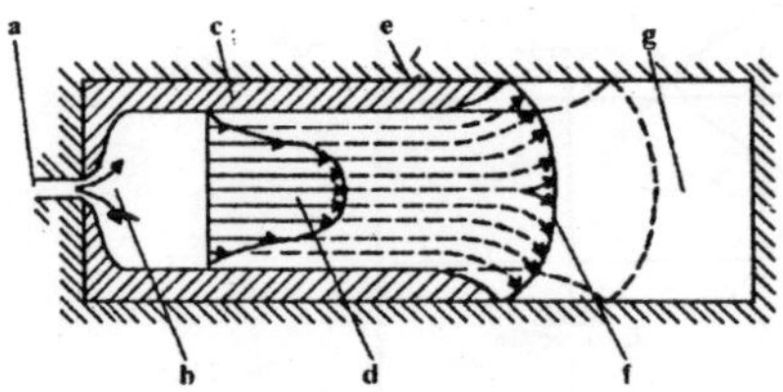

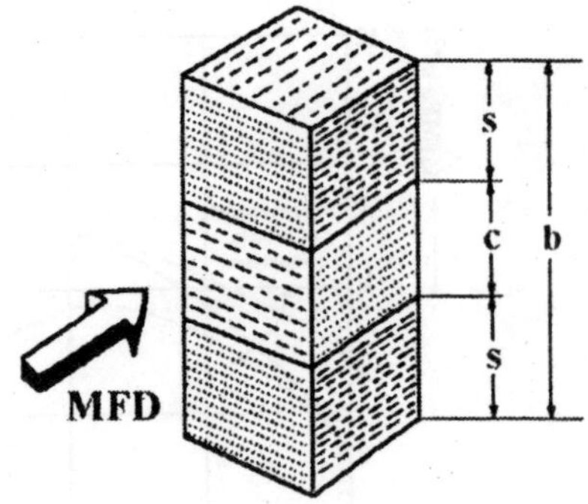

Figure 3.433. Schematic representation of the filling of a mould cavity during injection moulding: a) gate; b) injected stream; c) solidified skin layer; d) velocity profile of the molten core; e) wall of the mold; f) flow front; g) cavity [1254].

Figure 3.434 (right). Schematic of the short-fiber-alignment in the three principal layers (c and d) in an injection moulded rectangular specimen [1254].

Table 3.130. Effect of various factors on the fiber orientation during injection moulding of short-fiber-reinforced thermoplastics [1265]

Factors	Fibre orientation in the layers[a]
Injection moulding conditions	
melt temperature	-
mould temperature	-
additional pressure	?
injection speed	+
Material characteristics	
matrix	++
fiber concentration	(-)
fiber length	(-)
Design parameters	
gate	++
mould geometry	++
wall thickness'	?

[a]) The evaluation of symbols mean: ++ - very strong; + - insignificant; ? - small; (-) – with limitation

are rarely perfectly aligned (Figure 3.434). The influence of others different factors on the fibers orientation during processing by injection molding is presented in Table 3.130.

K.Friedrich investigating the mechanical properties of poly(ethylene terephthalate) as a function of fibres weight fraction (w/w) founded that both the modulus, Figure 3.435, and tensile strength, Figure 3.436, are generally enhanced by the fibres pres-

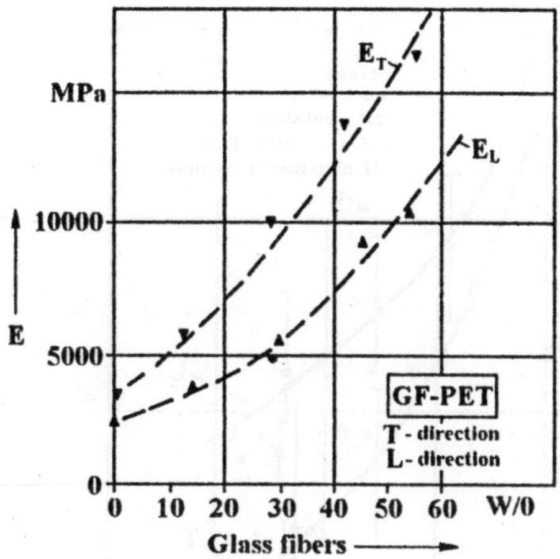

Figure 3.435. Dependence of the Young's modulus, *E*, on the short-glass fiber content in two directions (T and L; see Fig 3.438) for poly(ethylene terephthalate) filled with glass fibers (Rynite = GF-PET) [1256].

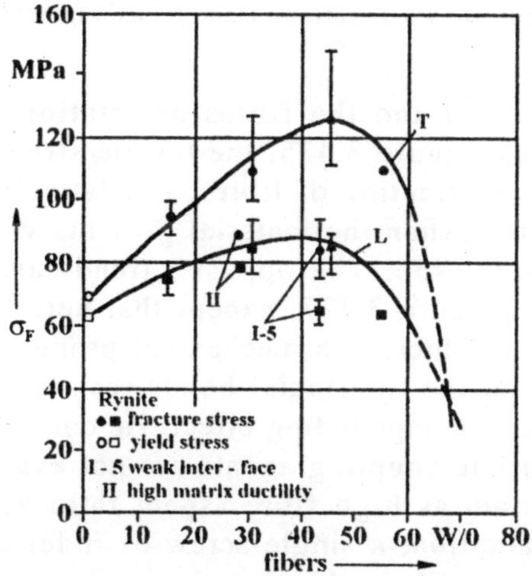

Figure 3.436. Dependence of the fracture strength σ_F on the contents of the short glass fibers (in two directions + T and L, see Fig 3.438) for the system shown in Fig 3.435 [1256].

ence, and the improvement is the highest when most of the fibres are oriented in the loading direction (T).

The short glass fibres, contained in the composite material, contribute to the increase both of modulus and tensile strength when-

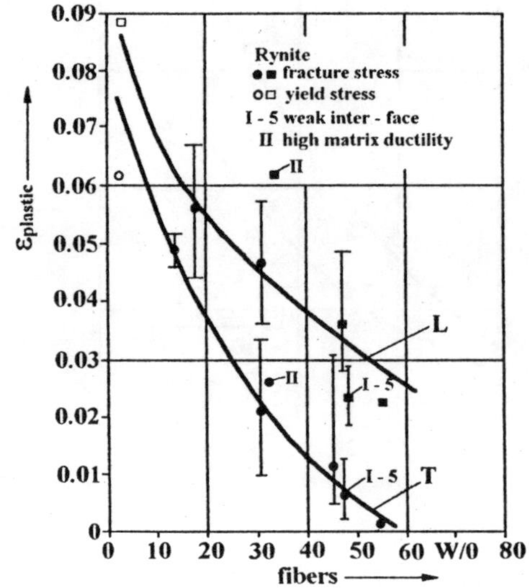

Figure 3.437. Dependence of the plastic strain $\varepsilon_{plastic}$ (strain at fracture) on the short- glass-fiber content (in two directions T and L, see Fig 3.438) for the system shown in Fig. 3.435.

ever the load direction and the fibres orientation are the same. Apart from modulus, Figure 3.435, the tensile strength in the two directions, for a given fraction of fibres, is relatively small for the highly ductile matrix, when the material presents very weak interfacial bonds, Figure 3.436. The opposite trends are observed for the strain at break, Figure 3.437, except that here also a poor fibre/matrix adhesion reduces the mechanical properties [1257].

D.M. Bigg treated the interfacial bonds resistance between the fibres and matrix as a compounding effect. He concludes that in the case where appropriate coupling agents are not available, it is not important to maintain as high fibre aspect ratio as possible in a moulded part. In addition, a single-screw extruder does less damage to glass or carbon fibres than a twin-screw one, under similar processing conditions when the polymer is in the form of pellets. When the polymer is in powder form, the twin-screw extruder can also be used.

The importance of the resistance of matrix/fibre adhesive bonds, for different types of matrix, is illustrated in Table 3.131.

Starting from the assumption that the factor K_c is a function of the adsorbed energy during the material fracture, K.Friederich initiated a series of investigations on neat polymers and their composites, using compact – tension specimens, Figure 3.438. This method

Table 3.131. Tensile strength of composite, s_c, and matrix, s_m, for 30 weight % (16 vol. %) short-glass-fiber-reinforced thermoplastic composites [1260]

No.	Polymer	Tensile strength, MPa		
		σ_c	σ_m	$\sigma_c - 0.846\sigma_m$
1	Nylon 6 - 6	183	84	120
2	Nylon 6	162	70	103
3	Nylon 6 - 10	148	60	97
4	ABS	134	44	97
5	PET	152	73	91
6	Acetal copolymer	136	61	84
7	Polycarbonate	130	63	77
8	Polysulfone	127	72	66
9	PPO	126	81	58
10	SAN	122	77	57
11	Polystyrene	88	56	41
12	Ionomer	60	28	36
13	Polyethylene (HD)	63	38	30
14	Polypropylene	55	32	28

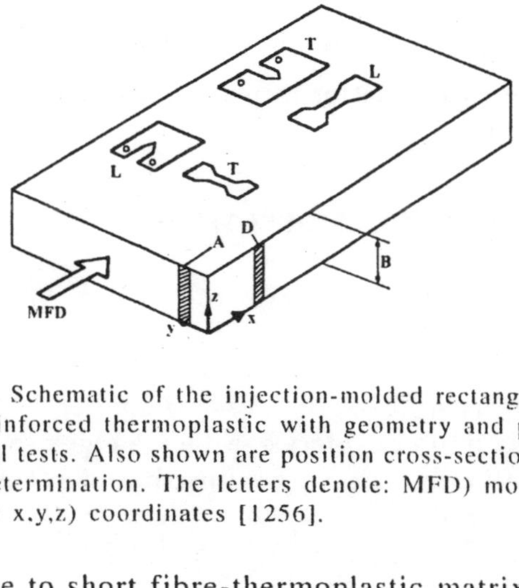

Figure 3.438. Schematic of the injection-molded rectangular plaque of a short-glass-fiber-reinforced thermoplastic with geometry and positions of specimens for mechanical tests. Also shown are position cross-sections (A and D) for fiber-orientation determination. The letters denote: MFD) mold-fill L) longitudinal; T) transverse; x,y,z) coordinates [1256].

is applicable to short fibre-thermoplastic matrix system as long as the general direction of the serrated crack remains perpendicular to the load axis applied. Typical results of fracture toughness K_c as a function of fibre fraction are given in Figure 3.439, for glass fibre reinforced PET [1261].

It was proved that the tendency and degree of variation in toughness are mainly function of the initial fracture toughness of the matrix, of the several microstructural effects related to the fibres, and to the fibre/matrix interface. K. Friedrich proposed the

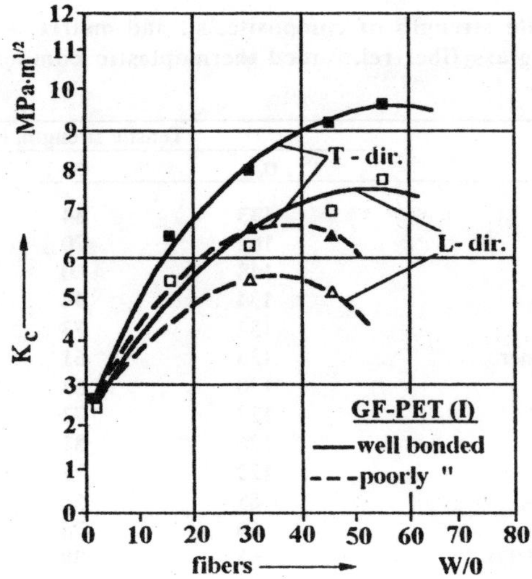

Figure 3.439. Dependence of the fracture toughness K_c on the short glass-fiber content (in two directions – T and L, see Fig 3.438) for the system shown in Fig 3.435 [1261].

following semi-empirical relationship to describe the composite's fracture toughness [1265]:

$$K_{c,c} = MK_{c,m} = (a + nR)K_{c,m} \tag{3.315}$$

where $K_{c,c}$ and $K_{c,m}$ – the fracture toughness of matrix and composite, respectively; M – microstructural efficiency factor; a – the matrix stress condition factor; n – the energy absorption ratio; and R – reinforcing effectiveness parameter.

The term, a takes into consideration the changes in the fracture toughness of the polymer matrix as a function of plaque thickness and/or the fibres presence:

$$a = K_{c,c(R=0)} / K_{c,m} \tag{3.316}$$

Normally the term a should be close to 1, *i.e.*, but other cases are also possible [1265,1266]:

a) $a < 1$ – occurs if the effective matrix fracture toughness in the composite is smaller than that of the neat-matrix polymer. This may be connected with the stress state transition (from plane stress to plane strain condition) or with a restriction of plastic deformation of the matrix, due to the fibres presence.

b) $a > 1$ – is probable when the fibres have a ductility enhancing effect (crazing, plastic deformation) on the surrounding brittle matrix, or if the measured $K_{c,m}$ is smaller than one necessary to express the real testifying condition applied.

Normalising the fracture toughness data of selected composites in the form:

$$\Delta K^*_{c,c} = \left(K_{c,c} - K_{c,c(R=0)}\right) / K_{c,m} \qquad (3.317)$$

and plotting them against R, a diagram, which reflects only significant differences between individual fibre/matrix combinations, is obtained, Figure 3.440. The term n from equation (3.315) can be obtained from the slope of the linear regression line fitted for the $\Delta K^*_{c,c} - R$ data pairs; n can be either positive or negative and its physical meaning is founded in the different energy absorbing mechanism during breakdown of the composites:

a) $n < 0$ – as long as the increase of the energy dissipation by the fibres during breakdown of the composite is greater than the loss of energy absorption by the matrix due to the incorporated fibres; and

b) $n > 0$ – if the opposite tendency becomes effective [1264].

H. Vass deduced an explicit equation for n by using the following assumptions:

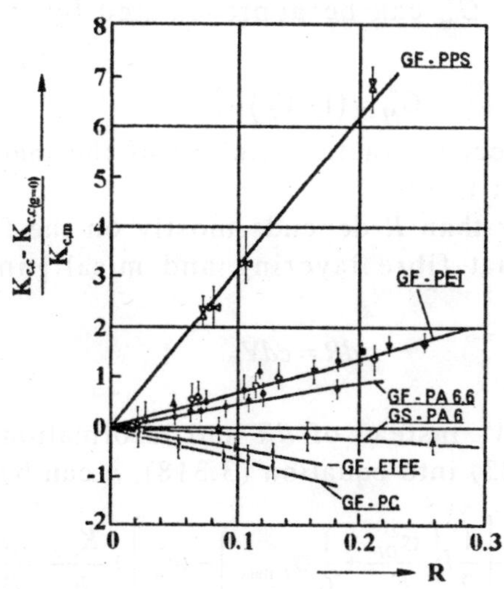

Figure 3.440. Relative changes in fracture toughness ($\Delta K^*_{c,c}$) of various thermoplastic matrix composites with increasing reinforcing effectiveness parameter R. [1254].

$$\frac{dG_{c,c}}{dR}=\frac{d}{dR}\left(\frac{K_{c,c}^2}{E_c}\right)=\frac{2K_{c,c}}{E_c}\cdot\frac{dK_{c,c}}{dR}-\frac{K_{c,c}^2}{E_c^2}\cdot\frac{dE_c}{dR} \tag{3.318}$$

From equation (3.315) it follows that:

$$dK_{c,c}/dR=nK_{c,m} \tag{3.319}$$

On the other hand, the fracture energy of the composite, $G_{c,c}$, is composed from those of the fibre, G_F, and matrix related term, G_M, the latter differs from that of the matrix fracture energy in the unfilled state [1265]. Therefore:

$$G_{c,c}=G_F+G_M^* \tag{3.320}$$

A high contribution to G_F comes from fibre/matrix debonding and fibre pull-out processes. This energy can be approximated by

$$G_F=V_F\frac{1}{2}l\left(\frac{\sigma_{DF}^2}{E_F}+\frac{1}{6}\sigma_{F\max}\right) \tag{3.321}$$

where σ_{DF} – the stress required for debonding initiation; E_F – the fibre modulus; $\sigma_{F\max}$ – the maximum tensile stress for fibre pull-out; $1/2l$ – the debonded or pull-out length, considered as being half of the fibre length (l); G_M^* can be approximated by the following expression:

$$G_M^*=(1-V_F)w_M^* \tag{3.322}$$

where: w_M^* is the specific fracture energy of the matrix material under the presence of fibres.

Supposing further than R depends mostly on the fibre volume fraction, V_F, and that fibre layering and misalignment can be neglected:

$$dR=cdV_F \tag{3.323}$$

Inserting now cdV_F instead of dR and information from equations (3.319) – (3.323) into equation (3.318), n can be derived:

$$n=\frac{1}{2K_{c,m}}\cdot\frac{1}{c}\left\{\frac{E_c}{K_{c,c}}\left[\frac{1}{2}l\left(\frac{\sigma_{DF}^2}{E_F}+\frac{1}{6}\sigma_{F\max}\right)-w_M^*\right]+\frac{K_{c,c}}{E_c}\cdot\frac{dE_c}{dV_F}\right\} \tag{3.324}$$

The first additive term determines whether n has a positive or negative value, *i.e.* fibres increase or decrease the fracture toughness of the composite. If w_M^* is higher than the specific energy connected to the fibre mechanism (debonding and pull-out) this term becomes negative, as is, for instance, the case of polycarbonate, PC, or poly(tetrafluoroethylene), PTFE. Based on the equation (3.324) upgrading of fracture toughness can be achieved in the following ways:

– increase in fibre length (LGF –reinforced thermoplastic composites);

– enhancement of fibre/matrix adhesion (higher σ_{DF} and $\sigma_{F\max}$ vales); and

– use of stiffer and lighter fibrous reinforcement, *e.g.* carbon fibres, since it increases the value of dE_c/dV_F more effectively.

Figure 3.441 summarises the main findings of the microstructural concept of fracture toughness.

The above outlined microstructural efficiency proved to be a valuable tool for upgrading the fracture toughness of short fibres reinforced semicrystalline plastics (SFRPs) by means of evaluation both of matrix and fibres – related failure mechanism [1267].

J-integral studies with SFRP become more and more popular in the characterisation of the fracture behaviour of both polymer matrices [1268–1270, 1271–1273] and their short fibre filled versions [1274,1254].

Figure 3.442 shows J *vs.* Δa curves of polypropylene (PP)/ elastomer blend and its related 40 wt% GF reinforced composite. It is clearly that fibre loading results in an enbrittlement of the material since J_c decreases from 9.8 to 5.0 kJ/m² for the T-cracked specimen. The relatively high difference obtained in J_c of the blend for L- and T-cracked specimens is due to a skin-core structure, which implies a massive segregation if the elastomer across the plaque thickness as demonstrated [1274].

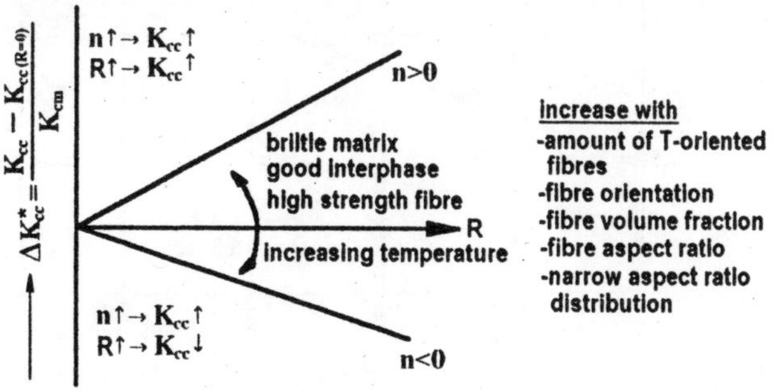

Figure 3.441. Schematic representation of principal $\Delta K_{c.c}^*$ versus R curves indicating upgrading and deterioration of composites' fracture toughness. [1274].

In Figure 3.443 the effect of the interfacial bond quality is depicted for 20 wt% GF – reinforced PP polymer. GF – reinforcement without surface treating yields J_c values in the 9.1–11.0 kJ/m^2 range, depending on the crack direction, while the use of coupled GF is associated with marked decrease of J_c, 1.9 kJ/m^2. This fact can be attributed to an operative amount of energy dissipation in the damage zone, which is significantly larger for the uncoupled then for the coupled GF – containing grade. Curves $I_C - V_F$ and K_C – R for PP homopolymer based short glass fibres composite, SGF-composites, show that $J_{c,c}$ values practically do not change as a function of V_F and R [1274]. A steep decrease in $J_{c,c}/J_{c,m}$ is expected because:

$$G_c = J_c = K_c^2/E \tag{3.325}$$

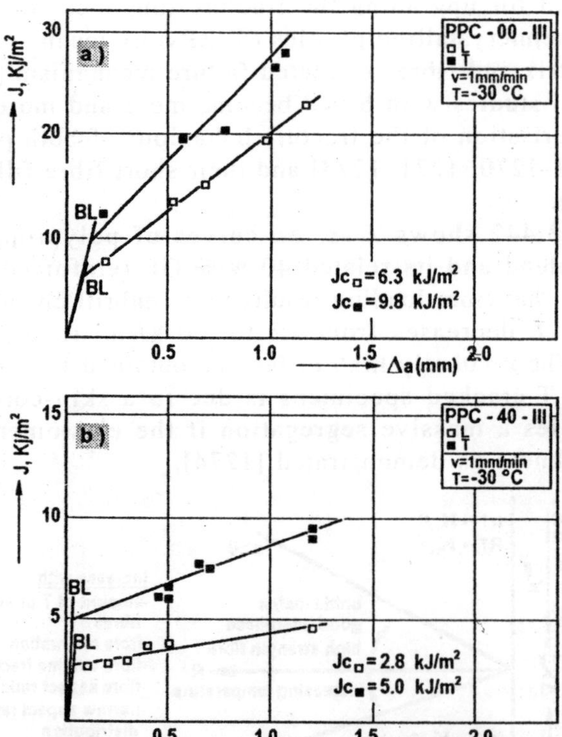

Figure 3.442. J – Δa curves established by the multiple specimen method for PP/elastomer/PE ternary blend (PPC-00-III) and its 40% wt.(= 19%wt.PPC-40-III) GF reinforced composite. Tests carried out at T = –30 °C nm/min. cross-head speed and BL designed blunting lines [1274].

$$\frac{J_{c,c}}{J_{c,m}} = \frac{K_{c,c}^2(R)}{E_c(R)} \cdot \frac{E_M}{K_{c,m}^2} \tag{3.326}$$

if

$$E_c = \sum_1 E_i V_F = R \cdot E_F + (1 - V_F) \cdot E_M \tag{3.327}$$

where: E_i is the E-modulus of the i^{-th} component; and V_F the volume fraction of glass fibres.

Inserting equations (3.327) and (3.315) into equation (3.326):

$$\frac{J_{c,c}}{J_{c,m}} = \frac{(a+nR)^2 E_M}{RE_F + (1 - V_F) E_M} = \frac{(a+nR)^2}{R\frac{E_F}{E_M} + 1 - V_F} \tag{3.328}$$

Based on equation (3.328) increasing R is connected with a steep decrease in $J_{c,c}/J_{c,m}$, due to the E_F/E_M multiplying term in the denominators. Analogous course was established when $J_{c,c}/J_{c,m}$ was

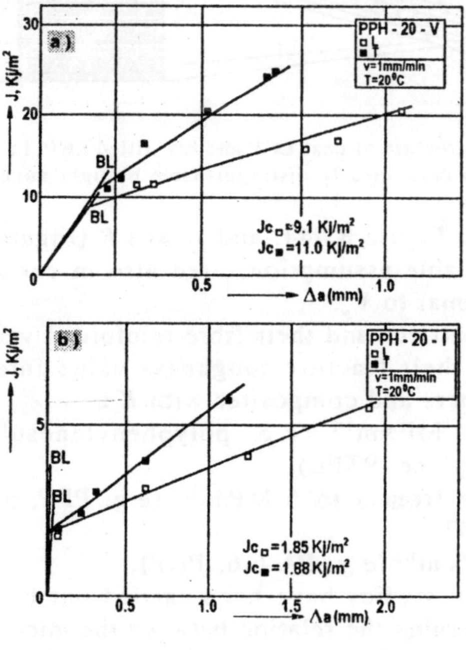

Figure 3.443. J – Δa curves for PP homopolymer composites reinforced by 20 wt% (=8 vol%) GF without (a) and with (b) PPH-20-I coupling agent, [1274].

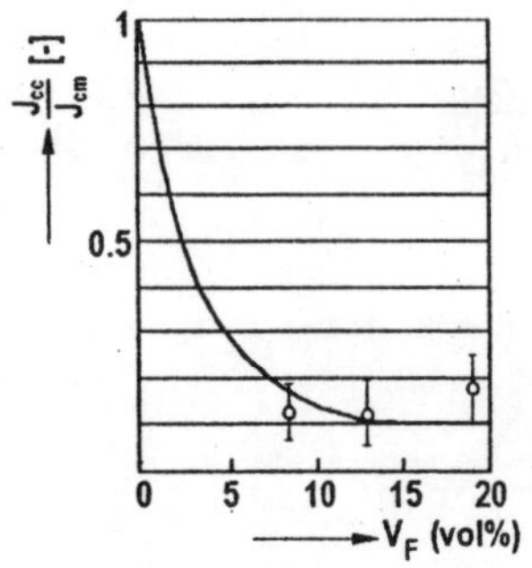

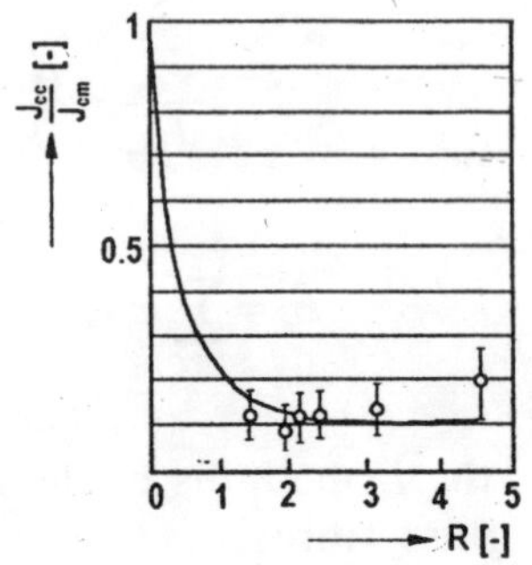

Figure 3.444. $J_{c,c}/J_{c,m}$ versus v_F and for coupled GF reinforced PP homopolymers. Testing carried out at $T = RT$ at $v = 10°$ mm/min; $J_{c,m}$=15 kJ/m^2 [1274].
Figure 3.445 (right). $J_{c,c}/J_{c,m}$ versus R for coupled GF reinforced PP homopolymers. Testing carried out at $T = RT$ at v = 10° mm/min, $J_{c,m}$=15 kJ/m^2 [1274]

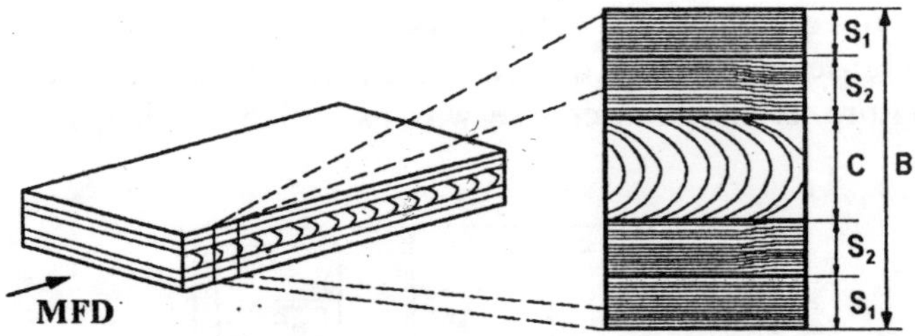

Figure 3.446. Layer-structure of neat LCP and low-filled LCP. In high-filled LCP there is only one skin layer clearly distinguishable by light microscopy, [1263].

plotted against V_F, Figure 3.444, and versus R (Figure 3.445) supporting the reasonable assumption, used also in equation (3.323), that R is proportional to V_F.

The polymer matrices and their fibre-reinforced versions can be ranked based on their fracture toughness value into three main groups, *i.e.* polymers and composites with K_c:

– less than 3 MPa·m$^{1/2}$ (*e.g.* polyphenylenesulphide, PPS, polytetraflouroethylene, PTFE);

– in the range from 3 to 6 MP·m$^{1/2}$ (*e.g.* PPP, poly(butylene terephthalate), PBT;

– above 6 MPa·m$^{1/2}$ (*e.g.* PA-6,6, PET).

Very interesting studies have been carried out by H. Voss and K. Friedrich concerning the relation between the microstructure and some mechanical properties of the liquid-crystal polymers reinforced with short fibres. The investigated composites were obtained as injection-moulded plates [1261–1262]. All materials, filled or unfilled, showed a distinct three-layer structure, Figure 3.446, [1263]. It was established that the dimensions different layers mainly are responsive by the differences between the materials. Thus, in the skin layer of the neat and low filled materials there was also a visible substructure, characterised by a change in orientation of the flow lines, as shown in Figure 3.446. The dimensions of the layers are given in Table 3.133. These results demonstrate the great influence of fillers (material and content) on the relative proportion of several layers. The data concerning the tensile tests, especially tensile strength, σ_f, strain to failure, ε_f, and elastic modulus, E, for both testing directions are collected in Table 3.133. The high

Table 3.132. Layer dimensions [1263]

<table>
<tr><th rowspan="2"></th><th rowspan="2">B
(mm)</th><th colspan="2">S
(mm)</th><th rowspan="2">C
(mm)</th><th rowspan="2">C/B</th></tr>
<tr><th>Inner</th><th>Outer</th></tr>
<tr><td>LCP</td><td>1.83</td><td>0.45</td><td>0.29</td><td>0.35</td><td>0.19</td></tr>
<tr><td>LCP + 40% talc</td><td>1.91</td><td colspan="2">0.82</td><td>0.27</td><td>0.15</td></tr>
<tr><td>LCP + 25% GF</td><td>1.84</td><td colspan="2">0.58</td><td>0.68</td><td>0.37</td></tr>
<tr><td>LCP + 5% Gr + 1% MoS₂</td><td>1.89</td><td>0.24</td><td>0.35</td><td>0.71</td><td>0.38</td></tr>
<tr><td>LCP + 2% PTFE</td><td>1.90</td><td>0.19</td><td>0.32</td><td>0.88</td><td>0.46</td></tr>
</table>

Table 3.133. Results of tensile tests [1263]

	σ_f (MPa)		ε_f (%)		E (MPa)	
	T	L	T	L	T	L
LCP						
B	99	29	2.0	1.9	7.8	1.8
S	120	31	2.0	2.1	8.2	1.9
C	85	36	1.6	2.0	8.1	2.0
LCP + 40% talc						
B	107	59	2.2	2.1	10.3	6.9
S	111	55	2.2	2.0	11.2	4.5
C	69	59	2.5	2.2	6.0	7.0
LCP + 25% GF						
B	111	50	1.8	1.7	11.4	5.4
S	148	41	1.3	1.8	15.8	3.2
C	60	53	1.3	1.5	7.5	4.5
LCP + 5% Gr + 1% MoS_2						
B	117	41	1.9	3.0	9.3	2.7
S	128	41	2.2	3.5	9.3	2.5
C	118	37	2.1	2.2	8.6	2.9
LCP + 1% PTFE						
B	110	21	1.6	0.9	10.0	2.7
S	149	15	1.5	1.0	14.9	1.7
C	50	21	1.1	1.0	4.6	2.7

anisotropy in mechanical behaviour is mainly reflected in a reduction of the tensile strength and elastic modulus if the fracture occurs parallel to the mould-filling direction (L-configuration). The strain to failure is less influenced by the loading direction. The high anisotropy in mechanical behaviour results from the high molecular orientation, parallel to the mould-filling direction in the skin layers. This is very clearly reflected if the skin layers are tested separately, after machining thinner specimens out of these regions. For the core layers the anisotropy is not as drastic as for the skin layers, but still exhibits some direction dependence (*i.e.* higher tensile strength and elastic modulus for the fracture perpendicular to MFD. This pointed to a preferred molecular orientation parallel to

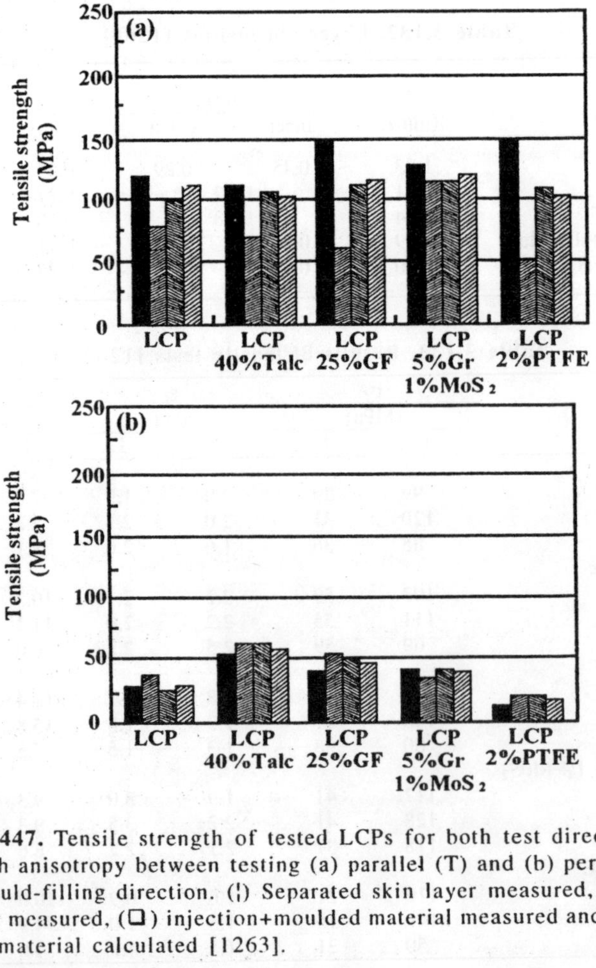

Figure 3.447. Tensile strength of tested LCPs for both test directions. Visible is the high anisotropy between testing (a) parallel (T) and (b) perpendicular (L) to the mould-filling direction. (¦) Separated skin layer measured, (▨) separated core layer measured, (☐) injection+moulded material measured and (▩) injection-moulded material calculated [1263].

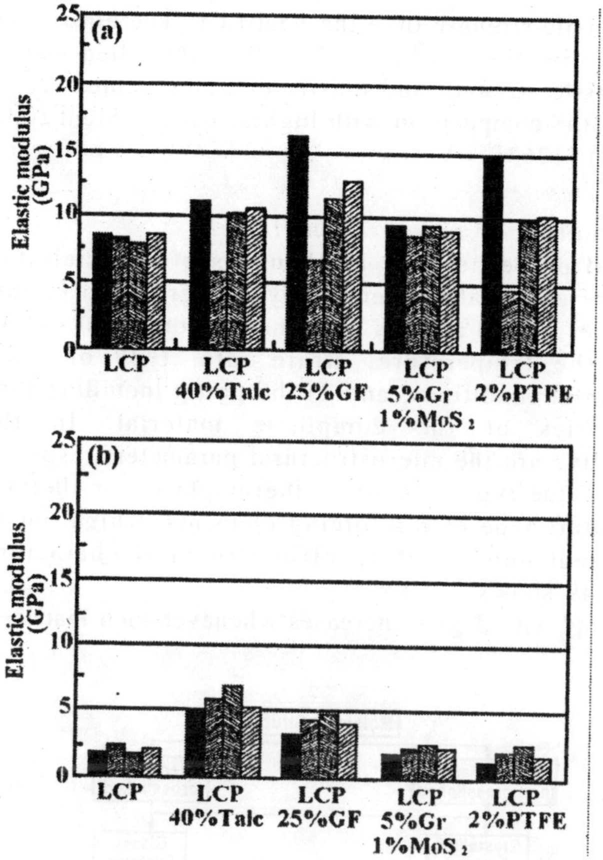

Figure 3.448. Elastic modulus of tested LCPs for both test directions. The elastic modulus shows nearly the same characteristics as the tensile strength, but here also differences between the materials' behaviour in the T-direction are visible. (¦) Separated skin layer measured, (▨) separated core layer measured, (□) injection-moulded material measured and (▥) injection-moulded material calculated [1263].

the mould-filling direction also in the core layer.

A summary of the measured tensile strength and elastic-modulus data of the injection-moulded material and of the specimens machined out of the skin and core layers is given in comparison with the calculated values in Figures 3.447 and 3.448, for both testing configurations (T/L). The agreement between the calculated and measured data is good. The differences in tensile strength as a result of the filling materials are mainly significant in the L-configuration, *i.e.* fracture parallel to the MFD. Except for the polytetrafluoroethylene – filled LCP, the other filled materials have a higher tensile strength than the neat LCP. The elastic modulus in both tested directions is higher for all of the filled materials. The highest tensile strength and elastic modulus values are especially notable for the composition with highest filler content (talc or glass fibre-filled) [1263].

3.10.5. Fracture of composite materials

Mechanical properties, the mechanisms of materials failure and fracture are essential influenced by two groups o factors: 1) the external ones, which reflects the concrete conditions of solicitation, such as: the temperature, strain rate, type of loading and environmental conditions; and 2) the group including the intricate characteristics of the composite material. In this case, representative are the microstructural parameters, especially those concerning, the type of matrix (thermoplastic or thermoreactive) and the nature/type of reinforcing elements, which, on their turn, bring a great number of specific structural characteristics, as Figure 3.449 shows.

The complexity degree increases whenever such matrices are used

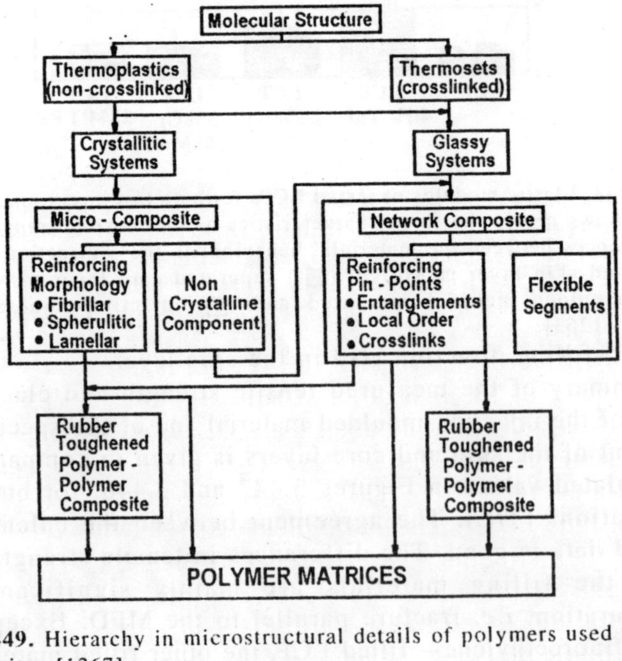

Figure 3.449. Hierarchy in microstructural details of polymers used as matrices for composites [1267].

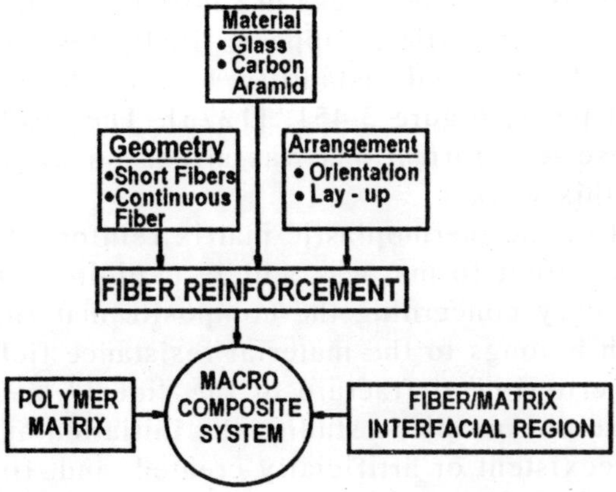

Figure 3.450. Basic phases in macrocomposite systems [1267].

for obtaining the fibres reinforced composites. The fibres presence implies a variety of new parameters, concerning both their chemical and supramolecular structure, and the morphological characteristics, such as: fibre length, their orientation and volume fraction. The arrangement of the fibres may also vary in different layers across the thickness of a component thereby affecting its fracture behaviour, Figure 3.450.

The composite materials currently as structural components can be supposed to the most different solicitation conditions, starting from the chemical attack to mechanical or electrical stresses, or the direct contact with other bodies. If these stressing conditions reach

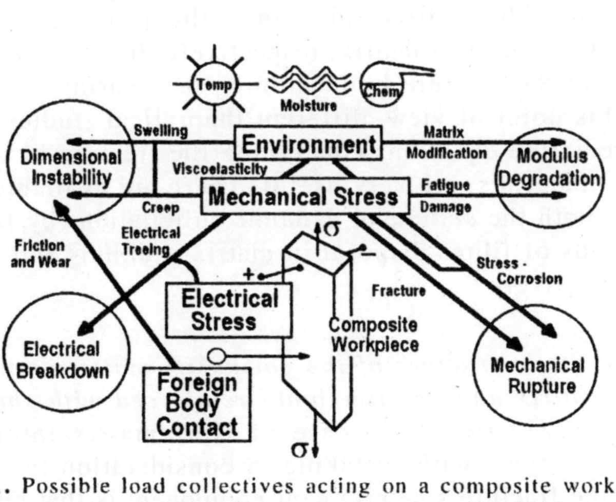

Figure 3.451. Possible load collectives acting on a composite work piece and resulting failure mechanism and changes in the material [1275].

critical values, the fatigue cracks are initiated within the composite material or wear particles appear, finally resulting a degradation modulus, dimensional instability, or even the total fracture of the structural parts, Figure 3.451, [1275]. The mechano-chemical nature of these transformations was already proved in the previous parts of this work.

The model of the thermoplastic matrix reinforced with short fibres, in its different forms, was widely exploited, in the elaboration of the theory concerning the composite materials fracture. A first approach belongs to the material resistance field, which uses the classical criterion of fracture of the "ideal" material, without faults; the fracture mechanic studies the initiation and propagation of a fault, preexistent or artificially created, and, finally, the failure mechanic investigates the material behaviour until to the apparition of a microscopic fault, which corresponds to the "catastrophic" fracture initiation.

All these theories furnish complementary information, thus contributing to the fracture process understanding and offer models within which practically all the concrete systems, supposed to the experimental analysis, can be framed.

3.10.5.1. The static fracture of the composite materials with thermoplastic matrix reinforced with short fibres

The fracture estimation of composites based on thermoplastic matrix reinforced with short fibres represents a complex problem, even in the case of the unidirectional composites supposed to a traction effort on the fibres direction, since the process can occur at interface, in fibre, or in matrix, respectively. In addition, the matrix should suffer the fracture by tensioning or shearing.

From this point of view, different theoretical studies, which offer the predictive equations concerning the fracture of this type of composite materials, are very useful. There are available equations describing both the static and dynamic solicitation (by fatigue), for the two kinds of fibres disposal in matrix, i.e. uni- and multidirectional, respectively.

3.10.5.1.1. Static fracture of the composite materials with thermoplastic matrix and unidirectional reinforced with short fibres

H.L. Cox firstly treated, a case of fibre/matrix interaction, of elastic/elastic type, without taking in consideration the fibre ends. He found the fracture's determinant component is that characterised by the smallest/weakest deformation [1276]. This approximate model was improved by A. Kelly, which using the concept of critical length and starting from the mixture's rule, proposed the following equation for calculating the fracture effort [1278]:

$$\sigma_c'' = \left[1-(1-\beta)\frac{l_c}{l}\sigma_f'' \cdot V_f + \sigma_m'(1-V_f)\right] \quad (3.329)$$

where: σ_m' - the effort in the matrix, corresponding to strain ε_f''; $\beta\sigma_f''$ - the average effort in a fibre at a distance equal to $l_c/2$ from the fibre end (l_c – critical length). According to this equation, about 95% from the fracture effort of the composites with continuous fibres can be attained in the case of the composites with short fibres.

A more realistic model is that proposed by P.E.Chen. Using the finite elements theory, the author takes into consideration the interactions between fibres. He found a plateau for the fracture effort of the unidirectional composites. The maximal effort supported for a epoxy/glass fibres composite does not overtake 60% from that one corresponding to the same matrix, but reinforced with continuous fibres.

The systems which present a poor plastic deformation, with influences in the fibre ends, are not correctly described by equation (3.329). For this case, J.R. Outwater developed the theory of mechanical friction, in order to characterise the polymers' behaviour. He replaced the matrix flowing effort with a transfer effort, which depends by a friction coefficient.

Based on the assumption that the deformation governs the fracture and realising an analysis of the deformations to fracture of the two components, the matrix and fibre, three distinct cases are found, Figure 3.452. The most frequent case corresponds to the situation when the matrix fracture deformation is lower that than of fibre; in this case, a multiple fracture of matrix occurs before of the sudden fibres fracture. The stress–strain curve presents a discontinuity, at the deformation level, in the moment of matrix failure, Figure 3.452a. The final fracture takes place at $l > l_c$:

$$\sigma_c'' = \sigma_f'' V_f\left(1-\frac{l_c}{2l}\right) \quad (3.330)$$

$$\varepsilon_m'' < \varepsilon_f'' \quad (3.331)$$

The second case describes the behaviour of composites with metallic matrix, Figure 3.452 b, when:

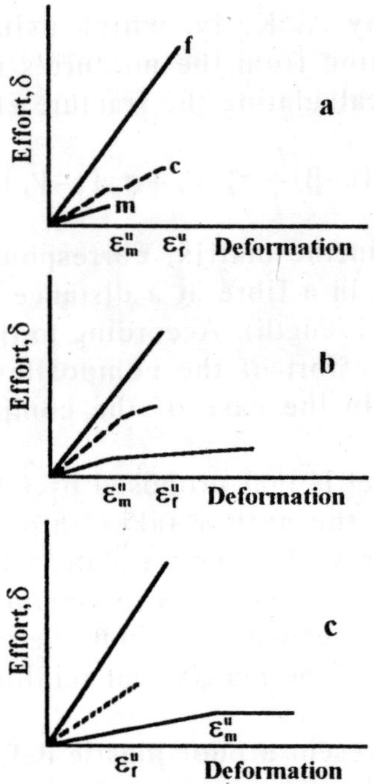

Figure 3.452. Various cases of mechanical behaviour that characterise the unidirectional composites [1279].

$$\varepsilon_{my} < \varepsilon_f^u < \varepsilon_m^u \tag{3.332}$$

The fracture occurs when the deformation value reached the fibre fracture deformation, while the matrix is plastically deformed. This is the case of elasto-plastic interaction (a = 1), illustrated in Figure 3.453, and it is described be the following equations:

$$\sigma_c^u = \sigma_f^u \cdot V_f \cdot \left(1 - \frac{l_c}{2l}\right) + \sigma_{my}\left(V_m - \frac{V_f}{2n^2 s}\right) \tag{3.333}$$

$$\varepsilon_c^u = \varepsilon_f^u \tag{3.334}$$

For the continuous fibres $l \to \infty$, and the classical mixture's rule is valuable:

$$\sigma_c^u = \sigma_f^u V_f + \sigma_m^u \left(1 - V_f\right) \tag{3.335}$$

The third case, illustrated in Figure 3.452 c, adequately describes the case of reinforced polymers:

$$\varepsilon_f^u < \varepsilon_{my} \tag{3.336}$$

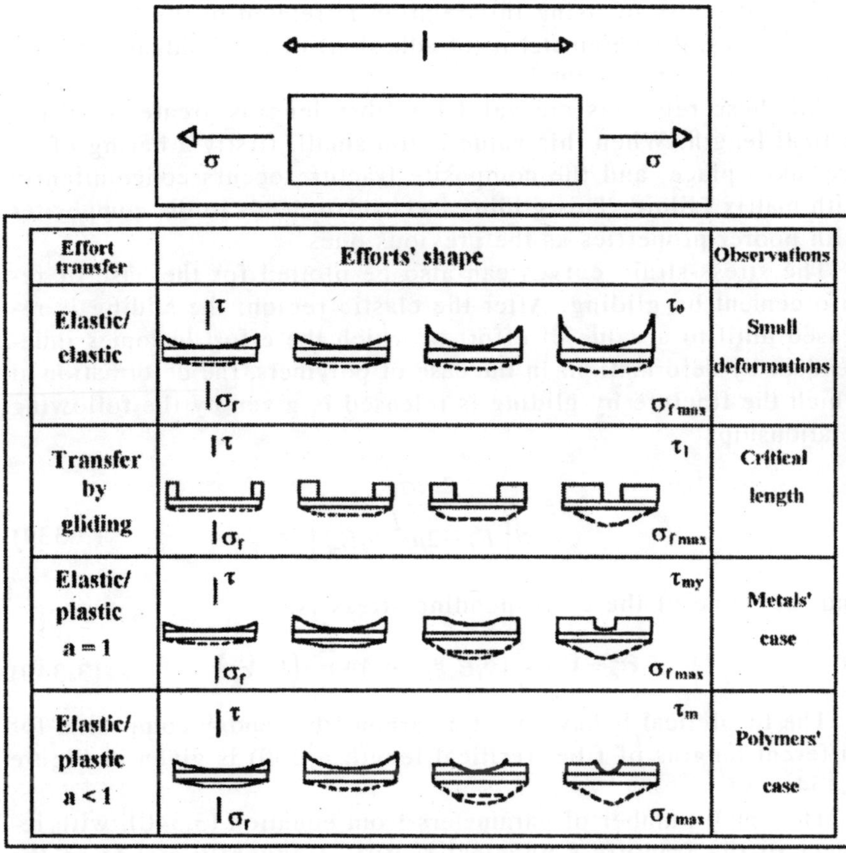

Figure 3.453. Diffrent type of load transfer mechanisms [1276].

The matrix elastically behaves until the fibres deformation to fracture, Figure 3.452 c, According to the equations developed by M.R. Piggot, in the case of a poor adhesion ($a < 1$) the fracture effort of the composite material is:

$$\sigma_c'' = \sigma_f'' V_f + E_f \varepsilon_f'' V_m - \frac{V_f}{s}\left[\frac{\sigma_f'' - \left(\frac{2a\tau_{my}}{n}\right)^2}{4\mu\left(\nu_l E_m \varepsilon_f'' - \sigma_r\right)} + \frac{2a\tau_{my}}{n^2}\right] \quad (3.337)$$

$$\varepsilon_c'' = \varepsilon_f'' \quad (3.338)$$

The difficulty of using this relation is related to the fact the in certain cases, the parameter *a* (of adhesion) or the residual tensions, σ_r, can not be determined.

All these relations are valid for fibre lengths greater than the critical length. When this value is too small, firstly a baring of fibre takes place, and the composite fracture occurs concomitantly with matrix failure. This gliding mechanism leads to the composites with poorer properties as the previous ones.

The stress-strain curves can also be plotted for the case of reinforcement by gliding. After the elastic region, the sliding is released until to a value of effort to which the effort becomes independent by deformation. In the case of polymers, the deformation at which the fracture by gliding is released is given by the following relationship:

$$\varepsilon_{ip} = \frac{2\mu l \sigma_r}{d\left(E_f - 2\mu \frac{l}{d} v_l E_m\right)} \tag{3.339}$$

and the value of the corresponding stress is:

$$\sigma_c^u = V_f \mu \frac{l}{d}(v_l E_m \varepsilon_l - \sigma_r) + \bar{\sigma}_m (1 - V_f) \tag{3.340}$$

The theoretical behaviour of a carbon fiber-epoxy composite, for different lengths of fiber (critical length ≈ 200) is given in Figure 3.454.

The small number of parameters from equation (3.340), with respect to the fracture's physical reality, frequently makes as the theoretical predictions to differ by the experimental results. From this reason, many authors adapt the basic models for their concrete purposes [1280–1282].

M.G. Bader, T.W. Chou, and J.J. Quigley proposed a new

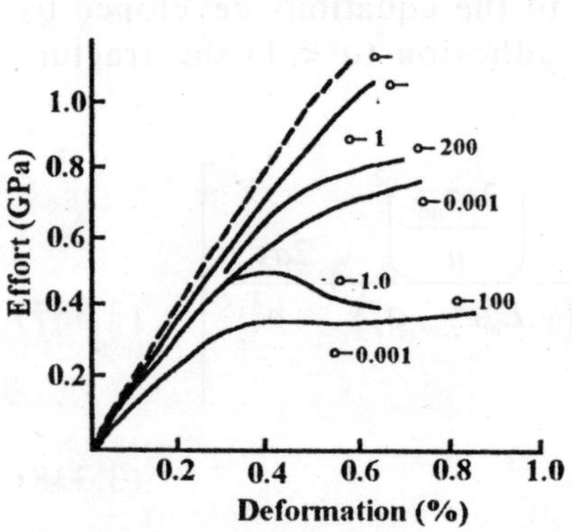

Figure 3.454. Stress – strain theoretical curve for electroplastic transfer. Effect of adhesion coefficient [1279].

theory, which assumes that the faults occur anytime at the fibre ends, their accumulation generates a critical zone, weakened, where the fracture is activated [1283].

3.10.5.1.2. Static fracture of the composite materials with thermoplastic matrix and multidirectional reinforced with short fibres

The fracture of unidirectional composite materials on a different direction than that of fibres one depends by the value of the angle θ between the fibres orientation and effort direction. Under limiting conditions, three distinct cases can be considered: 1) low values of θ, when the fibre is broken; 2) intermediate values of θ, when the fracture occurs by matrix shearing at the interface; and 3) $q \to 90°$, when the matrix fracture between fibres takes place.

For the composites with continuous fibres, a great number of criteria are available, such as those of deformation or of the maximal effort, and the Tsai/Hill criterion or Tsai/Wu criterion. In accordance with these criteria, the composite materials will be broken when one of the mechanical components reaches a limit value (experimentally determined, in the majority of cases). The Tsai/Hill criterion defines the upper limit of the effort, above which the fracture occur:

$$\frac{\sigma_1^2}{x^2}+\frac{\sigma_2^2}{y^2}+\frac{\tau_{12}^2}{s^2}-\frac{\sigma_1^2\sigma_2}{x^2}=k \qquad (3.341)$$

where: x, y, and s the fracture efforts at longitudinal and transversal traction, and shear, respectively. In the case of solicitation to compression the relations are more complicated. The Tsai/Hill criterion better describes the experimental data of the thermoplastic material reinforced with glass fibres (Figure 3.455).

Usually, the fibres are not unidirectional oriented in composite materials. Fibres orientation varies in function of the pieces thickness, processing method, and mould geometry. Accepting a random orientation of fibres, or a known and measured orientation, the laminates theory was appealed in order to describe the behaviour of this type of composites. The method allows the determination of the weakest layers, which will be firstly fractured [1285]. The Figure 3.456 illustrates a good agreement between the theory and practice, considering, for comparison, the data furnished by P.E. Chen [1285] and J.K. Lees [1286].

N.S. Sridharan performed an analysis of the layers progressive fracture of an hybrid composite, containing long and short fibres,

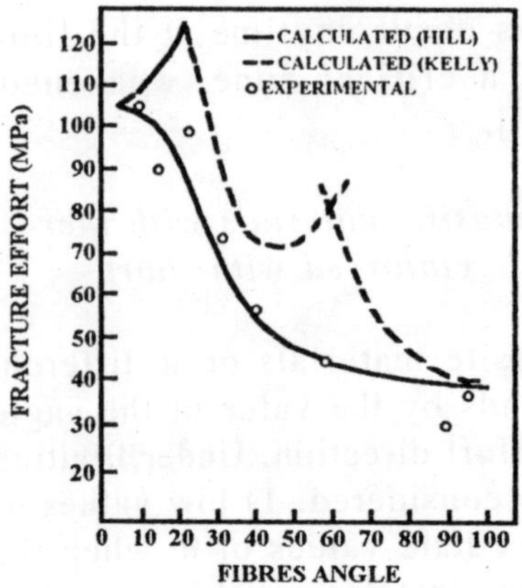

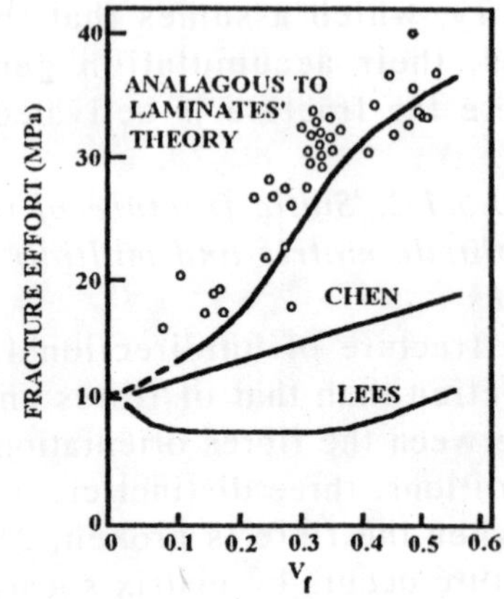

Figure 3.455. Fracture effort in function of fibers orientation (PMMA with 29% glass fibers) [1284].

Figure 3.456. Comparison between theory and experimental data on the influence of fibers orientation [1287].

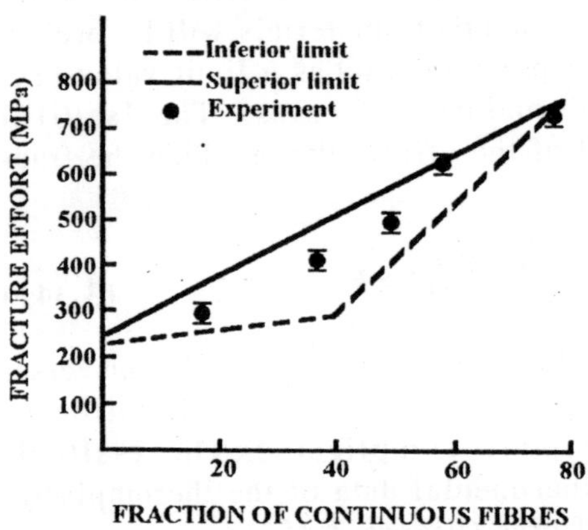

Figure 3.457. Fracture effort in function of GF content for a hybrid composite [1287].

calculating the fracture efforts, as is illustrated in Figure 3.457 [1287].

3.10.5.2. Dynamic fracture (by fatigue) of the composite materials with thermoplastic matrix and reinforced with short fibres

In this field, the investigations were devoted to elucidate the polymer and composite materials behaviour under exploitation conditions (non-notched specimens) as well as regarding the crack propagation (notched specimens). In the first place, the attention was focused on the composite materials with thermoreactive matrix, widely used in the automobiles industry. The test results, performed

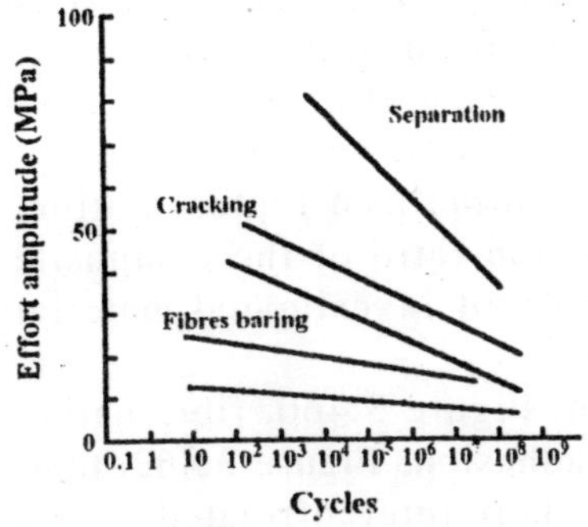

Figure 3.458. Endurance curves showing the fracture stages of a composite material reinforced with short fibers [1296].

on the non-notched specimens are plotted in the form of durability (endurance) curves.

Fracture by fatigue micromechanisms includes three successive stages of material deterioration. The first stage is manifested at the interface, in this zone the fracture occurring at relative small efforts. For instance, in the case of unsaturated polyesters reinforced with glass fibres, in the form of mats, it represents about 10% from the total fracture stress of the material, Figure 3.458.

The main fault, located at the microstructural level of composites, consists in the perpendicularly oriented fibres to the effort direction, which concentrate the mechanical energy, initiate the crack, whose propagation finalises the fracture [1288, 1289]. The fibers length and the matrix nature are very important characteristics. It was found that the stabilisation of fatigue resistance could be realised by increasing the ratio *l/d* [1290, 1291]. Using electron fractography, the role of the fibres' ends in the microcracks' development, under the conditions of a significant plastic deformation of the matrix was confirmed. The fragments containing broken fibres constitute faults that affect the fatigue resistance, the fracture starting even at very low efforts [1292, 1293]. From this reason, the fatigue limit is not very clear observable for these materials.

The fracture by fatigue is described as the crack propagation through material, occurring by so-called "fiber avoidance made". The experiments made at repeated traction, on different thermoplastic material or composites reinforced with glass or carbon fibres proved that a microcrack, initiated in matrix, will propagate avoiding the fibre, without causing its fracture. This finding proves that the cumulated deformation during the dynamic fracture is the same order of magnitude as the final deformation in the case of monotonous testing, for the composite materials filled with glass fibres [1294, 1295].

The modelling of the endurance of the thermoplastic materials reinforced with short fibres was mainly realised in the form of empirical equations, based on concrete experimental data [1296]. The typical endurance curves for glass or carbon fibres reinforced nylon 6,6 are depicted in Figure 3.459. A normalised representation, with respect to shows that the degradation ratio of the composite material is 10%, irrespective of the nature of investigated material [1296].

The influence of the fibre orientation, Figure 3.460, fibre nature and content; Figure 3.461, fibre/matrix adhesion, Figure 3.462, temperature, Figure 3.463, and especially their intercorrelated action determines the fatigue level in the composite material.

3.10.5.3. The application of fracture mechanics to the composite materials

The linear-elastic fracture mechanics describes the crack behaviour in isotropic and perfect elastic materials. Theoretically, the reinforced polymers are not isotropic materials and are

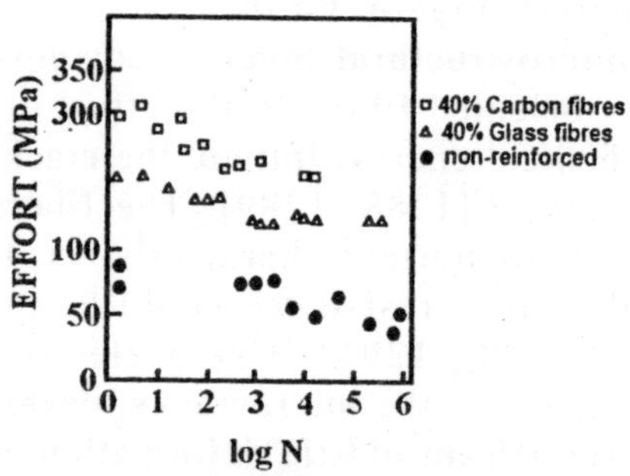

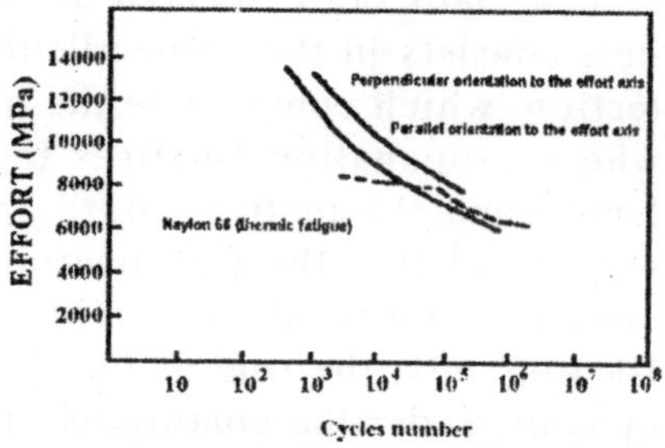

Figure 3.459. Endurance curve of Nylon-6,6 reinforced with short fibers [1296].
Figure 3.460. Influence of fibers orientation on the endurance curve of Nylon-6,6 reinforced with short fibers [1296].

characterised by a visco-elastic behaviour. In reality, the deviations with respect to the theoretical hypothesis are relatively small, *i.e.* macroscopic homogeneity and reduced anisotropy. This means that with some adaptations, generally, this theory may also be applied in the case of composites.

The theory leads to the toughness definition, K_{Ic}, and to the critical energy of crack propagation, G_{Ic}, the solving of the problems related to the efforts and plane deformations being of first importance. These parameters are particularised in the case of the ther-

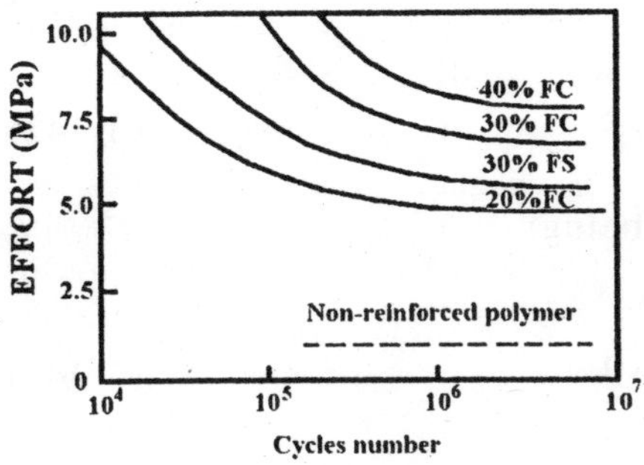

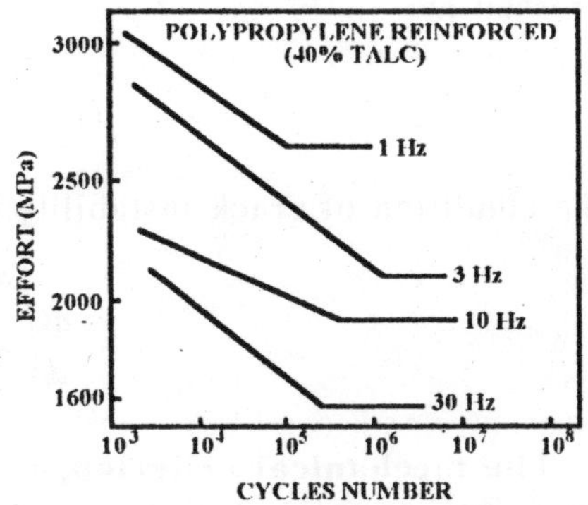

Figure 3.461. Influence of nature and content of fibres on the endurance curve [1–96].

Figure 3.462. Influence of solicitation frequency on the endurance curve for reinforced polypropylene (40% mineral talc).

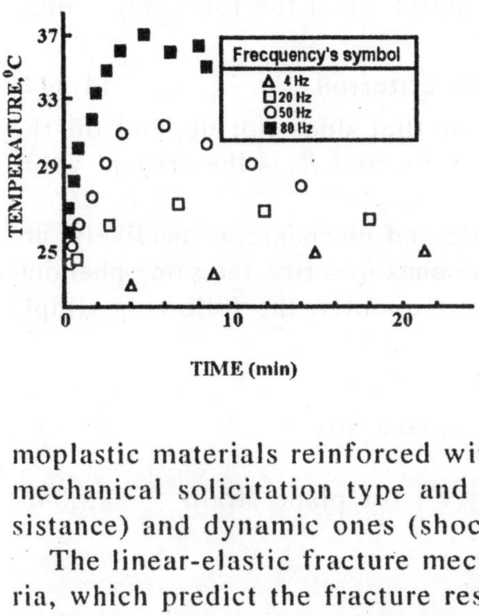

Figure 3.463. Variation of sample temperature during the first cycles of solicitation [1296].

moplastic materials reinforced with glass fibres, in function of the mechanical solicitation type and refers to static tests (flexion resistance) and dynamic ones (shock, fatigue).

The linear-elastic fracture mechanics developed two main criteria, which predict the fracture resistance of the different material, taking into consideration three parameters: 1) fault dimension; 2) its geometry; and 3) the applied effort.

The energetic criterion, according to which the crack development into a mechanically solicited composite releases a certain amount of elastic energy that is stocked in material and can be evaluated. When its level attains a critical value, G_c, the cracks will propagate until the material's fracture. The critical energy of propagation represents an intricate characteristic of material, and in the case of brittle fracture, the Griffith criterion can be expressed as:

$$G > G_c \quad (3.342)$$

the condition of crack instability being:

$$\frac{dG}{dA} > 0 \quad (3.343)$$

The mechanical criterion, according to which the crack presence into a stressed elastic material modifies the forces field in material. Thus, the real effort in the crack proximity is much higher than that corresponding to a far from crack zone.

The elasticity theory allows the calculation of the force fields. For instance, for a region of crack propagation, G.R. Irwin proposed a hypothesis that cracking starts when the following condition is fulfilled [1297]:

$$K_I = K_{Ic} \quad \text{Irwin's criterion} \quad (3.344)$$

where: K_I – stress intensity factor that does not depend on the nature of filler, or specimen geometry; and K_{Ic} – the critical value of K_I.

Even if the two criteria, energetic and mechanic, respectively, are based on different hypothesis, these ones describe the same phenomenon that of material fracture. Consequently, the following simple relations exist between G and K:

$$E \cdot G = K^2 \text{ for plane stress} \quad (3.345)$$

$$E \cdot G = K^2(1 - \nu^2) \text{ for plane strain} \quad (3.346)$$

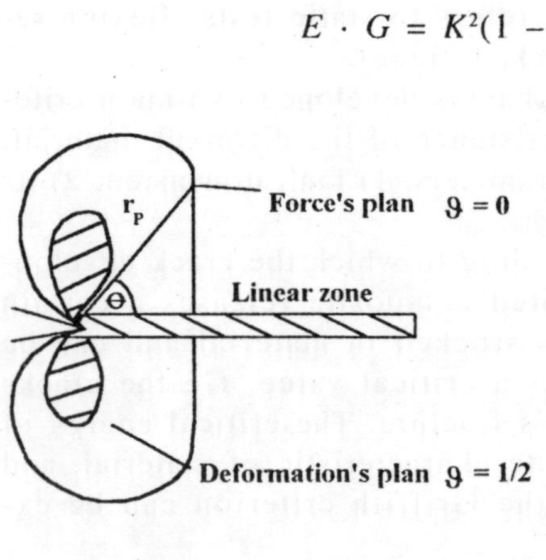

Figure 3.464. Model of plastic zone (Irwin and Dugdale) from crack region [1295].

where: E – the Young modulus of the material; and ν – the Poisson coefficient.

The elastic calculus shows that in the crack zone the theoretic effort is infinite. This means that on a certain distance around of the delimited region by crack, the material becomes plastic; therefore here the effort overtakes the flow limit effort, σ_y. Whether the dimension of this zone remains enough small, its evaluation becomes possible. Accepting that the plastic zone has an elongated shape, Figure 3.464, its dimension can be evaluated using Dugdale's method [1298]. The knowledge of the dimension of the affected zone allows the evaluation of the ductile or brittle character of material as well as toughness calculations. It also provides information concerning to the deformation state of material within the fracture region.

A crack, supposed to a certain manner of opening, contains at its centre the focus of its development and the lateral one, the centre of plane efforts development. The developed plastic zones differ each to other in function of the two deformation states and lead to different values of toughness. This is way, the knowing of the deformation state of the stressed material in the crack region, is so important.

In the case of polymers, an empirical rule is used which stipulates that the material will be deformed as a function of plane deformation, ε, under the following condition:

$$B > 4r_p \tag{3.347}$$

where: B – material thickness; and r_p – the radius of plastic zone.

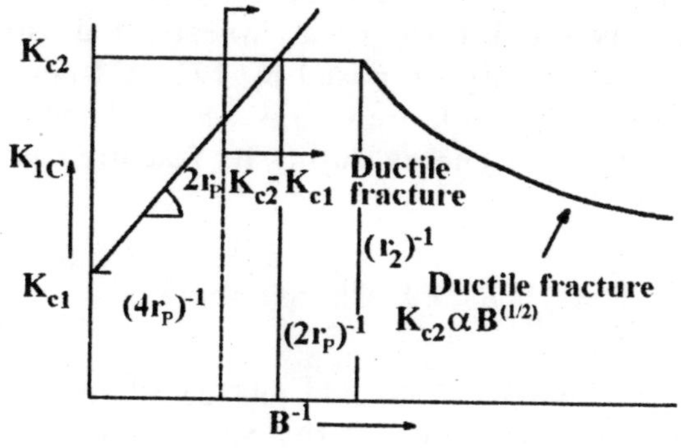

Figure 3.465. Influence of specimen thickness on toughness, K_{rc}, determination [1299].

The importance of sample thickness was established by J.G. Williams and it is illustrated in Figure 3.465 [1299]. If the thickness B is small, K_{Ic}, will be calculated for a state of mixed efforts, starting from K_{c1} and K_{c2}, which represent the toughness corresponding to the plane deformation and plane efforts, respectively. Only the toughness value of the plane deformations, K_{c1} is validated by the linear–elastic mechanics of fracture, since the values K_{Ic} calculated for different geometry of samples lead always to lower values of material toughness.

Depending on the concrete type of solicitation, one or other of the two criteria are used, allowing either the determination of K_{Ic} or G_{Ic}, respectively. Three cases are of peculiar importance from this viewpoint, namely: 1) static fracture; 2) shock fracture; and 3) dynamic fracture (fatigue).

In the case of static fracture, for a simple solicitation (traction) a cast plate with infinite dimensions that contains a crack with the length equal to $2a$, with a normal stress to the crack, σ, is considered. The expression for toughness calculation is:

$$K_{Ic} = \sigma_c^u \sqrt{\pi \cdot a} \tag{3.348}$$

where: σ_c^u - fracture stress; K_{Ic} – material toughness.

For a more general case:

$$K_{Ic} = \sigma_c^u y \sqrt{a} \tag{3.349}$$

where: y – the shape factor, which takes into account the specimen geometry and is tabled for the usual shapes.

The fracture by shock testing was investigated using a methodology, which was initially proposed by J.C. Williams [1299], taking into consideration the elastic energy accumulated within notched specimen. The absorbed energy during its fracture can be expressed as:

$$U = G_{Ic} \cdot B \cdot D \cdot \phi \text{ Charpy shock} \tag{3.350}$$

where: G_{Ic} – represents the critical energy of propagation; B and D – geometrical dimensions of the specimen; f - shape factor, especially determining the notch length. Varying the notch length and recording the fracture energies, the critical energy of propagation is obtained from the slope of the strength line obtained in graphical representation U vs. $BD\phi$. However, this method is not applicable to the ductile or semiductile materials.

An interesting method is that of instrumented shock, which allows the toughness calculation, K_{Ic}, at dynamic solicitation.

The fatigue case constitutes another face of fracture that presents a peculiar practical interest. Under the effect of repeated solicitation, a crack continuously propagates through the material. In this case, instead of the single factor, K_I, as the case of static fracture is, an interval $\Delta K = K_{max} - K_{min}$ must be considered, its limits corresponding to the maximal and minimal effort of the solicitation cycle. The crack propagates until attains a critical length, which equilibrates the material toughness, causing the sudden fracture of the stressed material. In order to describe this evolution, P.C. Paris and G.C. Sih proposed a model which was found to be valid for many materials [1300].

$$\frac{dA}{dN} = (\Delta K)^m \tag{3.351}$$

where: dA/dN – crack growth for every cycle; $\Delta K = K_{max} - K_{min}$; m – material constant.

In the case of composite materials, the application of mechanical fracture principles, and the approaching manner in this field evolved from the definition of same parameters, which represent criteria of fracture process evaluation, until to the establish of some precise models of deformation and fracture, in function of the matrix nature (ductile or brittle), filler type (particles or fibres), and, for fibres, their geometry and orientation.

Thus, it was defined a toughness factor, starting from polymer tenacity and limit yielding effort. The use of this method imposes the testing at low temperatures [1301]. For the reinforced polymers with spherical particles, a series of interesting interpretations concerning the toughness are offered [1302]. In the case of reinforcing with short fibres, there is a linear relation between the toughness and the fibre volume content. In the same time, a maximal value of G_{Ic} is found, since between the two parameters, *i.e.* K and G, the following relation exists: $K^2 = E{\cdot}G$ [1303]. The evaluation of resistance to cracking of the composite materials reinforced with unidirectional short fibres evidences that the best behaviour is obtained for the fibres length equal to the critical length, Figure 3.466, [1304]. The understanding of the fracture mechanism of the com-

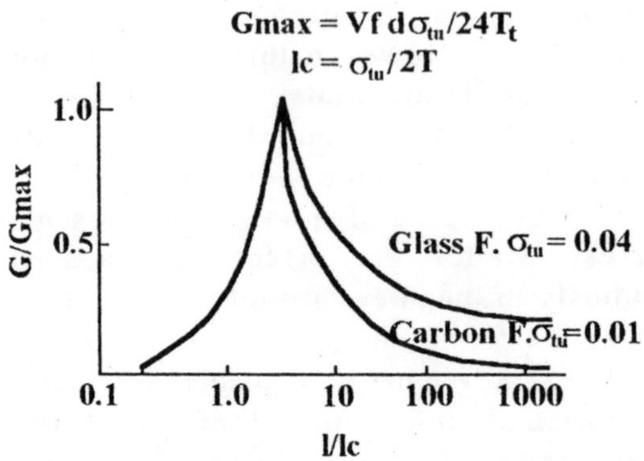

Figure 3.466. Theoretic effect of the filling ratio on the fracture energy [1304].

posites reinforced with glass fibres implies the knowing of its evolution in the crack region. The damaged zone, by its dimension, determines the material resistance. Fibres orientation represents an major factor. It was established that the fracture energy varies with a factor equal to 10 with fibres orientation, for the continuous composites, of epoxy/glass fibre type, tested to flexion. In the case of short fibres based composites this factor is lower [1305]. For a polyamide plate, obtained by injection, reinforced with 18% short glass fibres, an toughness anisotropy ratio equal to 1.7 was found [1306].

The investigation of the damaged zone, in the crack region, for the discontinuous systems was approached by A.C. Garg [1307]. He evaluated the crack dimension, starting form the models proposed by E.M. Waddauops [1308] and J.M. Whitney [1309], equivalent to the plastic zones analysed by D.S. Dugdale [1298]. The proposed model is in good accordance with the experimental data. Good results are also obtained studying the damaged zones for a series of reinforced thermoplastic materials, under the assumption that their dimension is equivalent to value of longest fibres. Figure 3.467 illustrates the good agreement with the experimental data [1310]. The model does not take into account the fibres orientation and outlined that the radius of the notch bottom does not represent a critical parameter for the reinforced systems.

A complete understanding of the cracking phenomenon implies to take into consideration of many factors of influence, the knowing of the damaged zone characteristics in function of matrix type, and fibres geometry and orientation. From the most important factors are those intricate of material, in the first place the composite's microstructure. From this point of view the matrix nature is very important. There are two classes of polymers widely used as matrix,

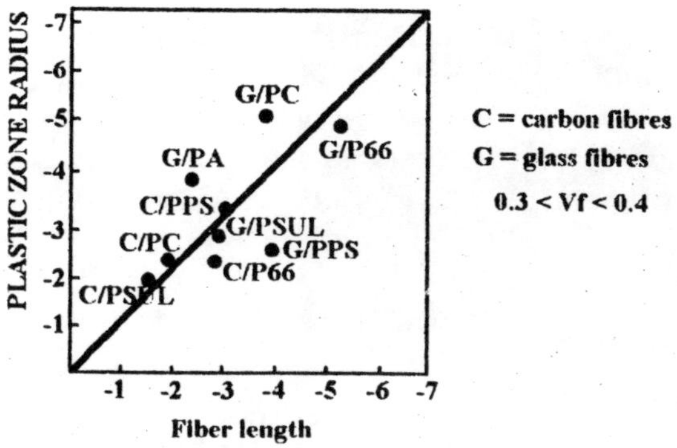

Figure 3.467. Radius of critical zone calculated in function of a length that overtakes with 5% fibers length for different reinforced thermoplastic materials [1310].

i.e. thermoplastic and thermoreactive polymers. The last ones are more tough, rigid, and brittle, due to their crosslinked structure. The thermoplastic polymers are characterised either by randomly coiled chains (amorphous phase) or aligned ones (crystalline phase). The morphology of crystalline polymers is influenced by their chemical structure and technological conditions of processing. At the molecular level, beside of chemical structure, the molecular weight plays a major role. With the increase of the polymer average molecular weight, the crystallinity degree and the material toughness increase, too.

The behaviour to fracture of a rigid polymer matrix can be positive influenced by the introduction of some finely divided particles, which increase material's toughness. As Figure 3.468 shows, these particles can determinate large zones of local deformation, in the proximity of the centres of stress/tension high concentration, as the cracks tips are, causing an important absorption of energy [1255, 1265], Fig. 3.468

Many polymers, such as PE, PP, PVC, PS, PA-6,6, PC, due to their durability, PEEK (polyetherethreketone), by its thermostability, and polyepoxy and polyimide resins are widely used as matrix in polymer composites.

Of equal importance is the second phase, rigid and highly resistant, such as long or short fibres, which fundamentally change the profile of the new material properties. The short fibres of carbon, aramide, or glass ones allows the material processing on the same principles and equipment as in the case of polymer. The following properties are enhanced by the fibres presence: tensile strength,

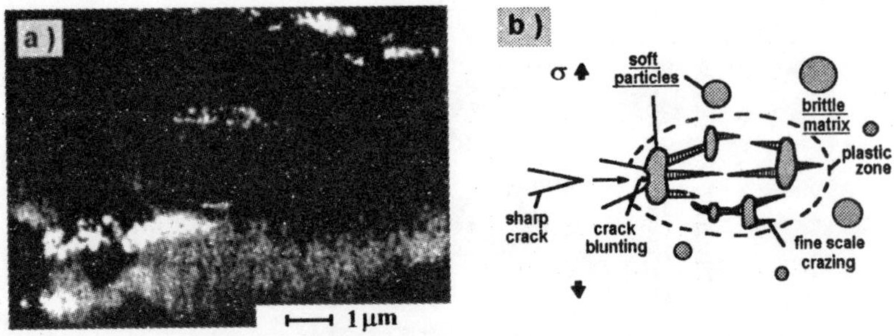

Figure 3.468. (a) TEM micrograph of crazes (C) in PET, toughness modified using Ionon-particles (dark zones); (b) Mechanisms of energy absorption by particles toughness into of brittle polymer matrix, rigid to the crack propagation [1255].

stiffness, and dimensional stability. As Figure 3.469 shows, using the casting under pressure, due to the flow conditions in mould, the reinforced material with short fibres is stratified.

Using long fibres, the composite material presents a laminate structure, with unidirectional disposing of fibres, under the same or different angles of individual layers [1311]. The appearance of such laminate, in the tridimensional space, is presented in Figure 3.470 [1312]. It is characterised by a symmetric arrangement, with a modification of the fibres direction between 0 and 90° with respect to the reference axis, which usually correspond to the stress action axis.

The correlation of some microstructural characteristics both with the toughness factor, K_c, and the integral J_{Ic} – as a parameter of the crack field, and using the methodology described in ref. [1313] gave the possibility of fracture prediction, following typical mechanisms in accordance to the more or less ductile, or brittle character of matrix. For instance, in the case of isotactic polypropylene, form the relation morphology (evaluated by spherulite diameter D) – toughness, it was deduced that the value K_c decreases with the spherulite diameter, concomitant with the increase of stiffness, which is controlled by the fraction of SiO_2, used as filler, Figure 3.471 [1313]. Correspondingly, the fracture behaviour is modified, passing from the mainly ductile fracture surface to brittle, Figure 3.472, [1256]. The prediction of the mechanisms and fracture properties can subsequently be correlated with the average molecular weight and the isotactic PP fraction.

In the case of the polymers with high toughness, the determination of toughness profiles is more rigorous achieved using J-integral

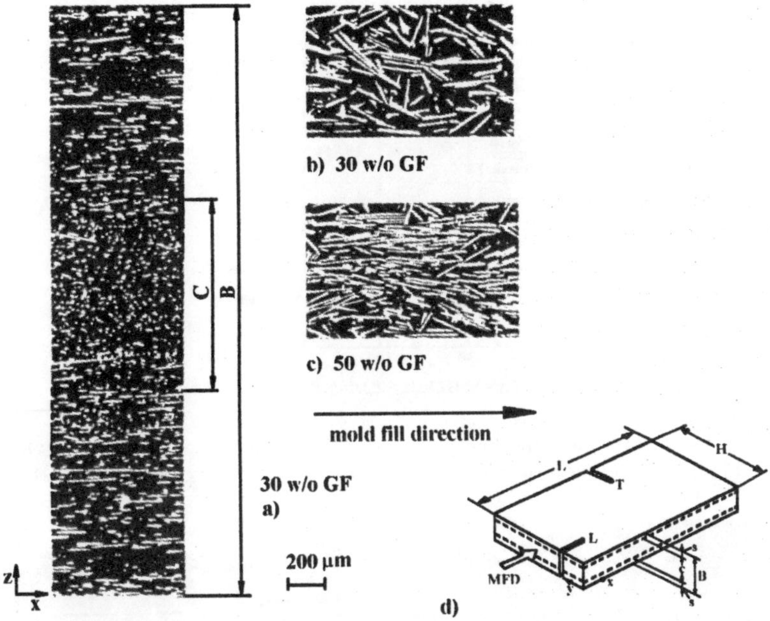

Figure 3.469. SEM-micrographs of polished surfaces of glass-fiber reinforced LCP. The layered structure across plate thickness is obvious in micrograph (a). The degree of fiber alignment parallel to the MFD is visible in micrograph (b) for LCP-30 GF and in (c) for LCP-50 GF. The high orientation in the latter one is obvious. (B-C = 25) [1255].

Figure 3.470. Tridimensional representation of the polished surfaces of a CF-epoxy laminate [0, 90, 0, 90] [1312].

method. In Figure 3.473 the crack's initial values, evaluated by $J_{Ic,}$ for the polymers with modified toughness, containing or not fibres [1313]. The values of toughness to fracture for different techno-polymers are summarised in Table 3.134 [1255].

In the case of reinforced thermoplastic polymers, for a given modulus E, the fracture toughness depends on the individual

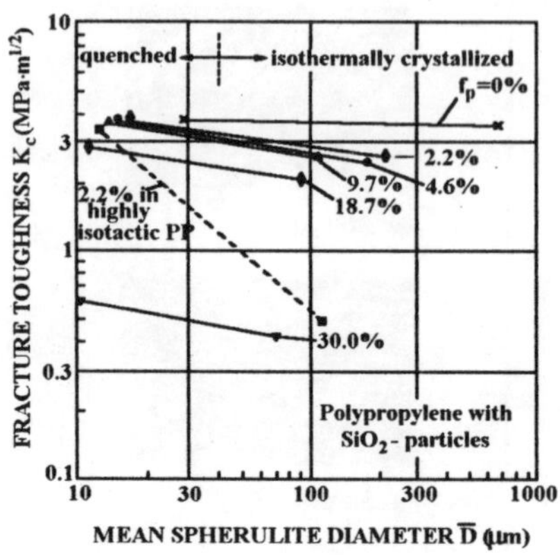

Figure 3.471. Fracture toughness, K_c, as a function of spherulite diameter, D, and volumic percentage of SiO_2 particles in isotactic polypropylene [1313].

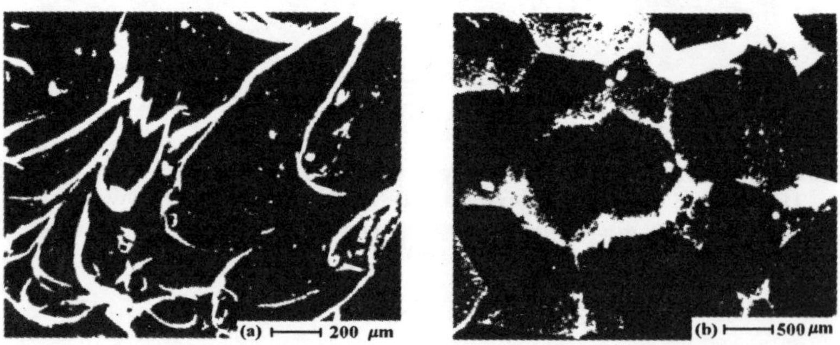

Figure 3.472. Fracture surfaces of: (a) fine-spherulitic polypropylene; (b) rough-spherulitic polypropylene; in both cases SiO_2 particles appear as impurities [1313].

mechanisms, acting in the damaged zone, Figure 3.474, namely: a) fibre fracture; b) separation fibre/matrix; c) fibre pull-out; and d) matrix deformation and fracture. The relative contribution of the individual mechanisms as well as the absolute value of the adsorbed energy are influenced by a series of factors, such as: 1) fibres characteristics (orientation, volume fraction, diameters size distribution; 2) matrix behaviour to deformation in correlation with different testing conditions; and 3) the quality of fibre/matrix bond.

Figures 3.475–3.480 illustrate the modification tendencies of fracture toughness in function of the fibres fraction of different polymer matrix and, correspondingly, the fracture behaviour as a function of composite microstructure and testing conditions. The

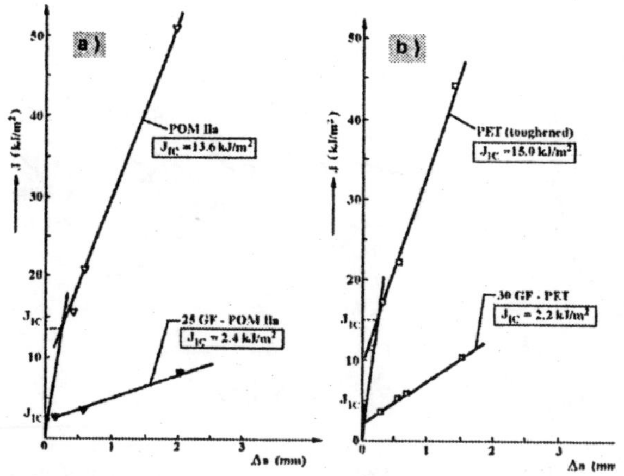

Figure 3.473. Curves of cracking resistance, as expressed by J_{ic} values for: (a) POM with PUR-particles; (b) PET modified with Ionon-particles [1313].

Table 3.134. Values of fracture toughness of different thermoplastic materials, measured at room temperature, RT, and a strain rate between 1 and 10 mm/min [1255]

Matrix	at RT	K_c (MPa m$^{1/2}$)	
		without fibers	with 30% GF
PC	very ductile	8.5	7.1
ETFE	very ductile	7.0	4.6
PA-6,6	ductile	7.5	6.9
PET II	ductile	7.5	8.0
PBT	ductile	5.0	6.2
PET I	ductile	2.3	7.5
PPS	very brittle	1.2	5.0

behaviour of different polymer composites can correctly be realised taking into account the already mentioned reinforcing factor, R, depicted in Figure 3.481 [1314,1255].

In the case of the composites based on high toughness matrix, *e.g.* PEEK, the reinforcement with fibres, in the best case will not negatively affect the composite properties. In the case of a weak bond fibre/matrix, for instance PTFE/very short glass fibres, a decrease of fracture toughness is expected. For the rigid matrix (PET at –60 °C), the fibres presence enhances the composite toughness as compared to the ductile matrix.

The fracture of continuous reinforced composites is characterised by the apparition of the main crack, perpendicularly oriented to the stress axis, under the condition as the fibres to have the same alingnment with respect to this axis, 90°. In the case of a parallel

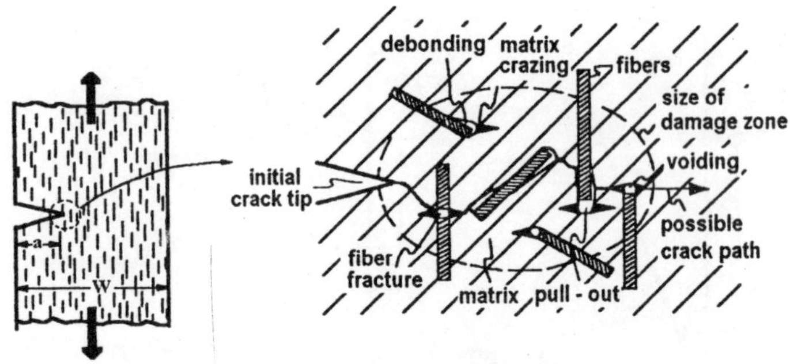

$$K_c = \sigma_c \cdot \sqrt{a} \cdot f\left(\frac{a}{W}\right)$$

$$K_c = \sqrt{G_c \cdot E}$$

$$G_c = \sum V_{fi} \cdot W_{fi} + (1 - V_f) \cdot W_m^*$$

Figure 3.474. Schematic representation of prediction of fracture mechanisms of the thermoplastics materials reinforced with short fibers and the corresponding equations for determination of K_c parameter, of the relation between fracture toughness and fracture energy as well as of the fracture energy, G_c, as source of partial contributions of adsorbed energy through individual mechanisms [1255].

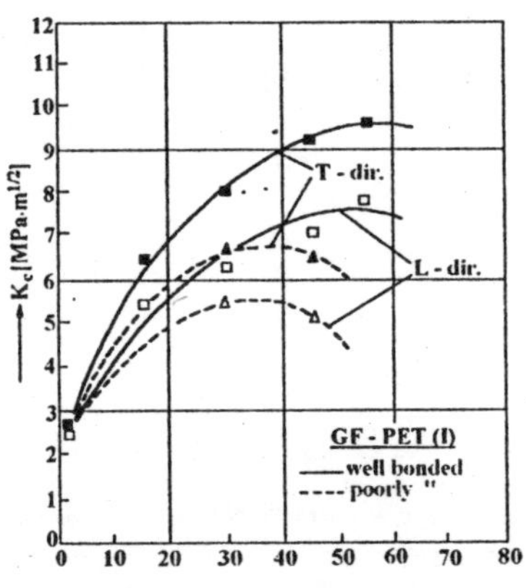

Figure 3.475. Toughness variation with fibres content in the case of PET plaques obtained by casting under pressure [1314].

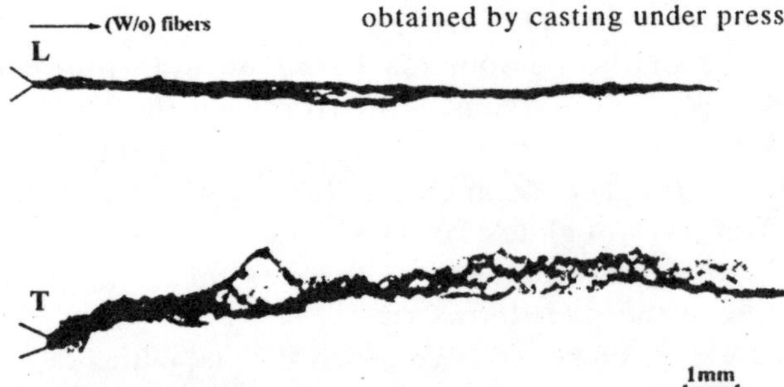

Figure 3.476. Progress of crack (roëntgenographic) in function of crack direction in the plaques cast under pressure [1255].

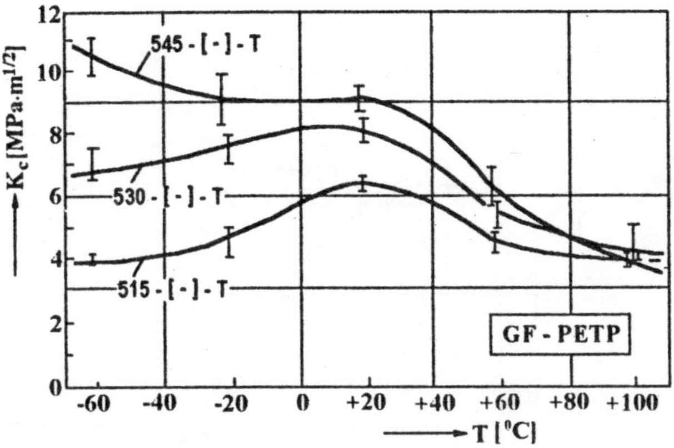

Figure 3.477. Fracture toughness of GF-PET as a function of testing temperature (515=15% wt; 530=30% wt; 545=45% wt-short fibers [1255].

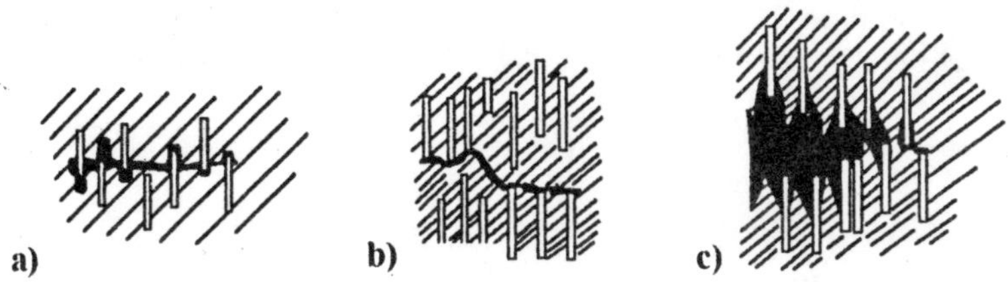

Figure 3.478. Schematic representation of crack propagation for thermoplastic materials reinforced with short fibers in various testing conditions: (a) low temperature; (b) room's temperature, local ductile matrix; (c) high temperature.

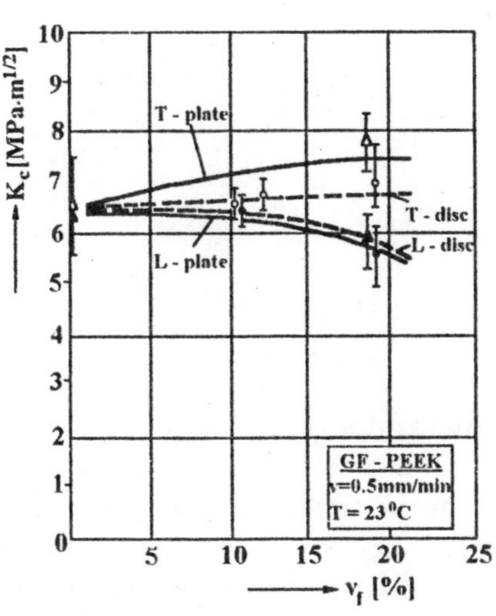

Figure 3.479. Fracture toughness of poly-ether-ether ketone (PEEK) [1255].

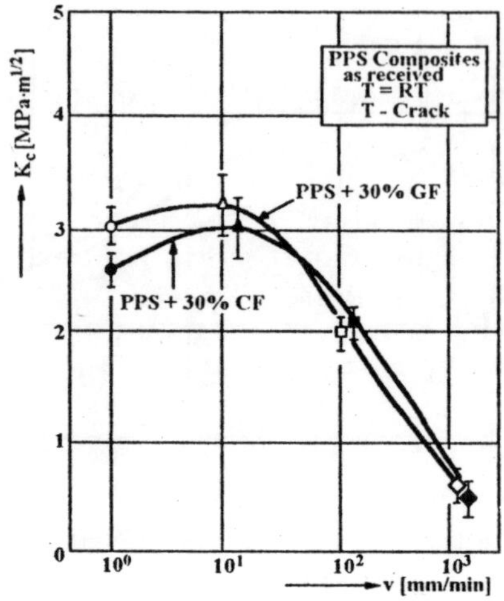

Figure 3.480. Fracture toughness as a function of loading rate, v, of poly-ether sulfone (PES) reinforced with glass and carbon fibers [1255].

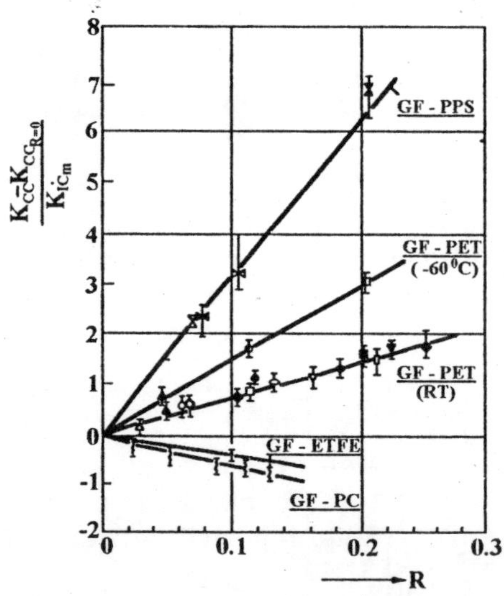

Figure 3.481. Influence of effective reinforcing factor on the relative variation of fracture toughness for different thermoplastic materials [1255].

orientation of the fibres, a multiple fracture occurs, the sample being fractured in many individual fragments, Figure 3.482, [1255]. Th tracture toughness of some polymer composites with high anisotropy is given in Table 3.135.

The evolution of the fracture process of the laminates with a

Figure 3.482. Fracture behaviour of unidirectional laminates (left 0°; middle 45°; right 90° – direction) [1255].

Table 3.135. Values of notched and unnotched laminates of CF-PES and CF-PI [1255]

	Material					
	CF-PI			CF-PES		
Laminate	$[0]_8$	$[90]_8$	[0,90,0,90]	$[0]_8$	$[90]_8$	$[0,90,0,90]_8$
Unnotched						
σ_F (MPa)	910	20	530	1000	46	490
ε_R (%)	0.99	0.25	1.20	1.09	0.69	1.14
E (GPa)	91	7.9	45	90	7.2	43
Center notched						
K_c (MPa$\sqrt{m}$)	51.8	1.1	30.1	56.9	2.1	27.8
G_c (kJ/m^2)	29.5	0.15	20.1	35.9	0.62	17.9

thermoreactive matrix (epoxy) or thermoplastic (PEEK) reinforced with carbon fibres, CF, was investigated, in order to clarify the mechanisms of rate effects on interlaminar fracture toughness. It was found that that the composite interlaminar fracture energy, G_c, is sensitive to strain rate. The studies initiated by A.J. Smiley proved a severe reduction in mode I and mode II toughness of CF/epoxy and CF/PEEK composites occur at high strain rates, Figure 3.483.

The fracture surfaces were investigated by scanning electron microscopy (SEM) to establish the mechanisms associated with toughness reduction [1315–1317]. The CF/PEEK fracture surfaces

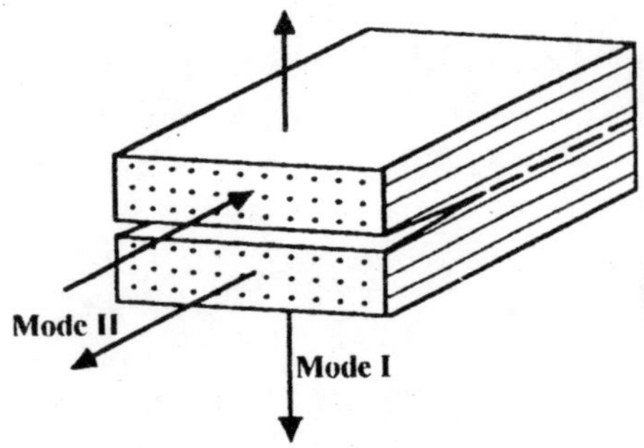

Figure 3.483. Illustration of the opening (mode I) and forward shearing (mode II) modes of crack surface displacements [1317].

displayed a larger extent of plastic deformation at low rates than at high rates, consistent with reduction of fracture toughness at high rates. The CF/epoxy fracture surface did not display any noticeable change in microscopic deformation behaviour that would explain the rate sensitivity [1315,1316].

To investigate the rate-dependent fracture mechanisms and the translation of matrix toughness to the composite interlaminar fracture toughness, K. Frieshich and co-workers realised the fracture testing of the materials at a large range of cross-head speeds under the opening (mode I) and shearing (mode II) modes of fracture, illustrated in Figure 3.483 [1317]. To relate the fracture toughness to the local behaviour at the crack tip, crack tip displacement rates for the mode I ($\dot{y}_{CT}$) and mode II ($\dot{u}_{CT}$), were determined as a function of cross-head speed (δ) and geometry parameters:

$$\dot{y}_{CT} = 1.5\xi^2 \qquad \text{(Mode I)} \qquad (3.352)$$

$$\dot{u}_{CT} = \frac{24\delta h a^2 \xi}{2l^3 + 3a^3} \qquad \text{(Mode II)} \qquad (3.353)$$

where: ξ - dimensionless distance from the crack tip, $\xi = x/a$, $x = 0.25$ mm and a is the crack length; h and l - semi-thickness and semi-span, respectively, of the end-notch flexure (ENF) geometry [1316,1317]. The obtained results are depicted in Figure 3.484.

It is observed that in the double logarithmic plats of G_{IC} against $\dot{y}_{CT}$ and G_{IIC} against $\dot{u}_{CT}$, over a wide range of low and medium crack tip displacement rates there is no remarkable effect of rate on fracture energy. Above a critical rate, the mode I and mode II toughness of both composites drops rapidly to an apparent minimal level.

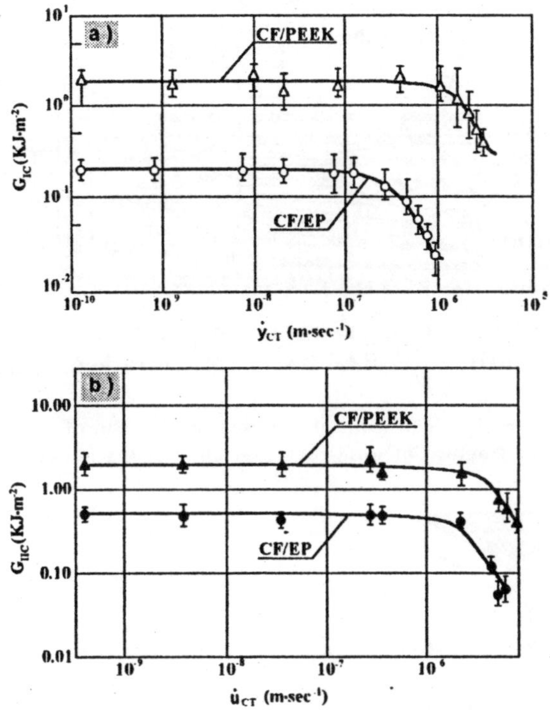

Figure 3.484. Fracture energy plotted against crack tip displacement rate in CF/PEEK and CF/Epoxy [1315-1317]: (a) log-log scale for mode I; (b) log–log scale for mode II.

A schematic illustration of the principal micromechanism of energy absorption in CF/Epoxy and CF/PEEK materials is shown in Figure 3.485. The succession of events that take place is the following: 1) crack bringing by fibres of fibre bundles, resulting in 2) fibre fracture; 3) formation of the fracture surface of the main crack; 4) formation of the side cracks; and 5) plastic deformation and/or microcracking of the matrix around the fibres.

The actual contribution of this microfailure mechanisms to the interlaminar fracture energy of the composites tested under particular conditions are a function of the number of events taking place, the real area of fracture surface formed, and the size (length and width) of the damage zone (DZS) around the main crack. The larger the latter becomes the more side cracks have to be expected, and the more energy is consumed by plastic deformation of the polymer matrix material.

Quantitatively, the interlaminar fracture energy of the compos-

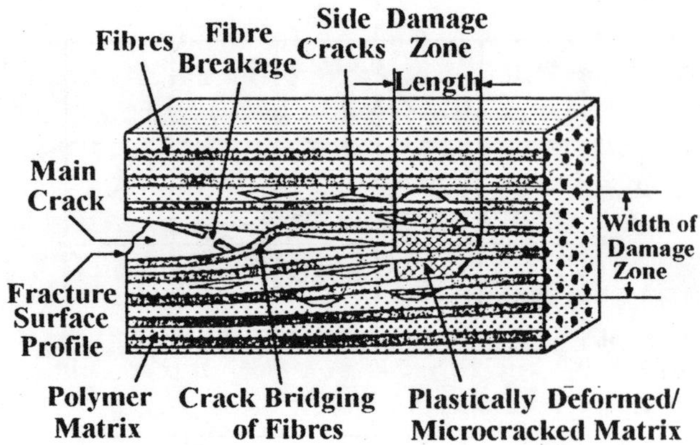

Figure 3.485. Schematic illustration of possible mechanism of energy absorption during interlaminar fracture of unidirectional composites [1317].

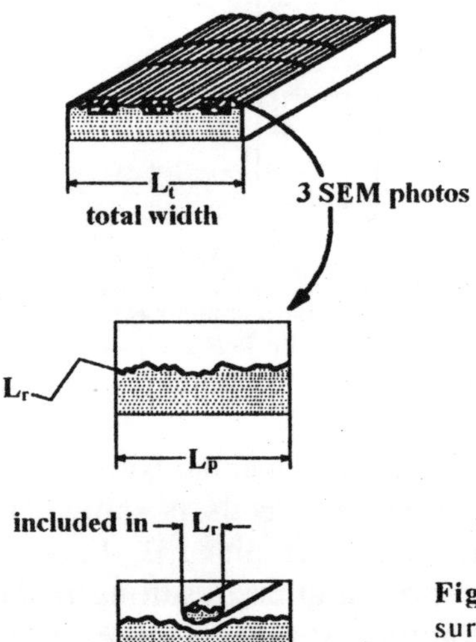

Figure 3.486. Schematic illustration of fracture surface profile determination [1317].

ites (G_{cc}) can be assumed to be proportional to the following quantities:

$$G_{cc} \approx \left(L_r/L_p\right)(\text{DZS})\left[W_m^* V_m + W_{sc}\rho_{sc} + \left(W_{fd/b}V_{fd/b} + W_{ff}V_{ff}\right)\right] \quad (3.354)$$

where: L_r – profile length; L_p – projected length; L_r/L_p – real fracture surface profile; (DZS) – the damage zone profile; W_m^* –

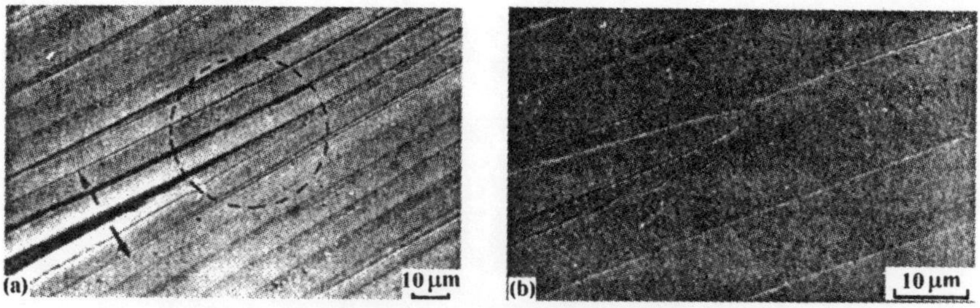

Figure 3.487. Slowly opened mode I crack (in situ) in CF/PEEK (dashed circle marks the approximate size of the damage zone around the crack tip [1317].

the specific deformation/fracture energy of the matrix in the presence of fibres; V_m – the matrix fraction involved in the deformation and fracture process; W_{cc} – the specific energy for side cracks formation, along fibre/matrix interfaces (f - interfacial bond quality; r_{sc} – the amount of side cracks in the damage zone; $W_{fd/b}$ – the specific energy of fiber debonding and bringing; W_{ff} – the specific energy for fibre fracture; $V_{fd/b}$ – the volume fraction of debonded and bridging fibres; V_{ff} – the volume fraction of fractured fibres.

The real fracture surface profiles, L_r/L_p, are always greater than 1 and are mostly affected by the ductility of the matrix material, Figure 3.486. A quantity which is hardly to determine is the damage zone size (DZS). In general, the DZS is larger for lower crack-opening rates, especially pronounced in the start part of the crack growth (the length of which is itself a function of the crack opening displacement rate). The DZS should increase with increasing ductility of the polymer matrix.

The above presented mechanism is sustained by the data of electronic fractography. Figure 3.487 a evidences the crack tip region around a mode I intralaminar crack in CF/PEEK composite that was opened slowly. The damaged zone around the crack tip, as indicated by dashed circle, extends over five to six fibre layers and has some distance ahead of the crack tip. The damage also includes microcracks in the matrix, Figure 3.487 b, microcracks between the matrix and the fibres in the interfacial region, plastically deformed matrix material and cracks adjacent to the main crack tip. The plastically deformed matrix between the fibres is easily observed in Figure 3.488.

In the CF/epoxy system, mod I cracks were studied at two dif-

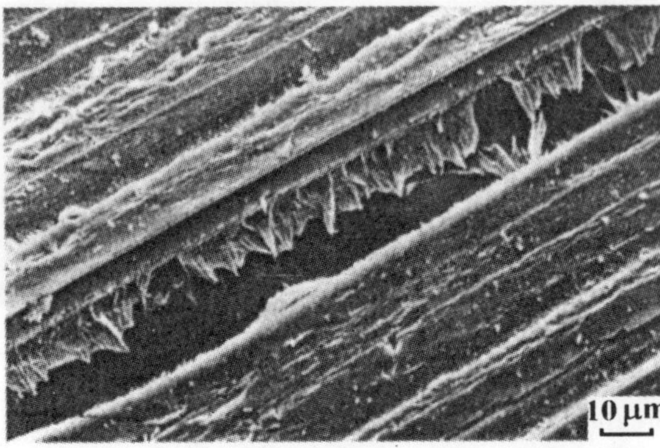

Figure 3.488. Very slowly opened mode I crack in CF/PEEK, giving evidence for significant deformation around the main crack.

ferent crack opening rates. Damage zone is observed around and ahead of the crack tip, Figure 3.489 a. The damage zone is smaller than for CF/PEEK case. Two or three fibre layers around the main crack are involved in the fracture process that induces the formation of microcracks in the matrix, between fibres and the fibre/matrix interfaces. The failure of the epoxy matrix between the fibres takes place by microcrack formation, as is illustrates in Figure 3.489 b. Even under very high magnification, Figure 3.490, no microcracks are visible. Separation of the material takes place primary along fibre/matrix interfaces. The above-mentioned results show that the damage zone extends into resin around the fibres, as it is schematically illustrated in Figures 3.491.

Figure 3.492 shows details of the damage mechanisms ahead of a mod II crack of CF/PEEK composite. Under the action of shear stress ahead of the crack tip, tensile cracks form in the matrix at 45°. In the regions within the damage zone where the stress intensity is highest, *i.e.* very close to the crack top, these cracks extend completely between fibre/matrix interface and open up. Finally, the material separation between the cracks leads to the full development of the hackled mode II fracture surface. The damage zone in CF/PEEK under low velocity/mode II conditions is extended over four to six fibre layers.

Figure 3.493 displays the mode II crack top zone for CF/Epoxy composite. At very higher magnification it is observed that crack formation is similar to that observed in CF/PEEK composite. A significantly smaller damage zone is observed in the more brittle CF/

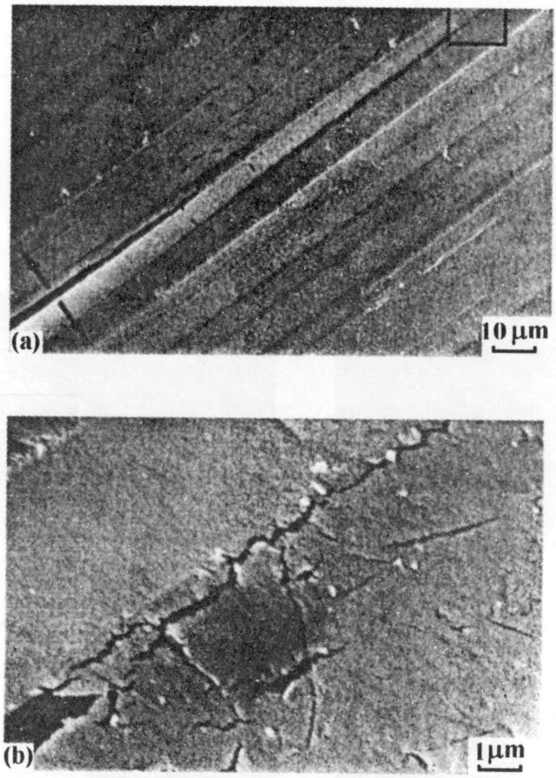

Figure 3.489. Slowly opened mode I crack in CF/epoxy (in situ) [1317].

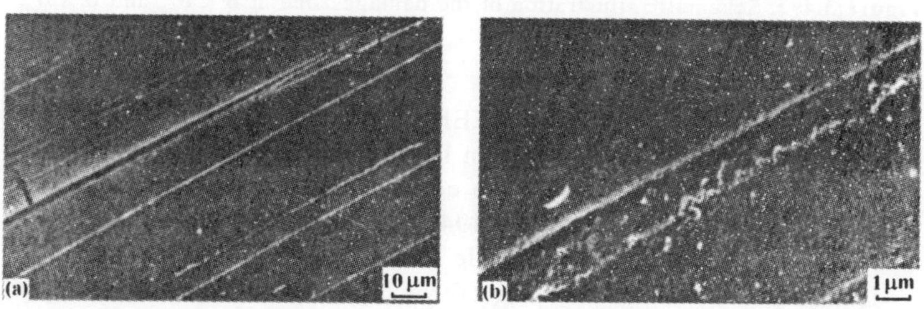

Figure 3.490. Fast propagated crack in CF/epoxy (mode I, in situ), showing (a) a smooth and sharp crack shape, and (b) no microcracking or damage in the polymer matrix ahead of the crack tip [1317].

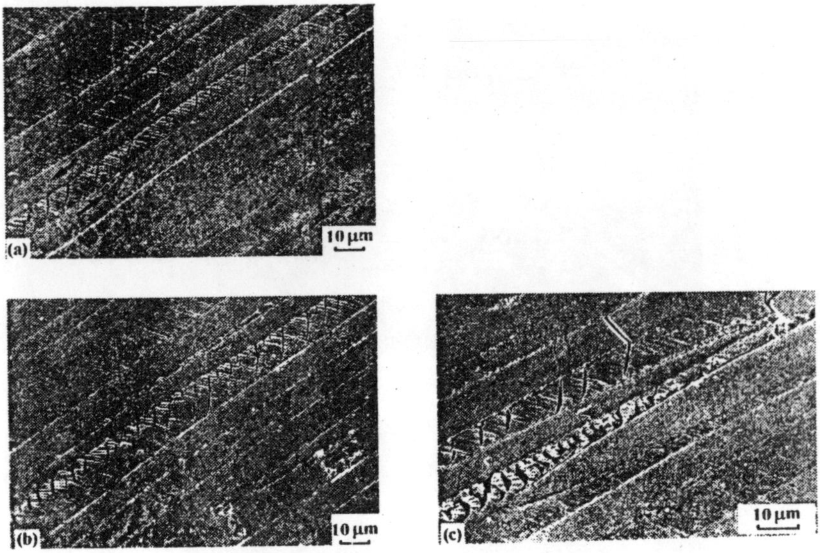

Figure 3.491. Higher magnifications of the damage zone ahead of a mode II crack in CF/PEEK: (a) 0.25 mm ahead: (b) 0.12 mm ahead: (c) 0.06 mm ahead [1317].

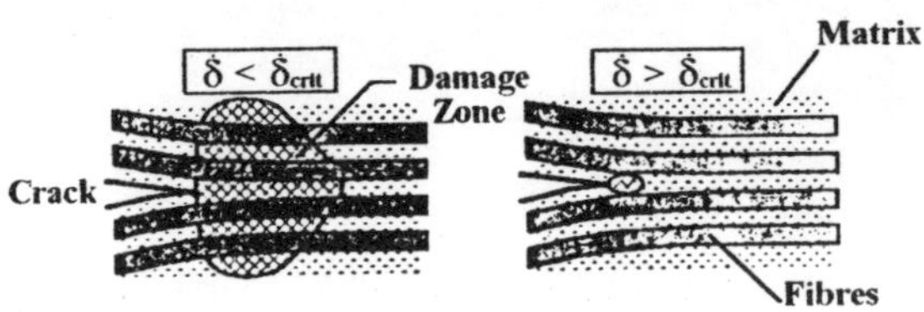

Figure 3.492. Schematic illustration of the damage zone at $\delta < \delta_{crit}$ and $\delta > \delta_{crit}$ [1317].

Epoxy, then in the ductile CF/PEEK.

The fracture surfaces profile in function of the stressing manner (mode I or mode II), for the two composites comparatively investigated, is depicted in Figure 3.494. Figure 3.495, exemplifies the counting of broken fibres on mode I and mode II fracture surfaces.

It was proved that the long fibres assure to the composites a higher resistance that the short ones, and are characterised by specific fracture properties. The microstructure and fracture mechanical behaviour of injection-moulded longer glass fibres reinforced polypropylene (Verton, aspect ratio ≈ 320) were studied as a function of the fibre volume fraction and compared to that of a shorter

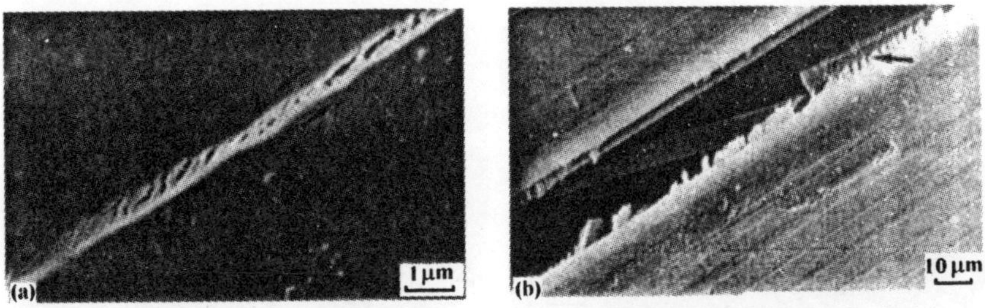

Figure 3.493. In situ mode II crack in CF/epoxy: (a) very high magnification of the region ahead of the crack tip; (b) crack with shear cusps as the edges (arrow) and a bridging fibre in the interior of the crack [1317].

fibre-filled polypropylene (aspect ratio ≈ 70). The toughness was measured using instrumented-notched Izod and falling weight impact tests, as well as compact tension specimens. It was found that the addition of longer fibres generally increased the toughness of the material, although more significant increases were seen in the impact tests than were seen in the compact tension tests. In the latter case, a correlation between toughness improvement and microstructural details was performed on the basis of the microstructural

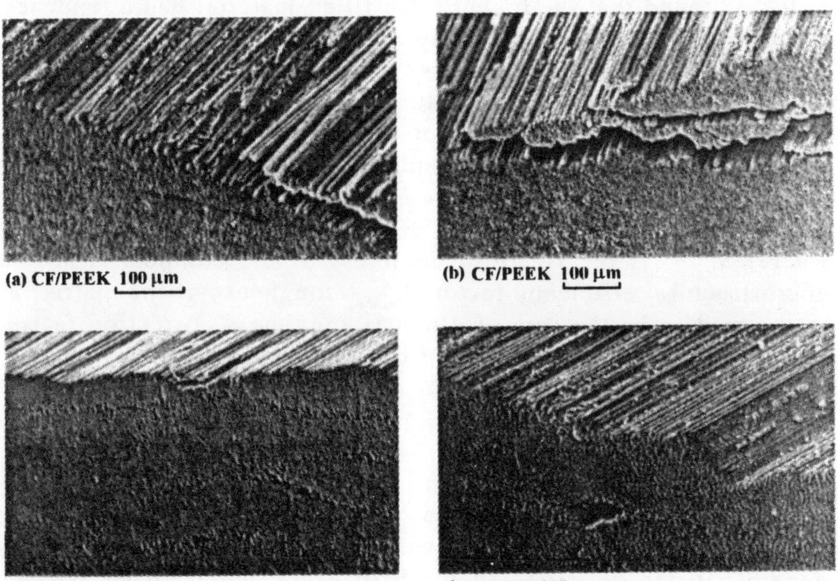

Figure 3.494. Examples of fracture surface profiles:(a) mode I; (b) mode II [1317].

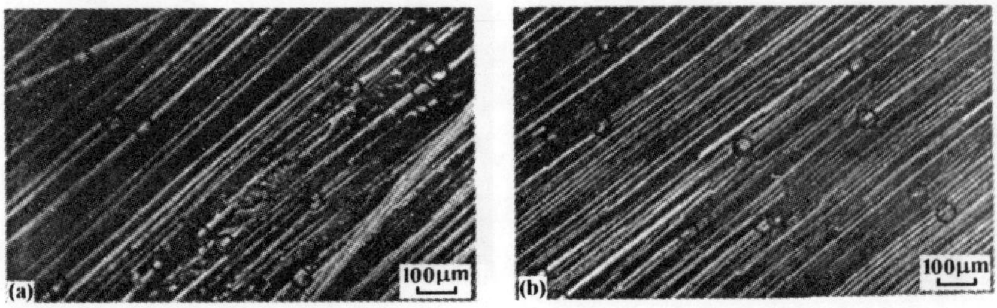

Figure 3.495. Examples of fibre breakages on fracture surfaces. (a) CF/PEEK, mode I (unstable propagation region), and (b) CF/epoxy, mode II (unstable propagation region) [1317].

efficiency concept, a semi-empirical approach of the form $K_{c,C} = (a^* + nR)K_{c,M}$, where $K_{c,C}$ and $K_{c,M}$ are fracture toughnesses of the composite and matrix, respectively; a^* is a matrix stress correlation factor; n is a scaling parameter; and R is a fibre reinforcement effectiveness factor.

The long fibre material performed better when energy related fracture toughnesses were considered (impact tests) than where the strength-related K_c toughness was considered.

It was found that the longer fibre-filled material had a tendency to form fibre bundles, especially at higher loadings and that this probably caused the effectiveness of the fibre reinforced to drop of considerably between 30% and 40% wt.

Additionally, the usefulness of the fracture toughness – microstructure correlation given by equations 3.355–3.358 was already shown [1319].

$$K_{c,C} = MK_{c,M} \tag{3.355}$$

where: $K_{c,C}$ – is the fracture toughness of the composite; M – a microstructural efficiency factor; $K_{c,M}$ – toughness of the matrix. M is assumed to be the sum of a matrix toughness correction factor, a^*, and a mainly fibre-dependent contribution, nR, *i.e.*:

$$M = a^* + nR \tag{3.356}$$

In this equation, a^* takes into account the factor that the matrix may not fail in the same way in the composite as when being total separately. If it does, the value of a^* may be less than 1. In general, it can be expected for brittle matrices that a^* will be greater than 1, while for ductile ones a^* may be less than 1. "n" is essentially a scaling factor which should remain constant as long as the temperature, testing conditions, and composite system remain constant. If the changes in the components of the composite system take place, *e.g.* other fibre/matrix bond quality, other fibre material or changes in the fibre aspect ratio, this should also result in different values of the "energy absorption ratio", *n*. *R* is a reinforcement effectiveness parameter, which takes into account the geometrical features of the composite microstructure and has the following expression [1319]:

$$R = f_{pe,S} V_{f.S} \frac{2S}{B} + f_{pe.C} V_{f.C} \frac{C}{B} \tag{3.357}$$

where: $V_{f.S}$ and $V_{f.C}$ – the volume fraction of fibres in the surface and core region, respectively; *S* – the thickness of the surface region; and *B* – the total thickness of the sample. The parameters $f_{pe.S}$ and $f_{pe.C}$ are the effective orientation factors for the surface and core layers, respectively. These are approximated using the relation:

$$f_{pe} \cong \alpha\left[1 + \tanh(\beta f_p)\right] \tag{3.358}$$

where: $\alpha = 0.5$ and $1 < \beta < 5$ [1319].

The values of *R* for all the materials and test directions are shown in Table 3.136.

The fibres orientation with respect to the stress direction essentially influences the fracture mechanism, as it was proved for a composite with semicrystalline matrix, PET reinforced with short fibres, having the structural characteristics presented in Figure 3.496 [1267]. It can be observed that different fibre orientation normally exists across the thickness of injection moulded plaques of short fiber reinforced thermoplastics.

This fact gives rise to various fibre-related failure mechanisms during breakdown under fatigue, static or impact loading conditions. In regions where fibres are preferentially oriented parallel to the fracture plane (longitudinally, L – cracks) failure develops by fibre/matrix debonding under normal loads, associated with craze initiation of fibre ends, Figure 3.497.

Fracture in composites where the major fibre orientation is transverse to the advancing crack front (T – cracks) involves the mechanisms such as fibre/matrix debonding due to the high stress concentrations at the fibre ends, Figure 3.498. The above-presented mechanisms are sustained by the data of electronic fractography [1267].

Concentrated efforts were devoted to investigation of the fracture behaviour of some composites with thermoplastic matrices of PPS, PEEK and liquid crystal polymers [1261, 1262, 1267, 1320].

Table 3.136 Normalised toughness, effective orientation and reinforcing effectiveness factor *R* for different GF-PP materials and crack directions (Note $f_{p,e\parallel}$ has to be used when the fibres in the layer are nearly parallel to the crack, and the opposite is true for $f_{p,e\perp}$ [1319]

Material code	Test direction	$\frac{K_{c,C}}{K_{c,M}}$	$f_{p,e\parallel}$		$f_{p,e\perp}$		*R*
			Surface	Core	Surface	Core	
$10L_1$	L	2.5	0.38	-	-	0.58	0.016
	T	2.7	-	0.42	0.63	-	0.023
$20L_1$	L	3.8	0.3 1	-	-	0.70	0.043
	T	4.2	-	0.30	0.69	-	0.066
$30L_1$	L	4.5	0.29	-	-	0.78	0.063
	T	5.2	-	0.22	0.71	-	0.082
$40L_1$	L	5.6	0.23	-	-	0.78	0.085
	T	5.3	-	0.22	0.77	-	0.128
$40S_1$	L	3.4	0.22	-	-	0.80	0.071
	T	5.4	-	0.20	0.79	-	0.149
$10S_2$	L	1.6	0.32	-	-	0.75	0.012
	T	2.0	-	0.25	0.68	-	0.025
$20S_2$	L	2.3	0.25	-	-	0.79	0.026
	T	2.8	-	0.22	0.75	-	0.062
30S2	L	2.7	0.22	-	-	0.81	0.037
	T	3.4	-	0.19	0.79	-	0.099
$40S_2$	L	3.1	0.18	-	-	0.85	0.050
	T	4.9	-	0.16	0.82	-	0.144

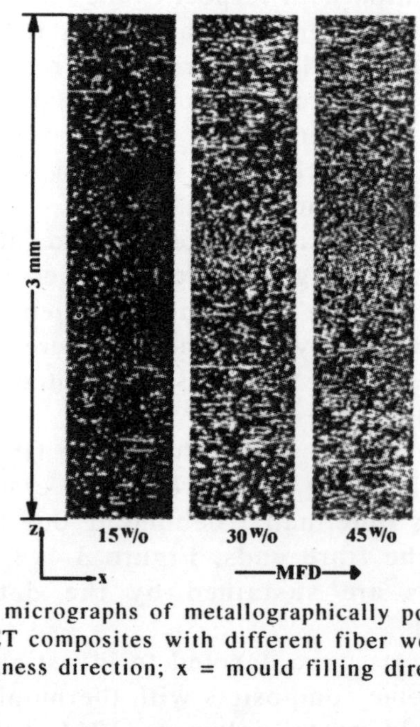

Figure 3.496. SEM micrographs of metallographically polished surfaces of the short glass fiber/PET composites with different fiber weight fractions (w/o = weight %; z = thickness direction; x = mould filling direction MFD) [1267].

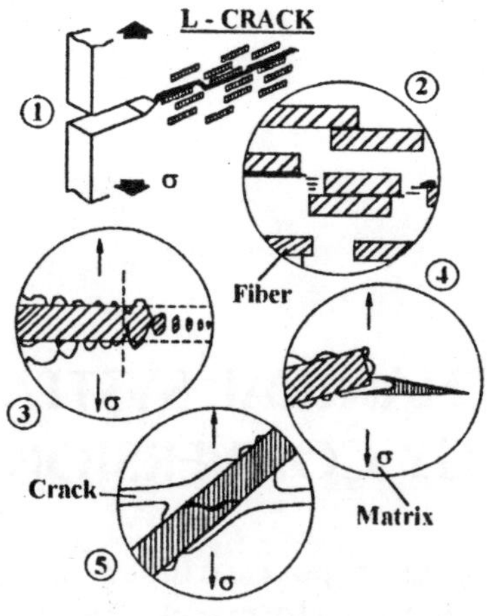

Figure 3.497. Schematic of possible failure mechanisms near L – oriented fibers: (1) tendency to form a straight crack profile; (2) coalescence of crazes or cracks between ends of fiber or fiber bundles; (3) void formation around fibers and rupture of the polymeric interphase; (4) craze initiation at fiber ends; (5) debonding along misoriented fibers and eventually fiber fracture due to bending. [1267].

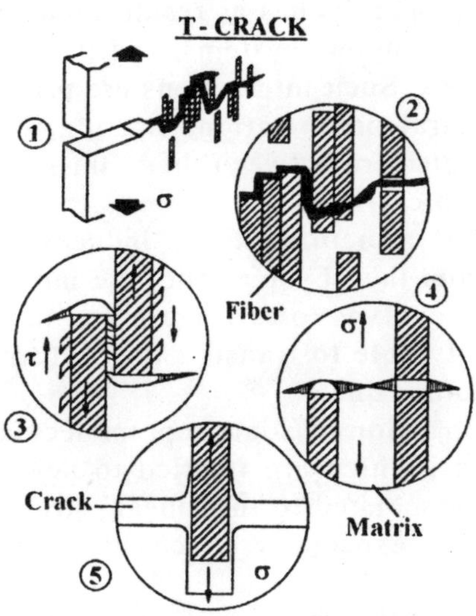

Figure 3.498. Schematic of possible failure mechanisms near T- oriented fibers: (1) tendency to form zig-zag crack profiles and crack bifurcations; (2) short-circuiting of cracks between cracks at fiber ends or broken fiber sites; (3) formation of crazes and cracks at fiber ends and along their well bonded interfaces at the onset of fiber pull-out; (4) as (2) in more detail; (5) fiber debonding and pull-out in case of a poorly bonded system [1267].

4

CHEMOMECHANICAL SYSTEMS FOR ENERGY CONVERSION

The modern technologies impose stringent demands of macromolecular systems up to high performance. The fundamental criterion in the achievement of this desideratum is the elucidation and efficient exploitation of the relation between their structure and performance properties. The nature enlightened such systems, offering practically perfect models.

The elucidation of the mechanisms that determine the complex interactions structure-properties has as result the discovering of the technical potential of biological systems, and its materialisation in performance applications. Such interactions are practically released at all levels of structural organisation, starting to the molecular one, and overlap and interfere between them, thus generating qualitatively new macroscopic effects.

The hierarchy of the structural levels induces an hierarchy of the motility forms, from the physical one, the most simple, to the chemical motility that evolves toward more and more complicated forms, the only motility able to transmit the "information" from a structural level to another one.

It stimulates the apparition of some supramolecular and morphological new formations, which are implied to new functions. The biological motility is associated to the "ideal" structures, perfected from prolonged periods, especially concerning the entropic pathway.

The relation between the structure and properties of the advanced macromolecular systems is frequently investigated by analogy with biological systems. One its turn, biomaterials are investigated by using common techniques to polymer science. Consequently, the transfer of some concepts and principles that characterise the biological systems to the technical macromolecular systems should stimulate the movement ability of the last ones,

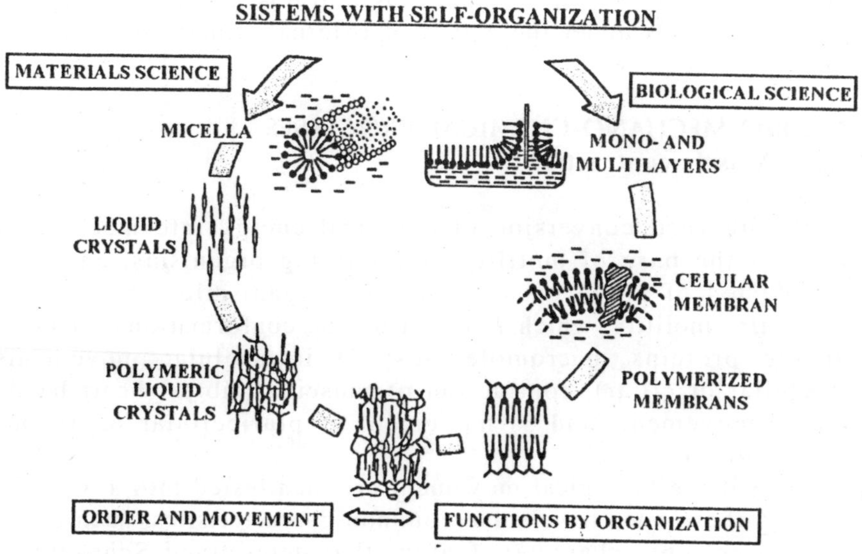

Figure 4.1. Macromolecular systems with self organisation ability [1].

through self-organisation, Figure 4.1 [1]. As is illustrated in this scheme, both in the field of biological science and in that of polymer materials, the self-organisation principle is essential for the release of motility's mechanisms. In the first case, the lipids are the fundamental structural components for the obtaining of functional membranes, able of chemo-mechanical functions. In the second case, structures with pronounced anisotropy of their properties, of the liquid-crystal type, are characterised by interesting mechanical and/or optical properties. The principle order and motility by self-organisation assures the performance of both biological membranes and polymeric liquid crystals.

Starting from the assumption that the living systems are composed, in the last instance, by macromolecules "biomacromolecules", similar to the materials offered by polymers science and engineering, it is clear that a careful corroboration of the knowledge of the two domains might furnish new methods, in measure to permit the decoding of mechanisms specific to living bodies. By using this methodology, interesting results have been obtained in the designing of technical devices, able of locomotion. Nowadays, noticeable results have been obtained in the synthesis of some macromolecular systems, with pre-established structures.

These systems are able to suffer high strains under the action of extenr stimulias under the action of external stimuli and generate force or execute work

4.1. BIO-MECHANO-CHEMICAL PROCESSES

4.1.1. Muscle contraction

The isothermal conversion of chemical energy into mechanical work is the base of motility of all living organisms, and it is manifested of all levels of structural organisation, namely: 1) molecular motility, which is based on the conformational changes of the proteins macromolecules; 2) intracelular movements (flagellar or ciliar), contraction of muscular fibres, heart beats, plants movement; and 3) movement of pluricellular organisms (locomotion).

Even if the biological movement are manifested into a wide diversity, these ones have some common characteristics, such as: 1) the antientropic character, feature that determined Schrödinger to define the life as "a continuous production of negative entropy" or of order; 2) occur with energy consumption; 3) self-regulation and self-control character; 4) tend to optimise the relations of living system with the medium within which this one evolves.

The muscular contraction implies the most important, most versatile as performances motility mechanism in the living life, and practically the only one able of evolutionary perfecting. Its support, the skeletal muscle, presents a complex structure, having at list six levels of hierarchy, Figure 4.2.

At the cellular level, the striated fiber is characterised by a high structural self-organisation as well as of its components. Consequently, striated fiber constitutes a suitable contractile model for the analysis of its technical potential, regarding the conversion of chemical energy into mechanical work. Its investigation is usually done by common methods used in polymer science, such as optical and electronic microscopy, X-ray diffraction; these ones are often used in combination with those typical to immunochemistry and enzymology fields [3].

The release of mechanical energy by muscle contraction is the result of a motile biosystem, of protein constitution, namely actin-myosin complex. This one is able of chemomechanical transformations that occur far from equilibrium and are enzymatically caused, external commanded by a trigger agent, Ca^{2+}, and transmitted to a macrofag, ATP, located in its proximity. In turn, ATP is partially

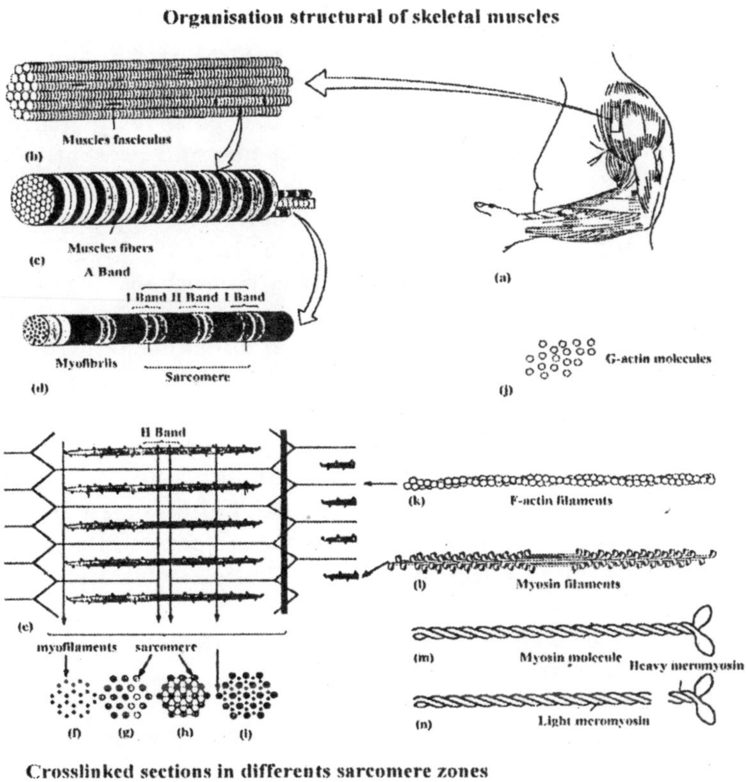

Figure 4.2. Structural organisation of the skeletal muscle [2].

converted, by splitting of a chemical bond with the concomitant release of a large amount of energy, in accordance with the following expression:

$$E + S \underset{k_{-1}}{\overset{k_1}{\rightleftharpoons}} X \underset{k_{-2}}{\overset{k_2}{\rightleftharpoons}} E + P$$

where: E – contractile and enzymatic macromolecule; S – the ligand and substrate of enzymatic activity; X – the complex of the two components; P – the product of complex splitting; k_i – the corresponding rate constants.

In order for such a system to be functional and efficient in time, it is necessary for the pathway X – E to be covered with a lower energetic consumption than that corresponding to the E–X pathway. The reversibility of these transformations is compulsory; this requirement is assured by adequate kinetic adjustments, on the different segment of the cycle (for instance, the ratio k_2/k_{-2}), related to the capacity of self-regulating and self-control of the contractile apparatus [4, 5].

The structure of striated muscle fiber has been investigated both by microscopic and X-ray diffraction methods, and presently it is demonstrated. Its most important parts are the following ones: sarcolemma (cell membrane), the fibrillar contractile apparatus, and sarcoplasm (cytoplasm of muscle fibres) [6–23].

Sarcolemma – it was proved by electron microscopy that sarcolemma is a complex consisting of the plasmatic membrane of the muscle fiber, covered of two dense optical zones, separated from each to other by a third lighter zone of about 200–300 Å. Externally a fine network of collagen appears, and the dense layer is connected to the Z membranes.

Contractile fibrillar apparatus – is composed by fascicles of myofibrils with diameter of about 1 μm, parallel arranged along of muscular fibre, in fact, constituting the muscle "contractile motor". This formation results by repeating of the basic contractile unit of the muscle, the sarcomere (b in Figure 4.2). Seen in longitudinal section (with an electron microscope), a myofilament shows several distinct bands, each of which has been given a special letter. The lightest (least electron dense) band is known as the I band and consists mostly of actin units. The wide, dark band, known as the A band, is composed primarily of myosin. In the center of the I band is an electron dense line, known as the Z-line. In the middle of the A band is another dense line known as the M line. The sarcomeres are delimited each to another by Z line. The length of vertebrates' sarcomere in relaxed state is of about 2.3 μm.

In principle, microfibrils are composed by two types of myofilaments, i.e. the "thick" myosin filament, ϕ = 100 Å and L = 1.5 μm, and the "thin" actin filaments, ϕ = 50 Å and L = 2.0 μm respectively. In cross section, under very high magnification, both A and I bands can be seen to be hexagonal networks, Figure 4.3. These networks are apparently ordered and fixed at the M- and Z-lines. In the region where the A and I bands overlap (sometimes known as the H band) the two hexagonal networks intermesh so that each myosin filament is surrounded by six actin filaments. These networks appear to be anchored to (and through) the cell membrane in two ways. At the ends of fibrils, special structures

Table 4.1. Myofilaments components and their function in skeletal muscle

Protein	$M \cdot 10^{-3}$	Structure and function	Ref.
		Thick filaments	
Myosin	500	- peptide responsible for force generation ; it is composed of a globular head with both ATP and actin binding sites, and a long tail involved in its polymerization into myosin filaments; - it contains 6 polypeptide chains; two heavy chains, HC (200 kDa), and four light chains (20 kDa each); - the two HC are wound into a helix, composing filamentary part of molecule "light meromyosin" LMM; - "heavy meromyosin" (HMM) is composed by two globular heads ($HMM\text{-}S_1$), one for each HC chain, and its connection part with light meromyosin, $HMM\text{-}S_2$; each $HMM\text{-}S_1$ also contains two LC (regulatory and essential, respectively); - $HMM\text{-}S_1$ is the myosin part directly implied in contraction.	24 - 28
Protein C	140	- ellipsoidal protein of about 35 nm, composed by a single polypeptide; fixed on LMM on the two sides of zone H, straightened toward actin filaments	29 - 31
Proteins of M line		- a complicated network structure, composed at list from three proteins; implied both in enzymatic functions and in maintaining the neatly ordered striation pattern; it appear to anchor the myosin network to the actin network	32 – 34
		Thin Filaments	
G-Actin	46	- conservatory protein, globular in appearance, with two approximately equal domains, containing the sites for binding ATP and Ca^{++}; - by polymerisation G-actin monomers are linked into long chains; -two actin chains are wound into a helix.	35 – 37
Tropomyosin (TM)	70	-filamentary protein (40 nm length and 2 nm thick), based on two polypeptide chains that compose a double helix, attached to the complex	38 - 40
Troponin (Tn-T, Tn-I, and Tn-C)	81 (40, 23, 18)	-apparently the major regulator of force production: its three subunits lie in the groove of each actin filameny blocking the myosin binding site, in the absence of Ca^{++}	
β-Actin	70	- favours G-actin polymerisation, determining the growth of actin filaments toward the sarcomere centre	41
α-Actin	100	- appears to act as a molecular ruler regulating the length of actin filaments ("ending factor").	

anchor the terminal actin filaments to the membrane. The Z and M lines also appear to be connected to the cell membrane. Apart from myosin and actin, several other proteins take part in filament formation Their most important characteristics and assembling possibilities are given in Table 4.1.

The assembling of individual proteins in filaments reversibly occurs, being influenced by the medium factors and/or certain enzymes. In essence, the contractile process implies the shortening of sacromere, caused by the sliding of the thick filaments along the thin ones. At the molecular level, the displacement of constitutive

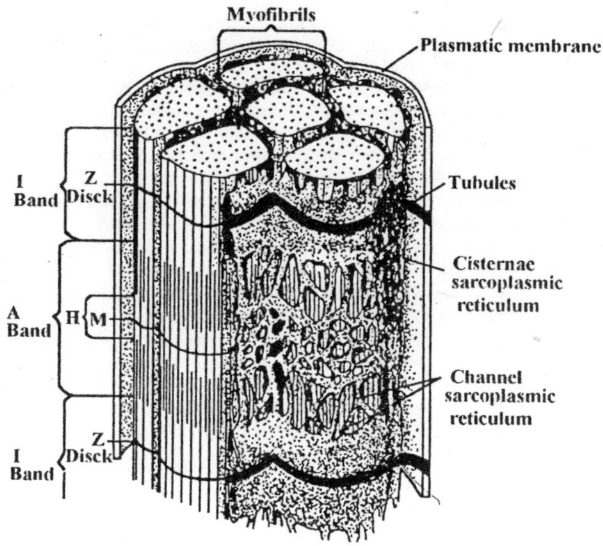

Figure 4.3. Tridimensional representation of 6 myofibrils [2]. The lumen of transversal tubules is in connection with extracellular medium and the material exchange is achieved in the filaments' proximity.

elements of the contractile apparatus is both the "motor" of muscle contraction and its manifestation.

Sarcoplasm – contains multiple nuclei, located peripherally beneath sarcolemma, sarcoplasmic reticulum, mithocondria, which are specialised into an intense oxidative mechanism, working into a medium rich in hemoglobin that fixes the oxygen. All these peculiarities just reflect the specialisation of muscle fiber with a view to perform the contractile function. Sarcoplasmic reticulum and T tubules is a membranous system of longitudinal tubules, in the zone of H band, and terminal cisterns (flattened reservoirs for Ca^{++}) that forms closely meshed network around each myofibril.

In the zone of Z membrane, invaginations of sarcolemma regularly occur, constituting the secondary system of T tubules or transversal, which is not connected to the sarcoplasmic reticulum. Functionally, these ones intervene by a "trap" system in the caption of Ca^{++} ions released during the muscle excitation and in metabolic exchanges of the fibrillar apparatus; T tubules system conveys the excitation from the sarcolemma to the fibrillar apparatus, causing the fibres contraction, by releasing Ca^{++}.

The sliding model of muscle contraction, according to which the

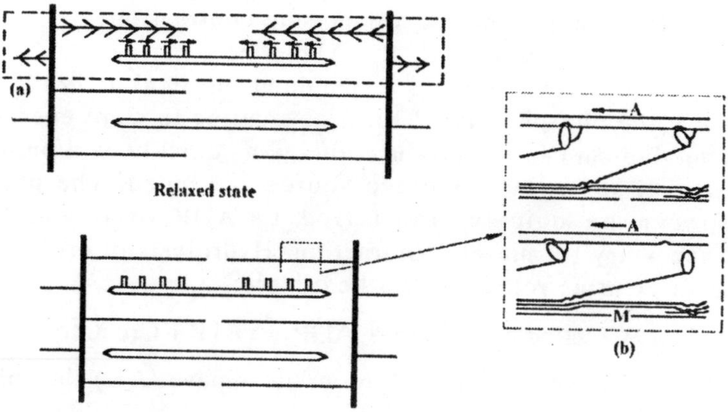

Figure 4.4. Model of muscle contraction based on sliding filaments [2]: (a) The bands H and I are shortened during contraction; the length of thick and thin filaments remain constant; the arrows indicate the myosin and actin-G polarity; (b) Schematic representation of the interactions between myosin head and thin filaments during contraction.

electric energy pass in chemical energy and the last on is converted in mechanical energy is practically unanimous accepted today, Figure 4.4. The electrical signal, in the form of nervous impulse, releases a complex of chemical reactions, followed by the mechanical process of contraction.

The energy require during contraction mainly results from the hydrolysis reaction of adenosine triphosphate, ATP, which implies the following steps:

$$\begin{array}{c} AM + ATP \rightleftharpoons AM{\cdot}ATP \rightleftharpoons M^{*}ATP + A \rightleftharpoons M^{**}ADP + P_a + A \\ \downarrow\uparrow \\ AM + ADP + P_a \rightleftharpoons AM{\cdot}ADP + P_a \end{array}$$

where: A – actin; M – miosin; AM – actin-myosin complex (the head S_1 attached to the actin filaments; P_a – inorganic phosphate; M* ... M** -mechano-excited states, charged with potential energy.

In its simplest form, actin (A) combines with myosin (M) and ATP to produce force, adenosine diphosphate (ADP) and inorganic phosphate. Scientists now agree that ATP serves at least two functions in skeletal muscle systems: first, ATP disconnects actin from myosin, and second, ATP is hydrolyzed by the myosin molecule to produce the energy required for muscle contraction.

Into a living muscle cell, the ATP concentration practically remains constant, even during a sustained physical activity. Muscle cells use the phosphorylated form of creatine to store energy. Normal metabolism can not produce energy as quickly as a muscle cell can use it, so an extra storage source is needed. The phosphate group can be quickly transferred to ADP to regenerate the ATP necessary for muscle contraction. Hydrolysis of creatine phosphate to creatine releases 10.3 kcal/mole:

$$\text{Creatine phosphate} + \text{ADP} \rightleftharpoons \text{ATP} + \text{Creatine}$$

Secondary, in the presence of adenilate kinase (AK) the following reaction occur:

$$2\,\text{ADP} \xrightleftharpoons{AK} \text{ATP} + \text{AMP}$$

where: AMP is adenosine monophosphate.

Under the effort conditions this enzyme intervenes for maintaining ATP concentration at constant level.

The ATP insufficiency or disappearance causes muscular rigidity. Whenever the level of ATP is overtaken into the muscular cell, and the concentration Ca^{++} intracellular, Ca^{++}_{ic}, is no further controlled, a rigor state is entered whereby actin and myosin interact to form a very stiff connection. Therefore, the role of ATP in muscle contraction is that of dissociating of actin-myosin complex but not of its formation. The flux of Ca^{++} ions is of peculiar importance in the release of movement and in sarcomere it is controlled by a sequence of events.

In the case of skeletal muscle, the contraction is electrically initiated, by the nervous impulse transmitted from the central nervous system to the motor effectors via, the neuromuscular junction, which is synapse-like contact between the nerve terminal and muscle fiber. A single action potential travelling down the axon of the motor neuron triggers the release of acetylcholine, ACh, at all branches of axon, (Figures 4.3 and 4.4). The interaction of Ach with the receptors of post-synaptic membranes, releases a sequence of events (for example, the depolarisation of postsynaptic membrane, the increase of Ca^{++} concentration in sarcomere, and the interaction of Ca^{++} with troponin C, Tn-C) triggering a contraction of the muscle fiber. A problem of first interest is that of the mechanism through which the Ca^{++} is released or transferred to sarcomere. It is clear that this process rapidly occurs, since the contraction must simultaneous be released for the all cells of stimulated muscle.

The postsynaptic membrane is densely packed with nicotinic acetylcholine receptors. The nervous impulse induces a rapid depolarization and of transverse tubules, Figures 4.5 and 4.6. A question arises here, namely whether: 1) the depolarization of transverse tubules directly causes the conformational changes of the terminal cisterns, which induces the release of Ca^{++}; or 2) the depolarization causes the release of a transmitter, which once bound by receptor induces the opening of Ca^{++} channels. In this moment the second assumption is considered more plausible. It im-

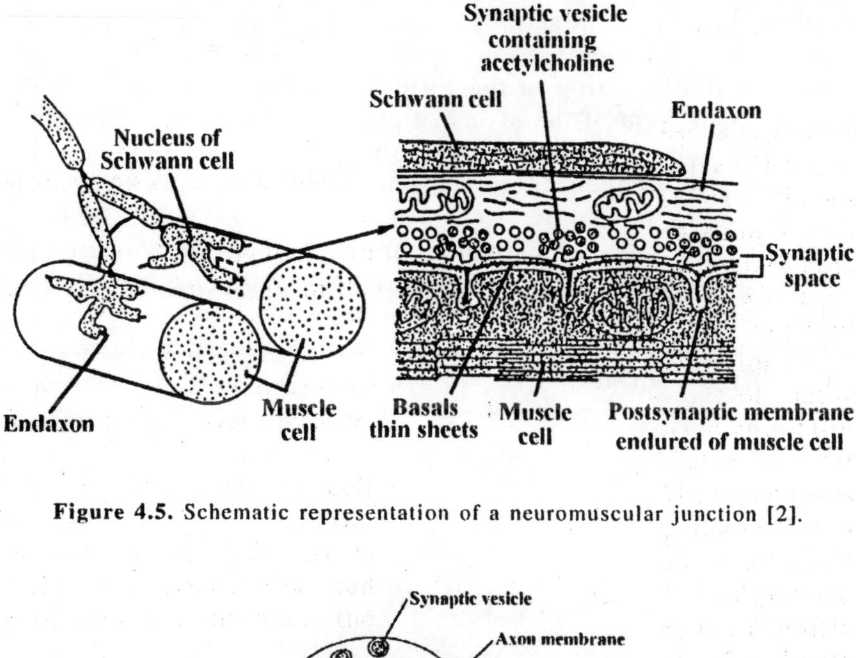

Figure 4.5. Schematic representation of a neuromuscular junction [2].

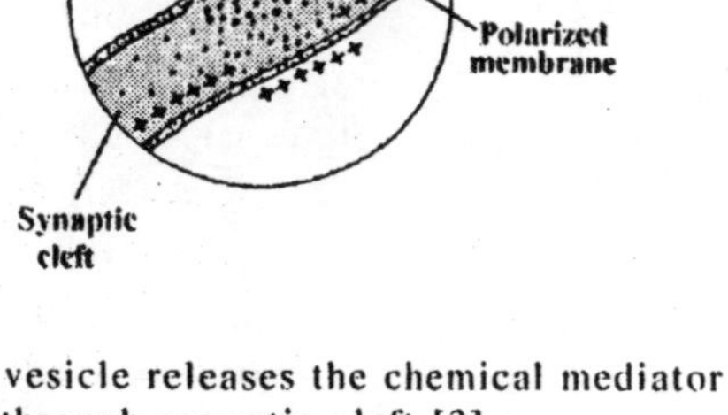

Figure 4.6. Synaptic vesicle releases the chemical mediator acetylcholine (ACh) that diffuses outside through synaptic cleft [3].

plies the stimulation into the sarcolemma of an enzyme, phosphorilase-C, that is associated with the formation of inisitol-1,4,5-triphosphate (IP_3) from phosphodilinositol-4,5-triphosphate (P_1P_2).

It was established the IP_3 may diffuse on a distance of about 20 nm from sarcoplasmic reticulum to interact with the receptors that open the Ca^{++} channels. In this sense the operational aspect of the mechanism involving IP_3 is proved. However, it is not clear yet whether the propagation rate of the considered sequences has the magnitude required of the rapid answer to the stimuli of the skeletal muscle. IP_3 fulfils this condition in the case of smooth muscles, which are specialised for prolonged, slow contraction, and no need for massive contractions requiring large releases of Ca^{++}.

Ca^{++} is a fundamental agent in the regulation of muscle contraction, essentially acting in the following elementary steps: 1) the phosphoriled form of the myosin light chain binds the actin in order to stimulate Mg-ATP-ase, which release the required energy of the contractile system; 2) the phosphorilation of myosin light chain is catalysed by myosin-light-chain kinase, MLCK; 3) MLCK is activated by the Ca^{++}-kalmodium complex, CM; 4) the formation of Ca^{++}-CM complex is conditioned by the concentration, therefore the release of Ca^{++} from terminal cisterns as well as its flux along the membrane constitutes the regulation factor of the muscle contraction; 5) the contraction is stopped either under the action of a

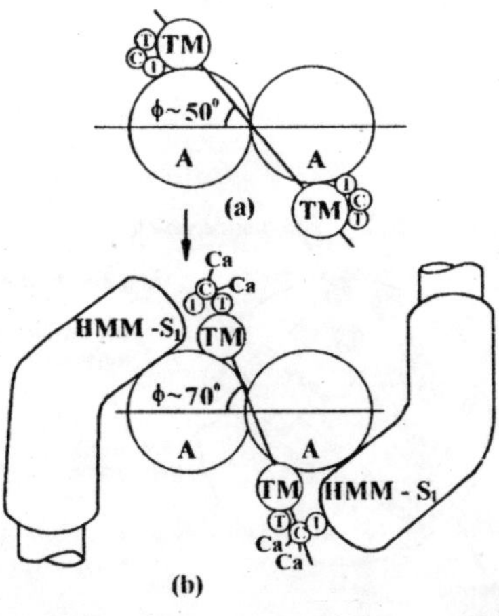

Figure 4.7. Structural model of proteins complex, consisting from: myosin head (HMM-S1), actin (A), tropomyosin (TM), and the three troponins (Tn - T, I, and C, respectively). (a) "blocked" position - relaxation; (b) "free" access of HMM-S1 to actin filaments - contraction.

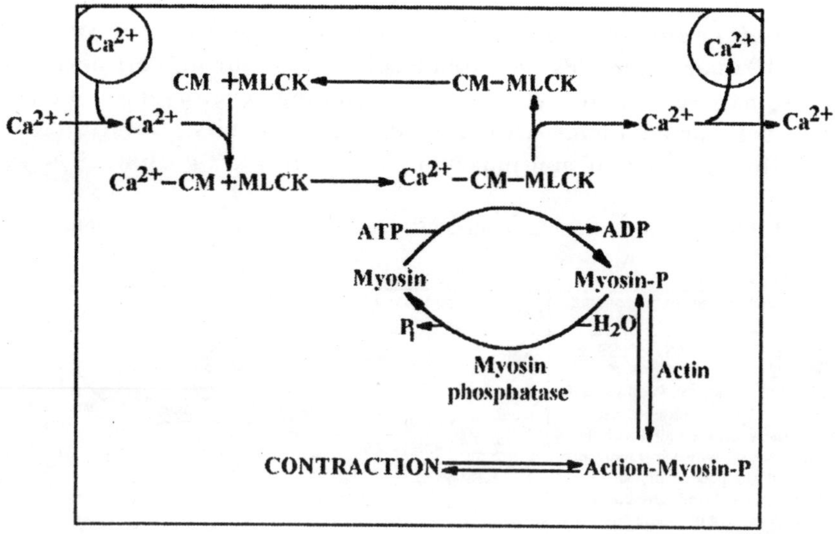

Figure 4.8. Schematic representation of the control of smooth muscle contraction [2]. Thick arrows indicate the tension accumulation and thin arrows its release, respectively. Mg-ATP-ase activity is maximal in the actin-myosin-Pa complex (CM – calmodulin, MLCK – light myosin kinase).

myosin-phosphatase or by the transport of Ca^{++} in the extracellular space, Figure 4.7. The intercorrelation between the sequences 1 – 5 is illustrated in Figure 4.8 [2].

The smooth muscle contraction implies a great number of biochemical steps that can be successively controlled and confers to it the capacity of developing various tensioning degrees, which is maintained for prolonged periods.

4.1.2. Non-muscular contractile systems

The organisation of the non-muscular contractile apparatus is mainly based on the ordered disposal of microfilaments (actin filaments), AF, the myosin being located in their proximity, in dispersed state. In non-muscular cells the gravimetric ratio myosin/actin is always subunitary (~0.1), while in the striated fibre this ratio is equal to 1.7. In this case, similar to the muscle contraction, the sliding mechanism is responsive for local motility. Small actin-myosin aggregates exist in some parts ofthe cell and cause localised contraction. Any living animal or vegetal cell contains a motile system, actin-myosin complex and/or microtubules, specialised for strictly determined functions. These ones can be either of high generality, such as: elongation and contraction of *pseudodia during amoeboid movement,* contracting ring of microfilaments pinches an animal cell in two during cell division, or its involvement in cytoplasmic streaming (cyclolysis) found in plant cell, or characteristics to some types of cells.

Different types of non-muscular movements are presented in Table 4.2.

Table 4.2. Non muscular contractile systems. Functioning principles

Biosystem and the corresponding type of movement	Structure and function of motile system
Cytoplasmic streaming (Cytolysis) (flowing of entire cytoplasm around the space between vacuole and plasma membrane in cells) a) the shear force, parallel with the contact surface between ectoplasmic actin and endoplasmic myosin generates cytoplasmic streaming; or b) the movement of motile endoplasmic organelles.	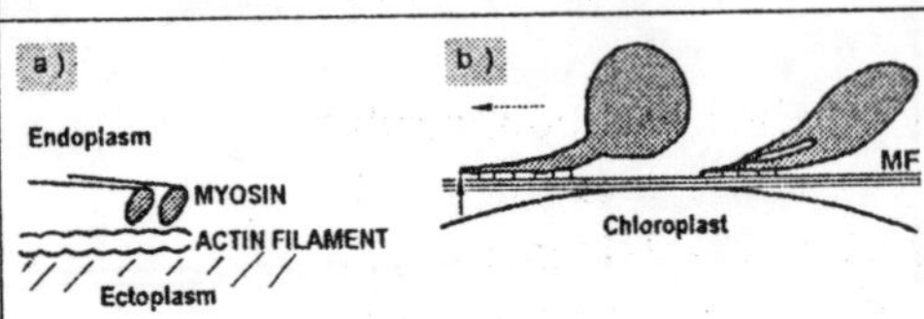
Cellular displacement - is characteristic for pseudopodia during amoeboid movement.	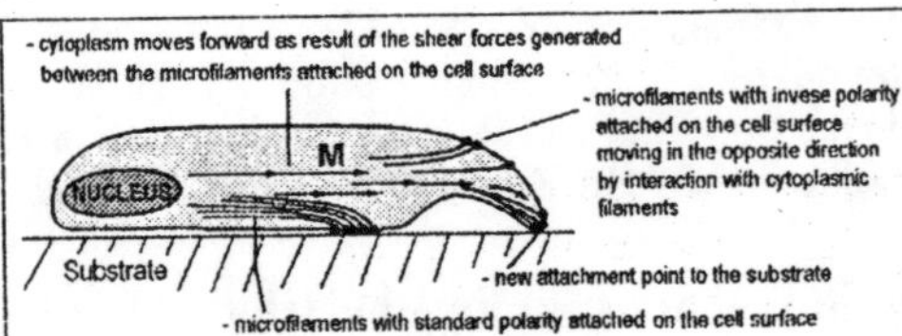
Cilia and flagella movement (locomotor organelles found in eukaryotes that are formed from a specialised arrangement of microtubuli)	
CILIA - work like oars, alternating power with recovery strokes; - the force is generated in a direction perpendicular to the axis of cillium; - tubulin and dyenin the proteins involves in force generation show many analogies with myosin-actin couple; - in this case the normal activity requires low concentration of Ca^{++}.	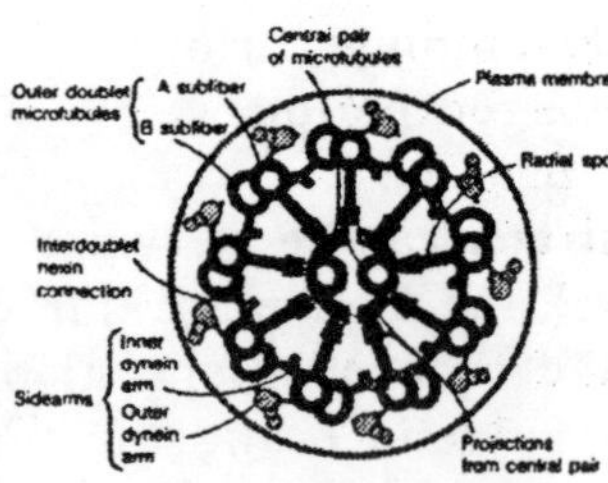
FLAGELLA - undulating motion that creates force in the same direction as the axis of flagellum ; - its motility does not depend directly by ATP but by a protons gradient on one and other side of the membrane.	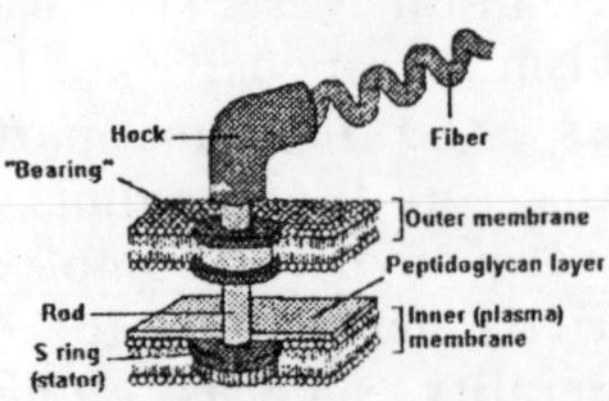

4.1.3. Contractile mechano-chemical systems of equilibrium

Very interesting contractile formations – known as the spasmoneme – exist to some unicellular eukaryotic organisms (*Vorticella, Carchesium*). These fibers, ϕ = 2–3 nm, are made of the contractile protein spasmin and are not acto-myosin type. The spasmoneme is variously described as being contained in an evagination of the cell body that extends the length of the stalk or as being extracellular. Individual spasmin fibers, separate from each other to extend longitudinally below the pellicle over the surface of the body. The presence of a system of longitudinal tubules and a Ca^{++} pomp associated to the spasmoneme filaments, was proved.

The movement arises by direct participation of Ca^{++} but not of ATP. It is accepted that the filaments are formed by polarised and longitudinally disposed macromolecules, with a high affinity for Ca^{++}. Whenever Ca^{++} ions are bonded on the spasmin sites, the ordering degree diminishes and the filament is characterised by a high configurational entropy; when the Ca^{++} ions live the contractile apparatus, this one is extended and self-organises, by hydrogen bonds formation between the molecules, and its entropy increases. The ATP role is that of furnishing the required energy of the ionic pomp to "sequester" the Ca^{++} ions; consequently, ATP induces the elongation of spasmoneme, not its contraction. There is not enzymatic activity, the protein represents only the motile support the movement being external controlled.

The main particularity of this system concern the fact that the spasmoneme is never in the relaxed state; it extends actively, producing a mechanical work, both during extension and contraction; in this way this biosystem constitutes the only example of "equilibrated" biochemical transformation.

4.2. MECHANO-CHEMICAL PROCESSES OF ENERGY CONVERSION BASED ON ARTIFICIAL SYSTEMS

The already described biomechano-chemical processes represent clear examples of reciprocal energies conversion. This ability is mainly related to their soft composition, based on "wet" components, such as liquids or gels. Presently, a growing number of workers, taking inspiration from nature, have begun to see opportunities in these materials. The technical design of some analogous systems implies, in the first instance, the elimination of rigid materials that work in dry medium, such as metals, ceramics, and plastics. From this viewpoint, it was imposed the adaptation of new materials, able to avoid the disadvantages of usual materials. The new materials have to work in reversible cycles of high yield of energy conversion, to be kind-hearted, quietly, no wastes producing, to enumerate only some advantages conferred by the direct conversion of chemical energy in other forms, by using of macromolecular supports.

A model, which harmonises into an artificial macromolecular system, the specific conditions of the natural motile systems is that of polymer gels. These ones are soft and wet materials and can change both their size and shape, thereby converting chemical energy directly into mechanical work. Generally, an elastic macromolecular network and a fluid filling the interstitial spaces between them compose the polymer gel. It does not flow like a polymer solution, but rather has the properties of a solid. For instance, if one of its sides is perturbed mechanically, the disturbance propagates directly along the macromolecular chains to the other side. Answering to many external stimuli (for instance the changes of medium factors) the gels are able to modify adequately their size and shape.

A rheological intricate peculiarity marks polymer gels' behaviour as chemo-mechanical systems, namely: even if it are easily deformed due to the high internal friction, in their bulk, under the action of hydrostatic pressure these ones are incompressible.

The simplest artificial mechano-chemical system is that able to realise the energy conversion by swelling of a polymer network, using only one liquid component. Thus, a poly(vinyl alcohol)-based gel, crosslinked with glutaric aldehyde and immersed in pure water, realises the mechano-chemical cycle as Figure 4.9 shows.

On the pathway (1–2) the non tensioned macromolecular net-

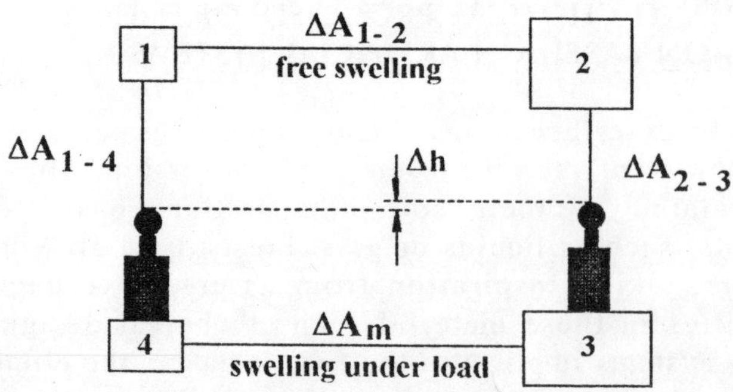

Figure 4.9. Schematic illustration of the chemomechanical cycle [69].

work swells, retaining the liquid at constant chemical potential; an additional load attached to it determines further its deformation on the partway (2–3); under the action of the external applied force, the network is also deformed on the route (1–4); finally, the network swelling occurs on the route (4–3). Correspondingly, one can write [69]:

$$\Delta A_{(1-2)} + \Delta A_{(2-3)} = \Delta A_{(1-4)} + \Delta A_m \tag{4.1}$$

$$\Delta A_m = \Delta A_{(1-2)} + \Delta A_{(2-3)} - \Delta A_{(1-4)} \tag{4.2}$$

$$\Delta A_m = \Delta A_c + mg\Delta h \tag{4.3}$$

where: ΔA_m is the free energy exchanged by the conversion of mechano-chemical energy; ΔA_c – the free energy changed during gel's swelling under the action of external load, with its weight equal to m; and $mg\Delta h$ – the mechanical work effectively produced, W. Under these conditions, the yield of energy is given by the following relation:

$$\eta = \frac{W}{-\Delta A_c} \tag{4.4}$$

In some gels, the bonds that hold the macromolecular network together are classic covalent bonds, but in other they are more subtle. Briefly, a macromolecular gel can be obtained, starting either from low molecular weight components or polymer chains, by: 1) chemical reactions; 2) both under the action of some precipitating reagents in the homogeneous solution of the macromolecular compound or of incompatible solvents; 3) under the presence of high energy radiation; 4) by alternating the freeze-thaw cycles, on the mechanochemical route.

Since in 1948, W. Kuhn [70], J.W. Breitenbach [71], and A. Katchalsky [72] independently found that water-swollen macromolecules can convert chemical energy directly into mechanical work under isothermal conditions. They exploited the principle of reversible contraction and dilatation that is based on the reversible ionisation of suitable groups (for example carboxylic acid groups) of a polyelectrolyte, by alternating addition of alkali and acid, respectively.

The changes of polymer gels volume can also be induced by modification solvent composition, salt concentration, temperature, light or UV, electric and magnetic fields, or by redox reaction.

The macromolecular systems thermodynamically able to convert the chemical energy directly into mechanical work or, conversely, of transforming mechanical into potential chemical energy (in fact passing into a mechano-activated state) are denoted as "mechanochemical systems". These ones are capable of performing cyclically. The first part of cycle is called "chemo-mechanical", and presently in literature on can find many models illustrating it [72–75]. The difficult part, which for the time being limits the applicability at large scale of these systems is related to the release of reversibility by adequate kinetic sequences.

Such kind of chemomechanical "engine" must revert after each cycle to its initial state, working further on repetitive and reproductive manner. In the first part of cycle, the variation of the chemical potential of the substrate is coupled to the release of mechanical work. Therefore the reactant falls from a high chemical potential to a lower one. It is necessary to introduce at least one feedback system to realise a self-regulating chemomecanical system in which the synthetic macromolecules involved are sensitive to changes of the environment (such as pH, ion species, ionic strength, temperature, reactive chemical species, etc.) and undergo important conformational changes. In essence, by analogy with the actual engines, a macromolecular system working on this principle should be defined as a "molecular-mechanochemical engine" [76].

The artificial mechanochemical systems, implying synthetic polymers, are capable not only of generating mechanical energy on the macroscopic level, but also of transforming the information as a "signal" or "receptor", whereby semimicroscopic deformation or strain play an essential role for displaying functions such as those of a switch or a sensor. Such kind of behaviour, located at the molecular–supramolecular boundary can also be found in many forms of biological movements, such as those occurring at the level of nervous cells, ions channels, enzymatic reactions, hormones release, etc.

The mechanochemical systems may be classified according to the principle of reaction:1) changes in ionisation state (hydrogen-ion transfer, ion exchange, chelation); 2) redox reaction; 3) steric isomerisation; 4) phase transition or order–disorder transition; 5) polymer–polymer association or aggregation; 6) electrokinetic processes [73].

Irrespective of the gel and stimuli nature, a molecular mechanochemical engine implies a receptor, usually located at the level of the functional groups, and a functional part, at the molecular and supramolecular level, Figure 4.10.

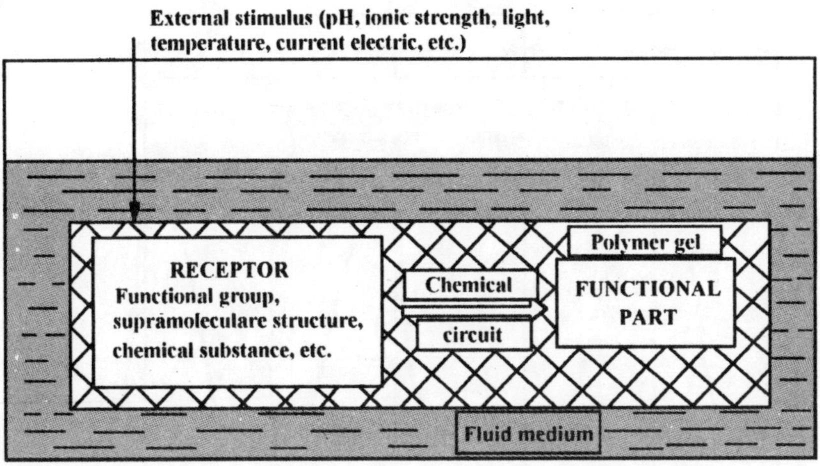

Figure 4.10. Generalised representation of the functional parts of a stimuli-responsive polymer gel.

A small change of the physical or chemical stimuli in the surrounding medium of the mechanochemical system is perceived by the receptor (chemical compound, functional group of the macromolecular chain, intricate characteristic of the polymer) that adequately answers converting it into a chemical signal. On its turn, this one is transmitted through a "chemical circuit" to the functional part of the system, assuring to the molecular engine the self-regulation ability.

The reversible conversion of the system properties caused by the conformational changes is described by the curves having approximately the same shape, irrespective of the stimulus nature, Figure 4.11. The plotted curves differ each to another by the amplitude and the frequency of the changes of measured property. Presently, the efforts are focused on the increase both of amplitude of dimensional changes and developed force, and shortening of mechanochemical cycle.

4.2.1. Mechanochemical systems based on the change of ionisation state of the system

By changing reversible the ionisation degree of the functional groups attached to the macromolecular chains or by exchanging monovalent ions with polyvalent ones, reversible contractions and dilatations occur, due to the conformational modifications of the macromolecular substrate, Figure 4.12. If an additional load is attached at one end of the membrane or filament gel, during

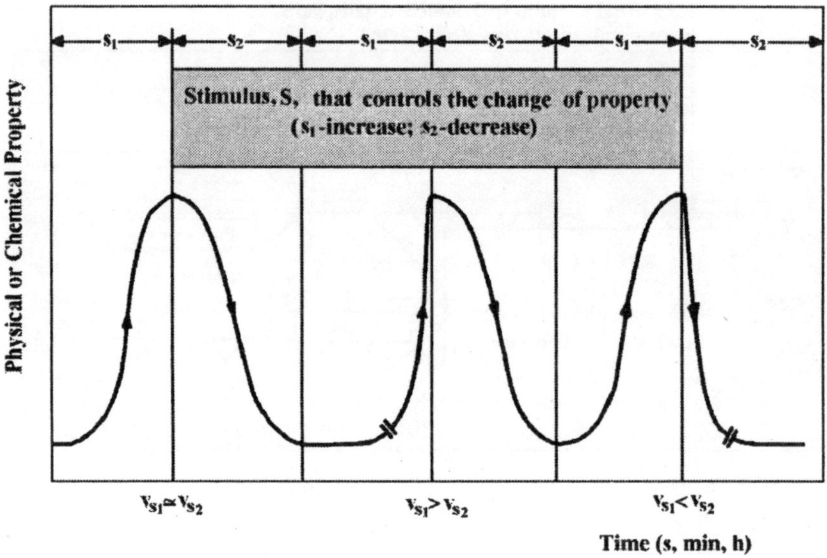

Figure 4.11. General evolution of the property *vs.* time curves: *S* - external stimulus (*S*1 - mechanochenocal cycle' part where the value of stimulated property increases; *S*2 - mechanochenocal cycle' part where the value of stimulated property decreases; *vS*1, *vS*2 - the speeds of stimulated property change).

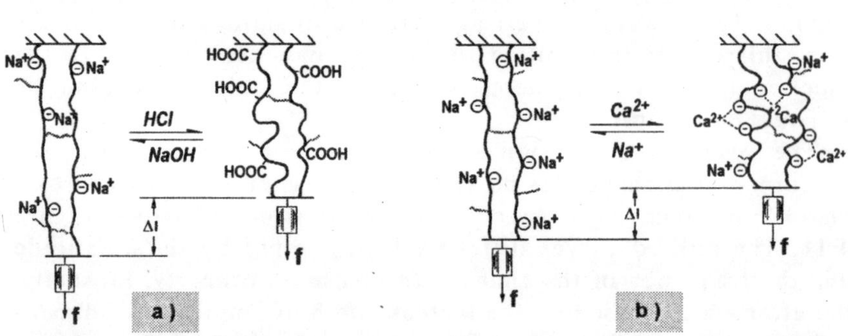

Figure 4.12. Schematic representation of principle of dimensional changes of polyelectrolyte fibers or films in the case of: (a) hydrogen ion transfer; (b) ion exchange or chelation.

dimensional changes it is lifted up and down, and chemical energy is thus transformed into mechanical work.

Presently the attention is paid to improving the polymer gel properties, mainly to reducing the response time to the stimuli action, and to increasing the developed force/weight ratio. Interesting results were obtained modifying the poly(acrylonitrile) fine fibers. Thus in the first step the fibres were annealed at high temperature, in the presence of oxygen, to induce crosslinking. In the next step, the fibres were saponified in boiling 1 N NaOH to convert the original cyano groups into ionizable carboxylic groups. The chemical structure of thus obtained gel fibres is the following one [77]:

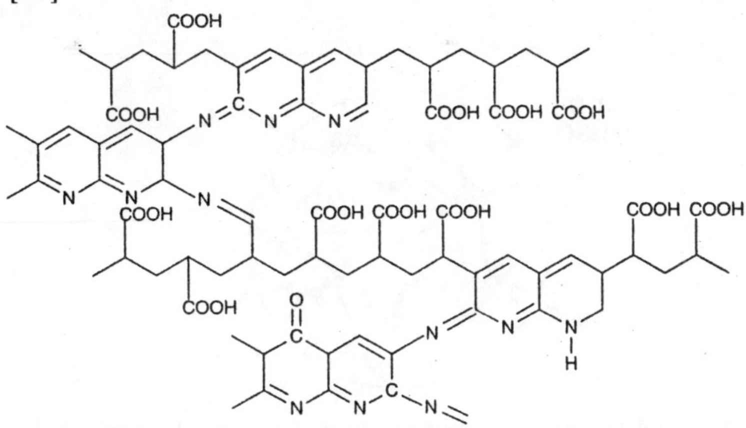

Bundles of about 60 gel filaments with such structure and ϕ = 25 μm immersed alternatively in alkaline and acidic solutions responded by elongation/contraction cycles under a constant load of 0.75 kg/cm^2 (isotonic contraction). Some dynamic characteristics of the described contractile system are presented in Figure 4.13 and Table 4.3.

An interesting chemomechanical behaviour was found in the

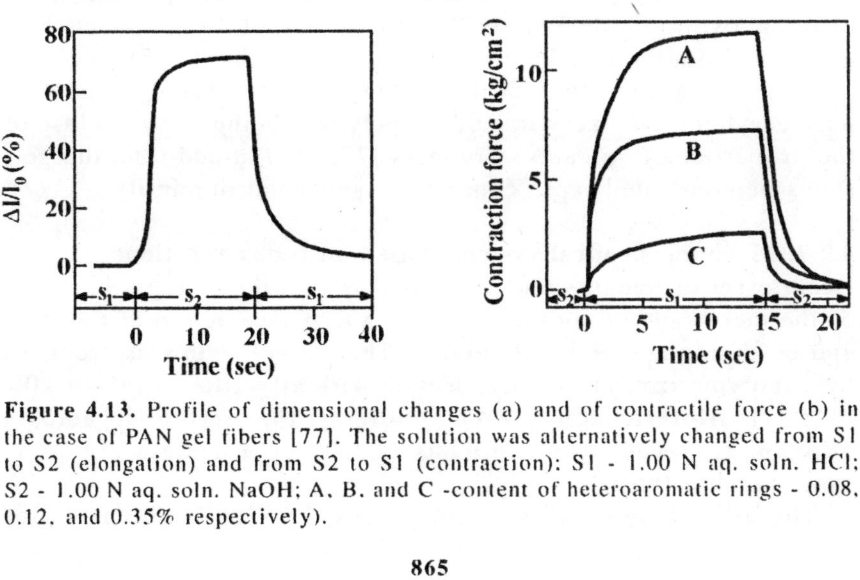

Figure 4.13. Profile of dimensional changes (a) and of contractile force (b) in the case of PAN gel fibers [77]. The solution was alternatively changed from S1 to S2 (elongation) and from S2 to S1 (contraction): S1 - 1.00 N aq. soln. HCl; S2 - 1.00 N aq. soln. NaOH; A, B, and C -content of heteroaromatic rings - 0.08, 0.12, and 0.35% respectively).

Table 4.3. Comparison between the performances of skeletal and artificial muscles

Parameter	Skeletal muscle	"Artificial muscle"		
		PAN	PVA-PAA	PMVE
l/l_0, %	30	80	30	> 100
Time of response, s	$10^{-2} - 10^{-3}$	2	0.12 – 0.33	> 1
Maximal contractile force, MPa	0.4 – 1	1.2	0.22	3 – 10

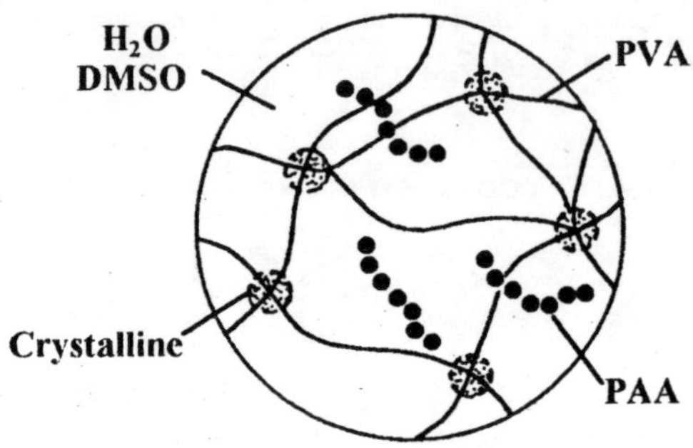

Figure 4.14. Schematic representation of the PVA-PAA gel [78].

case of composite PVA-PAA gels obtained by crylolitic treatment of homogenous solutions of poly(vinyl alcohol), with more then 95% saponified groups, and poly(acrylic acid), PAA, into water-DMSO mixture. X-ray analysis allowed the description of the gel strips as an interpenetrated network, where the PVA chains during cryolisis generated crystalline domains, which act as joining points of the network, while PAA is the stimuli-responsive polymer, Figure 4.14. Both the tensile strength and strain exhibited by thus prepared gels are more than 50 times higher as compared to those of poly(acrylamide-*co*-acrylic acid) copolymer, being enough close of the properties of vulcanised rubbers [77, 78]. In addition, the gels strips demonstrated rapid contraction and good durability.

4.2.2. Mechanochemical systems based on redox reactions

Even lesser efficient, some systems in which the changes in length of the membrane are produced by an alternative action of reducing and oxidizing agents is of interest. Thus, a gel strip was produced by copolymerization of vinyl acetate with allylalloxan (about 20% of allylbarbituric acid), then oxidizing the barbituric acid to alloxan, followed by alcoholysis of the vinyl acetate groups to vinyl alcohol [80]:

The alloxan and dialuric acid groups, implied in the oxidation

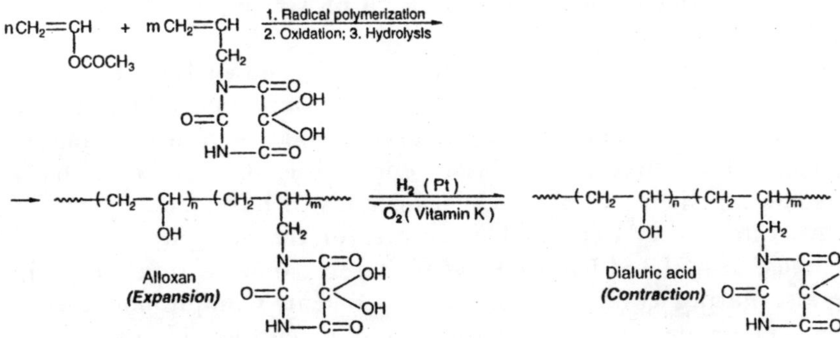

and reduction processes, constitute the redox system fixed to the gel. The reduction of an aqueous solution of alloxan to dialuric acid with the use of hydrogen gas was possible only in the presence of platinum. Since the alloxan from copolymer could not leave the gel, a soluble redox system, namely: 2-methyl-naphthoquinone/2-methyl-naphtohydroquinone (vitamin K component) was added to the solution. In this way, the soluble redox system transferred the redox action to the fixed system in the interior of the gel.

The gel strips showed a reversible change in length. During the reduction step, the original length was reduced by about 20%, and during oxidation step it recovered its initial length.

It was also investigated the behaviour of some chelates, mainly based on polyallylamine and copper. On the reduction stage, in the presence of catalyst, the copper ions change their oxidation state ($Cu^{2+} + H_2 \rightarrow Cu^{+}$); therefore the co-ordinative bonds are split and the strip elongation occurs. On the oxidation stage ($Cu^{+} + O_2 \rightarrow Cu^{2+}$) the strip contracts to the original length.

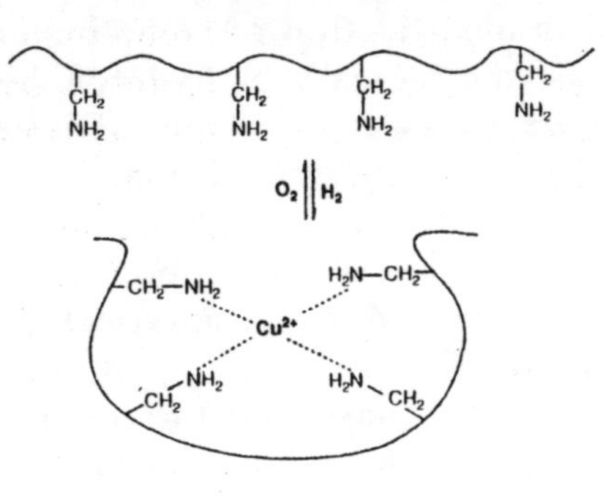

4.2.3. Mechanochemical systems based on steric photo-isomerization

The importance of UV irradiation is well known both in release and occurrence of life phenomena and in macromolecular synthesis. Furthermore, it is more and more used as information stocking and transmission source as well as in controlling of reversible change of some physical and chemical properties of polymers.

An intensively exploited principle for the designing of motile systems, stimulated by UV light is stereo-isomerization, occurring by alternating UV irradiation/dark or heat. Complementary and even independently, other factors can compete in this process. The conformational transformations thus stimulated as molecular level release the motility at the macroscopic level, mainly in the case of polymer gels containing in the backbone, attached, or physically incorporated such kind of functional groups. A series of photo-isomerizable compounds and the mechanisms of isomerization is given in Table 4.4.

The first mechanism is exemplified by the system composed from poly(methacrylic acid) and Crysofeine G, in aqueous media [87]:

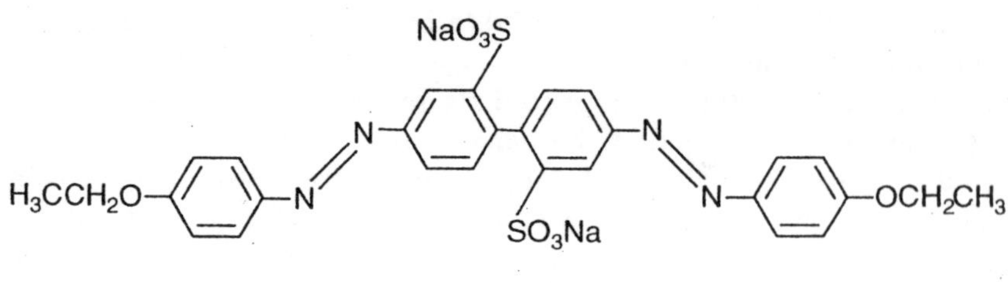

By UV irradiation the stable *trans* form passes into *cis* form, which is reflected in modification of some important properties of the system, such as: the increase of hydrophobicity, the decay of viscosity, and the weakness of intermolecular interactions. On the initial stage, due to the *trans* conformation of the azobenzene groups, the dye is disposed alongside of the extended macromolecular chains; the *cis* form, obtained by irradiation, limits the dye-polymer interactions and poly(methacrylic acid) adopts a coiled conformation.

A second mechanism is based on the modifications caused by the effect of UV waves on the intramolecular interactions between the pendant azobenzene groups bonded on a poly(styrene-*co*-maleic anhydride) copolymer.

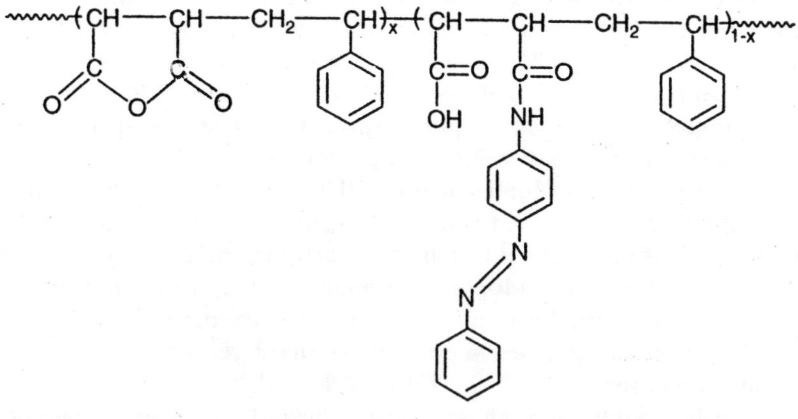

Table 4.4. Types of conformational changes stimulated by UV irradiation

Type of reaction. Examples	Conformational changes of poly chain (containing or interacting photo-ionizable groups) during I irradiation
1) *trans-cis* isomers 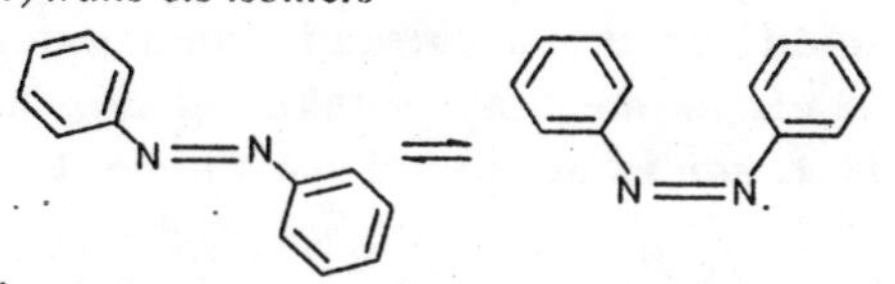	1 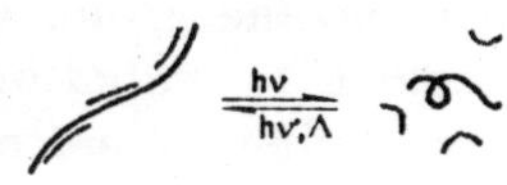
2) zwitter ions formation 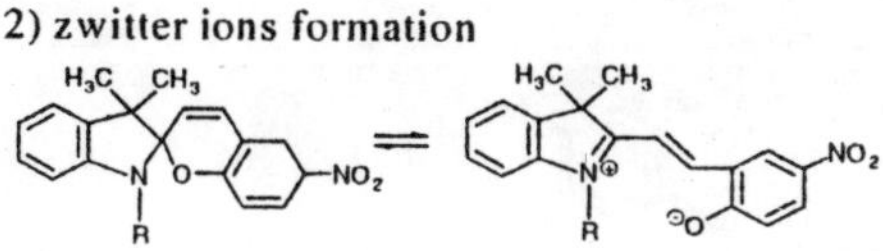	2 a 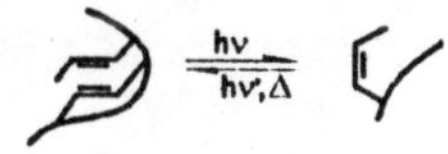 b 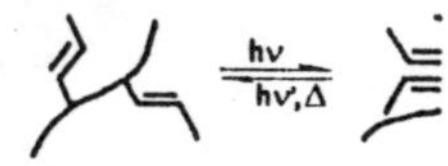
3) radicals formation 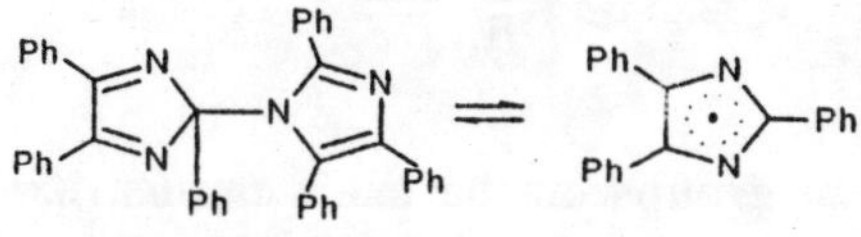	3 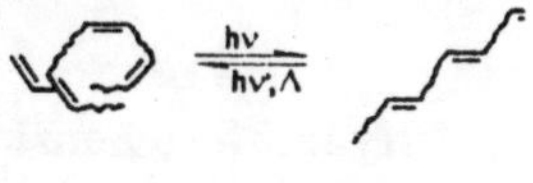
4) ionic dissociation 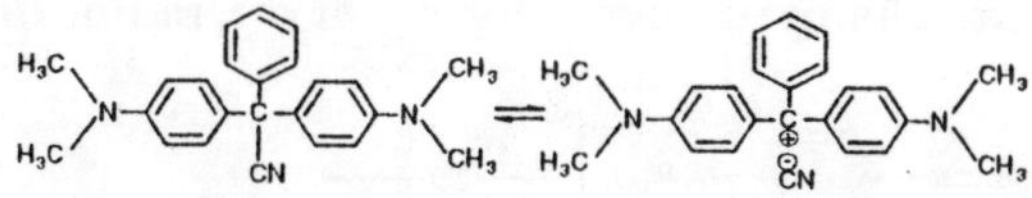	4 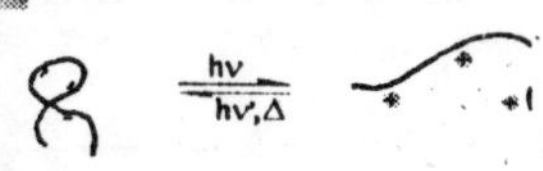
5) rings splitting and remaking 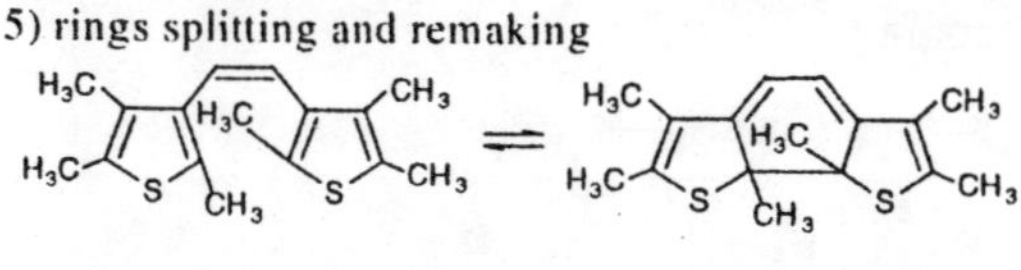	5 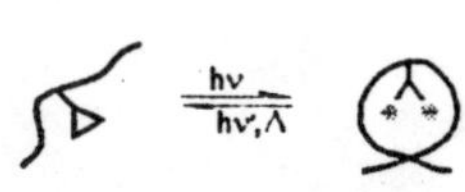

Under the alternative action *hν* ↔ *dark*, the viscosity of this copolymer in dioxane modifies in a reversible way. Thus, it is reduced with about 20% during UV irradiation. It seems that the chains shortening is related to the difference between the dipolemoment values corresponding to the to states (for *cis:* μ = 3.1 D; and for *trans:* μ = 2.6 D, respectively).

The *cis* ↔ *trans* isomerization will induce strong dipoles in the side azobenzene groups, these ones tend to orient parallel suffering, in low polar solvents, a mutual attraction; consequently, the macromolecular chain adopts a compact coiled conformation, case 2b in Table 4.4. In the presence of the polar solvents, the change of hydrophobe properties becomes the main reason of the conformational changes, case 2a in Table 4.4.

A similar mechanism characterises the polymers or polymer gels containing pendant units of spyrobenzopyrane derivatives [91].

The third mechanism implies the incorporation of *cis* ↔ *trans* stereoisomers into the polymers main chain. The photo-induced dimensional variation of these photochromic groups determines implicitly the dimensional changes of polymer gel. In this sense, the azobenzene group was widely used; as it was already mentioned under UV irradiation the *trans* → *cis* isomerization takes place, the recovering of more stable *trans* form is achieved by thermal or photochemical treatment [92];

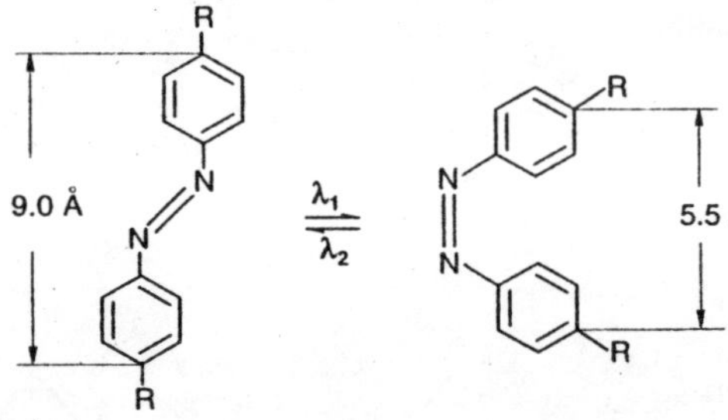

Polyamides, bearing azobenzene groups on the main chains, have been from the first macromoleculare supports used destinated to investigate the conformational changes occurring by UV irradiation [93, 94]:

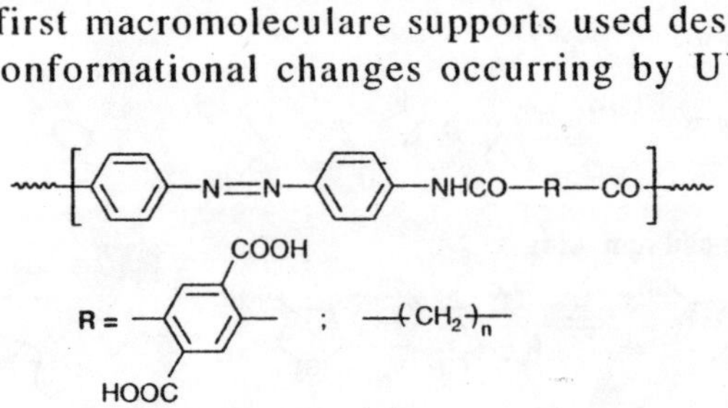

Other more complex systems have been obtained, consisting by a photoresponsive crown ethers which combines within the same molecule both a crown ether and a photoresponsive chromophore. The polymer changes its conformation in response to irradiation, which in turn induces a change in the complexation ability with suitable alkali metal ions, for instance Cs^+ [95]:

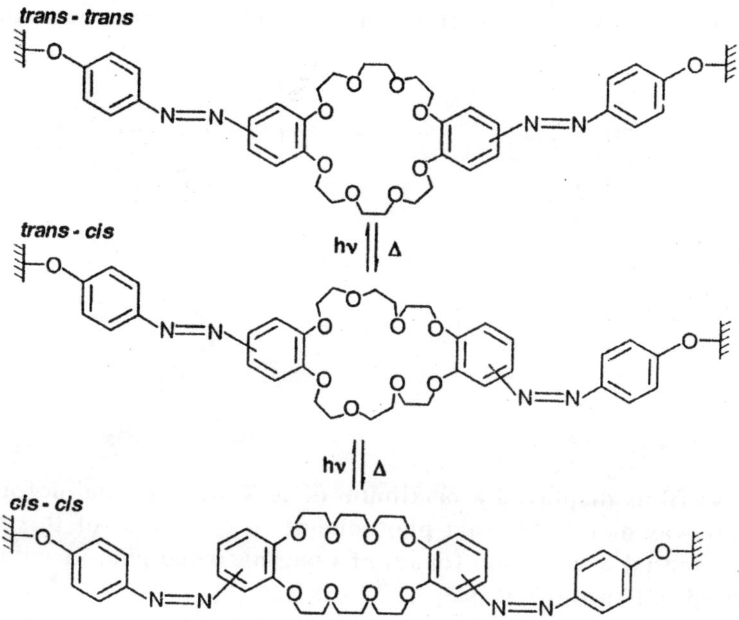

The fourth mechanism implies the presence of electrostatic repulsions between the charges generated by photoirradiation of a compound with a suitable structure, for example triphenylmethane derivatives. These ones have been introduced as pendant groups, by the copolymerization of the corresponding imide derivatives with N,N'-dimethylacrylamide [96]:

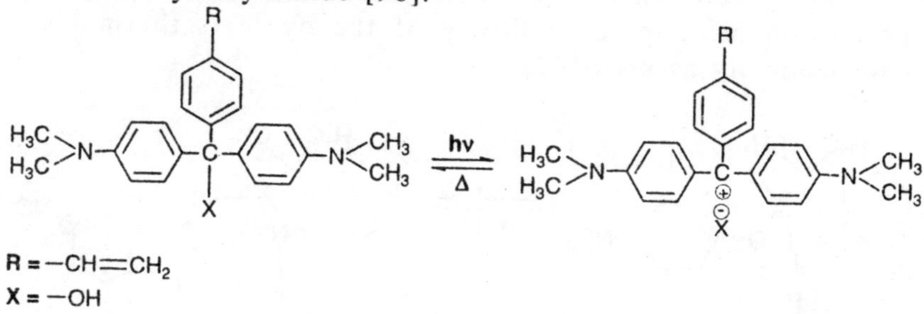

Under the action of UV light (λ = 270 nm) the coplymer solution becomes green and its viscosity increases from 0.55 to 1.6 dl/g; this behaviour is reproductible. The system response to the applied stimulus was evaluated by its capacity to recover half of the value of initial viscosity, i.e. 3.1 min.

Membranes based on poly(γ-glutamic acid) containing N,N-dimetil amino groups, working on the same mechanism, have been synthetised by A. Anviram. This copolymer undergoes photo-ionization in the presence of carbon tetrabromide or carbon tetrachloride upon UV irradiation:

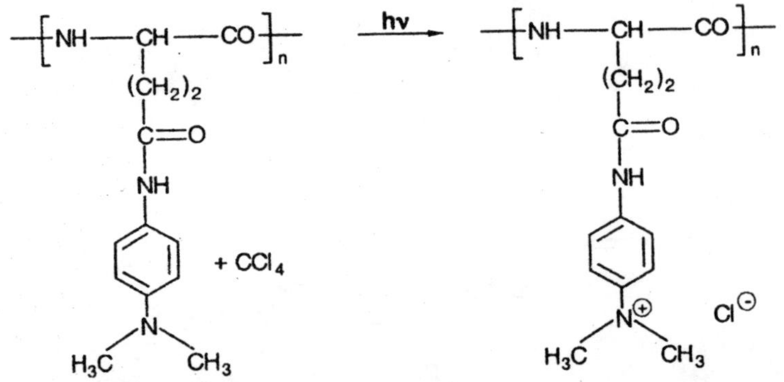

These films displayed a maximum of 35% dimensional deformations. It was mentioned that photochemical behaviour of this type has a potential application for relief imaging if the photoirradiation is carried out through masks.

The induction alongside of the polymer chains of some intense dipoles, capable to release specific intermolecular interactions, constitutes the fifth mechanism of conformational changes by UV leads by UV irradiation that can be used for mechanical work generation. Spirobebzopyranes, by UV irradiation split the 1,2 bonds of the pyrane ring concomitantly with the partial rotation of the structural unit obtained; this unit is characterised by an important dipole moment. The reversibility of the cycle is thermally or photochemically assured [97]:

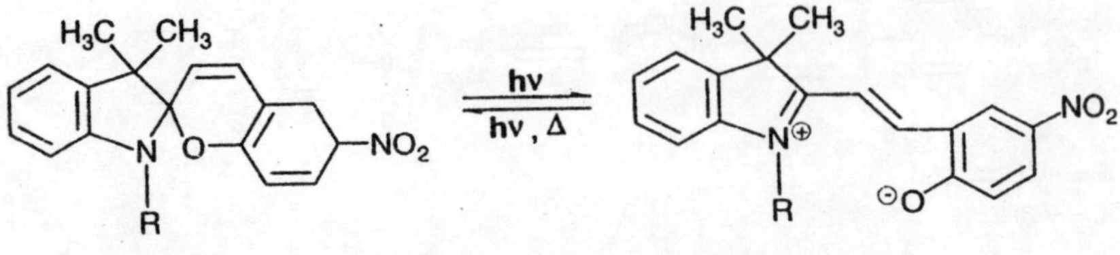

Such compounds, bonded as pendant groups in macromolecular chains of poly(dimethyl methacrylate) type, generates in non polar solvents (benzene) photo-active gels, with reversible response to the repeated *hν* ↔ *dark* cycles.

The amplification of the conformational changes of the polymer chains used as solutions/gels or even in solid state can be in such way performed that the release of mechanical energy to become perceptible to the macroscopic scale. Such effect was simulated using gels containing functional groups that give rise to photo-induced electrostatic repulsions [98, 99]. A disk-like polymer gel (ϕ = 10 mm, *h* = 2 mm) based on a copolymer with the following structure:

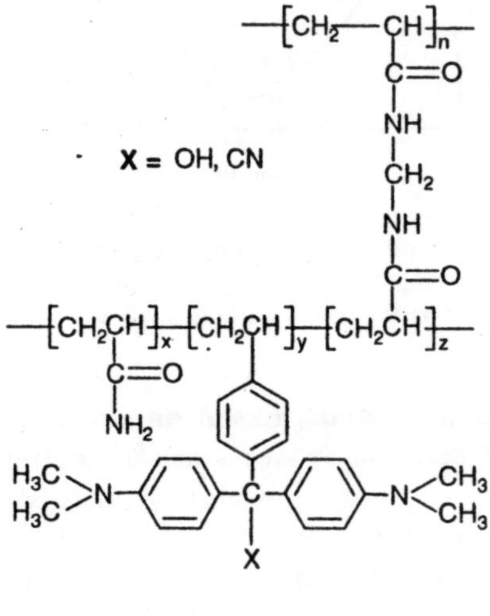

gives a reversible swelling, expressed by 3 times increase of its initial weight into 1 hr. In dark the gel contracts, regaining after 10 hr the initial weight, Figure 4.15. The increase of gel sensitivity to the variations of pH was achieved by replacing the -OH groups with –CH groups. Thus, the weight of gel increased more than 18 times upon UV irradiation.

This principle can also be used for the modification of polymeric films, namely: either by their physical penetration with chromophores [100, 101] or by covalent bonding of the last ones to the adequate functional groups of the macromolecular substrate [102, 106].

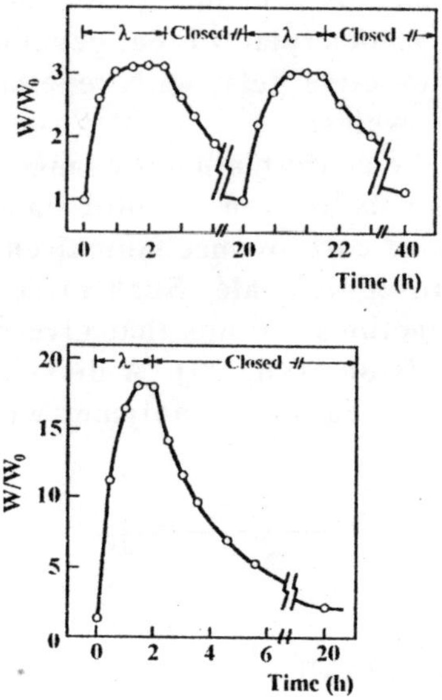

Figure 4.15. Photostimulated dilation and subsequent contraction of the polyacrylamide gel containing different percents of hydroxy triphenymethane pendant groups: (a) 3.7%; (b) 9%(UV irradiation (λ = 270 nm at 250 C, [90].

4.2.4. Mechanochemical systems based on phase transition

Macromolecular heat-sensible systems are characterised by negative temperature coefficients during water dissolution (LCST, lower critical solution temperature). Performance polymer gels based on this polymers are usually obtained by modification of the corresponding linear supports by grafting, copolymerization with bifunctional monomers, or crosslinking under the action of high-energy radiations (Co^{60}) followed by swelling.

This group comprises substituted polyacrylamides and their substituted derivatives and atactic poly(methyl vinyl ether), obtained at room temperature by cationic mechanism, which at this temperature is water insoluble and shows phase transition an 38 °C.

It was found that the decrease of response time is directly related to the synthesis method of the network. Thus, at high temperature a spongy-like gel is obtained, more sensible to the temperature than the homogeneous one, prepared at a temperature bellow LCST, Figure 4.16. Good results have been obtained, spinning poly(methyl vinyl ether) aqueous solutions at 40°C, into a $CuCl_2$ coagulation bath, followed by fibers reticulation under the action of γ radiations, at 23 °C and a radiation dose of 100 kGy. The gel fibrils shrink from 400 μm across at 20 °C to 200 μm at 40 °C, Figure 4.16. An additional load attached to this system converts the contractile energy developed into o tension of 98 kPa, three times higher than that developed by skeletal muscle, Table 4.3 [98].

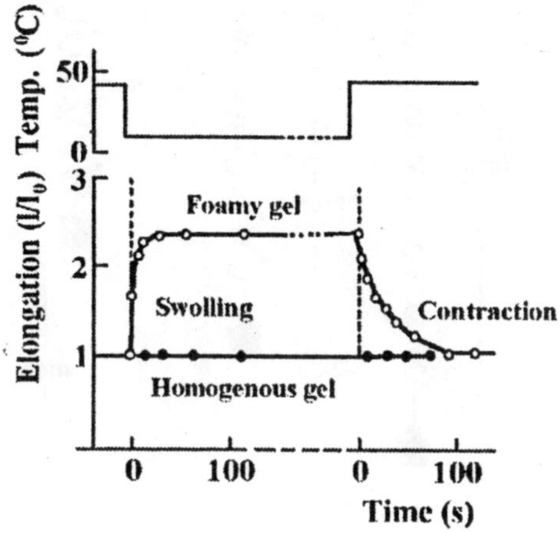

Figure 4.16. Chemomechanical response at temperature variation of the PVME gel crosslinked by γ radiation [98].

The order–disorder transitions are responsive for the conversion of chemical energy into mechanical work in the case of protein based networks. It was established that modifying the temperature and to certain values of pH, the conformational transitions from α-helix to elastic spiral takes place, Figure 4.17 [99, 100].

Since polypeptides' chemical melting temperature is associated with ionic strength and pH value, these factors controls this temperature, allowing the elastic bands to shrink and perform mechanical work, by lifting the attached load. At 37 °C the polypeptide matrix, equilibrated into a buffer solution, suffers reversible contraction and elongation by changing the pH from 4.3 to 3.3 and conversely. It was established that in case of this system, the contribution of electrostatic repulsions is insignificant (the average distance between two carboxylic groups is higher than 4 nm, therefore twice higher than the average distance between the negative species in buffer solution.

Other polypeptide networks, for example based on l-glutamic acid and dl-cystine residues contracted in their isoelectric region and extended in both acidic and basic media.

It worth to be mentioned that the first chemomechanical engine,

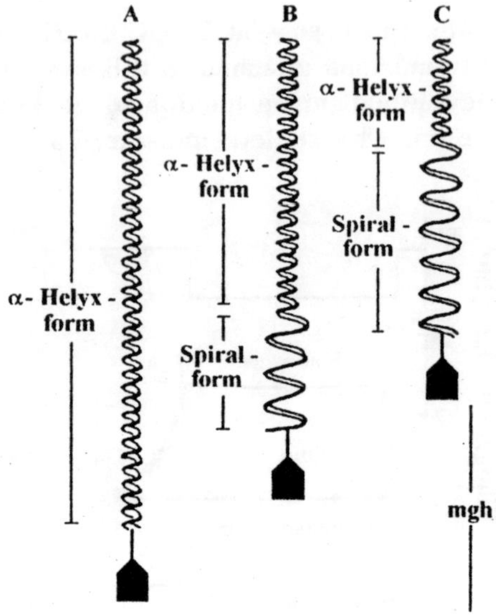

Figure 4.17. Elastic contraction that accompanies proteins conformational change form α-helix to spiral form [99].

constructed by A. Katchalsky, functioned on chemical melting and crystallisation of crosslinked collagen fibers by their treatment with salt solutions such as, LiBr, KSCN, or urea.

4.2.5. Mechanochemical systems based on polymer association

The study of interaction between two "complementary" macromolecules and the aggregation of the resulting complexes to a supramolecular structure in solution is of special interest, because biological phenomena, such as enzymatic processes and muscle contraction are based on the specific interactions of protein–protein type.

The finding of synthetic macromolecular systems, capable of reversible associations, accompanied by conformational changes, allowed the conversion of free energy into mechanical work. Thus, poly(ethylene glycol), PEG, of molecular weight (M_w) 2000 undergoes reversible complexation-dissociation with poly(methacrylic acid), PMMA, in aqueous medium, in by changing the temperature, Figure 4.19.

The pronounced change of viscosity (decrease), in the temperature range from 25 to 45 °C, proves the association of PEG-2000 with PMMA, curve 2 in Figure 4.19. The viscosity recovers com-

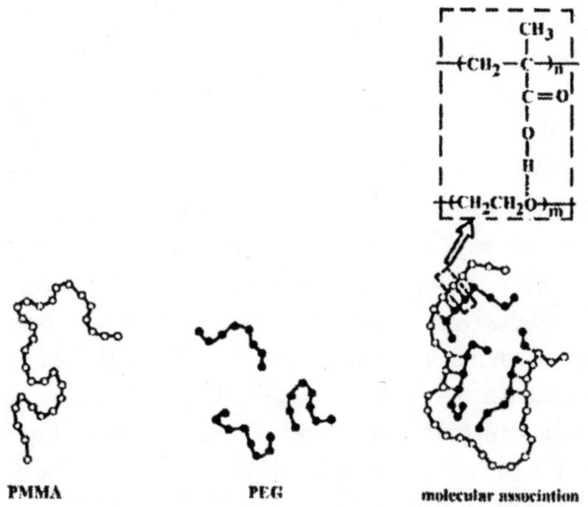

Figure 4.18. Representation of the reversible complexation-dissociations of poly(ethylene glycol), PEG, with poly(methacrylic acid), PMAA, in an aqueous media by changing the temperature [73].

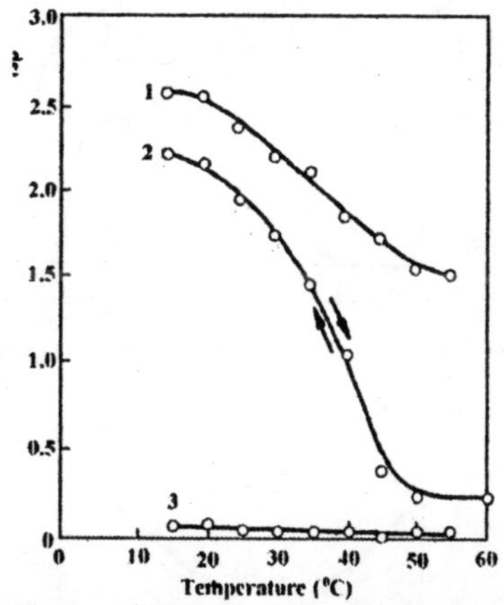

Figure 4.19. Temperature dependence of reduced viscosities of PMAA abd PMAA-PEG complexes in water [103]: 1) PMAA; 2) PMAA-PEG, Mw of PEG = 2000; 3) PMAA-PEG, Mw of PEG = 20 000. [PEG]/[PMAA] = 1.0 (per repeating unit); PMAA = 0.05 g/100 ml.

pletely on lowering the temperature. On the other side, a solution prepared from the same PMAA but with PEG having a molecular weight of 20 000 exhibits low viscosities over the entire range of temperature, curve 3 in Figure 4.19, indicating the formation of a stable complex. In addition, the use of PEG having a molecular weight lower than 1000 did not affect the variation of PMMA viscosity with temperature.

Y. Osada and its co-workers investigated this thermoreversible complexation, in order to proce chemical energy conversion into mechanical work [104]. Thus, a crosslinked PMAA membrane under a load of 490 mg was suspended in water at 10 °C. An aqueous solution of PEG was then added to the embedding fluid after constant length bad been attained. As Figure 4.20 shows, the membrane contracts sharply with rising the temperature (isotonic contraction), especially in the region of 20–30 °C. The dimensional changes are reversible and the work done per contraction by one gram of contractile substance is 5×10^{-3} cal. The internal stress of the membranes was measured at constant length for this system (isometric contraction), and it was found that the stress developed in PMAA membrane was 4–6 kg/cm³, which is almost the value found in natural muscle, Figure 4.20, [105].

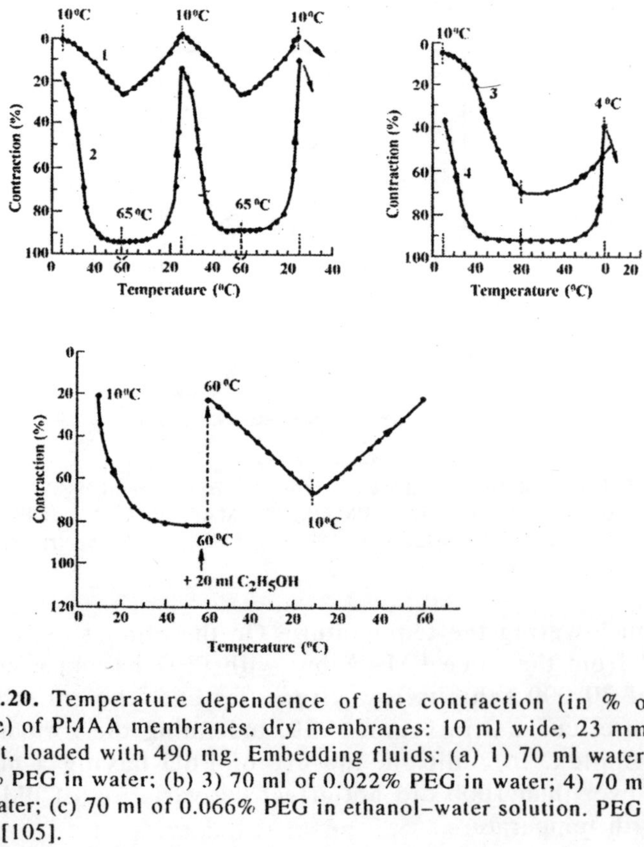

Figure 4.20. Temperature dependence of the contraction (in % of the dry membrane) of PMAA membranes, dry membranes: 10 ml wide, 23 mm long, 4.7 mg weight, loaded with 490 mg. Embedding fluids: (a) 1) 70 ml water; 2) 70 ml of 0.066% PEG in water; (b) 3) 70 ml of 0.022% PEG in water; 4) 70 ml of 0.11% PEG in water; (c) 70 ml of 0.066% PEG in ethanol–water solution. PEG molecular wt. 2000 [105].

4.2.6. Electro-activated mechanochemical systems

Electric modulation of the polymer gels contractile state is a real challenge for the scientist dealing with this field. It was found that a water swollen polymer-gel, containing ionizable groups, inserted between a pair of electrodes which is connected to a DC source starts to shrink, releasing the water droplets. The contraction occurs as a result of an electrostatic interaction between the charged macromolecules and the electrode, giving extensive desolvation (dehydration). In the case of polyanion gels contraction takes place only in the region of the anode and the water droplets was observed only at the edge of the gel near the cathode. The situation is reversed if a positively charged polymer gel is tested, i.e. contraction occurs near the cathode and water evolves near the anode. The fluid (water) transport to the electrodes takes place as result electrically induced migration of hydrated microcounter ions from gel.

Gel contraction can therefore be associated with the decrease in hydration power of macro-ions and microcounter ions due to the interaction with the electrodes. Thus, in the case of the polyanion gels, hydrated H_3O^+ ions travel to the cathode together with the water molecules, where they are reduced releasing hydrogen gas and water droplets from the gel:

$$2H_3O^+ + 2e^- \rightarrow H_2\uparrow + 2H_2O$$

Near the anode the following reaction takes place:

$$4OH^- \uparrow O_2- + 2H_2O + 2e^-$$

The contraction rate depends on the size of polymer gel supposed to action of the electric field. Whenever small gel particles, based on crosslinked PAA-Na, have been electro-activated using DC (46 V·cm^{-1}, 0.3 mA), contraction suddenly occurred accompanied by a volume reduction of 96%. The rate of volume change was found to be proportional both to the value of the applied current and the particles size square, Figure 4.21 [107,108]. Other factors that influence the contraction rate are the nature and concentration of the present salts, the nature and concentration of water miscible organic solvents, and temperature. No contraction was observed if an AC in the range of 6 Hz to 5 kHz was applied at 20 V.

A progress in modulating the electro-activated contraction was obtained by using of a composite material composed by a polyelec-

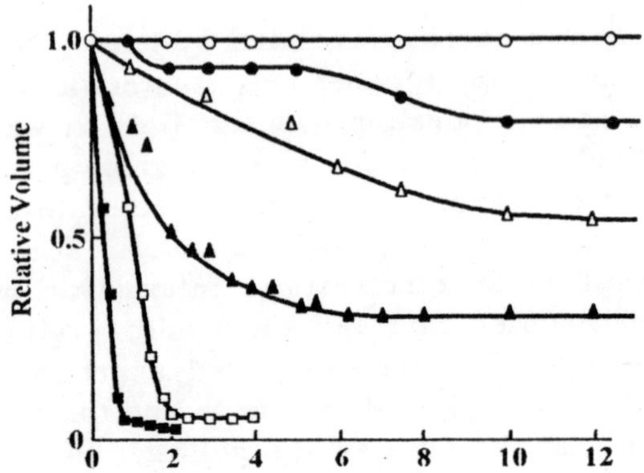

Figure 4.21. Variation in time of gel microparticle diameter with applied DC [107]. Electric potential: ▲ 4.2 V (32.3 V/ cm); □ 4.8 V (36.9 V/cm); ■ 6 V (49.2 V/cm). initial diameter of particles, 180 mm.

trolyte gel and an electro-conductive polymer. De Rossi investigate the chemomechanical behaviour of this kind of material, synthesised in two steps: 1) firstly the homogeneous aqueous solution consisting from a mixture of PVA and PAA was thermally treated in order to obtain a crosslinked hydrogel; followed in the second step by the pyrrole polymerization in gaseous phase in the presence of ferric chloride, $FeCl_3$, as catalyst [109]. The composite gel modifies reversible its dimensions as response to the potential difference between composite and counter electrode. It seems that the following electrochemical reactions occur at the polypyrrole electrode (PPy):

$$PPy + HO^- \rightarrow (PPy^+OH^-) + e^+$$

$$PPy + Cl^- \rightarrow (PPy^+Cl^-) + e^-$$

However, the water electrolysis is not excluded in this case. The two reactions determine the decay both of pH and ionic strength of solution at the PPy electrode. The change of local pH causes the sample contraction. The change of electrodes polarity causes the sense modification of the two above mentioned reactions, and the corresponding change of pH causes sample swelling.

A modulation of the ionic-gel contraction is possible by electrochemical control of the amount of microcounter ions changed (released and introduced) in conductive polymer (PPy) applying small differences of potential between the two electrodes

[110]. It was found that a microgel, with average particles size of 3 μm, obtained by complexation of poly(methacrylic acid), PMAA, with Ca^{2+} ions, placed on PPy electrode, rapidly changes its morphology concomitantly with pH modification under the action of electrical field. The dissolution and remaking of gel particles is correlated with the pH modification, whose values govern the chelation degree of PMAA. Thus, at low value of pH the PMAA-Ca^{2+} chelate is destroyed and the polymer becomes soluble; by increasing the pH the chelate is remade.

The electrode potential of the PPy film, deposed on a optical transparent electrode of $PtSnO_2$ (OTE) was controlled between –0.6 and 0.4 V against reference electrode Ag/AgCl. Gel's dissolution starts after 5 s and completes within 15 s. When the potential is fixed at –0.6 V with respect to the Ag/AgCl electrode the gel is remae in 10 seconds. In this case, the process is reversible and occurs without concomitant water electrolysis. Figure 4.22 schematically presents the proposed mechanism of morphologic changes of PMAA-Ca^{2+} microgel that is in direct contact with PPy film deposed on OTE. It can be seen that the anions are transferred in the PPy film proximity by electrochemical oxidation and neutralisation. The film doping and is controlled.

Similar, the effect of electric field on the cultures of animal cells, deposed on the same type of electrode, was investigated. It was found that the cells proliferation and the morphologic changes are affected by applying small differences of potential between electrodes. The phenomenon was ascribed to the structural changes of cellular cytoskeleton, in last instance, to the modification actin-myosin aggregates.

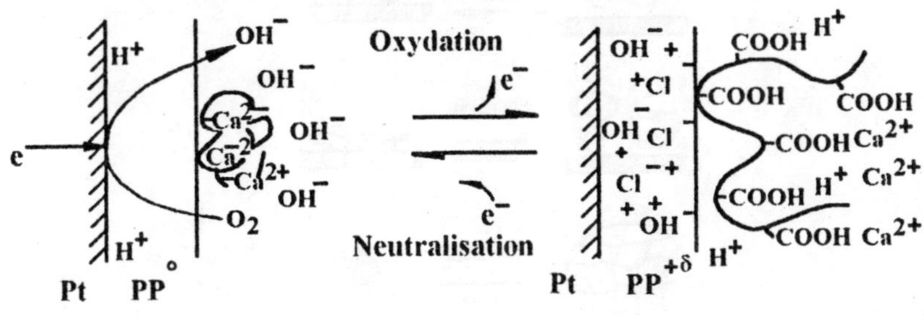

Figure 4.22. Mechanism of the electrically controlled morphologic changes of the PMMA-Ca2+ microgels, which are in direct contact with a PPy electrode [111].

4.3. APPLICATIONS. INTELLIGENT POLYMERS

The term "intelligent polymers" refers to soluble, surface coated or crosslinked polymers, which exhibit relatively large and sharp physical or chemical changes in response to small physical or chemical stimuli [112]. In the present days, the interest in "intelligent" polymer systems is focused on aqueous polymer solution, interfaces and hydrogels. "Intelligent" polymers are also called "smart" or "stimuli-responsive" polymers. In any case, in this class enters those polymer systems able to adapt their behaviour in relation to the surrounding medium. The manner within which smart polymers respond to the action of external stimuli is depicted in Figure 4.23. There are many different stimuli, which have been applied to the intelligent polymers and they are listed in Table 4.5.

In order as the conformational changes induced by the external stimuli that occur at the molecular level to become macroscopic perceptible, the composition of the environmental medium is in such manner adjusted that the smart polymer to be in the proximity of phase transition. In this way small changes of stimuli will cause sharp phase transitions, which acts as an amplifier of the external signal [114]. Two typical examples of such phase transitions for soluble polymers and hydrogels are illustrated in Figure 4.24. The most frequent responses of to the stimuli action are listed in Table 4.6.

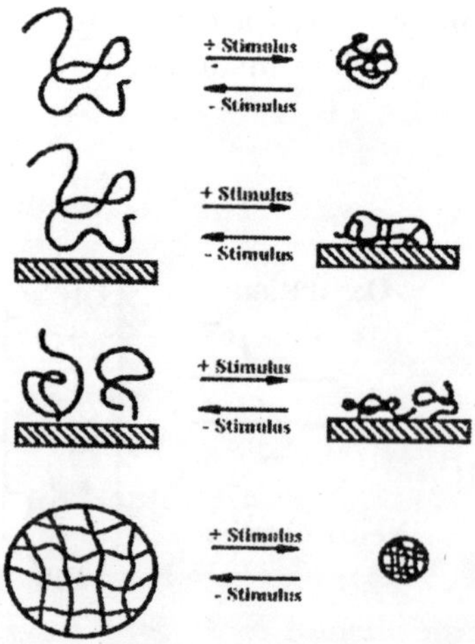

Figure 4.23. Schematic representation of "intelligent" polymer systems in solution, on surfaces and as hydrogels [113].

Table 4.5. Some of environmental stimuli applied to the smart polymers [113]

• temperature	• electric fields
• pH	• magnetic fields
• ions	• stress
• solvents	• recognition
• light or UV	• reactants

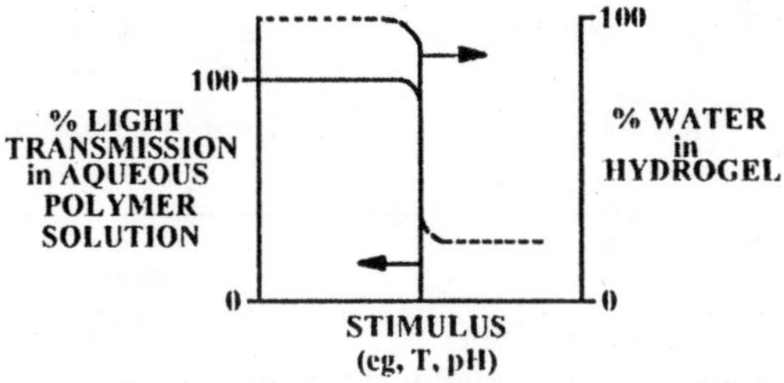

Figure 4.24. Typical responses to a stimulus for aqueous-based "intelligent" polymers in solution and as a hydrogel [114].

Table 4.6. Intelligent polymers responses can cause changes in [113]:

♦ phase	♦ surface energies
♦ shape	♦ reaction rates
♦ optics	♦ permeation rates
♦ mechanics	♦ recognition
♦ electric field	

There are a number of possible molecular mechanisms that can cause such sharp, sometimes discontinuous transitions in polymer systems, and they are listed in Table 4.7. and were reviewed in Chapter 4.2.

The most important applications of "intelligent" polymers are in correlation with polymers physical state, i.e. in aqueous solution, grafted or coated onto a solid support or as hydrogels; this is way, in this part the applications of smart polymers are treated in accordance with their physical state. The efficiency of the artificial chemomechanical systems can be enhanced by their combination with biomolecules, enabling the conjugation of many biomolecules to the same polymer backbone, thereby providing the opportunity for significant amplification of biological activity, Table 4.8. A conjugated system intelligent polymer-biomolecules is illustrated in Figure 4.25. The biomolecules may be conjugated to pendant groups along a polymer backbone, or to a terminal end of the polymer. In either case, the smart polymer may be a soluble polymer, a grafted or physically-adsorbed polymer on a solid substrate, or a polymer chain segment within a gel.

Significant amplification of biological action in these conjugated systems can be achieved by immobilization of a great number of "biomolecules" on the same macromolecular chain, (Figure 4.26). The biomolecules may also be physically-entrapped within hydrogel, either permanently, as in the case of large proteins, or tempo-

Table 4.7. Some molecular mechanisms of sharp transition in natural and synthetic polymer systems [113]

- ionisation or neutralisation
- ion exchange
- ion–ion repulsion and attraction
- release or formation of hydrophobically bound water
- helix-coil transition
- onset or inhibition of chain mobility
- crystallization or melting
- isomerization between hydrophobic and hydrophilic forms
- counter-ion movement in an electric field
- electron transfer redox reactions

Table 4.8. Examples of molecules which may be immobilised on or within "intelligent" polymers [113]

Proteins/Peptides	*Drugs*
Enzymes	Anti-thrombogenic
Antibodies	Anticancer agents
Antigens	Contraceptives
Cell adhesion molecules	Drug antagonists
"Blocking" proteins	Peptide, protein drugs
Saccharides	*Ligands*
Sugars	Hormone receptors
Oligosaccarides	Cell surface receptors
Polysaccharides	Avidin, biotin
Lipids	*Nucleic acids, nucleotides*
Fatty acids	Single or double-stranded DNA, RNA
Phospholipids	(e.g. anti-sense oligonucleotides)
Glycolipids	
Other	
Conjugates or mixtures of the above	Therapeutic isotopes
Labels	"Stealth" molecules
Chromophores	

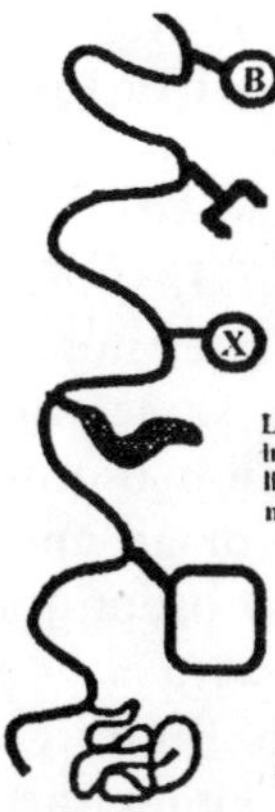

Figure 4.25. Schematic illustration of a variety of natural or synthetic biomolecules which have been conjugated to an "intelligent" polymer backbone. In some uses only one biomolecule may be conjugated, while in other uses more then one biomolecule is needed [115].

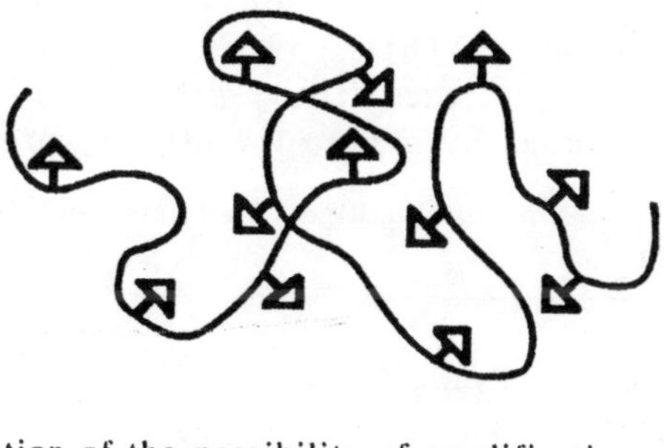

Figure 4.26. Illustration of the possibility of amplification of the action of a particular biomolecule (e.g. a drug or a ligand or an enzyme), by conjugating many biomolecules to the same "intelligent" polymer molecule. The polymer may be in solution, on a surface, or in a hydrogel [115].

rary, as in the case of small drug molecules.

There are many diverse biomedical and biotechnological applications of environmentally-sensitive "smart" polymeric biomaterials, containing or not immobilised biomolecules, Table 4.9.

4.3.1. Applications of stimuli-responsive soluble polymers

Soluble, stimuli-responsive polymers in aqueous solutions can be precipitated at specific environmental conditions. Even if such systems can not be used for chemical energy conversion into mechanical work, it can be useful as temperature of pH indicators, or as "on-off" light transmission switches. Of peculiar interest are the applications of the conjugated systems smart polymers – biomolecules, some of them being listed in Figure 4.27.

In the first example, a biomolecule is conjugated to the polymer, and them it is phase-separated from the solution by a small change in the environmental conditions. For instance, an enzyme in a bioprocess may be easily phase-separated and recycled, also permitting recovery of the product at the same time.

In the second example, a recognition biomolecule or receptor ligand such as a cell receptor peptide or an antibody is conjugated to an intelligent polymer and used in a precipitation-induced affinity separation process. When mixed with a complex solution, the conjugate will selectively complex its binding partner and then it can be readily and cleanly separated under the action of adequate stimulus (e.g. a small change of temperature), which causes the complex precipitation. A.S. Hoffman and co-workers applied the thermally-induced affinity precipitation process to recover IgG from solution using a poly(N-isopropyl acrylamide)-protein A conjugate, and are currently using it to recover CD-44 cell receptors from membrane lysates [116].

In the third example, the affinity precipitation principle is extended to an immunoassay, where the affinity precipitates the con-

Table 4.9. Some applications of "intelligent" polymers and their combinations with biomolecules [113]

A. *Polymers in solution*

- optical indicators (sensors, switchers)
- precipitation separations
- affinity precipitation (separations, sensors, and diagnostics)
- phase-transfer catalysis
- binding to an stimulating cells (cell separations, endo- or exocytosis, lysis, etc.)

B. *Polymers on surfaces (Physically adsorbed-desorbed polymers, graft polymers, gels surfaces)*

- wettability changes
- cell and protein attachment/detachement
- bioactive surfaces (immobilised enzymes)
- affinity separations
- permeation switchers in microporous membranes
- optical indicators (sensors, switches)

C. *Homogeneous or heterogeneous hydrogels*

- separations (size or affinity)
- drug delivery (pulsed, cyclic, controlled release)
- immobilized enzymes, cells (bioprocesses, implants, therapeutic devices)
- permeation switches (molecular pores)
- robotics

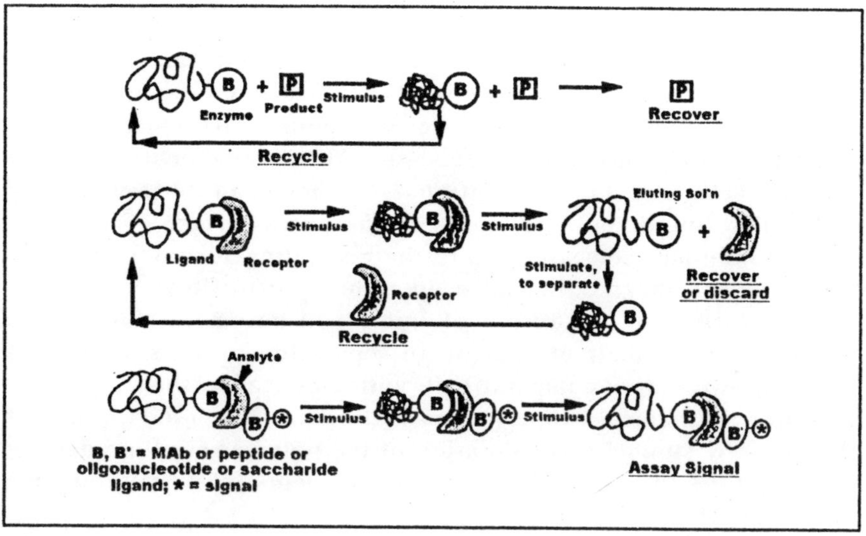

Figure 4.27. The use "intelligent" polymers-biomolecule conjugates in application including: bioprocesses and downstream separations, affinity separations, sensors, diagnostics and environmental water processing [115–117].

jugate/complex of (polymer-firstantibody/antigen/second, labeled antibody [117]. Such "intelligent" polymer-based affinity separations are potentially more efficient than traditional methods, such as affinity chromatography or the ELISA immunoassay.

When soluble smart polymers are mixed with lysosome or cell suspensions, they may be phase-separated by a stimulus, and may interact with lysosomal or cell membranes by hydrophobic interactions [118]. After the conjugate interacts with the membranes and phase-separation take place, a gel may be formed and cells can be reversibly cultured on surfaces [119].

Another kind of "intelligent polymer molecule in solution is based on the incorporation of the key chemical groups of a know recognition sequence from a natural biomolecule, randomly along a water-soluble polymer backbone, and in the same ratio as in the natural recognition molecule [120]. Under the action of external stimulus the conformational changes of polymer chain can enhance the possibility of recognition.

4.3.2. Applications of stimuli-sensitive polymers deposed on solid surfaces

Stimuli-responsive polymers can be chemically-grafted or

physically-adsorbed onto solid polymer supports; they can rapidly change surface film thickness, wettability, or surface charge in response to small changes in stimuli such as, temperature, pH, or specific ionic concentration. These responses are usually much faster than for solids as hydrogels since the surface coating is very fine. In such way, the permeation "switches" can be prepared by depositing "intelligent" polymers on the pores surface of a porous solid membrane, and stimulating their swelling to block the pore flow or, conversely, collapse to open the pore to flow.

The grafting or adsorption of intelligent polymers onto porous membranes and their utilisation in separation processes shows a series of advantages comparatively with classical methods, such as: 1) better mechanical properties as hydrogels; 2) a good control of flow rate by stimulated modulation of the polymer swelling degree; and 3) allows the separation of the components from water/organic solvent mixtures.

Basically, the principle of chemomechanical swelling and shrinking of the pores, in the case of inorganic porous supports grafted with stimuli-sensitive polymers is related to the specific interactions between the micro and macromolecular solutes contained in the permeant and the polymer chains grafted onto the membrane, giving rise to significant conformational changes of the grafted polymer. Such interactions can also be stimulated by changing the solution pH and ionic strength, or by addition of metal ions or polymers able capable of forming complexes with the grafted polymers. For example, considering a porous membrane onto PMAA was grafted, at low pH the polymer is mostly unionised, assuming a compact conformation, and therefore the composite membrane exhibits high permeability, Figure 4.28a. In contrast, if the permeate is of high pH, PMAA is almost completely ionised and assumes an extended conformation. The membrane's pores are covered by polymer chains and the apparent permeability is low, Figure 4.28 b. The polymer can also form chelate complexes with di- and trivalent metal ions, Figure 4.28 d, or form polymer complexes with complementary polymers, like PEG, through cooperative hydrogen bonding, Figure 4.28 c.

On the same principle, composite membranes based on the polyacrylamide gel deposited in the pores surface of a ceramic membrane of a silice and alumina were synthesised. Thus, acrylamide was polymerized with N, N'-methylenbisacrylamide, which acts as crosslinker [122]. In the same purpose, it was used of a flat innorganic support based on porous glass ($\phi_{m,pores} \cong$

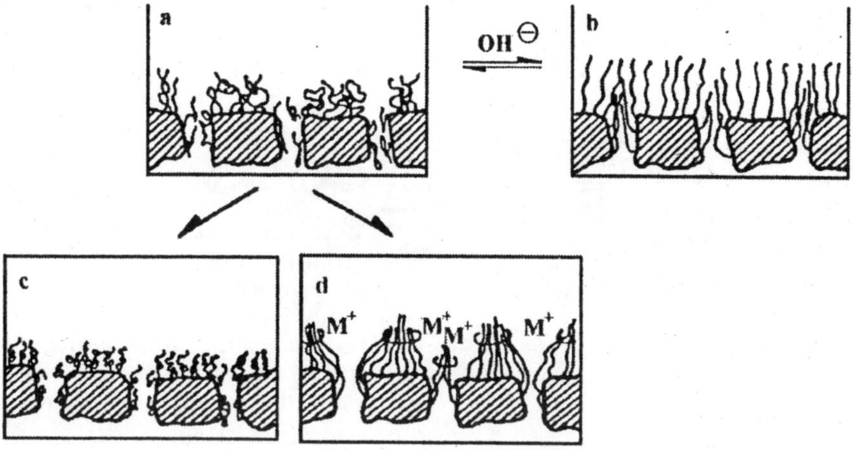

Figure 4.28. Schematic illustration of a "chemical valve" membrane prepared with graft polymer on a porous support [121].

6000 Å) with crosslinked poly(N-isopropylacrylamide) by a special manner [123].

If proteins or cells are exposed to "intelligent" polymer surfaces, which are in swollen or collapsed state, they usually will preferentially adsorb on the more hydrophobic surface composition, Figure 4.29 [124].

One potential use of a such system is to reversibly culture cells on a chemically-grafted LCST polymer surface; after the cells have been cultured for a while, they can by detached simply by reversing the stimulus, and converting the surface to the more hydrophilic condition [125–127].

The peptide cell receptor ligand, RGD, has been conjugated to special temperature-sensitive polymer compositions (having LCSTs close to ambient temperature) which have been matched to specific polymer support surfaces, for thermally-reversible cell culture based of temperature stimulated, polymer sorption-desorption, Figure 4.30 [128].

4.3.3. Applications of stimuli-responsive hydrogels

The most extensive work on "intelligent" polymers has been carried out on stimuli-responsive hydrogels, particularly those based on pH-sensitive and thermally-sensitive monomers. Their most important applications are related to desalting and/or dewatering of protein solutions, delivery of drugs, "on-off" immobilised enzyme reactors, microrobotics and artificial muscle.

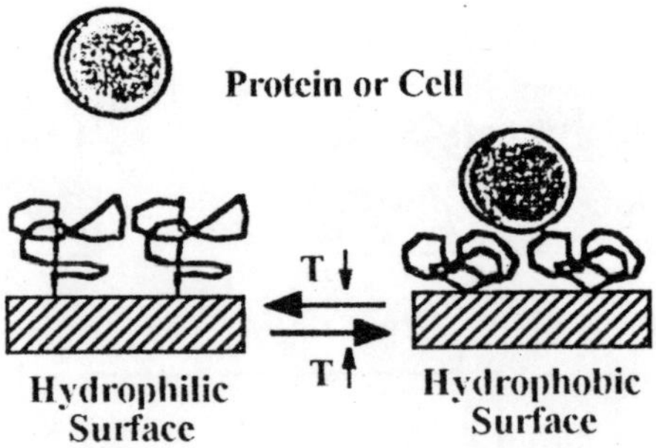

Figure 4.29. Illustration of the use of a stimulus to convert a surface coated with an "intelligent" polymer from hydrophilic to hydrophobic, and thus from a protein- or cell-repelling surface to one which is more attractive to the protein or the cell [124].

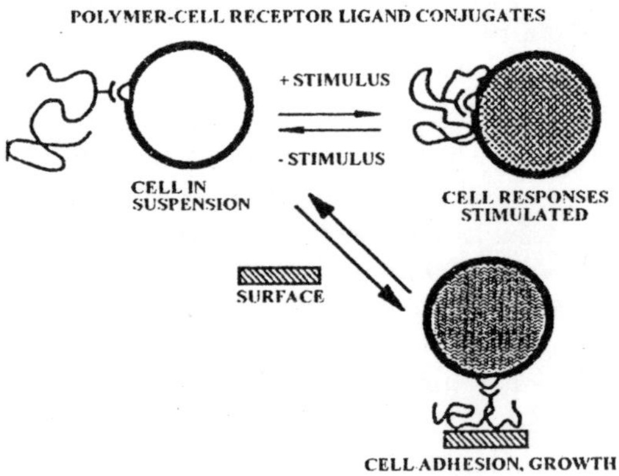

Figure 4.30. Schematic illustration of a cell receptor ligand-"intelligent" polymer conjugate binding to the cell membrane receptor, and stimulating various cell responses, including reversible cell culture on a surface, when stimulated to precipitate or redissolve [128].

Some of these applications are illustrated in Figure 4.31.

When enzymes are immobilised within smart hydrogels, then cyclic changes in external stimuli can lead to "on-off" activity of the enzyme due to the cyclic collapse and re-swelling of the hydrogel pores. This action can also be used to enhance mass transport of substrate *into* and product *out* of the immobilised enzyme hydrogels.

If the enzyme is immobilised within a pH-sensitive gel, and the enzyme substrate reaction produces a local micro-environmental pH change, then the actual "stimulus" for the resultant swelling or deswelling of the gel is the substrate concentration in the external solution. This type of gel can be used either as a biosensor or as a permeation switch, e.g., to permit release of a drug such as insulin through a swelling gel as response to an increase of a systemic metabolite such as glucose [129].

The pore size in an "intelligent" polymer hydrogel can be controlled by the environmental conditions and the hydrogel composi-

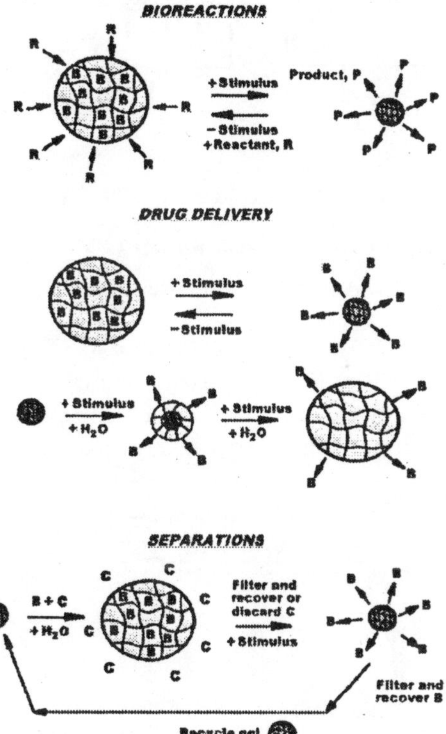

Figure 4.31. Schematic illustration of various applications of "intelligent" hydrogels in medicine and biotechnology [129).

tion. Thus, such gels can be used to separate molecules on the basis of size. This process can be used to selectively recover a desired compound or to selectively remove an undesirable compound such as toxin.

4.3.3.1. Chemomechanical valve for water circulation

O. Hirosa and co-workers constructed a thermally-sensitive self-regulating valve, using a poly(methyl vinyl ether) gel (PMVE). Whenever inside of valve is running cold water, the gel-made sphere from left side compartment is in swollen state (PMVE is

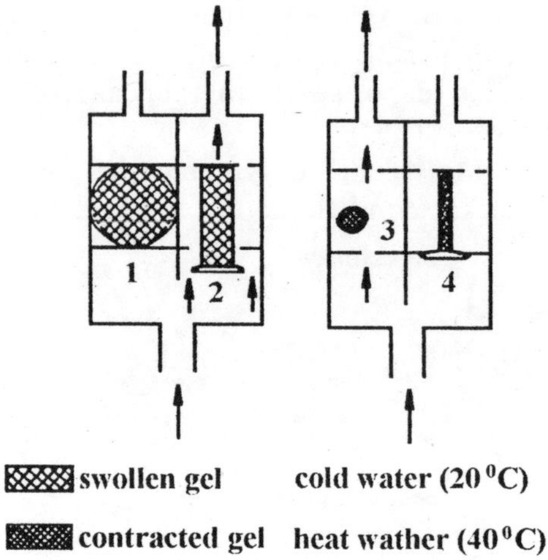

Figure 4.32. Automate thermo-sensible valve [130].

characterised by LCST) and blocks the water flowing through this way, Figure 4.32 [130]. The opposite situation is happened when worm waters enters inside; the sphere contracts allowing the water flow through this part while in the right compartment the water passing is blocked due to the contraction of the-gel piston from this compartment. Thus, always the cold water runs through the right compartment and the worm water through the left one, respectively.

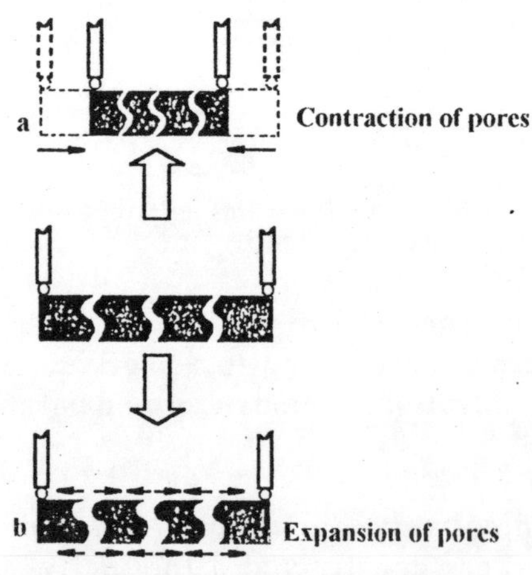

Figure 4.33. Schematic representation of chemomechanical contraction of a PMAA membrane by polymer-membrane complexation: (a) isotonic contraction); (b) isometric contraction [131].

4.3.3.2. Chemical valve

The function of the above-described device is based on the isotonic contraction of polymer gels. If the chemomechanical contraction is developed isometrically, the contractile stress appearing in the membrane should expand the pores through which water and dissoluted components permeate. Such systems are widely used in ultrafiltration membranes having chemomechanically expanding and contracting pores. A schematic illustration is given in Figure 4.33.

A PMAA membrane was treated with small amounts of PEG, of various molecular weights. PMAA membrane exhibits a marked increase in water permeation as soon as the membrane is contacted with the PEG solution and it maintains a high permeability for prolonged periods. If the membrane is rinsed with alkali solution, pH about 8, to dissociate the complex and remove the PEG, it returns to the initial low water permeability [131]. The best results were obtained when PEG having the molecular weight of 3000 was used for complexation. It was established that the complexation takes place not only on the surface, but also inside the swollen membrane due to the penetration of PEG molecules through the expanded pore channels.

Thus, the chemomechanical PEG treated PMAA membrane can behave as a "chemical valve", expanding and contracting the pore size. Flow tests with hemoglobin and albumin solutions were conducted on this chemical membrane. Figure 4.34 shows that both

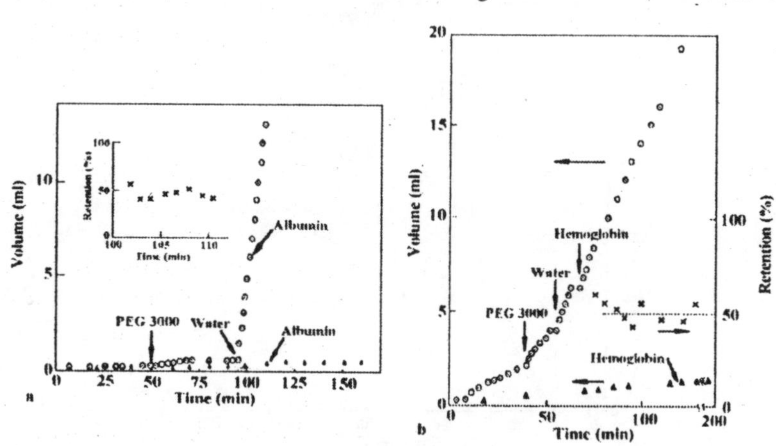

Figure 4.34. Effect of chemomechanical contraction of PEG treated PMAA membranes on albumin (a) and hemoglobin (b) permeation [131]: (a) PEG added: M_w=3000; 5×10^{-2} mol×l^{-1}; albumin added: 2.5×10^{-2} % solution; (b) PEG added: M_w = 3000, 5.8×10^{-2} mol×l^{-1} solution; hemoglobin added 2.5×10^{-3} solution. In both cases the transmembrane pressure was 0.2 kg/cm^2.

protein solutions permeate the membrane without decreasing the high permeation velocity owing to the expansion of pores, whereas in the case of an untreated PMAA membrane, the proteins clog the membrane rapidly and hardly permeate.

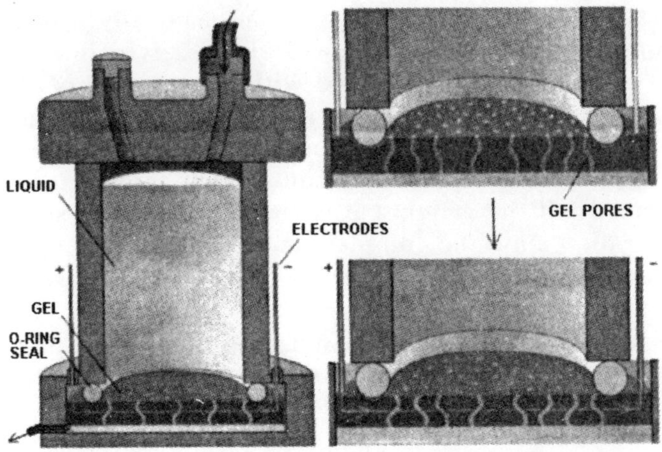

Figure 4.35. Prototype of electroactivated "chemical membrane" and its working principle [133].

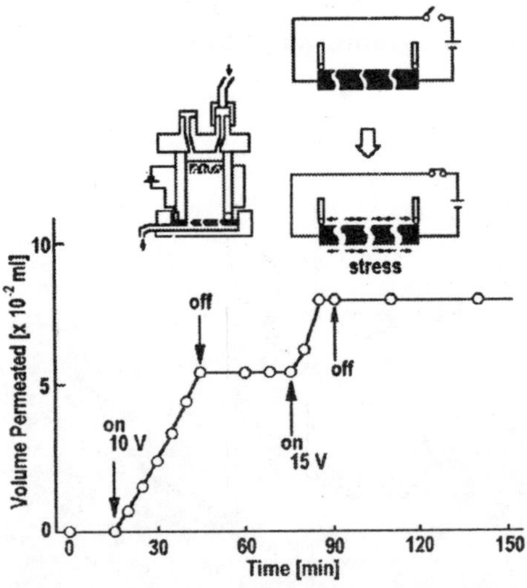

Figure 4.36. Apparatus for electrodriven "chemical valve" membrane (top) and the change in permeability of a polyacrilic acid-polyvinyl alcohol composite membrane as the DC source is switched on and off [133].

4.3.3.3. Electro-activated chemical valve

Y. Osada and M. Hasabe designed a chemical valve that contracts and expands its pores as response to the electrical current. Under the normal conditions, the gel is swollen and impermeable to liquids. When an electric current passes through the gel, its larger internal pores open up, allowing liquid to flow through. If the gel is very homogeneous, its pore size can be controlled so precisely that it passes small molecules while blocking large ones, Figure 4.35 [132]. The effect of applied current on the quantity of water that penetrates a poly(acrylamido-2-methylpropanesulfonic acid), PAMPS, by applying a DC of 6.5 V, is illustrated in Figure 4.36 [133].

4.3.3.4. Extraction and separation processes using chemomechanical systems

E. Cussler proposed an interesting application that uses the difference of adsorption parameters between the ionised and unionised forms of polyelectrolyte gels refers to extraction of pigments particles from a large volume of aqueous dispersion [134, 135]. Thus, a diluted aqueous solution, containing the dispersed

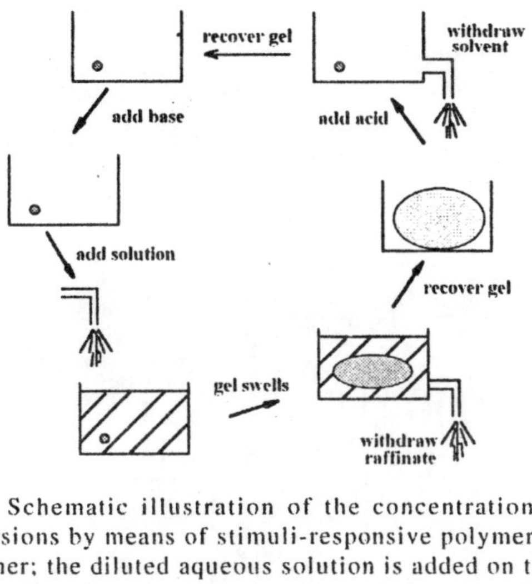

Figure 4.37. Schematic illustration of the concentration process of different diluted dispersions by means of stimuli-responsive polymer gels. The cycle starts in left up corner; the diluted aqueous solution is added on the small gel particles. The gel swells by water absorption but it excludes the voluminous solute. After removing of thus concentrated solution, over the swollen gel particles a small amount of acid is added and the adsorbed water is released due to the gel contraction. After adding of a small amount of base, for carboxylic group ionisation, the gel particles may enter into a new cycle [135].

pigment, was effectively concentrated using a partially ionised polyacrylamide gel. The sequence of steps implied in this process is schematically presented in Figure 4.37. In the swelling step, the small particles of pigment can not penetrate the gel pores, and the gel retains only the mobile component (water). In the second step the swollen gel particles are removed and placed into another vessel. Here a small amount of acid is added, the gel particles suddenly collapse, releasing the retained water. The process is repeated until the pigment solution reaches the desired concentration.

Thermally-sensitive hydrogels were also used as dehydration reagents. In this case, the water absorption and desorption is led in the proximity of phase-transition temperature of used gels. Polymer gels based on N-isopropylacrylamide have been used for concentrating albumin, gelatine or dextrane solutions [135].

The change of reciprocal interactions between the polymer gels and different organic compounds was also exploited by means of thermally-sensitive gels. In the vicinity of transition temperature, the hydrophile/hydrophobe equilibrium drastically modifies, and this behaviour stays on the base of separation and concentration of hydrophobic organic substances, such as non-ionic surfactants from residual waters. Thus, nonylphenyl-polyethylene glycol was removed from residual water using a PMVE gel fixed on alumina by g irradiation. The schematic diagram of this process is presented in Figure 4.38. It was found that the adsorbed amount of surfactant at 40 °C is of about 40 times higher than that retained at 5 °C.

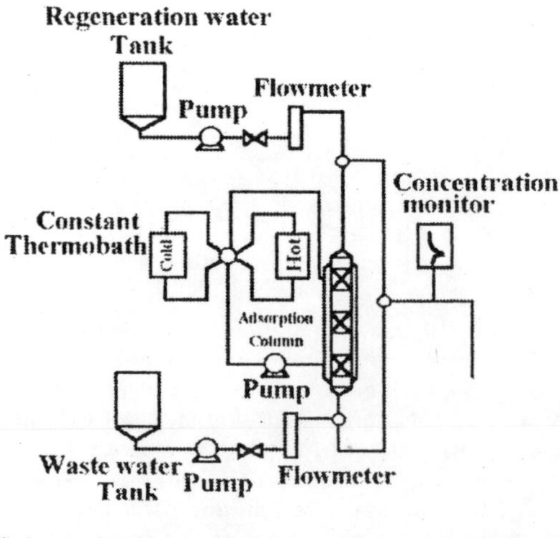

Figure 4.38. Schematic diagram of reversible systems of adsorption [135].

It worth to be mentioned that the separation processes based on thermally-sensitive chemomechanical gels are more convenient than the classical ones, being less expensive.

4.3.3.5. Drug delivery systems and control of enzymatic activity based on chemomechanical systems

The functional, from the therapeutic viewpoint, drug release systems suppose the achievement of some "active devices", capable to perceive the organs requirements and to release the corresponding dose of medication in the proper moment and in the specific place.

The controlled release of drugs can be realised with many types of active devices, such as: mechanical and osmotic pumps, systems of controlled diffusion, chemically-controlled systems composed from bio- or non-biodegradable polymers, magnetically-controlled systems, and not in the last instance chemomechanical systems, Figure 4.39 [136, 137].

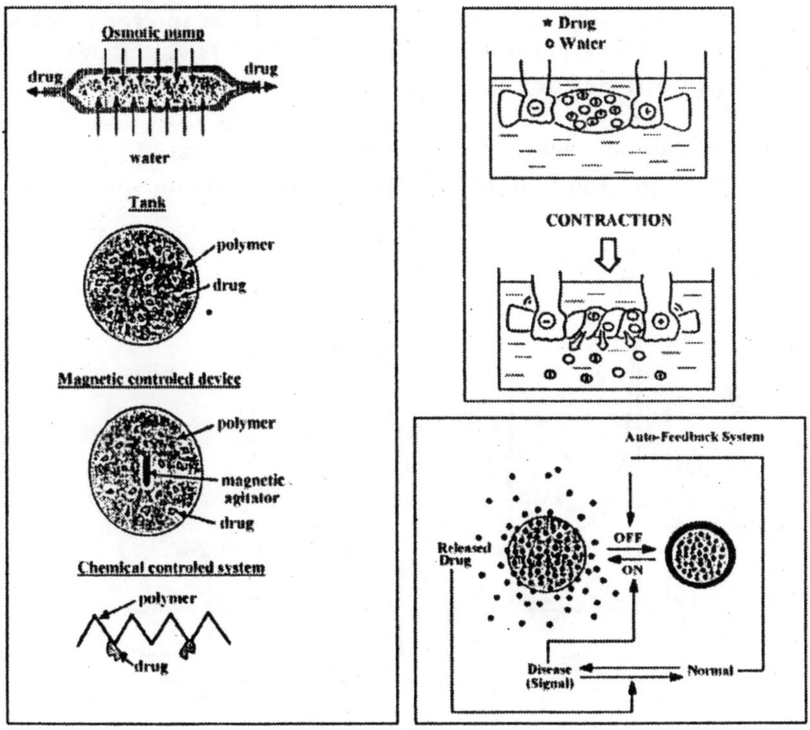

Figure 4.39. Comparative schematic representation of the classical systems of drugs controlled release and the corresponding chemomechanical systems [137].

In the case of drugs immobilisation within smart polymer gels, which show reversible swelling/shrinking cycles at external stimuli (pH, temperature, electric current, etc.), these ones are released in bloodstream due to the contractile forces developed, which release the fluid swelling the polymer network that also contains the dissolved drug.

R.A. Siegel and co-workers have developed a simple, gel-based system protecting acid-sensitive medication from the hostile environment of the stomach. Their gels shrinks when exposed to low pH, but in more alkaline environment of the intestines it expands and becomes permeable, allowing the encapsulated drug to diffuse under the proper conditions.

W. Kim and co-workers prepared a thermally-sensitive gel based on poly(N-isopropylacrylamide), NIPAm, and butyl methacrylate and used it for determination indometacine and insulin release rates, controlled by temperature changes, Figure 4.40, [139, 140].

Hydrogels able to release drugs or biomolecules in response to variations in electric field. Thus, Y. Osada and co-workers studied the electro-activated release of pilocarpin from microparticles of sodium salt.of poy(acrylic acid) gel, Na-PAA [141]. It was found that insulin diffuses out of a gel made of weakly crosslinked polyelectrolyte gels when electric current is turned on, but the flow ceases immediately when it is turned off. This kind of gel could be the basis on an implantable insulin pump with no moving parts [142].

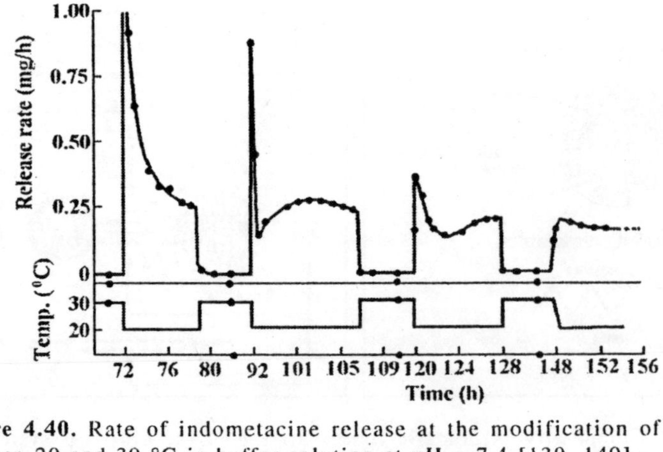

Figure 4.40. Rate of indometacine release at the modification of temperature between 20 and 30 °C in buffer solution at pH = 7.4 [139, 140].

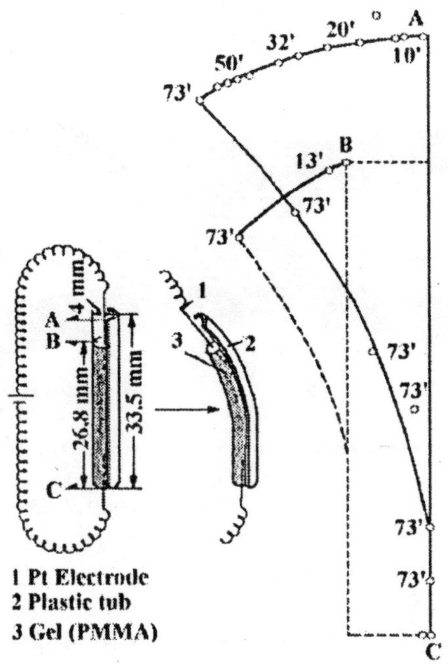

Figure 4.41. Schematic illustration of a polymer composite functioning like a "bimetal" and the profile of its "bending". Gel: poly(methacrylic acid); plastic: poly(vinyl chloride); electric field: 0.25 V/cm; the numbers indicate time (min) elapsed after electric field was imposed [143].

4.3.3.6. Sensors

Y. Osada and M. Hasabe constructed a polymeric switching device functioning like a "bimetal", from a laminated polymer gel/plastic structure [106]. The contractile force developed in the gel by an electric stimulus through a pair of platinum-wire electrodes bends the "nonshrinkable" poly(vinyl chloride) tube, as illustrated in Figure 4.41, The rate and extent of bending were enough large. Thus, the composite device consisting of crosslinked poly (methacrylic acid) (degree of swelling 3 times) and poly(vinyl chloride tube) bent as much as 24° after 30 min, and 43° after 70 min. It recovers the original linear form after dipping it into water.

In the case of the next described sensor, the change of the applied force induces the modification of pH, which in monitored by ^{19}F-NMR spectroscopy. The sensor consists from a non-permeable but flexible membrane, encapsulating o polyelectrolyte gel soaked by an ionic solution, Figure 4.42.

The polyelectrolyte gel is hydrated into an aqueous mixture con-

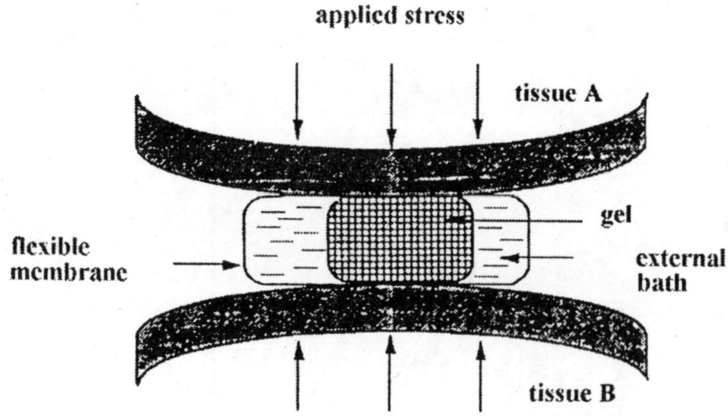

Figure 4.42. Schematic representation of the sensor used for measuring the contact force, consisting from a flexible and impermeable membrane that contains a polyelectrolyte gel embedded by a ionic solution. A and B - surfaces of two tissues in contact with the sensor [143].

taining Na^+, Cl^-, and difluoromethylalanine. The last compound functions as a sensible indicator of pH, allowing the intra and extracellular measurement of this parameter using ^{19}F-NMR spectroscopy. Since the majority of ionizable groups of the polymer chains are dissociated at pH = 7.5, at equilibrium the concentrations of mobile ions [Na^+], [HO^-], [H_3O^+], [Cl^-] and the two forms of pH indicator are different within the fluid from interstitial compartment and the external one, respectively. If a higher force is applied to the sensor, the polymer gel will exclude a part form interstitial liquid into external bath, all the mobile ions being, including the H^+ ion, redistributed between the interstitial liquid and external bath. The new attained equilibrium is reflected in the change of NMR spectrum of the pH-indicator compound. Both empirical and mathematical methods have been used for the sensor calibration. This sensor could be used in clinical applications, including the monitoring of contact tension between tumour an adjacent tissues, tumour and bone, between muscle and fascia, etc.

It is well known the human tissues are capable to convert the mechanical energy into electrical energy, using as transducer soft and wet piezoelectric materials. Starting from the assumption that piezoelectricity should be one of intricate properties of certain animal and vegetal tissues, consisting from a network of proteins and polysaccharins the interstitial space being filled with water, Y. Osada and co-workers designed an artificial system of mechani-

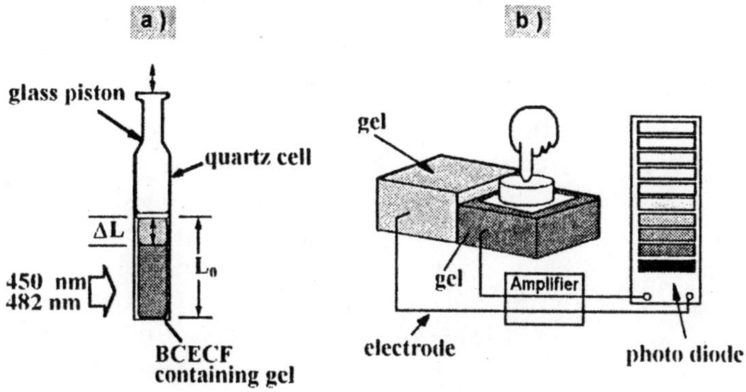

Figure 4.43. Schematic illustration of the devices used for following the pH changes during gel deformation (a) and of tactile system able to illuminate the diodes (b) [143].

cal energy conversion into electrical energy. Thus, two gel pieces – made from a poly(acrylamide-*co*-acrylic acid) copolymer (molar composition 1:1), crosslinked with 2% molar N,N'-methylene-bisacrylamide – were placed into a cell, being in direct contact each to another. One of the gels was gentle compressed using a glass piston. The cell volume was higher than gel volume, and during deformation the gels fitted the all cell space. No water release was observed during compression.

In order to avoid any interference in measuring the electrostatic potential of gel, the pH changes have been followed using 2,7-bis-(carboxyethyl)carboxyfluorescein (BCECF), which changes its absorption peaks from 450 nm and 482 nm in accordance with pH changes; the absorption ratio at the two wavelengths (A_{450}) and (A_{482}) varies linearly in the pH range from 2 to 4. In Figure 4.43 the device used for measuring the pH of the stressed gel is pre-

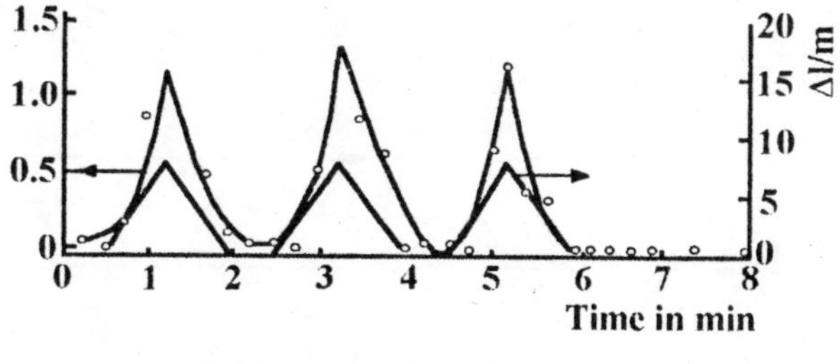

Figure 4.44. Profile of time-changes of the electric potential at the interface of two gels, one deformed and the second one in free state, respectively. Upper curve: electric potential; lower curve: deformation. Gel dimensions: 15×10×10 mm [143]. Swelling degree, 167.

sented. The potential difference between the deformed and non-deformed gels was measured by inserting a pair of very thin platinum-wire electrodes in each gel piece. Since, no water is released during deformation, the constant variation of pH was associated with the change of ionization degree of carboxyl groups with the stress applied. Being compressed in one direction the gel is constrained to expand laterally; consequently its free energy increase and the macromolecular chains entropy diminishes, respectively, which is compensated by an increase of ionization degree.

Mechanically-induced ionisation was observed only to the weak polyelectrolyte gels, capable to sensible modify the ionization state in function of chains conformation. In this experiment, when the gel is deformed the extra-protons migrates into non-deformed gel through the interface until the Donnan equilibrium is attained, when a difference of potential was observed, Figure 4.44. The experiment from Figure 4.43 b demonstrates that this system of tactile sensing, based on polymer gels, can light the cascade of bulbs proportionally with the applied stress.

By their similitude with the receptor from living organisms these mechano-electrical systems, composed from polymer gels, open new possibilities of investigation of the artificial perception, with potential application in robotics or prosthesis.

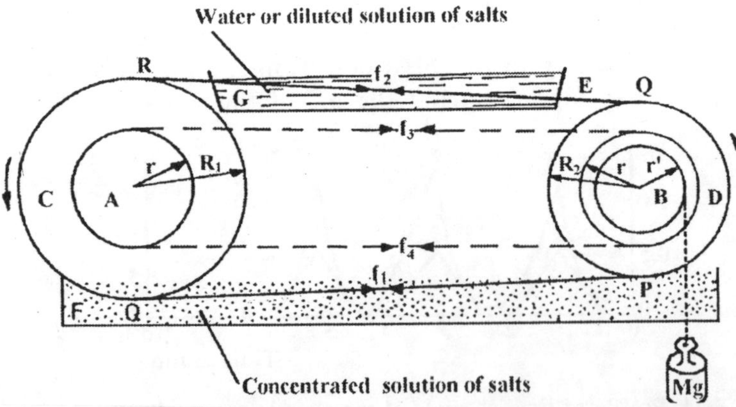

Figure 4.45. Schematic illustration of a chemomechanical engine according to A Katchalsky [144].

4.3.3.7. Devices with moving parts based on chemomechanical systems

4.3.3.7.1. Micromotors

A. Katchalsky constructed the first engine continuously working on the principle of alternative contraction and relaxation of contractile fibers. The motile system of this engine consists from partially crosslinked collagen fibers. These fibers alternatively treated with concentrated solutions of salts such as LiBr, KSCH, or urea, and water, respectively, exhibit rapid and reversible contractions and relaxation. The chemomechanical behaviour is relate to reversible chemical meeting of the fibers and crystallisation, (Figure 4.45).

At Weizmann Institute several prototypes functioning on this principle have been constructed. Figure 4.46 shows photographs of a family of experimental engines. In Figure 4.45 a closed loop of fiber E is wound around pulleys C and D, respectively. A part of

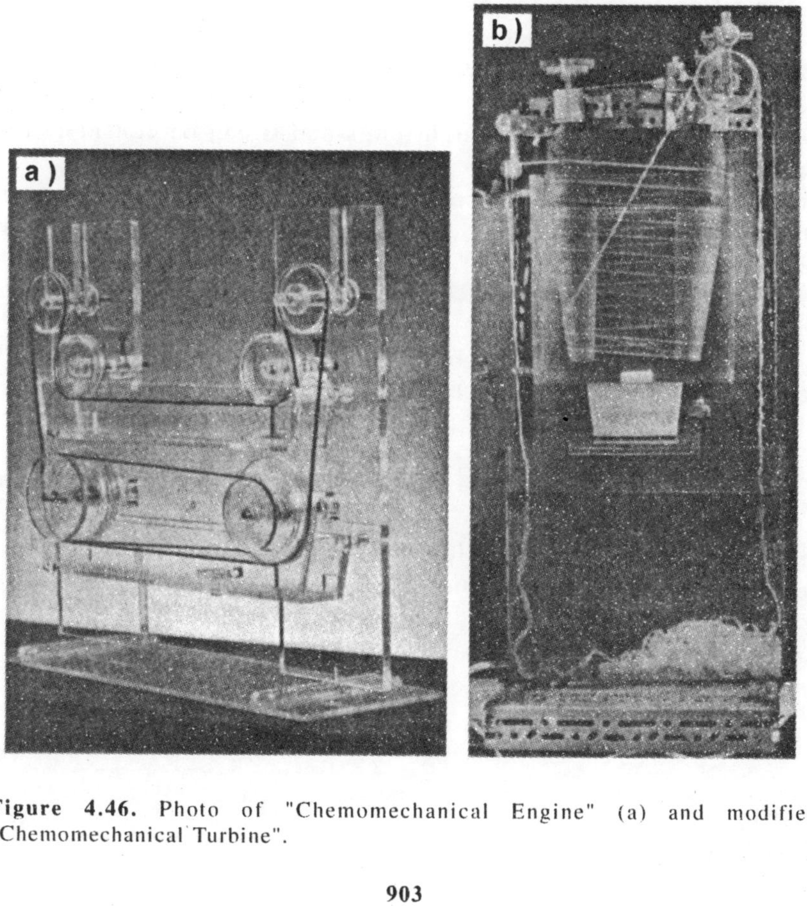

Figure 4.46. Photo of "Chemomechanical Engine" (a) and modified "Chemomechanical Turbine".

fiber loop is laced into the inferior bath containing a strong salt solution and another part is introduced into the superior bath, containing water. The pulleys of transmission A and B of equal diameters are connected through an unelastic loop and fixed to the main pulleys, C and D, respectively. If pulley C is dipped into a strong salt solution (such as 8 – 10 M aqueous LiBr), the collagen undergoes contraction and exerts equal forces on the rims of pulleys C and D. Since the radius of C is larger than that of D, a net rotary moment acts on the compound pulley C-D and causes o counterclockwise rotation. The rotation brings a new part of the collagen belt into the salt solution, washing out the salt and relaxing the collagen. While the engine is working, salt is transferred from the concentrated solution to the dilute one.

The mechanical efficiency was calculated into an elegant manner, starting from the following hypothesis [144]:

1. The force f_1 that acts on the fibers from bath F is constant on the portion O–P, and similarly, f_2 is constant on the portion Q–R in bath G;

2. The two baths are enough longs and it can be considered that the fibers attain the equilibrium length, $\bar{l}_P$ in the point R, at the emergence from salt bath; correspondingly, they attained the equilibrium length, $\bar{l}_R$ in the point R, at the emergence from water bath;

3. $\bar{l}_R$ = constant, on the R–O portion, and $\bar{l}_P$ = constant, on the portion P – Q, respectively.

The mechanical work performed by engine when a point of fiber loop performs o complete rotation is given by the following expression:

$$W = 2pr'nMg \tag{4.5}$$

where r is the radius of pulley B which lifted the attached load M; n - the number of rotations effectuated by the pulley B; g – gravitational acceleration. Since the pulleys B and D are rigid coupled, concomitantly a mass of collagen. m, is transferred by the pulley D into bath F:

$$m = \frac{2\pi R_2}{\bar{l}_P} \tag{4.6}$$

The mechanical work performed per unit mass of collagen transferred $\overline{W}$:

$$W = \frac{2\pi r' \nu Mg}{2\pi R_2 / \bar{l}_P} = \nu \cdot M_g \cdot \bar{l}_P \cdot \frac{r'}{R_2} \tag{4.7}$$

When the engine works at constant speed, the net moment that acts any of pulleys A-C and B-D is equal to zero. Neglecting the friction forces, it can be written:

$$\begin{aligned} R_1(f_1 - f_2) + r(f_3 - f_4) &= 0 \\ R_2(f_1 - f_2) + r(f_3 - f_4) + r'Mg &= 0 \end{aligned} \tag{4.8}$$

therefore:

$$(R_1 - R_2)(f_1 - f_2) = r'Mg \tag{4.9}$$

By replacing equation (4.9) in equation (4.7):

$$\overline{W} = \frac{R_1 - R_2}{R_2}(f_1 - f_2)\bar{l}_P \tag{4.10}$$

In order for the engine to operate steadily, the amount of fiber that enters into the salt bath for each rotation of wheel C should be equal to the amount which leaves the bath with wheel D, or:

$$\frac{2R_1}{\bar{l}_R} = \frac{2R_2}{\bar{l}_P} \tag{4.11}$$

Replacing now equation (4.11) in equation (4.10):

$$\overline{W} = (f_1 - f_2)(\bar{l}_R - \bar{l}_P) \tag{4.12}$$

and the efficiency, η, of the conversion of mechanical energy into mechanical work is given by:

$$\eta = \frac{\overline{W}}{\Delta n_s \cdot \Delta\mu_s + \Delta n_W \cdot \Delta\mu_W} = \frac{(f_1 - f_2)(\bar{l}_R - \bar{l}_P)}{\Delta n_s \cdot \Delta\mu_s + \Delta n_W \cdot \Delta\mu_W} \tag{4.13}$$

where Δn_S and Δn_W is the number of moles of salt and water, respectively, transferred by the collagen mass unit; $\Delta\mu_S$ and $\Delta\mu_W$ – chemical potential differences in the two baths.

The maximal power delivered was about 0.03 W/g of collagen, corresponding to a mechanical efficiency of about 40%. This value is practically identical with the maximal power generated, per 1 g of contractile proteins, of frog's sartorius muscle [145].

The contraction of polymer gels under the action of electric current and the reswelling after its turning off was exploited by T. Hayashi and co-workers which constructed a micromotor having as active parts spherical gels of sodium salt of poly(acrilic acid), PAA-Na. In principle, this device converts the swelling and shrinking of gel balls into a rotational movement of an eccentric shaft, which is in contact with the balls. Four such polymer balls with diameter of 2 mm are placed on the internal surface of a cylindrical container at equal distance each to another. The shaft is also placed inside of container, slightly eccentric with respect to the container center. An aqueous solution of Na_2SO_4 (0.01 mole/l) is introduced in container covering the gel balls. The DC is supplied using two platinum electrodes; one of them being placed on the internal circumference of container and the second one covering a half of shaft circumference. In such way, all the time half of gel balls are under electrical field action (continuously shrinking) and another half is out of electrical field (continuously swelling). The gel ball shrinks under the action of DC up to 50 –70% from its initial volume. The swelling and shrinking degrees of the gel balls are different, depending on the contact period between gel ball and electrodes. Figure 4.47 schematically presents this device [146].

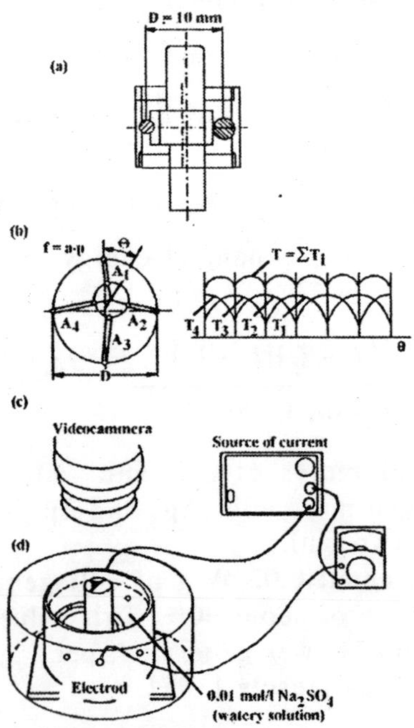

Figure 4.47. Constructive principle and experimental arrangement of T. Hayashi micromotor [146]: (a) cross section of turning device; (b) functional principle of the gel-motor; (c) experimental arrangement.

A torsion moment is generated by the interaction between the swelling/shrinking balls and the eccentric shaft, which varies in time, its minimal value, T_{min}, being given by the following expression:

$$T_{min}=a/\tan(\pi/n)p\Delta r \tag{4.14}$$

where: a – surface of contact between gel ball and shaft; p – the pressure exerted at the contact gel ball-shaft; Δr – shaft's and eccentricity; n – angle of rotation.

The process efficiency can be calculated from the relation:

$$\eta = 1-(\mu r + \mu_s r_s + T_0/w)\Delta r \tag{4.15}$$

where: μr and $\mu_s r_s$ – friction coefficients of the gel spheres and shaft, respectively; T_0/w – the ratio between static friction and weight.

The shaft rotation was followed in time, using a video camera, by measuring the total angle of rotation. Even if the shaft slowly rotates, the presented example clearly demonstrates the possibility of construction of micromotors based on electro-activated conversion on mechanical energy, Figure 4.48.

The interaction of positively charged surfactants with negatively charged gels was used by Y. Osada to construct an inchworm-like device that moves by repeatedly curling and straightening itself, called by author "gel looper". The system is driven by an ensemble of electro-activated interactions between the surfactant molecules and negatively charged sites on the gel [146].

Thus, a thin strip of gel based on poly(acrylamide-*co*-acrylamido-2-methylpropanesulfonic acid) is hanged from a plastic road

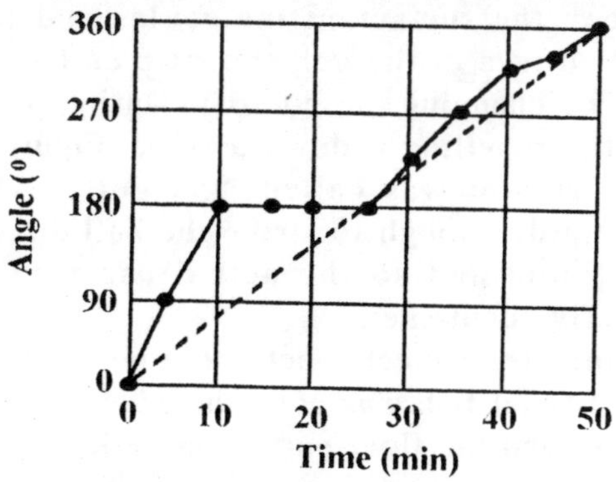

Figure 4.48. Experimental results for a complete rotation of the shaft [146].

by means of metal hooks while immersed in an aqueous solution containing Na_2SO_4 and surfactant, i.e. cetylpyridinium chloride, $C_{16}PyCl$. Parallel carbon electrodes, placed above and bellow the gel strip controls its motion. The polarity was changed at each 2 s. When a voltage of 20 V is applied across the electrodes, the surfactant molecules migrate toward the negative electrode. On their way they encounter the negatively charged surface of the gel and attach themselves to it, causing the gel to contract. Since the surfactant molecules alight preferentially on the gel side that faces the positive electrode, the gel contracts mainly on that side and curls up. When the polarity of the electric field is reversed the surfactant molecules are released (going back into the solution), and the gel strip straightens. A sawtooth pattern on the top of supporting rod ensures that the looper moves forward with each cycle of bending and straightening, Figure 4.49.

The maximal speed of this linear actuator in about 25 cm/min at an applied DC of 20 V (15 mA/cm^3). The moving speed depends on the applied current, surfactant concentration and solution ionic strength. The maximal speed corresponds to the concentrations of 3×10^{-2} mole/l Na_2SO_4 and 1×10^{-2} M cetylpyridinium chloride, $C_{16}PyCl$, respectively.

4.3.3.7.2. Microrobots

Y. Osada demonstrated the ability of intelligent gels to act on their surrounding. Thus, he prepared a gel strip from the same material as the "gel looper" and hanged it at one end. The strip of gel was immersed into an aqueous solution, having the same composition as in the case of "gel looper" bath. In this case, the one ėlectrode was placed in the front of gel strip and the second one behind of it. Under the influence of electric current, due to the asymmetrical interactions between the surfactant molecules and the polymer negatively charged surface of the gel, the strip curls first one way and then the other, by changing the polarity. During this transition, it strikes a golf ball, propelling it down a slope, Figure 4.50 [112]. This is way, this actuator was called "gel golf". Although the "club" material is sturdy enough to strike the ball directly, it mast be submerged in liquid and so the actual use would probably encased in a protective container.

On the same principle of electro-activated bending of ionizable polymer gels, an artificial fish was constructed. Its body was made from a PAA-PVA gel and the float part from an inert polymer. The fish's body had the following dimensions: 30 mm length, 10 mm

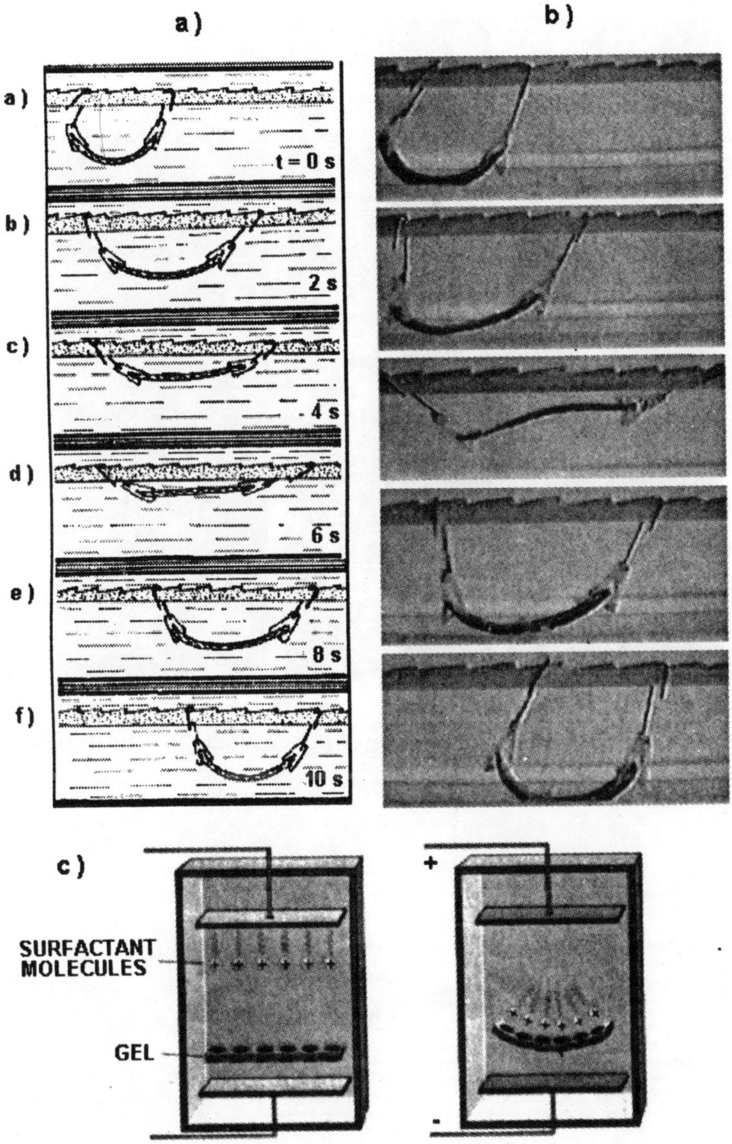

Figure 4.49. "Gel Looper", an inchwormlike device that moves by repeatedly curling and straightening itself under the cooperative interactions at the gel surface with surfactant molecules that drift through the surrounding solvent under the influence of an electric field [147]: schematic representation; (a) schematic illustration: a – d) stretching process; e – f) curling process; (b) and photos showing its movement; (c) schematic illustration of the interaction of surfactant molecules with the gel strip under the influence of an electric field [147].

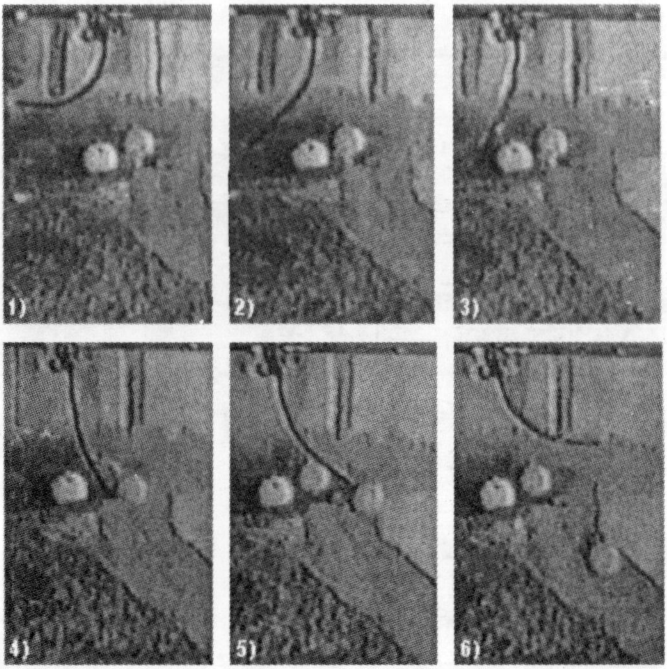

Figure 4.50 "Gel Golf" demonstrating the ability of an intelligent polymer gel to act on its surrounding [112].

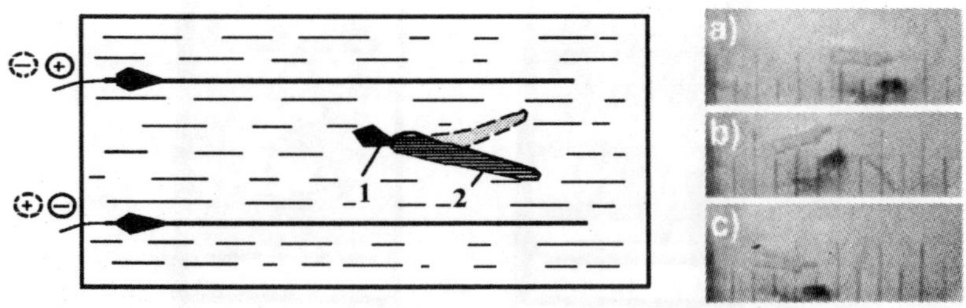

Figure 4.51. 'Gel fish'. Schematic representation of the electroactivated movement of the gel: 1) plastic-made floating part; 2) fish's body (hydrogel active part). In the right hand part, the photos showing the progressive change in motion of the gel fish at a voltage of 10 V. Electrode distance: 70 mm, 0.25 Hz. Photographs taken every 8 s.

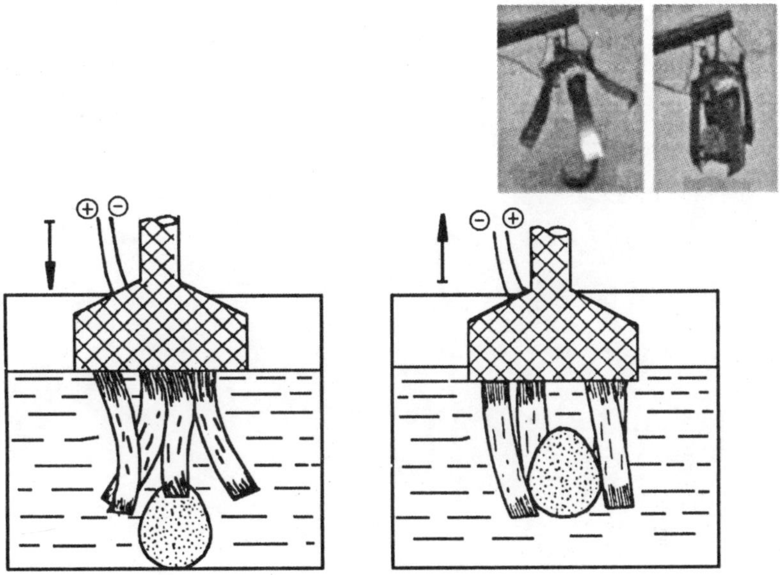

Figure 4.52. Schematic representation of an end-effector gripper before lifting (a) and lifting an egg (b) under electrical activation using four finges made of PVAPAA hydrogel [148]. Photos in the right hand upper corner show an end-effector gripper lifting 10.3-g rock under 5-V, 25-mW activation using four 0.1-g fingers made of ion-exchange polymer gel-metal composite (IPMC).

width, and 1 mm thickness. The gel fish swims into an aqueous solution containing Na_2CO_3 by changing the polarity of electric field by means of a pair of electrodes disposed on one and other side of it. Fish swimming speed was about 2 cm/s by applying a potential difference of 50 V, Figure 4.51, [148].

A electro-activated microrobot that mimes the human hand was constructed. It has four rectangular fingers with the following dimensions: 60 mm length, 6 mm width, and 7 mm thickness. In order to bend the fingers a pair of electrodes was disposed on one and other side of each finger. The robot can easily lift a fragile object, for example an egg. The fingers are closed by applying a DC of 50V into an aqueous solution containing Na_2CO_3. By reversing the polarity the gel curls in the opposite direction releasing the egg, Figure 4.52.

A series of other, relatively recent, data concerning the smart gels behaviour have been obtained by the authors of this work [149–151].

Bibliography

691. J.W.Hearle and R.H.Peters, *Fiber Structure,* Eds. Manchester and London,The Textile Insitute, Butherworths,1964
692. W.G.Protasov and N.K.Baramboim, *Plast. Massy,* **2** (1969) 8.
693. K.Bischoff, *Rev. Roum. Chim.,* **13** (1962) 205.
694. S.A.Arakov, N.F.Bakeev, and V.A.Kabanov, *Vysokomol. Soedin.,* **A 15** (5) (1973) 1154.
695. N.K.Baramboim and V.N.Gordiev, *Vysokomol. Soedin.,* **2** (1960) 197.
696. Cl.Vasiliu Oprea, Cl.Negulianu, M.Popa, and Cr.Simionescu, *Plaste und Kautschuk,* **6** (1977) 604.
697. Cr.Simionescu, Cl.Vasiliu Oprea, Cl.Negulianu, and M.Popa, *Plaste und Kautschuk,* **10** (1977) 689.
698. S.Gurevici, *Teor. Prikld. Mekh.,* **9** (1978) 117.
699. J.Constable, J.C.Williams, and D.J.Burns, *J. Mech. Eng. Sci.,* **12** (1970) 1.
700. N.J.Mills and N.Walker, *Polymer,* **17** (1976) 335.
701. Cl.Vasiliu Oprea and M.Popa, *Bull. I.P.Iasi,* **27**(31) (1981) 115.
702. Yu.S.Urzimtsev, *Mekhanika Polimerov,* **3** (1967) 467.
703. T.Hancock, *Personal Narrativeon the origin and progress of the Caoutchuk in India,* Rubber Manufacture in England, London, 1857.
704. A.Mallok, *Proc. Roy. Soc.,* **46** (1889) 133.
705. C.Checheveau and F.J.Heim, *Franklin Institute,* **142** (1896) 187.
706. D.G.Geletian, D.G.Sarkissian, Yu.K.Kabalian, and N.D.Zakharov, *Kauciuk I Rezina,* **31**(11) (1972) 17.
707. T.Kempermann, *Gummi, Asbest, Kunststoffe,* **27**(8) (1974) 566.
708. B.N.Dynsburg, N.K.Baramboim, and B.P.Kozarenko, *Vysokomol. Soedin.,* **A17**(9) (1975) 2023.
709. W.Louer, *Kautsch., Gummi, Kunststoffe,* **28** (1975) 536.
710. P.Compagnon and C.R.LeBras, *Acad. Sci.Paris,* **212** (1941) 616.
711. G.L.Slonimskii and E.V.Rezstova, *Vysokomol. Soedin.,***1**(4) (1959) 534.
712. Cl.Vasiliu Oprea and M.Popa, *Mekhanika Compoz. Mat.,***4** (1982) 679.
713. P.J.Flory, *Principles of Polymer Chemistry,* Cornell Univ.Press, Ithaca, 1953.
714. A.Tager, *Physical Chemistry of Polymers,* MIR Publish., Moskva, 1960.
715. P.H.Lindenmayer, *J. Macromol. Sci.,* **B8** (1973) 361.
716. K.V.Ovicinikov, G.S.Markova, and V.A.Kargin, *Vysokomol. Soedin.,* **A11** (1969) 329.
717. G.S.Y.Yeh, *J. Macromol. Sci.,* **B6** (1972) 462.
718. E.W.Fischer, J.W.Wendorf, M.Dittenmayer, G.Lieser, and T.Voight-Martin, *Polymer Preprints, Div. Polymer Chem. Inc. Amer. Chem. Soc.,* **15**(2) (1974) 8.
719. V.V.Jinescu, *Proprietatile fizice si termomecanice ale materialelor plastice,* Vol.1 and 2, Ed.Tehnica, Bucuresti, 1979.
720. W.Pechhold, E.Liska and A.Baumgartner, *Koll. Z. Z.Polym.,* **250** (1972) 1017.
721. A.S.Lodge, *Rheol. Acta,* **7** (1968) 379.
722. C.F.Curtiss and R.B.Bird, *J. Chem. Phys.,* **74** (1981) 2016.
723. M.Doi and S.F.Edwards, *J. Chem. Soc. Faraday Trans.* **II** (1978) 1789, 1802, 1818.
724. J.D.Ferry, *Viscoelastic Properties of Polymers,* 3rd Ed., J.Wiley & Sons Inc, N.Y., 1980.
725. J.V.Mointrie, *J. Appl. Polym. Sci.,* **16** (1972) 2901.
726. C.D.Han, *J. Appl. Polym. Sci.,* **17** (1973) 1403.

727. R.T.Balmer, *J. Appl. Polym. Sci.*, **18** (1974) 3127.
728. B.Wunderlich, *Macromolecular Physics*, Vol.2, Academic Press, N.Y., 1976.
729. A.Windle, *MRS Bulletin*, Nov.16/Dec.31 (1987) 18.
730. F.Bueche, *Physical Properties of Polymers*, Intersci. Publ. J.Wiley, N.Y., London, 1962.
731. V.R.Allen and T.G.Fox, *J. Chem. Phys.*, **41** (1964) 337.
732. R.Z.Tudose, T.Volintiru, N.Asandei, M.Lungu, E.Merica, and Gh.Ivan, *Reologia compusilor macromoleculari. Reologia starii lichide.* Vol. 2, Ed.Tehnica, Bucuresti, 1983.
733. O.L.Castellanos, *Rev. Plast. Mod.*, **361** (1986) 51.
734. S.M.Aharoni, *J. Appl. Polym. Sci.*, **16** (1972) 3275.
735. C.D.Hann, *Rheology in Polymer Processing*, Academic Press, N.Y., 1976.
736. H.M.Laun, *Rheol. Acta*, **18** (1979) 478.
737. J.W.Hill and J.A.Cuculo, *J. Macromol. Sci. Rev. Macromol.Chem.*, **C14**(1) (1976) 107.
738. A.P.Saveliev, L.N.Malisev, L.N.Braginschi, and K.S.Minscher, *Plast. Massy*, **6** (1973) 56.
739. K.K.Chee and Y.H.Yap, *J. Appl. Polym. Sci.*, **20** (1976) 1313.
740. M.J.Folkes and D.A.M.Russell, *Polymer*, **21** (1980) 1252.
741. R.S.Lenk, *J. Appl. Polym. Sci.*, **22** (1978) 1781.
742. E.Boudreaux and J.A.Cuculo, *J. Appl. Polym. Sci.*, **27** (1982) 301.
743. J.F.Agassant, P.Avenas and J.Ph.Sergent, *La mise en forme des matières plastiques*, II-ème edition, Technique & Documentation-Lavoisier, Paris, 1986.
744. C.D.Hann and R.R.Lamonte, *J. Appl. Polym. Sci.*, **16** (1972) 3307.
745. F.N.Cogswell, *Polym. Eng. Sci.*, **12** (1972) 64.
746. A.G.Gibson and G.A.Williamson, *Polym. Eng. Sci.*, **25** (1985) 980.
747. J.A.Brydson, *Flow properties of polymer melts*, 2nd Ed., G.Godwin Ltd.London, 1981.
748. N.P.Krasnikova, E.V.Kotova, G.V.Vinogradov, and I.Pelzbauer, *J. Appl. Polym. Sci.*, **22** (1978) 2081.
749. K.A.Norh and A.Keller, *Polymer*, **32**(14) (1991) 2512.
750. J.R.C.Pereira and R.S.Porter, *J. Polym. Sci.; Polym. Phys. Ed.*, **21** (1983) 1147.
751. N.P.Cheremissinoff, in *Elastomer Technology Handbook* (Cheremissinoff N.P.Ed.), CRC Press, Boca Raton, Ann Arbor, London, Tokyo, 763, 1993.
752. V.A.Kargin and T.I.Sokolova, *Dokl. Akad. Nauk SSSR*, **108** (1956) 662.
753. 753.V.A.Kargin and T.I.Sokolova, *Zh. Fiz. Khim.*, **31** (1957) 1328.
754. A.P.Saveliev, G.I.Shilov, L.N.Malyshev, and V.A.Braghinskii, *Plast. Massy*, **12** (1971) 25.
755. C.Rauwendaal, *Polymer Extrusion*, Hanser Publish., Munich, 1986.
756. D.Poller and A.M.Kotliar, *J. Appl. Polym. Sci.*, **9** (1965) 501.
757. C.T.Kiang and J.A.Cuculo, *J. Appl. Polym. Sci.*, **46** (1992) 55.
758. P.Desai and A.S.Abhirman, *J. Polym. Sci.; Polym. Phys. Ed.*, **26** (1988) 1657.
759. A.K.Kulshreshta, M.V.S.Rao, and N.E.D.Dwetts, *J. Appl. Polym. Sci.*, **30** (1985) 3423.
760. M.J.Napolitano and A.Moet, *J. Appl. Polym. Sci.*, **34** (1987) 1285.
761. W.R.Krigbaum, C.K.Liu and D.J.Yang, *J. Polym. Sci.; Polym. Phys. Ed.*, **26** (1988) 1711.
762. A.G.Gibson and A.G.Williamson, *Polym. Eng. Sci.*, **25** (1985) 968.
763. Y.Chan, J.L.White, and Y.Oyanagy, *Polym. Eng. Sci.*, **18** (1978) 268.
764. Cl.Vasiliu Oprea and Al.Savin, *Premier Seminaire Franco-Roumaine sur les polymeres*, Iasi, 1993.

765. C.Vasile, C.Calugaru, A.Stoleru, M.Jablinovschi, and M.Rusu, *Comportarea termica a polimerilor,* Editura Academiei Romane, Bucuresti, 1980.
766. M.F.Vallat and D.J.Plazek, *J. Polym. Sci.; Polym. Phys. Ed.,* **26** (1988) 545.
767. M.F.Vallat, D.J.Plazek, and B.Bhushan, *J. Polym. Sci.; Polym. Phys. Ed.,* **26** (1988) 555.
768. B.Ke, *J. Appl. Polym.Sci.,* **24** (1962) 624.
769. A.Rudin and H.P.Schreiber, *J. Polym. Sci.,* **44** (1960) 261.
770. F.E.Divine, *J. Appl. Polym. Sci.,* **9** (1965) 369.
771. K.Komuro, Y.Todany, and N.Nagata, *J. Polym. Sci.,* **B2** (1964) 6443.
772. J.W.Teh, A.Rudin, and H.P.Schreiber, *Plast. Rubber Process. Appl.* **4**(2) (1984) 157.
773. R.S.Lenk, *Plastics Rheology,*, MacLaren and Sons Ltd., London, 1968.
774. G.V.Vinogradov and A.Y.Malkin, *Rhoelogy of Polymers,* MIR Moscow, 1980.
775. H.Tahakashi, T.Matsuoka, and T.Kurauki, *J. Appl. Polym. Sci.,* **30** (1985) 4669.
776. Cl.Vasiliu Oprea and Al.Savin, *Proceeding of IUPAC-Macro'83,* Bucuresti, 1982, Sect.II, 120.
777. V.K.S.Kumar, G.C.Das, and D.Gangopadhyay, *Man Made Textiles India,* **XXV**(10) (1982) 483.
778. R.T.Conley and R.A.Gaudiane, *Thermal Stability of Polymers,* Vol.II., M.Dekker, N.Y., 1970.
779. Gh.Rozmarin and E.Twartochlieb, *Fibre Poliesterice. Chimie si Tehnologie,* Editura Tehnica, Bucuresti, 1975.
780. Cl.Vasiliu Oprea and Al.Savin, *Paper presented at Symp.of Res.Centre R.Valcea,* Romania, 1983.
781. K.B.Abbas and R.S.Porter, *J. Appl. Polym. Sci.,* **20** (1976) 1289.
782. G.Perez, *Some effects of the rheological properties of poly(ethylene terephtalate) on spinning line profile and structure developed in high-speed spinning,* Chap.12, in *High –speed Fiber Spinning Scince and Engineering Aspects,* (A.Ziabicki and H.Kawai Eds.), J.Wiley & Sons, N.Y., 1965.
783. J.L.White and H.Yamane, *Pure Appl. Chem.,* **57**(10) (1985) 1441.
784. Cl.Vasiliu Oprea, A.Savin, and M.Popa, *Report Research Polytechnic Institute Iasi, 14,734,* 1984
785. S.A.Jabarin and E.A.Lafgren, *Polym. Eng. Sci.,* **24**(13) (1984) 1056.
786. Cl.Vasiliu Oprea and Al.Savin, *Paper presented at Symp.of Research Centre R.Valcea,* Romania, 1985.
787. C.Sanmo, C.Xi, and Z.Yu, *Huadany Fangzi Gongxueyuan Xuebao,* **2** (1982) 65.
788. J.M.Besnoin and K.Y.Choi, *J. Macromol. Sci. Rev. Macromol. Chem. Phys.,* **C29** (1989) 55.
789. D.Becker, H.Zimmermann, and C.Kalbig, *Faserforsch. u. Textiltech.,* **22**(5) (1976) 229.
790. S.Oprea and Al.Savin, *Materiale Plastice,* **28**(3-4) (1991) 129.
791. H.Inata and S.Matsumura, *J. Appl. Polym. Sci.,* **30** (1985) 3325.
792. F.C.Wampler and D.R.Gregory, *J. Appl. Polym. Sci.,* **16** (1972) 3253.
793. V.Marinoiu, C.Stratula, C.Patrascioiu, and C.Marinescu, *Metode numerice aplicate in ingineria chimica,* Editura Tehnica, Bucuresti, 1986.
794. W.H.Press, B.P.Flannery, and W.T.Vaterling, *Numerical Recipes,* Cambride University Press, Cambridge, 1986.
795. A.Savin and M.D.Bucevschi, *Poster at The 2nd Mediteranian School on Science and Technology of Advanced Polymer Based Materials,* Capri, Italy, 1991.
796. S.Oprea, Gh.Cristian, A.Savin, I.Druta, I.M.Popa, and S.Ciumasu, *Report Research,*

Polytechnic Institute, No. 2,321, 1988.
797. A.M.Basedow and K.H.Ebert, *Makromol. Chem.*, **176** (1975) 745.
798. M.S.Doulah, *J. Appl. Polym. Sci.*, **22** (1978) 1735.
799. M.M.Kuharcic and N.K.Baramboim, *Vysokomol. Soedin.*, **A14**(11) (1972) 843.
800. S.Matsuzawa, K.Yamamura, Y.Umeda, Y.Inoue, H.Kitamua, and S.Y.Pak, *Macromol. Chem.*, **180** (1979) 2009.
801. G.Akay, *Rheol. Proc. Int. Cong. 8th*, 1980, 1351.
802. J.D.Culter, J.L.Zakin, and G.Patterson, *J. Appl. Polym. Sci.*, **19** (1975) 3235.
803. H.G.Müller, *Angew. Makromol. Chem.*, **57** (1977) 99.
804. A.Nakano and Y.Minoura, *J. Appl. Polym. Sci.*, **19** (1975) 2749.
805. S.H.Agarwall and R.S.Porter, *J. Appl. Polym. Sci.*, **25** (1980) 173.
806. K.Yamamura, K.Nakada and S.Matsuzawa, *Makromol. Chem.*, **177** (1976) 927.
807. F.S.S.Yu, J.L.Zakin, and G.Patterson, *J. Appl. Polym. Sci.*, **23** (1978) 851.
808. H.Staudinger and V.Heuer, *Ber. Dstch. Chem. Ges.*, **67B** (1934) 1159.
809. J.Frenkel, *Acta Physicochim. SSSR*, **19** (1944) 51.
810. J.W.Morris and R.Shnurman, *Nature*, **160** (1967) 674.
811. H.Grohn and F.Krause, *Plaste und Kautschuk*, **11** (1964) 2.
812. F.Krause and H.Grohn, *Plaste und Kautschuk*, **11** (1964) 6.
813. Cl.Vasiliu Oprea and M.Popa, *Plaste und Kautschuk*, **32**(9) (1985) 337.
814. Cl.Vasiliu Oprea, M.Popa, and S.Petrovan, *J. Appl. Polym. Sci.*, **32** (1986) 5557.
815. D.H.Fischer and F.Rodriguez, *J. Appl. Polym. Sci.*, **15** (1971) 2975.
816. R.Y.Ting and R.C.Little, *J. Appl. Polym. Sci.*, **11** (1973) 3345.
817. R.S.Seright and J.H.Maerker, *Polym. Prep.(Am.Chem.Soc.Div.Polym.Chem.)* **22**(2) (1981) 30.
818. H.G.Müller and J.Klein, *Makromol. Chem. Rapid Commun.*, **1**(1) (1980) 27.
819. H.G.Müller and J.Klein, *Makromol. Chem.*, **182**(2) (1981) 523.
820. H.G.Müller, J.Klein, and A.Rottliff, *Makromol. Chem.*, **182**(2) (1981) 529.
821. A.M.Basedow, K.H.Ebert, and H.H.Hunger, *Makromol. Chem.*, **180** (1979) 411.
822. V.D.Fihman and E.A.Paksver, *Khim. Volokna*, **3** (1980) 6.
823. H.U.Mashudanov, O.G.Volkova, L.A.Haimelina, and M.T.Primkulov, *Khim. Volokna*, **1** (1981) 25.
824. C.P.Buckley and M.Habibulah, *J. Appl. Polym. Sci.*, **26**(8) (1981) 2613.
825. R.J.Rigby and T.W.Mitchell, *J. Text. Inst.*, **63** (1972) 416.
826. M.Ballauf and B.A.Wolf, *Macromolecules*, **17**(2) (1984) 211.
827. P.H.Goldblat and R.S.Porter, *J. Appl. Polym. Sci.*, **20** (1980) 1193.
828. A.Casale, R.G.Reinwell, and R.S.Porter, *Rheol. Acta*, **10** (1971) 412.
829. Cl.Vasiliu Oprea and M.Popa, *Synthetic Polymers J.*, **1**(2-3) (1994) 1043.
830. M.Popa, *Contributii la cunoasterea unor reactii mecanochimice privind sinteza, prelucrarea si exploatarea copolimerului ternar pe baza de poliacrilonitril*, Ph.D. Thesis, Polytechnic Institute of Iasi, 1989.
831. E.H.Andrews, *Fatigue in Polymers*, in *Testing of Polymers*, Vol.IV, John Wiley & Sons, N.Y., 1969.
832. R.W.Herzberg and J.A.Manson, *Fatigue of Engineering Plastics*, Academic Press, N.Y., 1980.
833. J.A.Manson and R.W.Herzberg, *CRC Critical Review* in *Macromol. Sci.*, 433, 1973.
834. R.W.Herzberg, *Deformation and Fracture Mechanics of Engineering Materials*, 2nd Ed, John Wiley & Sons, N.Y., 1983.
835. K.Oberbach, *Kunststoffe-Kennwerte fur Konstrukteure*, Ed.C.Hanser, Munchen, 1975.
836. J.A.Sauer and G.C.Richardson, *Int. J. Fracture*, **16** (1980) 499.

837. J.A.Sauer and C.C.Chen. in *Adv.in Polymer Sci.*, **52/53**, Ed.Springer, Berlin, 1983.
838. A.J.Kinloch and H.J.Young. *Fracture Behaviour of Polymers*, Ed. Appl. Sci. Publ., London, 1983.
839. M.T.Kakemori. *Ann. Review of Mater. Sci.*, **14** (1984) 171.
840. E.J.Kramer, in *Developments in Polymer Fracture*, Vol.I, (E.H.Andrews Ed.), Ed. Appl. Sci. Publ., London, 1979.
841. L.Morbitzer, *Coll. Polym. Sci.*, **259** (1981) 832.
842. D.Yuwanda, R.Lew, and S.T.Balke, *36th Canadian Chemical Eng. Conference*, Saarnia, Ontario, Oct. 5-8, 1986.
843. G.M.Bartenev, *Dokl. Akad. Nauk SSSR*, **9**(8) (1955) 53.
844. G.M.Bartenev, *Mekhanika Polimerov*, **5**(8) (1966) 700.
845. G.M.Bartenev and B.M.Tulinov, *Mekhanika Polimerov*, **1**(8) (1977) 3.
846. G.M.Bartenev and E.M.Kartasov. *Plaste und Kautschuk*, **28**(5) (1981) 241.
847. A.I.Gubanov, *Fiz. Tverd. Tela, Phys. Chem. Glas*, **1**(8) (1981) 427.
848. G.M.Bartenev and B.M.Tulinov, *Phys. Chem. Glas*, **1**(8) (1978) 427.
849. S.N.Jurkov, V.R.Regel, and T.P.Sanfirova, *Vysokomol. Soedin.*, **6**(8) (1964) 1092.
850. B.Ya.Bilski, O.P.Pozdniakov, and B.P.Redkov, *Mekhanika Polimerov*, **5**(8) (1973) 835.
851. A.V.Amelin, O.P.Pozdniakov, V.R.Regel, and T.P.Sanfirova, *Fiz. Tverd. Tela*, **12**(8) (1976) 2528.
852. S.N.Jurkov, *Vysokomol. Soedin.*, **4**(8) (1962) 1703.
853. N.N.Petcianskaia and V.A.Stepanov, *Fiz. Tverd. Tela*, **7**(8) (1965) 2982.
854. S.N.Jurkov and S.A.Abassov, *Vysokomol. Soedin.*, **3**(8) (1964) 450.
855. V.R.Regel and A.M.Leksovski, *Mekhanika Polimerov*, **1** (1969) 70.
856. V.R.Regel, A.M.Leksovski, and I.Bolitekov, *Mekhanika Polimerov*, **2** (1972) 247.
857. O.V.Bonda, V. Zilvar, and J.Staverman, *J. Appl. Polym. Sci.; Polym. Phys. Ed.*, **14** (1976) 2313.
858. J.A.Sauer and G.C.Richardson, *Int. J. Fracture*, **16** (1980) 6.
859. N.P.Procopciuk, *Dokl. Akad. Nauk SSSR*, **25**(5) (1981) 453.
860. K.Friedrich, *Polym. Composite*, **3**(2) (1982) 65.
861. W.G.Lyons, *Text. Res. J.*, **40** (1970) 60.
862. J.C.Radon, *Int. J. Fract.*, **16** (1980) 533.
863. M.T.Kakemori and R.P.Kambour, *J. Mater. Sci.*, **16** (1981) 1108.
864. M.T.Kakemori and D.S.Matsuda, *J. Polym. Sci.; Polym. Phys. Ed.*, **20** (1982) 2027.
865. A.R.Bunsell and J.W.S.Hearle, *J. Appl. Polym. Sci.*, **18** (1974) 267.
866. Ch.Oudet and A.R.Bunsell, *J. Mater. Sci.; Letters*, **3** (1984) 295.
867. Ch.Oudet and A.R.Bunsell, *J. Appl. Polym. Sci.*, **29** (1984) 436.
868. P.Kassenbeck, *III-eme Congres International de Recherche Textile Lainiere*, CIRTEL, Paris, Sect. I, 1965.
869. A.A.Griffith, *Phil. Trans. Roy. Soc.*, **A221** (1951) 169.
870. A.D.Stuart and L.O.Anderson, *J. Am. Chem. Soc.*, **36** (1953) 27.
871. L.T.Smith, *Pure and Appl.Chem.*, **23** (1970) 235.
872. J.Murray and D.Hull, *Polymer*, **19** (1969) 451.
873. C.D.Prevorsek and J.M.Lyons, *J. Appl. Phys.*, **35** (1964) 3154.
874. C.D.Prevorsek, *J. Polym. Sci.*, (Symp.No.32), (1971) 343.
875. O.Malcomete, A.Grigoriu, and E.Grigoriu, *6th European Plastics Conference*, June 7-9, 1982, Paris.
876. O.Malcomete and E.Hagiu, *Ind.Usoara*, **7** (1982) 308.

877. I.Taraboanta, *Studiul comportarii chemoreologice a unor fibre sintetice de tip poli(alcool vinilic),* Ph.D.Thesis, Technical University Iasi, 1998.
878. J.W.S.Hearle, B.Lamos, and W.D.Cooke, *Fibre Failure and Wear of Materials on Atlas of Fracture, Fatigue and Durability,* Ellis Harword Limited, England, 44, 1989.
879. J.W.S.Hearle and P.M.Cross, *J. Mater. Sci.,* **5** (1970) 507.
880. Cl.Vasiliu Oprea and I.Taraboanta, *Polym. Plast. Techn. Eng.,* 1999 (in press).
881. R.D.Anondjwola and B.C.Gosvaani, *Textile Tes. J.,* **63**(7) (1993) 392.
882. A.Peterlin, *J. Macromol. Sci.,* **8** (1973) 277.
883. A.Peterlin, *J.Macromol. Sci., Phys.,* **B19** (1980) 40.
884. S.Turner, *Mechanical Testing of Plastics,* Illife, London, 1973.
885. Cl.Vasiliu Oprea, Al.Constantinescu, and P.Barsanescu, *Coll. Polym.Sci.,* **264** (1986) 590.
886. L.Geanau, A.Taranu, S.Seiler, and V.Popescu, *Rom. Pat. 100,965,* 1980.
887. B.W.Cherry and D.F.Kinder, *Mat. Sci. Eng.,* **49** (1981) 285.
888. C.S.Schollenberger, H.Scott, and G.R.Moore, *Rubber World,* **4** (1958) 549.
889. C.S.Schollenberger and F.D.Stuart, in *Advances in Urethane Science and Technology,* (K.D.Fresch and S.L.Reegan, Eds.), Vol. 2, Technonic Publish., Connecticut, 1973.
890. Z.Osawa and N.Nagashima, *Polym. Deg. Stab.,* **1** (1979) 311.
891. C.E.Hoyle and K.J.Kim, *J. Polym. Sci.; Polym. Chem.,* **25** (1987) 2631.
892. C.S.Schollenberger, L.G.Papos, J.C.Park, and V.V.Vicktory, *Rubber World,* **139** (1960) 81.
893. C.S.Schollenberger and K.Dinsbergs, *SPE Trans.,* **1** (1961) 31.
894. S.C.Schollenberger and J.D.Stuart, *J. Elastoplastics,* **4** (1972) 294.
895. L.V.Nevskii and D.G.Tarakanov, *Soviet Plastics,* **9** (1967) 47.
896. L.V.Nevskii, O.G.Tarakanov, and V.Belyakov, *Soviet Plastics,* **10** (1967) 23.
897. O.G.Tarakanov, L.V.Nevskii, and V.Belyakov, *J. Polym. Sci.,* (Part C), **23** (1968) 193.
898. L.V.Nevskii, O.G.Tarakanov, and A.P.Kafenganz, *Soviet Plastics,* **14** (1972) 46.
899. J.L.Gardette and J.Lemaire, *Polym. Deg. Stab.,* **6** (1984) 135.
900. G.G.Samoilov and E.E.Tomasevskii, *Fiz. Tverd. Tela,* **10** (1968) 1094.
901. T.B.Boboev and V.R.Regel, *Mekhanika Polimerov,* **5** (1969) 929.
902. K.Akimbekov, V.S.Kuksenko, S.Nezanudinov, A.I.Stutsker, and A.A.Yastubinskii, *Fiz. Tverd. Tela,* **14** (1972) 2708.
903. Cl.Vasiliu Oprea and Al.Constantinescu, *Paper presented at 11st Scientific Conference of Research Centre R.Valcea,* Calimanesti, Romania, 1985.
904. S.N.Jurkov and V.E.Korsukov, *Fiz. Tverd. Tela,* **15** (1973) 2071.
905. A.V.Savitskii, V.A.Malchevskii, T.P.Sanfirova, and L.P.Lozin, *Vysokomol. Soedin.,* **A16** (1964) 2130.
906. Cl.Vasiliu Oprea and Al.Constantinescu, *31st IUPAC Macromolecular Symosium, Merseburg,* (II/47), (1987) 118.
907. K.L.DeVries, D.K.Roylance, and M.L.Williams, *J. Polym. Sci.,* **A2**(10) (1972) 599.
908. H.W.Siesler, *Paper presented at 29th IUPAC Symposium on Macromolecules,* Bucuresti, 1983.
909. Cl.Vasiliu Oprea and Al.Constantinescu, *Materiale Plastice,* **20** (1986) 83.
910. Cl.Vasiliu Oprea and Al.Constantinescu, *Rev. Roum. Chim.,* **32** (1987) 407.
911. A.J.Kinloch and R.J.Young, *Fracture Behaviour of Polymers,* Appl.Sci.Publ.,London, 67, 1983.
912. Cl.Vasiliu Oprea, Al.Constantinescu, and P.Barsanescu, *Rom. Pat. 145,214* (1990).

913. Cl.Vasiliu Oprea and Al.Constantinescu, Al-III-lea Congres National de Chimie, ucuresti, Vol.I, (1988) 43.
914. H.H.Kausch, *Polymer Fracture*, 2nd Ed., Springer Verlag, Berlin, 1987.
915. A.Peterlin, *Polymer Preprints*, **18** (1977) 2
916. Cl.Vasiliu Oprea, M.Popa, and E.G.Ioanid, *Rom. Pat. 99,613* (1989).
917. Cl.Vasiliu Oprea and Al.Constantinescu, *Paper presented at 33rd IUPAC Symposium on Macromolecules*, Montreal, 1990.
918. F.Coiffier, R.Arnaud and J.Lemaire, *Makromol. Chem.*, **185** (1984) 1095.
919. 919.Cl.Vasiliu Oprea, Al.Constantinescu, and P.Barsanescu, *Plaste und Kautschuk*, **37**(11) (1990) 370.
920. M.C.Kenney, J.F.Mandell, and F.J.Magarey, *J. Mater. Sci.*, **20**(6) (1985) 2045.
921. R.A.Schultz, *Curso de Reologia Textil-Buletin del Instituto de Investigation Textil*, Univ. Politecnica, Barcelona, (1993) 552.
922. L.Vangheluwe, *Melliand*, **73** (1992) 117.
923. L.Vangheluwe, *Melland*, **74** (1994) 496.
924. L.Vangheluwe, *Text. Res .J.*, **63**(9) (1993) 552.
925. H.E.Andrews, *Fatigue in Polymers*, in *Testing of Polymers*, Vol.VI, J.Wiley & Sons, N.Y., (1969) 63.
926. W.R.Hertzberg and A.J.Manson, *Fatigue of Engineering Plastics*, Academic Press, N.Y., (1980) 45.
927. A.M.Mannich, *Text. Res. J.*, **62** (1992) 196.
928. O.Malcomete and E.Covrig, *Industria Usoara*, **9** (1984) 9.
929. O.Malcomete, *Merceologie*, Ed."Gh.Zane", Iasi, 1995.
930. A.Peterlin, *Structure and Properties of Oriented Polymers*, I.M.Ward Ed., Appl.Sci.Publ., London, 1975, 330.
931. A.Peterlin, *J. Macromol. Sci. Phys.*, **13**(19) (1981) 401.
932. A.Peterlin, *Polym. Sci., Symp.*, **32** (1971) 297.
933. A.Peterlin, *Text. Res .J.*, **42**(1) (1972) 20.
934. J.Wood and R.Postle, *Text. Res .J.*, **48** (1978) 68.
935. I.Alig, K.H.Feller, and R.Souberlich, *Acta Polym.*, **31**(5) (1980) 316.
936. I.Alig, *Dipl. Arbeit*, T.H. Merseburg, 1978.
937. Lu-Fu-Min, B.Groswami, J.E.Spruiell, and K.E.Duckett, *J. Appl. Polym. Sci.*, **30** (1985) 185.
938. J.H.Dumbleton, *J. Polym. Sci.*, **A2** (1968) 795.
939. W.Wegener, *Chemiefasern.*, **20 Nov.** (1970) 985(I).
940. W.Wegener, *Chemiefasern.*, **20 Dec.** (1970) 1072(II).
941. Cl.Vasiliu Oprea, I.Taraboanta, and L.Wangheluwe, *Bull. I.P.Iasi*, **XLIII**(1-4) (1997) 26.
942. W.W.Mosley, *J. Appl. Polym. Sci.*, **3** (1960) 266.
943. R.J.Samuels, *Structure and Polymer Properties*, Wiley, N.Y., 1977.
944. V.Subramanian and H.K.Santhi, *Sonic Methods in Testing Synthetic Fibres*, The Indian Text.J., Oct (1982) 232.
945. G.Blith and R.Postle, *Text. Res. J.*, **49** (1979) 601.
946. G.F.S.Hussain and R.Nachane, *Text. Res. J.*, **57** (1987) 592.
947. A.Ryan and R.Postle, *Text. Res .J.*, **51** (1981) 730.
948. J.L.Woo and R.Postle, *J. Appl. Polym. Sci.*, **18** (1974) 589.
949. J.G.M.Miltenburg, *Text. Res. J.*, **61** (1991) 363.
950. Y.Schimizu, R.Yamamoto, and H.Shimizu, *Text. Res. Inst.*, Nov. (1989) 684.
951. Cl.Vasiliu Oprea and M.Popa, *Bull. I.P.Iasi*, **XXVII(XXXI)**(1-2) (1981) 115.
952. S.Urjimtsev, *Mekhanika Polimerov*, **3** (1967) 467.
953. Cl.Vasiliu Oprea and M.Popa, *Symp: Modernizarea fluxurilor Tehnologice ale*

fibrelor sintetice, Vaslui, Romania, 23 Sept. 1987.

954. K.Ked and G.Menning, in *Verschleiss in der Kunstoffverarbeitung. Phanomene und Schutzmassnalmen,* Carl Hanser Verlag, Munchen-Wien, (1990) 1.

955. N.A.Krotova, *Developing researches in mechanochemistry and mechano-emission solid body field,* in *Mekhano-emisia I Mekhanohimia tverdovo tela,* Frunze, USSR, (1974) 9.

956. D.W.Krevelen, *Properties of Polymers,* 2nd Ed.Elsevier, Amsterdam, 1990.

957. R.P.Kambour, *Corrosion Fatigue,* Ed.Nàtional Assoc.of Corrosion Eng., Huston, 1972.

958. R.P.Kambour in: *Mechanism of Enviromental Sensitive Craking of Materials,* Ed. Metal. Soc., London, 1977.

959. N.K.Baramboim, *Mechanochimia vysokomolekulearnyh soedinenii,* Izd. Himia, Moskva, (1978) 293.

960. G.A.Gorohovskii, T.V.Dmitrievna and S.I.Kuzmina, in *Fiziceskaia Himia polimernih kompozitii,* Moskva (1974) 154-160, 166-169.

961. G.A.Gorohovskii, T.V.Dmitrievna, and V.O.Movcean, *Dokld. Akad. Nauk SSSR,* **A10** (1973) 936.

962. G.A.Gorohovskii, S.I.Kuzmina, and V.P.Gorsinov, *Dokld. Akad. Nauk SSSR,* **A6** (1974) 658.

963. G.A.Gorohovskii, P.A.Cernenco, and V.A.Vonseatskii, *Dokld. Akad. Nauk SSSR,* **A1** (1972) 204.

964. G.A.Gorohovskii and T.V.Dmitrievna, *Dokld. Akad. Nauk SSSR,* **B12** (1974) 1099.

965. G.A.Gorohovskii and T.V.Dmitrievna, *Fiziko-himiceskaia mehanika materialov,* **2** (1972) 43.

966. G.A.Gorohovskii and O.A.Potemkina, *Fiziko-himiceskia Mehanika materialov,* **3** (1972) 39.

967. G.A.Gorohovskii, P.A.Cernenco, and V.A.Smirnov, *Fiziko-himiceskaia Mehanika Materialov,* **5** (1972) 42.

968. G.A.Gorohovskii, *Vestnik Akad. Nauk SSSR,* **10** (1973) 43.

969. G.A.Gorohovskii, V.P.Gorsinov, and L.M.Graevskaia, *Dokld. Akad. Nauk SSSR,* **B3** (1972) 224.

970. N.G.Savciuc, D.F.Simonenco, N.N.Pjidaev, V.A.Volnitskii, and E.P.Mamulev, *Izv. Vuzov. Tehnologia legkoi Promaslenosti,* **4** (1969) 166.

971. A.J.Sviridionok, *Wear,* **25**(3) (1973) 301.

972. J.K.Lancaster, *Plastics and Polymers,* **41** (1973) 297.

973. J.Gzerny, D.Zurajda, and J.Coufal, *Symp. IUPAC,* Prague, Prelim.Report, Part 2, (1973) F/119 – F/134.

974. V.A.Mustufaev, *Kinetics and Mechanism Polyreacts.* Vol.V, Preprs.Budapest, (1969) 427.

975. K.L.DeVries, D.K.Roylance, M.L.Williams, *J. Polym. Sci.,* **B8** (1971) 605.

976. Sh.M.Vilek and V.I.Donskih, *Dokld. Akad. Nauk SSSR,* **153** (1963) 397.

977. S.B.Rotner and E.G.Lurie, *Mekahanika Polimerov,* **9** (1960) 33.

978. S.B.Rotner and E.G.Lurie, *Dokld. Akad. Nauk SSSR,* **169** (1966) 909.

979. E.G.Lurie and E.B.Rotner, *Plast. Massy,* **11** (1962) 47.

980. S.E.Bressler, S.N.Jurkov, E.N.Kazbekov, E.M.Slonimskii, and E.E.Tomasevskii, *J. Teort. Fiz.,* **29** (1959) 358.

981. S.N.Jurkov, E.E.Tomasevskii, and V.A.Zakrevskii, *Fiz. Tverd. Tela,* **3** (1961) 2841.

982. E.T.Konincik, D.M.Kafalovici, and I.L.Roih, *Plast. Massy,* **8** (1966) 72.

983. W.E.Morton and J.W.S.Hearle, *Physical Properties of Textile Fibres,* Ed.Text.Inst.,

London, (1993) 630.
984. Cl.Vasiliu Oprea and I.Taraboanta, Paper in press.
985. C.Sadrok, *Flow Properties of Dysperse Systhems*, Ed. Intersci. Publ., N.Y., 1953.
986. H.L.Frish and R.Simha, *Rheology*, Vol.I, Academic Press, N.Y., 1956.
987. D.Thomas, *J. Coll. Sci.*, **20** (1965) 267.
988. F.Bueche, *J. Chem. Phys.*, **40** (1964) 484.
989. P.Debye and Bueche, *J. Chem. Phys.*, **16** (1948) 575.
990. P.Debye, *J. Chem. Phys.*, **14** (1946) 636.
991. W.F.Busse and R.Longworth, *J. Polym. Sci.*, **8** (1964) 1053.
992. D.Gupta and W.C.Forsman, *Macromolecules*, **2** (1969) 304.
993. R.S.Porter and J.F.Johnson, *Trans. Soc. Rheol.*, **6** (1962) 107.
994. M.L.Miller, *The Structure of Polymers*, Reinhold Publish.Comp., N.Y., 1966.
995. F.Bueche, *Physical Properties of Polymers*, Ed. J.Wiley & Sons, N.Y., (1962) 234.
996. E.Passaglia, Y.T.Yang, and N.Y.Wegemer, *J. Polym. Sci.*, **47** (1960) 333.
997. B.Novakovskii and G.Meyerhoff, *Rheol. Acta*, **3** (1964) 233.
998. Y.Go, S.Matsuzawa, I.Kondo, and K.Nakamura, *Kobunshi Kagaku*, **24** (1967) 711.
999. R.S.Porter, M.J.R.Cantow, and J.F.Johnson, *Proceedings of the Fourth International Congress on Rheology*, Part 2, Intersci., N.Y., (1965) 479.
1000. F.Bueche, *J. Appl. Phys.*, **26** (1955) 738.
1001. L.Merker, *J. Polym. Sci.*, **22** (1956) 353.
1002. T.Takemura, *J. Polym. Sci.*, **28** (1958) 185.
1003. T.E.Newlin, S.E.Lovell, P.P.Saunders, and J.D.Ferry, *J. Coll. Sci.*, **170** (1962) 10.
1004. J.Furukawa, *Rubber Chem. Technol.*, **35** (1962) 1013.
1005. J.D.Ferry, *Viscoelastic Properties of Polymers*, J.Wiley & Sons, N.Y., (1961) 192.
1006. W.W.Grassley, *J. Chem. Phys.*, **43** (1965) 2696.
1007. M.C.Williams, *A.Ch.E.J.*, **12** (1966) 1064.
1008. M.Fixman, *J. Chem. Phys.*, **35** (1961) 889.
1009. R.A.Schultz, *Rheol. Acta*, **8** (1969) 349.
1010. R.S.Lenk, *J. Appl. Polym. Sci.*, **11** (1967) 1033.
1011. Iu.E.Ezner, *Vysokomol. Soedin.*, **A9** (1967) 765.
1012. M.Johnson and J.D.Ferry, *J. Coll. Sci.*, **7** (1952) 498.
1013. R.Simha and L.Ultracki, *J. Polym. Sci.*, **A2** (1967) 853.
1014. V.P.Budtov, *Vysokomol. Soedin.*, **12** (1970) 1355.
1015. H.Staudinger and H.Heuer, *Ber. dschen. Chem. Ges.*, **67B** (1934) 1159.
1016. H.Staudinger and E.Dreher, *Ber. dschen. Chem. Ges.*, **69B** (1936) 1091.
1017. J.W.Morish and R.Schnurmann, *Nature*, **160** (1947) 764.
1018. J.W.Morish and Schnurmann, *Rev. Sci. Instrum.*, **17** (1946) 17.
1019. R.Kigama and H.Kinoshita, *Rev. Phys. Chem. Japan*, **22** (1952) 1.
1020. A.B.Bestul, *J. Phys. Chem.*, **61** (1957) 418.
1021. A.B.Bestul and H.V.Belcher, *J. Apll. Phys.*, **24** (1953) 1011.
1022. A.B.Bestul, *J. Appl. Phys.*, **25** (1954) 1069.
1023. P.Goodman and A.B.Bestul, *J. Polym. Sci.*, **18** (1955) 235.
1024. A.B.Bestul, *J. Chem. Phys.*, **24** (1956) 1196.
1025. P.Goodman, *J. Polym. Sci.*, **25** (1957) 235.
1026. H.S.White and H.V.Belcher, *J. Res. Nat. Bur. Stand.*, **60** (1958) 215.
1027. A.B.Bestul and P.Goodman, *Phys. Rev.*, **94** (1954) 1425.
1028. S.A.McKee and H.S.White, *J. Res. Nat. Bur. Stand.*, **46** (1959) 18.

1029. L.S.Wood, *J. Appl. Phys. Suppl.*, **1** (1950) 202.
1030. G.Ngul and G.Moscz, *Freiberger Forsch.*, **196** (1961) 83.
1031. H.Grohn and F.Krause, *Plaste und Kautschuk,* **11**(1) (1964) 2.
1032. H.Grohn and F.Krause, *Plaste und Kautschuk,* **11**(1) (1964) 6.
1033. Cl.Vasiliu Oprea and C.Popa, *Graduate Diploma*, Inst.Polytechnic Iasi, 1966.
1034. D.Levinthal and P.V.Dowson, *J. Mol. Biol.*, **3** (1961) 674.
1035. R.E.Harrington and B.H.Zimm, *J. Phys. Chem.*, **69** (1965) 161.
1036. R.E.Harrington and B.H.Zimm, *J. Polym. Sci.*, **A2** (1962) 294.
1037. D.E.Moore and A.G.Parts, *Polymer,* **9** (1968) 52.
1038. A.D.Hershey and E.Burgi, *J. Mol. Biol.*, **3** (1971) 674.
1039. L.Forgacs and S.G.Mason, *J. Coll. Sci.*, **14** (1959) 473.
1040. O.L.Forgacs and G.S.Mason, *J. Colloid. Sci.*, **14** (1959) 457.
1041. W.Bartok and S.G.Mason, *J. Coll. Sci.*, **14** (1970) 2145.
1042. P.Thomas, *Ind. Eng. Chem.*, **32** (1940) 299.
1043. P.Alexander and M.Fox, *J. Polym. Sci.*, **12** (1954) 533.
1044. F.Sontag and E.Frenckel, *Koll. Z.*, **81** (1954) 135.
1045. Y.Minoura, T.Kasuya, S.Kawamura, and A.Nakano, *J. Polym. Sci.*, **A2**(5) (1967) 125.
1046. Y.Minoura, T.Kasuya, S.Kawamura, and A.Nakano, *J. Polym. Sci.*, **A1** (1967) 43.
1047. A.Nakano and Y.Minoura, *J. Appl. Polym. Sci.*, **15** (1971) 927.
1048. A.Nakano and Y.Minoura, *J. Appl. Polym. Sci.*, **16** (1972) 627.
1049. K.Goto and H.Fujiwara, *Kobunshi Kagaku,* **21** (1964) 716.
1050. K.Arai, K.Nakamura, T.Kamatsu, and T.Nagawa, *J. Chem. Soc. Japan Ind. Chem. Soc.*, **71** (1968) 1438.
1051. R.E.Harrington, *J. Polym. Sci.*, **A1**(4) (1966) 489.
1052. R.S.Porter, R.F.Klaven, and J.F.Johnson, *Rev. Sci. Inst.*, **36** (1965) 1846.
1053. R.S.Porter, M.J.R.Cantow, and J.F.Johnson, *J. Polym. Sci.*, **14** (1970) 879.
1054. N.I.Kaverina, *J Appl. Chem. USSR* (Eng.), **29** (1956) 1565.
1055. N.I.Kaverina and G.Semenida, *J. Appl. Chem. Phys. USSR* (Eng.)., **8** (1964) 2585.
1056. H.Shin, *Ph.D. Thesis,* MIT, 1965.
1057. K.I.Klimov and P.P.Zarundi, *Khim. Technol. Topliv Masel,* **4**(3) (1959) 37.
1058. Y.Go, Y.Kondo, E.Yambe, and T.Okamura, *Kobunshi Kagaku,* **25** (1968) 477.
1059. K.J.Mysels, *US Pat., 2,492,173,* (1947).
1060. B.A.Toms, *Proc. Int. Congress of Rheology,* Part 2, North Holland Publishing Co., Amsterdam, (1949) 130.
1061. J.W.Hoyt and A.G.Fabula, *US Naval Ordnance Test, Station Nawers, Report 8636,* Chinur Lake, 1964.
1062. G.E.Gadd, *Nature,* **212** (1966) 874.
1063. W.Hoyt, *Solution Properties of Natural Polymers, Spec. Publ. Nr.23,* The Chemical Soc.London, (1968) 207.
1064. R.C.Little, *Naval Research Laboratory,* Memorandum, Rep. 1957,Washington D.C., 1969.
1065. R.W.Peterson. *Ph.D.Thesis,* Harward Univ., 1969.
1066. A.G.Tabula, *Proc. Fourth Int. Congress of Rheology,* Part. 3, Intersci. Publish., N.Y., (1963) 456.
1067. R.G.Shaver and E.W.Merrile, *Am. Inst. Chem. Eng. J.*, **5** (1959) 181.
1068. C.D.Dever, R.G.Harbour and W.F.Seijert, *US Pat 3,023,760,* (1962).
1069. H.R.Craword, *US Pat.,3,230,919,* (1966).
1070. P.S.Wirk, E.W.Merrill, H.Schickley, and R.A.Smith, *J. Fluid Mech.* ,**30** (1967)

305.
1071. P.R.Kenis, *Nature,* **217** (1968) 940.
1072. P.R.Kenis, *Int. Shipbuilt. Progr.,* **16** (1969) 342.
1073. P.R.Kenis, *J. Appl. Polym. Sci.,* **15** (1971) 607.
1074. D.H.Fisher and F.Rodriguez, *J. Appl. Polym. Sci.,* **15**(12) (1971) 2975.
1075. J.Hand and M.C.Williams, *J. Appl. Polym. Sci.,* **13** (1969) 2499.
1076. J.Hand and M.C.Williams, *Nature,* **227** (1970) 369.
1077. P.G.Assarson, P.S.Leung, and G.J.Safford, *Polymer Preprints,* **10** (1969) 1241.
1078. R.W.Patterson and F.H.Abernatly, *J. Fluid Mech.,* **43**(4) (1970) 689.
1079. Iu.A.Buevici and N.I.Iakusin, *J. Prikld. i Techn. Fiziki,* **1** (1968) 56.
1080. J.W.Hoyt, *Drag Reduction* in *Encyclopedia of Polymer Science and Engineering,* Vol. 5, (1985) 129.
1081. B.Gampert and P.Wagner, in *IUTAM Symposium: The Influence of Polymer Additives on Velocity and Temperature Fields,* Univ. Essen, Germany, 1985.
1082. J.W.Hoyt, *Nature,* **211** (1966) 170.
1083. J.W.Hoyt, *Polymer Letter,* **9** (1971) 851.
1084. J.W.Hoyt, *Trans. ASME,* **94D** (1972) 258.
1085. D.R.Oliver and S.I.Bakhtiarow, *J. Non-Newtonian Mech.,* **12** (1983) 113.
1086. M.P.Tulin In: *Proceedings of the 3rd Int. Conf. on Drag Reduction,* (R.H.J.Moses and R.T.Moses Eds.), Bristol, Univ.of Bristol, U.K., 1984.
1087. J.W.Hoyt and J.J.Taylor, in *Flow – Visualization,* Vol. II (W.Merzrich Ed.), Aemisphere Publishing Co., N.Y., 1982.
1088. I.E.Elpiner, *Ultrazvuk. Fizikohimiceskoe I biologiceskoie deistvie,* Fizmatgiz, Moskva, 1963.
1089. E.W.Flosdorf and L.A.Chambers, *J. Am. Chem. Soc.,* **58** (1936) 1069.
1090. R.Prudhomme, D.Picard and R.Busnel, *J. Chim. Phys.,* **50** (1953) 107.
1091. R.Busnel, H.R.Asbach, and H.Bouzinger, *J. Chem. Phys.,* **50** (1953) 97.
1092. W.Demann and H.R.Asbach, *Techn. Mitt. Krupp.,* **1** (1940) 12.
1093. M.G.Sirotiuk, *Akust. J.,* **10** (1964) 465.
1094. M.A.Margulis and A.N.Altsev, *J. Fiz. Khim.,* **42** (1968) 1441, 1447.
1095. N.Miller, *Trans. Faraday Soc.,* **46** (1950) 5546.
1096. N.Sata and K.Nakasima, *Bull. Chem. Soc. Japan,* **18** (1943) 220.
1097. O.Nomoto and S.Okui, *J. Phys. Soc. Japan,* **3** (1948) 7.
1098. K.Negishi, *J. Phys. Soc. Japan,* **16** (1961) 1450.
1099. I.G.Polotski, *J. Obscei Khim.,* **17** (194) 1048.
1100. S.S.Urazovski and I.G.Polotski, *J. Obscei Khim.,* **10** (1940) 812.
1101. E.A.Neppiras and V.E.Nolting, *J. Proc. Phys. Soc.,* **64B** (1951) 1032.
1102. R.Prudhomme, *Bull. Soc. Chim. Biol.,* **39** (1957) 425.
1103. A.Weissler, H.Cooper, and S.Snyler, *J. Am. Chem. Soc.,* **72**(1950) 1769.
1104. A.Hengleim, *Naturwiss.,* **43** (1956) 277.
1105. E.C.Morboe and W.A.Weyl, *J. Appl. Phys.,* **21**(9) (1950) 937.
1106. M.Minnaert, *Phil. Mag.,* **16** (1933) 235.
1107. H.Rust, *Angew. Chem.,* **65**(10) (1953) 249.
1108. H.Fark, *Frequenz,* **6** (1952) 255.
1109. V.Griffing, *J. Chem. Phys.,* **20**(6) (1952) 939.
1110. W.T.Richards and A.L.Loomis, *J. Am. Chem. Soc.,* **43** (1927) 3086.
1111. A.Dognon and Y.Simonot, *J. Chim. Phys.,* **50** (1953) 94.
1112. V.D.Eltsefon and A.A.Berlin, *Vysokomol. Soedin.,* **5** (1963) 1562.
1113. E.Flosdorf and L.A.Chambers, *J. Am. Chem. Soc.,* **55** (1933) 3057.
1114. A.Szent-Gyorgyi, *Nature,* **131** (1933) 278.
1115. A.Szalay, *Z. Phys. Chem.,* **164A** (1933) 234.

1116. H.Freundlich and D.W.Gillings, *Trans. Faraday Soc.*, **34** (1938) 649.
1117. G.Schmid and O.Rommel, *Z. Electrochem.*, **45** (1939) 659.
1118. G.Schmid and O.Rommel, *Z. Phys. Chem.*, **185A** (1939) 97.
1119. G.Schmid, *Chem. Z.*, **64** (1940) 423.
1120. G.Schmid, *Z. Phys. Chem.*, **186A** (1940) 113.
1121. G.Schmid and E.Beuttenmüller, *Z. Electrochem.*, **50** (1944) 209.
1122. G.Schmid and W.Poppe, *Z. Electrochem.*, **53** (1949) 28.
1123. I.I.Zhukov and M.A.Klenokh, *Dokld. Akad. Nauk SSSR*, **68** (1949) 333.
1124. H.Sobue and H.Kawoi, *J. Soc. Chem. Japan*, **49** (1946) 36.
1125. H.Sobue and K.Ishikawa, *J. Soc. Text. Cell. Ind. Japan*, **5** (194?) 366.
1126. H.Mark, *J. Acoust. Soc. Am.*, **16** (1954) 183.
1127. M.A.K.Mostafa, *J. Polym. Sci.*, **33** (1958) 311.
1128. M.A.K.Mostafa, *J. Polym. Sci.*, **33** (1958) 323.
1129. H.Okuyama, *Z. Electrochem.*, **59** (1951) 565.
1130. N.Sata, H.Okuyama, and K.Cheyo, *Koll. Z.* **121** (1951) 46.
1131. M.A.K.Mostafa, *J. Polym. Sci.*, **22** (1956) 535.
1132. M.A.K.Mostafa, *J. Polym. Sci.*, **27** (1958) 473.
1133. M.A.K.Mostafa, *J. Polym. Sci.*, **28** (1959) 499.
1134. M.A.K.Mostafa, *J. Polym. Sci.*, **28** (1958) 519.
1135. M.A.K.Mostafa, *J. Polym. Sci.*, **33** (1958) 311, 323.
1136. D.W.Ovenall, G.W.Hastings, and P.E.M.Allen, *J. Polym. Sci.*, **33** (1958) 207.
1137. D.W.Ovenall, *J. Polym. Sci.*, **42** (1960) 456.
1138. E.A.Neppiras and B.E.Molingk, *Proc. Phys. Soc.*, **63B** (1950) 674.
1139. N.Sato and H.Okuyama, *Z. Electrochem.*, **58** (1954) 196.
1140. H.Okuyama and T.Hirose, *J. Appl. Polym. Sci.*, **7** (1963) 591.
1141. T.Hirose, *Technol. Rep. Osaka Univ.*, **10** (1960) 395.
1142. P.E.M.Allen, G.M.Burnel, G.M.Hastings, H.W.Melville, and D.W.Owenal, *J. Polym. Sci.*, **33** (1958) 213.
1143. S.Blawmik and P.Chowdhury, *J. Ind. Chem. Soc.*, **48** (1971) 699.
1144. H.H.Jellinek, *Degradation of Vynil Polymers*, Academic Press, N.Y., 1955.
1145. Y.Boyle and L.Lehman, *Rep. on Ultrasonics, Com. Res. Council*, 1923.
1146. V.C.Sorensen, *Am. Phys. Leipzig (V)*, **26** (1936) 121.
1147. A.Weissler, *J. Appl. Phys.*, **21** (1950) 171.
1148. A.Weissler, H.W.Cooper and S.Snyder, *J. Am. Chem. Soc.*, **72** (1950) 1771.
1149. A.Weissler, *J. Appl. Phys.*, **21** (1950) 172.
1150. H.W.Melville and A.J.R.Murray, *Trans. Faraday Soc.*, **46** (1950) 996.
1151. M.Kosino and I.Miyagawa, *J. Chem. Soc. Japan. Ind. Chem. Sect.*, **54** (1951) 193.
1152. H.W.W.Brett and H.H.G.Jellinek, *J. Polym. Sci*, **13** (1954) 44.
1153. G.Seini and G.Ostecoli, *Atti. Acad. Sci. Torino, Classe Sci. Fiz. Mat. Nat.*, **90** (1955-1956) 393.
1154. B.B.Thomas and W.J.Alexander, *J. Polym. Sci.*, **25** (1957) 285.
1155. N.H.Langton and P.Vaughon, *Plastics*, **23** (1958) 308.
1156. .N.H.Langton and P.Vaughon, *Brit. J. Appl. Phys.*, **13** (1962) 478.
1157. H.V.Bargstead, *Brit. J. Appl. Phys.*, **15** (1964) 773.
1158. S.Ono, *Rev. Phys. Chem. Japan*, **14** (1940) 25.
1159. H.H.W.Brett and H.H.G.Jellinek, *J. Polym. Sci.*, **21** (1956) 235.
1160. R.O.Prudhomme and P.Graber, *J. Chim. Phys.*, **46** (1949) 667.
1161. R.O.Prudhomme and P.Graber, *J. Chim. Phys.*, **46** (1949) 323.
1162. R.O.Prudhomme, *J. Chim. Phys.*, **47** (1950) 795.
1163. S.Akiya, O.Nomoto, and S.Okui, *J. Pharm. Soc. Japan*, **63** (1949) 113.

1164. W.Kuhn, *Helv. Chim. Acta,* **39** (1956) 1071.
1165. N.I.Nikitin and P.I.Klenkova, *Issledovanie v oblasti vysokomolekulearnih soedinenii,* Moskva, 1949, 138.
1166. A.A.Berlin, *Tezisy Dokld. naucinih konferntii MTIMP po itogam rabot,* (1956) 33.
1167. A.A.Berlin, *Dokld. Akad. Nauk SSSR,* **110** (1956) 401.
1168. A.A.Berlin and E.A.Penskaia, *Dokld. Akad. Nauk SSSR,* **110** (1956) 585.
1169. A.A.Berlin, E.A.Penskaia, and G.I.Volkova, *Mejd. Sym. po makromol. himii, SSSR,* Moskva, S.III-a, (1960) 334.
1170. F.Patat and W.Högner, *Makromol. Chem.,* **75** (1964) 85.
1171. W.Kuhn, E.Peterlin, and H.Mayer, *Z. Electrochem.,* **62** (1958) 296.
1172. W.Kuhn and H.Mayer, *Angew. Chem.,* **68** (1956) 345.
1173. W.Kuhn and H.Mayer, *Z. Phys. Chem.,* **3** (1955) 330.
1174. S.Oprea and Cr.Simionescu, *Vysokomol. Soedin.,* **8** (1966) 1133.
1175. S.Oprea and Cr.Simionescu, *Materiale Plastice,* **3** (1966) 177.
1176. P.Iu.Butjagin, A.A.Berlin, A.E.Kalmanson, and L.A.Blumenfeld, *Vysokomol. Soedin.,* **1** (1959) 865.
1177. G.Kanig, *Koll. Z.,* **173** (1960) 97.
1178. A.A.Berlin, *Biofizika,* **8** (1963) 28.
1179. P.Iu.Butjaghin, *Dokld. Akad. Nauk SSSR,* **140** (1961) 45.
1180. P.Iu.Butjaghin, I.V.Kolbanev, and V.A.Radtig, *Fiz. Tverd. Tela,* **5** (1963) 235.
1181. S.N.Jurkov, E.E.Tomasevskii, and V.A.Zakrevski, *Fiz. Tverd. Tela,* **3** (1961) 2841.
1182. N.I.Nikitin and T.I.Rudneva, *J. Prikld. Khimii,* **29** (1956) 243.
1183. H.H.G.Jellinek and S.Y.Foks, *Makromol. Chem.,* **104** (1967) 18.
1184. L.Mullins, *Kaucik I Rezina,* **10** (1970) 20.
1185. H.W.Greensmith, *J. Polym. Sci.,* **21** (1956) 175.
1186. E.H.Andrews, *J. Polym. Sci.,* **33** (1958) 41.
1187. L.Mullins, *Rubber Chem.Techn.,* **33** (1960) 315.
1188. E.H.Andrews, *J. Appl. Polym. Phys.,* **32** (1961) 548.
1189. L.Mullins and N.K.Tobin, *J. Appl. Polym. Sci.,* **9** (1965) 2993.
1190. J.A.C.Horwood, L.Mullins, and A.R.Payne, *J.I.R.I.,* **1** (1967) 17.
1191. S.S.Voiutski, V.G.Raevski, and S.M.Zagneatinskaia, *Kaucik I Rezina,* **7** (1964) 16.
1192. A.N.Kamenski, S.M.Iagneatinskaia, N.A.Pereghibova, N.N.Fadiman, and S.S.Voiutski, *Mehanika Polimerov,* **2** (1967) 291.
1193. E.M.Dannenberg, *Rubber Chem. Technol.,* **48** (1971) 411.
1194. G.Kraus, in *Reinforcement of Elastomers,* J.Wiley & Sons, N.Y., 1965, 123.
1195. J.B.Donnet and A.Voet, *Carbon Black: Physics, Chemistry and Elastomer's Reinforcement,* Marcel Dekker, N.Y., 1976.
1196. L.Mullins and N.B.Tobins, *J. Appl. Polym. Sci.,* **9** (1965) 1993.
1197. A.M.Bueche, *J. Polym. Sci.,* **25** (1957) 25.
1198. A.Cifferi and D.J.Ferry, *J. Appl. Polym. Sci.,* **30** (1959) 1498.
1199. A.F.Blanchard, *Rubber J.,* **153** (1971) 44.
1200. F.Bueche, in *Reinforcement of Elastomers,* J.Wiley & Sons, N.Y., 1965.
1201. G.Krauss, K.W.Rollman, and J.T.Gruver, *Macromolecules,* **3** (1970) 92.
1202. A.R.Payne and E.R.Whitacker, *Comn. Int. Coloq. Elastomer,* Paris, 1978.
1203. J.B.Donnet and E.Papierer, *Brit. Polym. J.,* (1973) 213.
1204. Z.Righi, *Adv. Polym. Sci.,* **36** (1980) 21.
1205. C.Stanescu, *Cercetari privind obtinerea si caracterizarea unor noi materiale compozite prin prelucrarea mecanica a amestecurilor binare de elastomeri,*

Ph.D.Thesis, Technical University "Gh.Asachi" Iasi, 1998.

1206. N.V.Veresotskaia, K.D.Berbis, and G.L.Slonimski, *Kauciuk I Rezina,* **12** (1963) 27.

1207. B.A.Dogatkin, *Himiceskaia Prom.,* **4** (1955) 55.

1208. K.W.Scott, O.Lorenz and C.R.Parks, *J. Appl. Polym. Sci.,* **8** (1964) 2909.

1209. I.A.Tutovski, L.A.Shumonov, and B.A.Dogatkin, *Eur. Polym. J., Suppl.,* (1969) 457.

1210. B.A.Dogatkin, G.B.N.Zacesova, and T.A.Sohim, *Kauciuk I Rezina,* **12** (1961) 15.

1211. V.F.Drozdovski and D.R. Razgon, *Kauciuk I Rezina,* **8** (1969) 12.

1212. P.Iu.Butjaghin, V.F.Drozdovski, D.R.Razgon, and N.V.Kolbanev, *Fiz. Tverd. Tela,***7** (1965) 941.

1213. E.V.Reztsova, P.F.Badenkov, and V.A.Kargin, *Dokld. Akad. Nauk SSSR,* **182** (1968) 5

1214. T.Volintiru and G.Ivan, *Technological Basis of Elastomer Processing,* Editura Tehnica, Bucuresti, 1980.

1215. G.Ivan, *Industria Usoara,* **16**(5) (1969) 290.

1216. G.Ivan, *Industria Usoara,* **16**(11) (1969) 746.

1217. G.Ivan, *Studii si Cercetari, Chimie,* **18**(11) (1970) 1133.

1218. G.Ivan, *Industria Usoara,* **18**(2) (1971) 83.

1219. G.Ivan, *Industria Usoara,* **18**(6) (1971) 338.

1220. G.Ivan, G.Bolohan, and M.Giurginca, *Rev. Gen. Caout. Plast.,***50** (1973) 413.

1221. G.Ivan, M.Ganea, and I.Languri, *Blends of two elastomers with one filler. Examination of the Fracture Surface by Electronic Microscopy.* Int.Symp.on Elastomer Reinforcement, Obernai, Strassbourg, 1973.

1222. T.Volintiru, E.Bugaru, and G.Ivan, *Mat. Plast.,* **13**(3) (1976) 170.

1223. 1223.F.M.Albert,M.Giurginca and G.Ivan,*Mat.Plast.,***18**,3 (1981) 179

1224. G.Ivan and M.Giurginca, *Intraction in the System Elastomer-Promotor-Carbon Black,* Int. Rubber Conf., Moscow, 1984.

1225. G.Ivan, E.Bugaru, R.Plesa, and P.Bujenita, *Mat. Plast.,***24**(1) (1987) 50.

1226. G.Ivan, E.Danila, V.Ionescu, and A.Toader, *Fine ground vulcanzed rubber as an elastic filler in elastomer matrix,* Symp. of ICPCMP "Polymers Composite Materials", Tg.Mures, Romania, 1988.

1227. G.Ivan, *Polymer Filler Interaction in Elastomer Reinforcement,* Sesiunea stiintifica a Institutului de Chimie Macromoleculara"P.Poni" Iasi,1991

1228. G.Ivan, E.Bugaru and P.Bujenita, *Influence of some oligomers containing functional groups on the processing behaviour of the diene elastomers,* Symp. "Stiinta Polimerilor", Bucuresti, 1992.

1229. M.Giurginca, G.Ivan, and J.M.Herdan, *Grafting antioxidants. VII. Graftinf of antioxidants containing amine and thiol groups onto epoxidized natural rubber,* a-XIX-a Sesiune de comunicari stiintifice, INCERCHIM.-Calimanesti, R.Valcea, Romania, 1993.

1230. E.V.Veitsman, *J. Phys. Chem.,* **56** (1982) 524.

1231. G.Backer, *Handbuch der Experimentalphysik,* Akad. Verlag, Leipzig, Bd.6 (1928) 70.

1232. T.L.Hill, *J. Chem. Phys.,* **17** (1949) 520.

1233. V.L.Vakula and L.M.Putykih, *Polymer Adhesion, Physico-chemical Principles,* Ellis Horwood Ltd., London, 1991.

1234. J.D.Anohode, *Surface and Interfacial Aspects of Biomedical Polymers,* J.Wiley & Sons., N.Y., 1985.

1235. F.N.Fowers, *J. Coll. Interface Sci.,* **78** (1980) 200.

1236. E.A.Raud, *Colloid J. (USSR)*, **47** (1985) 1200.
1237. J.R.Good, *J. Am. Chem. Soc.*, **74** (1952) 5049.
1238. A.N.Frumnin, *J. Phys. Chem. (USSR)*, **12** (1938) 337.
1239. Y.M.Malinski, *PhD Thesis*, Karpov Phys. Chem. Inst., Moskva, 1970.
1240. D.H.Raable, *J. Adhesion*, **1** (1969) 102.
1241. P.S.Theokaris, *Mech. Caract. Load Bear-m Fibre Composite Laminates*, London, 1985.
1242. Y.S.Lipatov, *Adsorption of Polymers*, Nauka va Dumka, Kiev, 1972.
1243. P.G.deGennes, *Ann. Chim.*,**77** (1982) 389.
1244. J.Klein and P.Pinars, *Macromolecules*, **15** (1982) 1129.
1245. Y.S.Lipatov, *USSR Adv. Chem.*, **50** (1981) 355.
1246. D.C.Douglas, *Macromolecules*, **10** (1977) 178.
1247. T.M.Birtstein, *Macromolecules*, **12** (1979) 315.
1248. S.A.Tsygonov and Y.A.Goldman, *USSR High Polymers*, **23A** (1989) 294.
1249. R.D.Calvert, *J. Appl. Polym. Sci.* **24** (1979) 357.
1250. A.E.Chalyk, *Diffusion in Polymers Systems*, Khimia Ed., Moskva, 1987.
1251. A.I.Malkov, *Autodiffusion in solutions and melts of polymers*, Univ.of Kazan Publ., 1987.
1252. E.P.Pledermann, *Bonding through coupling agents*, in *Molecular characterization of composite interface*, Plenum Press, N.Y., 1985.
1253. K.Friedrich, *MRS-Eurpe*, Nov. 1985.
1254. J.Karger-Kocsis, *Application of fracture mechanics to composite materials*, Composite Materials Series, Vol. 6, (Ed.K.Friedrich), Elsevier Sci. Publ., Amsterdam-Oxford-N.Y.-Tokyo, 1989.
1255. K.Friedrich and E.Reese, *Mikrostrukturelle Aspekte der Rissausbreitung in Polymere und Polymerverbundwerkstoffe, Werkstoffprüftagung* 31 May-2 Juni, 1989, Gefüge und Bruch, Material Kundt.-Tech., Reihe, Bd.9, 1990, Berlin-Stutgart (Bonntraeger).
1256. S.Fakirov, V.Djakovitch, B.Hardinghaus, O.Kalwies, K.Friedrich, and K.Barting, *Wiss. Zeitschrift Leuna-Merseburg*, **33**(2) (1991) 335.
1257. K.Friedrich, *Proc. Material Research Soc. Conf.*, Strassbourg, 1895.
1258. D.M.Bigg, *Polym. Comp.*, **6** (1985) 20.
1259. R.P.Hegler, *Kunststoffe*, **74**(5) (1984) 271.
1260. L.A.Golttler, *Polym. Comp.*, **5** (1984) 60.
1261. K.Friedrich, *Fracture Mechanical Behaviour of Short Fiber Reinforced Thermoplastics*, Fortsch.-Ber.VDI-Z, Reihe, (1984) 18.
1262. H.Voss and K.Friedrich, *J. Mater. Sci.*, **21** (1986) 2889.
1263. R.Schledjewski and K.Friedrich, *J. Mater. Sci., Letters*, **11**(1992) 840.
1264. K.Friedrich, *Compos. Sci. Technol.*, **22** (1985) 43.
1265. K.Friedrich, *Fracture Mechanical Behaviour of Short Fiber Reinforced Thermoplastics*, Fortsch.-Ber. VDI, Reihe 18, VDI –Verlag, Düsseldorf, Germany, 1984.
1266. H.Voss, *Aufbau, Bruchverhalten und Verschleisseigenschaften Kurzfaserverstärker Hochleistungsthermoplaste*, Fortschrifberichte VDI, Reihe 5,1 16, VDI-Verlag Düsseldorf, Germany, 1987.
1267. K.Fridrich and J.Karger-Kocsis, *Fractography of Unfilled and Short Fibre Filled Semicrystalline Thermoplastics*, in *Fractography of Polymers and Composites*, (Ed.A.C.Roulin-Moloney), Elsevier Applied Science Publishers, Barking, U.K., 1989, 437.
1268. S.Hashemi and J.G.Williams, *Polym. Eng. Sci.*, **26** (1986) 760.
1269. S.Hashemi and J.G.Williams, *Polymer*, **27** (1986) 384.
1270. Hahemi and J.G.Williams, *Plast. & Rubber Process. & Appl.*, **6** (1986) 363.

1271. S.Hashemi and J.G.Williams, *Testing of Polymers,* VI Int. Conf. on Deformation Yield and Fracture of Polymers, Cambridge, April 1985, The Plastic and Rubber Institute (Chameleon Press Ltd.) 1985, 531.
1272. J.Krey, R.Walter, and K.Friedrich, *Z. Werkstofftechnick,* **17** (1986) 226.
1273. B.Melve, *Testing of Polymers,* VI Int. Conf. on Deformation Yield and Fracture of Polymers, Cambridge, 1985, The Plastic and Rubber Institute (Chameleon Press Ltd.), 1985, 60.
1274. J.Karger –Kocsis and I.Csikai, *Polym. Eng. Sci.,* **27** (1987) 241.
1275. K.Friedrich, *Application of fracture mechanics to composite materials.Composite Materials Series 6,* (Ed.K.Friedrich), Elsevier, Amsterdam-Oxford-N.Y.-Tokyo, 1989, 427.
1276. H.L.Cox, *J. Appl. Phys.,* **3** (1952) 72.
1277. A.Kelly and W.R.Tyson, *J. Mech. Phys. Solids,* **13** (1965) 329.
1278. A.Kelly, in *Fibre Reinforcement,* Clarendon Press, Oxford, 1966.
1279. J.R.Outwater, *Mod. Plast.,* **33** (1964) 156.
1280. J.L.Kardos, J.C.Halpin, and S.L.Cheng, *The strong of oriented short-fiber reinforced composites,* Plenum Press, N.Y., 1980, 234.
1281. E.Mosaoumy, L.Kacir, and J.L.Kardos, *Effect of fiber ratio and orientation on the stress-strain behaviour of alligned short fiber reinforced, ductil epoxy,* Report Mat. Res. Lab. Washington University, St.Louis, Missouri, 1982.
1282. H.Fukuda and T.W.Chou, *Fibre Sci. Technol.,* **13** (1981) 243.
1283. M.G.Bader, T.W.Chou, and J.J.Quigley, in *New Development and Applications in Composites,* (Ed. D.Wilsdorf), N.Y., 1978.
1284. J.L.Kardos, *Structure, property relations in short fibre reinforced plastics,* CRC Critical Rew. **73** (1973) 419.
1285. P.E.Chen, *Polym. Eng. Sci.,* **11** (1971) 51.
1286. J.K.Lees, *Polym. Eng. Sci.,* **8** (1968) 186.
1287. N.S.Sridharan, *Short fiber reinforced composite materials,* (Ed.B.A.Sonders), ASTM, 1982, 145.
1288. M.J.Owen and R.Dukes, *J. Strain Analysis,* **2** (1967) 272.
1289. M.J.Owen, *J. Phys.; Appl. Phys.,* **7** (1974) 1214.
1290. R.E.Lavengood and L.E.Gulbransen, *Polym. Eng. Sci.,* **9** (1969) 305.
1291. J.W.Dally and D.H.Carillo, *Polym. Eng. Sci.,* **9** (1969) 434.
1292. A.T.Dibendefto and G.Salee, *Polym. Eng. Sci.,* **18** (1978) 634.
1293. A.T.Dibendetto and G.Salee, *Polym. Eng. Sci.,* **19** (1979) 512.
1294. J.F.Mandell, D.D.Huang, and F.J.MacGarry, (Ed.Sanders, ASTM), 1982.
1295. J.C.B.Bucknall, *Toughened Plastcs,* Appl. Sci. Publ. Mat. Sci. Series, 1977.
1296. J.F.Mandell, in *Composite reliability,* ASTM SIP, 1976, 580.
1297. G.R.Irwin, *J. Appl. Mech.,* **9** (1957) 139.
1298. D.S.Dugdale, *J. Mech. Phys. Solids,* **8** (1960) 100.
1299. J.G.Williams, *Polym. Eng. Sci.,* **17** (1977) 3.
1300. P.C.Paris and G.C.Sih, *Stress analysis of cracks,* in *Fracture toughness testing and applcations,* ASTM, STP, **381** (1965) 30.
1301. D.R.Moore and C.J.Hooly, *Plast. Rubber Proc. Appl.,* **1**(2) (1981) 121.
1302. C.B.Bucknall, *Adv. Polym. Sci.,* **27** (1978) 121.
1303. P.W.R.Beaumont, *J. Mater. Sci.,* **10** (1974) 1334.
1304. M.R.Pigott, *J. Mater. Sci.,* **9** (1975) 494.
1305. J.R.Ellis, *Polym. Eng. Sci.,* **14**(7) (1974) 157.
1306. L.Mascia and F.S.Speck, *BPF Reinforced Plast. Congress,* V/3, 1974.
1307. A.C.Garg, *Fibre Sci.Technol.,* **14** (1981) 293.
1308. M.E.Waddoups, *J. Composites Materials,* **5** (1971) 445.

1309. J.M.Whitney and R.J.Nuismer, *J. Composites Materials*, **8** (1974) 259.
1310. J.F.Mandell, A.Dorwish and F.J.McGarry, in *Test. method and design allowbles for fibrous composites*, ASTM, STP, 1981, 734.
1311. M.M.Schwartz, *Composite Materials Handbook*, McGrow Hill Book Comp., N.Y., 1983.
1312. K.Schulte, S.Kutler, and K.Friedrich, *DFVLR Report No. Df VLLB-FB-86-18*, 1986.
1313. K.Friedrich and U.A.Karch, *Fibre Sci. Technol.*, **18** (1983) 37.
1314. K.Friedrich, *Microstructural Efficiency and Fracture Toughness of Short Fibre/ Thermoplastic Matrix Composites, Composites Sci. Technol.*, **22** (1985) 43.
1315. A.J.Smiley and R.B.Pipes, *J. Comp. Mater.*, **21** (1987) 670.
1316. A.J.Smiley and R.B.Pipes, *Comp. Sci. Technol.*, **29** (1987) 1.
1317. K.Friedrich, R.Walter, L.A.Carlson, A.J.Smiley, and J.W.Gillespie Jr., *J. Mater. Sci.*, **24** (1989) 3387.
1318. L.A.Carlson, J.W.Gillespie, and B.R.Trethewey, *J. Reinf. Plast. Comp.*, **5** (1986) 170.
1319. D.E.Spahr, K.Friedrich, J.M.Scultz, and R.S.Bailey, *J. Mater. Sci.*, **25** (1990) 4427.
1320. J.E.O'Connor, A.Y.Lou, and W.H.Veever, *Polyphenylene sulphide a thermoplastic polymer matrix for high preformance composites*, in *Proc.5th Int.Conf.on Composite Materials (ICCMV)*, San Diego, July 29-Aug.1, The Metallurgical Soc.Inc., 1985, 963.
1321. G.R.Belbin, *Inst. Mech. Engrs.*, **198B** (1984) 71.
1322. J.Karger-Kocsis and K.Friedrich, *Polymer*, **27** (1986) 1753.
1323. D.P.Jones, D.C.Leads, and D.R.Moore, *Polymer*, **26** (1985) 1385.
1324. F.N.Cogswell, *Microstructure and properties of thermoplastic aromatic polymer composites, SAMPE Quarterly*, **14**(7) (1983) 33.
1325. R.B.Riby, *High temperature thermoplastic matrices for advaced composites*, Proc.27th Nat. SAMPE Symp., May, 1982, 747.
1326. R.W.Lang, M.Kiym, H.Tesch and H.Stutz, *Influence of constituent properties on interlaminar crack growth in composites*, Proc.7th Int. Conf. SAMPE EUROPE, Munchen, June 10-12, 1986, 261.
1327. P.W.R.Beaumont, J.K.Wells and P.D.Angtice, *Development of fracture maps for fibrous composites*, Proc.37th Annual Techn. Conf. Reinforced Plastics, Composites Institute, Jan.11-15, 1982, 20 A. 1.
1328. J.K.Wells and P.W.R.Beaumont, *J. Mater. Sci.*, **17** (1982) 397.
1329. K.Friedrich, H.Walter, H.Voss, and J.Karger-Kocsis, *Composites*, **17** (1986) 205.
1330. H.Voss and J.Karger-Kocsis, *Int. J. Fatigue*, **10** (1988) 3.
1331. T.Weng, A.Hiltner, and E.Baer, *J. Mater. Sci.*, **21** (1986) 744.
1332. Y.Ide and T.S.Chung, *J. Macromol. Sci. Phys.*, **B23**(4-6) (1984/1985) 497.
1333. T.S.Chung and P.E.McMahon, *J. Appl. Polym. Sci.*, **31** (1986) 965.
1334. D.C.Phillips and B.Harris, *The strenght, toughness and fatigue properties of polymer composites*, in *Polymer Engineering Composites*, (Ed.M.O.W.Richardson), Applied Science Publisher, London, 1977, 45.
1335. J.S.Wu, K.Friedrich, and M.Grasse, *Composites*, **20** (1989) 223.

Bibliography Part 4

1. ** *Max Planck – Geselschaft Berichte und Mitteilungen*, **1** (1989) 5.
2. T.M.Devlin, *Textbook of Biochemistry with Clinical Correlation*, John Wiley & Sons Inc., New York, 1991.
3. A.Despopoulos and S.Silbernagl, *Color Atlas of Physiology*, 4th Edition, Thieme Medical Publishers Inc., New York, 1991.
4. A.E.Bukatina, V.N.Morozov, and S.E.Snol, in *GMF*, (1977) 7.
5. A.E.Bukatina, in *GRI*, (1980) 212.
6. J.Hanson and H.E.Huxley, *Symp. Soc. exp. Biol.*, **9** (1955) 228.
7. F.Carlsen, G.G.Knappeis, and F.Buchthal, *J. Biophys. Bioch. Cytol.*, **11** (1961) 95.
8. L.D.Yates and M.L.Greaser, *J. Mat. Biol.*, **168** (1983) 123.
9. G.Hoyle, *Symp. Biol. Hung.*, **8** (1967) 34.
10. R.W.D.Rowe, *J. Cell. Biol.*, **57** (1973) 261.
11. M.Yamaguchi, M.Izumimoto, R.M.Robson, and M.H.Stromer, *J. Mol. Biol.*, **184** (1985) 621.
12. K.Wang and R.Ramirez-Mitchell, *J. Cell. Biol.*, **96** (1983) 562.
13. J.C.Kentish, *Muscle Res. Cell. Motil.*, **4** (1983) 1797.
14. K.Maruyama, M.Kuroda, and Y.Nomomura, *Biochim. biolphys. Acta*, **829** (1985) 229.
15. E.Veratti, *J. biophys. biochim. Cytol.*, **10** (1961), Suppl., 1-59.
16. K.R.Porter and G.E.Palade, *J. biophys. biochim. Cytol.*, **3** (1957) 269.
17. N.Garamvologyi, *Symp. biol. hung.*, **17** (1974) 61.
18. H.E.Huxley, *Nature*, **202** (1967) 1067.
19. H.C.Luttgau, *J. comp. Physiol.*, **120** (1977) 51.
20. H.S.Bennett, in *GHB* (1960) 137.
21. H.Dalen, P.Scheie, R.Myklebust, and T.Saetersdal, *J. Micros.*, **131** (1983) 35.
22. C.Franzini-Armstrong and K.R.Porter, *J. Cell. Biol.*, **22** (1964) 675.
23. C.Wittenberg and M.Keul, *Mecanismele miscarii in lumea vie*, Editura Academiei Romane, Bucuresti, 1987.
24. L.D.Yates and M.L.Greaser, *J. mol. Biol.*, **168** (1983) 123.
25. R.A.Cross, M.L.Bardsley, D.A.Ledward, J.V.Small, and A.Sobleszek, *Eur. J. Biomed.*, **145** (1984) 305.
26. R.Kassab, D.Mornet, P.Pantal, R.Bertrand, and E.Audemard, *Biochimie*, **63** (1981) 273.
27. G.S.Waller and S.Lowey, *J. Biol. Chem.*, **260** (1985) 14368.
28. D.Pette, J.Hendrikson, and M.Emmerich, *FEBS Lett.*, **103** (1979) 152.
29. F.A.Pepe, in *PSN*, (1979) 105.
30. R.Craig and J.Megerman, in *PSN*, (1979) 91.
31. F.J.Wilson and M.J.Inish, *Cell. Tiss. Res.*, **212** (1980) 213.
32. C.W.Heizmann, I.E.Bläuenstein, and H.M.Eppenberger, *Experimentia*, **34** (1978) 38.
33. E.E.Strehler, G.Pelloni, C.W.Heizmann, and H.M.Eppenberger, *J. Cell. Biol.*, **86** (1980) 775.
34. P.Anversa, G.Olivetti, P.G.Bracchi, and A.V.Lond, *Circ. Res.*, **41**(1981) 561.
35. E.Prochniewicz and T.Yanagida, *J. Biochim.*, **89** (1981) 1215.
36. P.W.Brandt, M.S.Diamond, and F.H.Schachat, *J. mol. Biol.*, **189** (1984) 379.
37. J.E.Morel and I.Pinset-Harstrom, *Biomedicine*, **22** (1975) 88.

38. J.E.Morel and I.Pinset-Harstrom, *Biomedicine,* **22** (1975) 186.
39. 39.U.Murakami and K.Uchido.*J.Biochem.,***98** (1985) 187
40. T.Obinata and R.Matsuda, in *EME* (1980) 275.
41. K.Maruyama, S.Kunimoto, S.Kimura, and K.Ohashi, *J. Biomed.,* **81** (1977) 245.
42. V.N.Mescereakov and L.G.Filatova, *Biofizika,* **26**(1981) 1057.
43. P.K.Hepler and B.A.Peleritz, *Ann. Rev. Plant. Physiol.,* **25** (1974) 309.
44. K.Seitz, in *Studies on Aquatic Vascular Plants,* red. J.J.Symoens, S.S.Hooseravin and E.E.Orlovskaia, *Roy. Bot. Soc. Belgium,* Brussels, (1982) 89.
45. V.Soran and G.Lazar-Keul, *Cytologia,* **43** (1978) 265.
46. K.Seitz, *Pflanzenphysiol.,* **95** (1979) 265.
47. E.Kamitsubo, *Protoplasma,* **74** (1972) 53.
48. B.A.Paleritz and P.K.Hepler, *J. Cell. Biol.,* **65** (1973) 29.
49. R.Nagai and T.Hayama, *J. Cell. Sci.,* **36** (1979) 121.
50. J.K.W.Khen and N.Kamiya, *Cell. Struct. Funct.,* **6** (1981) 201.
51. R.Nagai, *What's new Plant Physiol.,* **10** (1979) 45.
52. H.E.Huxley, *Nature,* **243** (1973) 445.
53. G.Haemmerli, *Eur. J. Cancer,* **16** (1980) 1.
54. K.E.Wolfarth-Bottermann, *Eur. J. Cancer,* **16** (1980) 10.
55. R.D.Allen, *Can. J. Bot.,* **58** (1980) 745.
56. D.L.Taylor and J.S.Condeelis, *Int. Rev. Cytol.,* **56** (1979) 57.
57. A.Grebecki, *Eur. J. Biol.,* **241** (1981) 163.
58. P.Satir, *Scient. Amer.,* **231** (1974) 44.
59. P.Satir and G.K.Ojakian, in *HF* (1979) 224.
60. W.Haupt, *Progr. Bot.,* **42** (1980) 143.
61. V.K.Pavlenko, in *GMF* (1977) 35.
62. U.W.Goolenough and J.E.Heuser, *J. Cell. Biol.,* **95** (1981) 798.
63. R.M.Macnab, in *HF* (1979) 207.
64. C.R.Calladine, *J. mol. Biol.,* **118** (1978) 457.
65. H.C.Berg, R.A.Anderson, *Nature,* **245** (1973) 380.
66. A.N.Glagoler, in *GMF* (1977) 66.
67. T.Weis-Fogh, in *BMS* (1975) 83.
68. N.J.Arronet, in *GMF* (1973) 222.
69. F.Horkay and M.Zriny, *Macromolek.Chem.; Macromolec. Symp.,* **30** (1989) 133.
70. W.Kuhn, *Experimentia,* **5** (1949) 318.
71. J.W.Breitenbach and H.Karllinger, *Monastsh. Chem.,* **80** (1949) 211.
72. W.Kuhn, B.Hargitay, A.Katchalsky, and H.Eisenberg, *Nature,* **165** (1950) 514.
73. Y.Osada, *Adv. Polym. Sci.,* **82** (1987) 1.
74. D.DeRossi, K.Kajiwara, Y.Osada, and A.Yamauchi, Eds., *Polymer Gel,* Plenum Press, New York, 1991.
75. M.Irie, *Adv. Polym. Sci.,* **94** (1990) 22.
76. Y.Osada and J.Gong, *Progr.Polym. Sci.,* **18** (1993) 187.
77. S.Umemoto, N.Okui, and T.Sakai, in ref. 74, 257.
78. T.Kurauchi,T.Shiga,Y.Hirose and A.Okada, in ref.74, 237.
79. M.Suzuki, in ref. 74, 231.
80. W.Kuhn, *Gazg. Chem. Ital.,* **92** (1962) 951.
81. R.Lovrien, *Proc. Nat. Acad. Sci.,* **57** (1967) 236.
82. N.Negishi, K.Ishihara, and I.Shinohara, *J. Polym. Sci.; Chem. Ed.,* **20** (1982) 1907.
83. L.Matejka and K.Dusek, *Makromol. Chem.,* **182** (1981) 3528.
84. M.Irie and K.Hayashi, *J. Macromol. Sci.; Chem.* **A 13** (1979) 511.
85. M.Irie, K.Hirano, S.Hashimoto, and K.Hayashi, *Macromolecules,* **14** (1981)

262.
86. S.Shinkai, T.Ogawa, Y.Kusano, O.Manabe, T.Goto, and T.Matsuda, *J. Amer. Chem. Soc.*, **104** (1982) 1960.
87. M.Irie and M.Hosada, *Makromol. Chem.; Rapid Commun.*, **5** (1984) 829
88. M.Irie, A.Menju, and K.Hayashi, *Macromolecules,* **12** (1981) 1176.
89. M.Irie and D.Kungwatchakun, *Makromol.Chem.; Rapid Commun.*, **5** (1981) 829.
90. M.Irie and D.Kungwatchakun, *Macromolecules,* **19** (1986) 2476.
91. H.S.Blair and T.K.Law, *Polymer,* **21** (1980) 1475.
92. R.S.Blair and H.I.Pogue, *Polymer,* **23** (1982) 779.
93. C.D.Eisenbach, *Polymer,* **21** (1981) 1175.
94. F.Agolini and F.P.Gay, *Macromolecules,* **3** (1970) 349.
95. E.K.Zimmerman and J.K.Stille, *Macromolecules,* **14** (1985) 1246.
96. S.Kohjiya, T.Hashimoto, S.Yamashita, and M.Irie, *Chem. Lett.*, (1985) 1479.
97. L.Matejka, M.Ilavsky, and O.Wichterle, *Polymer,* **22** (1981) 1511.
98. O.Hirasa, S.Ito, A.Yamauchi, S.Fujishige, and H.Ichijo, in ref. 74, 247.
99. 9D.W.Urry, R.D.Harris, and U.K.Prasad, *J. Am. Chem. Soc.*, **110** (1988) 3303.
100. D.W.Urry, B.Haynes, H.Zhang, and K.U.Prasad, *Proc. Natl. Acad. Sci.USA,* **85** (1988) 3407.
101. V.A.Kabanov, *IUPAC Macromol.Cem.*, **8** (1973) 121.
102. I.M.Papsivov, A.D.Antipina, and V.A.Kabanov, *Dokld. Akad. Nauk SSSR,* **199** (1971) 1364.
103. Y.Osada and M.Sato, *J. Polym. Sci.; Polym. Lett.Ed.*, **14** (1976) 129.
104. Y.Osada and M.Sato, *Polymer,* **21**(1980) 1057.
105. Y.Osada and Y.Saito, *Makromol. Chem.*, **176** (1975) 2761.
106. Y.Osada and M.Hasebe, *Chem. Lett.*, (1985) 1285.
107. Y.Osada and R.Kishi, *J. Chem. Soc. Faraday Trans.*, **85** (1989) 655.
108. R.Kinshi, M.Hara, K.Sawahata and Y.Osada, in ref. 74, 205.
109. P.Chiarelli, K.Umezawa, and D.DeRossi, in ref. 74, 195.
110. H.Shinohara, M.Aizawa, and H.Shirakawa, *J. Chem. Soc. Chem. Commun.*, **87** (1986) 453.
111. H.Shinohara, M.Yaoita, Y.Ikariyama, and M.Aizawa, in ref. 74,147.
112. Y.Osada and B.S. Ross-Murphy, *Scient.Amer.*, **5** (1993) 82.
113. A.S.Hoffman, *Makromol. Chem.; Macomol. Symp.*, **98** (1995) 645.
114. T.Tanaka, *Scient. Amer.*, **8** (1981) 82.
115. G.H.Chen and A.S.Hoffman, *Bioconj. Chem.*, **4** (1993) 509.
116. J.P.Chen and A.S.Hoffman, *Biomat.*, **11** (1990) 631.
117. N.Monji and A.S.Hoffman, *Appl. Biochem. Biotech.*, **14** (1987) 107.
118. H.G.Schild and D.A.Tirrell, *Langmuir,* **7** (1991) 1319.
119. X.S.Wu, et al., *J. Intell. Mtls. Syst. and Struct.*,**4** (1993) 202.
120. M.Tardieu, et al., *J. Cell. Physiol.*, **150** (1992) 194.
121. Y.Osada and K.Honda, *J. Membrane Sci.*, **27** (1986) 327.
122. M.Saeda, in ref. 74, 161.
123. S.Saito, in ref.74, 173.
124. H.Kawaguchi, et al., *Coll. Polym. Sci.*, **270** (1992) 53.
125. T.Okano, et al., *J. Biomed. Mater. Res.*, **27** (1993) 1243.
126. N.Yamada, et al., *Makromol.Chem.; Rapid Commun.*, **11** (1990) 571.
127. T.Takezawa, et al., *Biotech.*, **8** (1990) 854.
128. M.Miura and A.S.Hoffman, *Conference at Washington University*, Seatle, Washington, 1992.
129. G.Albin, et al., *J. Contr. Rel.*, **2** (1985) 153.
130. O.Hirasa, in ref.74, 128.

131. Y.Osada and Y.Takeuchi, *J. Polym. Sci.;Polym. Lett. Ed.*, **19** (1981) 303.
132. Y.Osada and M.Hasabe, *Chem. Lett.*, (1985) 1285.
133. Y.Osada, *Advanced Materials*, 3(2) (1991) 107.
134. E.Cussler, et al., *AICHE J.*, **30** (1984) 578.
135. E.Cussler and R.F.S.Freitas, *Chem. Eng. Sci.*, **42** (1987) 97.
136. T.Okano, in ref.74, 287.
137. Y.Osada, in ref.74, 215.
138. R.A.Siegel, in ref.74, 152.
139. W.Kim, et al., *J. Contr. Rel.*, **2** (1895) 153.
140. W.Kim, et al., *Pharm. Res.*, **8** (1991) 624.
141. Y.Osada and R.Kishi, *J. Chem. Soc. Faraday Trans.*, **85** (1989) 655.
142. T.G.Park and A.G.S.Hoffman, *Appl. Biochem. Biotech.*, **19** (1988) 1.
143. P.J.Basser, in ref.74, 271.
144. A.Katchalsky, et al., *Nature*, **210** (1966) 568; M.V.Sussmann and A.Katchalsky, *Science*, **167** (1970) 45.
145. F.D.Carlson, D.J.Hardy, and D.R.Wilkie, *J. Gen. Physiol.*, **46** (1963) 852.
146. T.Hayashi, et al., *Preprints 2nd Symp. Polym. Gels*, Tsukuba, 1991, 123.
147. Y.Osada and H.Okuzachi, *Nature*, **355** (1992) 242.
148. Y.Hirose, in ref. 74, 237.
149. Cl.Vasiliu Oprea and F.Dan, *Plaste und Kautschuk*, **41**(6) (1994) 265.
150. F.Dan and Cl.Vasiliu Oprea, *Rev. Rom. Chem.*, **7-8** (1995) 713.
151. Cl.Vasiliu Oprea, I.Marcu, and F.Dan, *Polym. Plast. Technol. Eng.*, **38**(4) (1999) 609.

Index